LONDON MATHEMATICAL SOCIETY LECTURE NOTE SERIES

Managing Editor: Professor M. Reid, Mathematics Institute,
University of Warwick, Coventry CV4 7AL, United Kingdom

The titles below are available from booksellers, or from Cambridge University Press at http://www.cambridge.org/mathematics

287 Topics on Riemann surfaces and Fuchsian groups, E. BUJALANCE, A.F. COSTA & E. MARTÍNEZ (eds)
288 Surveys in combinatorics, 2001, J.W.P. HIRSCHFELD (ed)
289 Aspects of Sobolev-type inequalities, L. SALOFF-COSTE
290 Quantum groups and Lie theory, A. PRESSLEY (ed)
291 Tits buildings and the model theory of groups, K. TENT (ed)
292 A quantum groups primer, S. MAJID
293 Second order partial differential equations in Hilbert spaces, G. DA PRATO & J. ZABCZYK
294 Introduction to operator space theory, G. PISIER
295 Geometry and integrability, L. MASON & Y. NUTKU (eds)
296 Lectures on invariant theory, I. DOLGACHEV
297 The homotopy category of simply connected 4-manifolds, H.-J. BAUES
298 Higher operads, higher categories, T. LEINSTER (ed)
299 Kleinian groups and hyperbolic 3-manifolds, Y. KOMORI, V. MARKOVIC & C. SERIES (eds)
300 Introduction to Möbius differential geometry, U. HERTRICH-JEROMIN
301 Stable modules and the D(2)-problem, F.E.A. JOHNSON
302 Discrete and continuous nonlinear Schrödinger systems, M.J. ABLOWITZ, B. PRINARI & A.D. TRUBATCH
303 Number theory and algebraic geometry, M. REID & A. SKOROBOGATOV (eds)
304 Groups St Andrews 2001 in Oxford I, C.M. CAMPBELL, E.F. ROBERTSON & G.C. SMITH (eds)
305 Groups St Andrews 2001 in Oxford II, C.M. CAMPBELL, E.F. ROBERTSON & G.C. SMITH (eds)
306 Geometric mechanics and symmetry, J. MONTALDI & T. RATIU (eds)
307 Surveys in combinatorics 2003, C.D. WENSLEY (ed.)
308 Topology, geometry and quantum field theory, U.L. TILLMANN (ed)
309 Corings and comodules, T. BRZEZINSKI & R. WISBAUER
310 Topics in dynamics and ergodic theory, S. BEZUGLYI & S. KOLYADA (eds)
311 Groups: topological, combinatorial and arithmetic aspects, T.W. MÜLLER (ed)
312 Foundations of computational mathematics, Minneapolis 2002, F. CUCKER *et al* (eds)
313 Transcendental aspects of algebraic cycles, S. MÜLLER-STACH & C. PETERS (eds)
314 Spectral generalizations of line graphs, D. CVETKOVIC, P. ROWLINSON & S. SIMIC
315 Structured ring spectra, A. BAKER & B. RICHTER (eds)
316 Linear logic in computer science, T. EHRHARD, P. RUET, J.-Y. GIRARD & P. SCOTT (eds)
317 Advances in elliptic curve cryptography, I.F. BLAKE, G. SEROUSSI & N.P. SMART (eds)
318 Perturbation of the boundary in boundary-value problems of partial differential equations, D. HENRY
319 Double affine Hecke algebras, I. CHEREDNIK
320 L-functions and Galois representations, D. BURNS, K. BUZZARD & J. NEKOVÁŘ (eds)
321 Surveys in modern mathematics, V. PRASOLOV & Y. ILYASHENKO (eds)
322 Recent perspectives in random matrix theory and number theory, F. MEZZADRI & N.C. SNAITH (eds)
323 Poisson geometry, deformation quantisation and group representations, S. GUTT *et al* (eds)
324 Singularities and computer algebra, C. LOSSEN & G. PFISTER (eds)
325 Lectures on the Ricci flow, P. TOPPING
326 Modular representations of finite groups of Lie type, J.E. HUMPHREYS
327 Surveys in combinatorics 2005, B.S. WEBB (ed)
328 Fundamentals of hyperbolic manifolds, R. CANARY, D. EPSTEIN & A. MARDEN (eds)
329 Spaces of Kleinian groups, Y. MINSKY, M. SAKUMA & C. SERIES (eds)
330 Noncommutative localization in algebra and topology, A. RANICKI (ed)
331 Foundations of computational mathematics, Santander 2005, L.M PARDO, A. PINKUS, E. SÜLI & M.J. TODD (eds)
332 Handbook of tilting theory, L. ANGELERI HÜGEL, D. HAPPEL & H. KRAUSE (eds)
333 Synthetic differential geometry (2nd Edition), A. KOCK
334 The Navier–Stokes equations, N. RILEY & P. DRAZIN
335 Lectures on the combinatorics of free probability, A. NICA & R. SPEICHER
336 Integral closure of ideals, rings, and modules, I. SWANSON & C. HUNEKE
337 Methods in Banach space theory, J.M.F. CASTILLO & W.B. JOHNSON (eds)
338 Surveys in geometry and number theory, N. YOUNG (ed)
339 Groups St Andrews 2005 I, C.M. CAMPBELL, M.R. QUICK, E.F. ROBERTSON & G.C. SMITH (eds)
340 Groups St Andrews 2005 II, C.M. CAMPBELL, M.R. QUICK, E.F. ROBERTSON & G.C. SMITH (eds)
341 Ranks of elliptic curves and random matrix theory, J.B. CONREY, D.W. FARMER, F. MEZZADRI & N.C. SNAITH (eds)
342 Elliptic cohomology, H.R. MILLER & D.C. RAVENEL (eds)
343 Algebraic cycles and motives I, J. NAGEL & C. PETERS (eds)
344 Algebraic cycles and motives II, J. NAGEL & C. PETERS (eds)
345 Algebraic and analytic geometry, A. NEEMAN
346 Surveys in combinatorics 2007, A. HILTON & J. TALBOT (eds)
347 Surveys in contemporary mathematics, N. YOUNG & Y. CHOI (eds)
348 Transcendental dynamics and complex analysis, P.J. RIPPON & G.M. STALLARD (eds)
349 Model theory with applications to algebra and analysis I, Z. CHATZIDAKIS, D. MACPHERSON, A. PILLAY & A. WILKIE (eds)

350 Model theory with applications to algebra and analysis II, Z. CHATZIDAKIS, D. MACPHERSON, A. PILLAY & A. WILKIE (eds)
351 Finite von Neumann algebras and masas, A.M. SINCLAIR & R.R. SMITH
352 Number theory and polynomials, J. MCKEE & C. SMYTH (eds)
353 Trends in stochastic analysis, J. BLATH, P. MÖRTERS & M. SCHEUTZOW (eds)
354 Groups and analysis, K. TENT (ed)
355 Non-equilibrium statistical mechanics and turbulence, J. CARDY, G. FALKOVICH & K. GAWEDZKI
356 Elliptic curves and big Galois representations, D. DELBOURGO
357 Algebraic theory of differential equations, M.A.H. MACCALLUM & A.V. MIKHAILOV (eds)
358 Geometric and cohomological methods in group theory, M.R. BRIDSON, P.H. KROPHOLLER & I.J. LEARY (eds)
359 Moduli spaces and vector bundles, L. BRAMBILA-PAZ, S.B. BRADLOW, O. GARCÍA-PRADA & S. RAMANAN (eds)
360 Zariski geometries, B. ZILBER
361 Words: Notes on verbal width in groups, D. SEGAL
362 Differential tensor algebras and their module categories, R. BAUTISTA, L. SALMERÓN & R. ZUAZUA
363 Foundations of computational mathematics, Hong Kong 2008, F. CUCKER, A. PINKUS & M.J. TODD (eds)
364 Partial differential equations and fluid mechanics, J.C. ROBINSON & J.L. RODRIGO (eds)
365 Surveys in combinatorics 2009, S. HUCZYNSKA, J.D. MITCHELL & C.M. RONEY-DOUGAL (eds)
366 Highly oscillatory problems, B. ENGQUIST, A. FOKAS, E. HAIRER & A. ISERLES (eds)
367 Random matrices: High dimensional phenomena, G. BLOWER
368 Geometry of Riemann surfaces, F.P. GARDINER, G. GONZÁLEZ-DIEZ & C. KOUROUNIOTIS (eds)
369 Epidemics and rumours in complex networks, M. DRAIEF & L. MASSOULIÉ
370 Theory of p-adic distributions, S. ALBEVERIO, A.YU. KHRENNIKOV & V.M. SHELKOVICH
371 Conformal fractals, F. PRZYTYCKI & M. URBAŃSKI
372 Moonshine: The first quarter century and beyond, J. LEPOWSKY, J. MCKAY & M.P. TUITE (eds)
373 Smoothness, regularity and complete intersection, J. MAJADAS & A. G. RODICIO
374 Geometric analysis of hyperbolic differential equations: An introduction, S. ALINHAC
375 Triangulated categories, T. HOLM, P. JØRGENSEN & R. ROUQUIER (eds)
376 Permutation patterns, S. LINTON, N. RUŠKUC & V. VATTER (eds)
377 An introduction to Galois cohomology and its applications, G. BERHUY
378 Probability and mathematical genetics, N. H. BINGHAM & C. M. GOLDIE (eds)
379 Finite and algorithmic model theory, J. ESPARZA, C. MICHAUX & C. STEINHORN (eds)
380 Real and complex singularities, M. MANOEL, M.C. ROMERO FUSTER & C.T.C WALL (eds)
381 Symmetries and integrability of difference equations, D. LEVI, P. OLVER, Z. THOMOVA & P. WINTERNITZ (eds)
382 Forcing with random variables and proof complexity, J. KRAJÍČEK
383 Motivic integration and its interactions with model theory and non-Archimedean geometry I, R. CLUCKERS, J. NICAISE & J. SEBAG (eds)
384 Motivic integration and its interactions with model theory and non-Archimedean geometry II, R. CLUCKERS, J. NICAISE & J. SEBAG (eds)
385 Entropy of hidden Markov processes and connections to dynamical systems, B. MARCUS, K. PETERSEN & T. WEISSMAN (eds)
386 Independence-friendly logic, A.L. MANN, G. SANDU & M. SEVENSTER
387 Groups St Andrews 2009 in Bath I, C.M. CAMPBELL *et al* (eds)
388 Groups St Andrews 2009 in Bath II, C.M. CAMPBELL *et al* (eds)
389 Random fields on the sphere, D. MARINUCCI & G. PECCATI
390 Localization in periodic potentials, D.E. PELINOVSKY
391 Fusion systems in algebra and topology, M. ASCHBACHER, R. KESSAR & B. OLIVER
392 Surveys in combinatorics 2011, R. CHAPMAN (ed)
393 Non-abelian fundamental groups and Iwasawa theory, J. COATES *et al* (eds)
394 Variational problems in differential geometry, R. BIELAWSKI, K. HOUSTON & M. SPEIGHT (eds)
395 How groups grow, A. MANN
396 Arithmetic dfferential operators over the p-adic Integers, C.C. RALPH & S.R. SIMANCA
397 Hyperbolic geometry and applications in quantum chaos and cosmology, J. BOLTE & F. STEINER (eds)
398 Mathematical models in contact mechanics, M. SOFONEA & A. MATEI
399 Circuit double cover of graphs, C.-Q. ZHANG
400 Dense sphere packings: a blueprint for formal proofs, T. HALES
401 A double Hall algebra approach to affine quantum Schur-Weyl theory, B. DENG, J. DU & Q. FU
402 Mathematical aspects of fluid mechanics, J. ROBINSON, J.L. RODRIGO & W. SADOWSKI (eds)
403 Foundations of computational mathematics: Budapest 2011, F. CUCKER, T. KRICK, A. SZANTO & A. PINKUS (eds)
404 Operator methods for boundary value problems, S. HASSI, H.S.V. DE SNOO & F.H. SZAFRANIEC (eds)
405 Torsors, étale homotopy and applications to rational points, A.N. SKOROBOGATOV (ed)
406 Appalachian set theory, J. CUMMINGS & E. SCHIMMERLING (eds)
407 The maximal subgroups of the low-dimensional finite classical groups, J.N. BRAY, D.F. HOLT & C.M. RONEY-DOUGAL
408 Complexity science: the Warwick master's course, R. BALL, R.S. MACKAY & V. KOLOKOLTSOV (eds)
409 Surveys in combinatorics 2013, S. BLACKBURN, S. GERKE & M. WILDON (eds)
410 Representation theory and harmonic analysis of wreath products of finite groups, T. CECCHERINI SILBERSTEIN, F. SCARABOTTI & F. TOLLI
411 Moduli spaces, L. BRAMBILA-PAZ, O. GARCIA-PRADA, P. NEWSTEAD & R. THOMAS (eds)
412 Automorphisms and equivalence relations in topological dynamics, D.B. ELLIS & R. ELLIS
413 Optimal transportation: theory and applications, Y. OLLIVIER, H. PAJOT & C. VILLANI (eds)

London Mathematical Society Lecture Note Series: 414

Automorphic Forms and Galois Representations

Volume 1

Edited by

FRED DIAMOND
King's College London

PAYMAN L. KASSAEI
McGill University, Montréal

MINHYONG KIM
University of Oxford

CAMBRIDGE
UNIVERSITY PRESS

University Printing House, Cambridge CB2 8BS, United Kingdom

Cambridge University Press is part of the University of Cambridge.

It furthers the University's mission by disseminating knowledge in the pursuit of education, learning and research at the highest international levels of excellence.

www.cambridge.org
Information on this title: www.cambridge.org/9781107691926

First published 2014

A catalogue record for this publication is available from the British Library

Library of Congress Cataloguing in Publication data
Automorphic forms and galois representations / edited by Fred Diamond, King's College London, Payman L. Kassaei, McGill University, Montréal, Minhyong Kim, University of Oxford.
volumes <1–2> cm. – (London Mathematical Society lecture note series ; 414, 415)
Papers presented at the London Mathematical Society, and EPSRC (Great Britain Engineering and Physical Sciences Research Council), Symposium on Galois Representations and Automorphic Forms, held at the University of Durham from July 18–28, 2011.
ISBN 978-1-107-69192-6 (v. 1) – ISBN 978-1-107-69363-0 (v. 2)
1. Automorphic forms–Congresses. 2. Automorphic functions–Congresses. 3. Forms (Mathematics)–Congresses. 4. Galois theory–Congresses. I. Diamond, Fred, editor of compilation. II. Kassaei, Payman L., 1973– editor of compilation. III. Kim, Minhyong, editor of compilation. IV. Symposium on Galois Representations and Automorphic Forms (2011 : Durham, England)
QA353.A9A925 2014
515′.9–dc23
2014001841

ISBN 978-1-107-69192-6 Paperback

Contents

List of contributors — *page* vi
Preface — ix

1 A semi-stable case of the Shafarevich Conjecture — 1
Victor Abrashkin

2 Irreducible modular representations of the Borel subgroup of $\mathrm{GL}_2(\mathbf{Q}_p)$ — 32
Laurent Berger and Mathieu Vienney

3 p-adic L-functions and Euler systems: a tale in two trilogies — 52
Massimo Bertolini, Francesc Castella, Henri Darmon, Samit Dasgupta, Kartik Prasanna, and Victor Rotger

4 Effective local Langlands correspondence — 102
Colin J. Bushnell

5 The conjectural connections between automorphic representations and Galois representations — 135
Kevin Buzzard and Toby Gee

6 Geometry of the fundamental lemma — 188
Pierre-Henri Chaudouard

7 The p-adic analytic space of pseudocharacters of a profinite group and pseudorepresentations over arbitrary rings — 221
Gaëtan Chenevier

8 La série principale unitaire de $\mathbf{GL}_2(\mathbf{Q}_p)$: vecteurs localement analytiques — 286
Pierre Colmez

9 Equations différentielles p-adiques et modules de Jacquet analytiques — 359
Gabriel Dospinescu

Contributors

Victor Abrashkin, Department of Mathematical Sciences, University of Durham, Durham, DH1, 3LE, UK.

Laurent Berger, UMPA, Ecole Normale Supérieure de Lyon, Lyon, 69007, France.

Massimo Bertolini, Diparetimento di Matematica, Università di Milano, Milano 20133, Italy.

Colin Bushnell, Department of Mathematics, King's College London, London, WC2R 2LS, UK.

Kevin Buzzard, Department of Mathematics, Imperial College London, London, SW7 2RH, UK.

Francesc Castella, Department of Mathematics, University of California, Los Angeles, CA 90095-1555, USA.

Pierre-Henri Chaudouard, Institut de Mathématiques de Jussieu, Université Paris Diderot (Paris 7), 75205 Paris Cedex 13, France.

Gaëtan Chenevier, Centre de Mathématiques Laurent Schwartz, École Polytechnique, Paris, 91128 Palaiseau Cedex, France.

Pierre Colmez, CNRS. Institute de Mathématiques de Jussieu, Paris, 75005, France.

Henri Darmon, Department of Mathematics and Statistics, McGill University, Montreal, Canada, H3A 0B9.

Samit Dasgupta, Department of Mathematics, Univeristy of California, Santa Cruz, CA 95064, USA.

Gabriel Dospinescu, UMPA – UMR, Ecole Normale Supérieure de Lyon, 69 364 Lyon Cedex 07, France.

Toby Gee, Department of Mathematics, Imperial College London, London, SW7 2RH, UK.

Kartik Prasanna, Department of Mathematics, University of Michigan, Ann Arbor, MI 48109-1043, USA.

Victor Rotger, Mathemàtica Aplicada II, Universitat Politècnica de Catalunya, Barcelona 08034, Spain.

Mathieu Vienney, UMPA, Ecole Normale Supérieure de Lyon, Lyon 69007, France.

Preface

The London Mathematical Society Symposium – EPSRC Symposium on Galois Representations and Automorphic Forms was held at the University of Durham from 18th July until 28th July 2011. These topics have been playing an important role in present-day number theory, especially via the Langlands program and the connections it entails. The meeting brought together researchers from around the world on these and related topics, with lectures on a variety of recent major developments in the area. Roughly half of these talks were individual lectures, while the rest constituted series on the following themes:

- p-adic local Langlands
- Curves and vector bundles in p-adic Hodge theory
- The fundamental lemma and trace formula
- Anabelian geometry
- Potential automorphy

These Proceedings present much of the progress described in those lectures. The organizers are very grateful to all the speakers and to others who contributed articles. We also wish to thank the London Mathematical Society and EPSRC for the financial support that made the meeting possible. We warmly appreciate the assistance and hospitality provided by the University of Durham's Department of Mathematics and Grey College. These institutions have helped to make the Symposia such a well established and highly valued event in the number theory community.

Fred Diamond
Payman Kassaei
Minhyong Kim

1

A semi-stable case of the Shafarevich Conjecture

Victor Abrashkin

Abstract

Suppose $K = W(k)[1/p]$, where $W(k)$ is the ring of Witt vectors with coefficients in an algebraically closed field k of characteristic $p \neq 2$. We discuss an explicit construction of p-adic semi-stable representations of the absolute Galois group of K with Hodge–Tate weights from $[0, p)$. This theory is applied to projective algebraic varieties over $\mathbb{Q}$ with good reduction outside 3 and semi-stable reduction modulo 3.

Introduction

In this expository paper we discuss the following result in the spirit of the Shafarevich Conjecture about non-existence of non-trivial abelian schemes over $\mathbb{Z}$.

Theorem 1.1. *If Y is a projective algebraic variety over $\mathbb{Q}$ with good reduction outside 3 and semi-stable reduction modulo 3 then $h^2(Y_{\mathbb{C}}) = h^{1,1}(Y_{\mathbb{C}})$.*

In particular, the above theorem implies that there are no such (non-trivial) abelian varieties Y (first proved in [13, 27]). Our result also eliminates a great deal of other varieties, e.g. all K3-surfaces.

The proof of Theorem 1.1 is given in [11] and is based on a:

- study of torsion subquotients of the Galois module $H^2_{et}(Y_{\bar{Q}}, \mathbb{Q}_3)$;
- modification of Breuil's torsion theory of semi-stable p-adic represenations with HT (Hodge–Tate) weights from $[0, p-1]$ over $W(k)$, where k is an algebraically closed field of characteristic p;

2010 *Mathematics Subject Classification*. 11S20, 11G35, 14K15.
Key words and phrases. p-adic semi-stable representations, Shafarevich Conjecture.

Automorphic Forms and Galois Representations, ed. Fred Diamond, Payman L. Kassaei and Minhyong Kim. Published by Cambridge University Press.

– formalism of pre-abelian categories (short exact sequences, 6-term Hom – Ext exact sequences, p-divisible group objects, devissage);
– study of the group of fundamental units in $\mathbb{Q}(\sqrt[3]{3}, e^{2\pi i/9})$ (via the computing package SAGE).

The strategy of the proof is very close to the strategy used in the following "crystalline case" of the Shafarevich Conjecture [23, 7].

Theorem 1.2. *Suppose X is a projective algebraic variety over $\mathbb{Q}$ with everywhere good reduction. Then*

(a) $h^1(X_{\mathbb{C}}) = 0$, $h^2(X_{\mathbb{C}}) = h^{1,1}(X_{\mathbb{C}})$ *and* $h^3(X_{\mathbb{C}}) = 0$;
(b) $h^4(X_{\mathbb{C}}) = h^{2,2}(X_{\mathbb{C}})$ *under Generalized Riemann Hypothesis (GRH).*

Part (a) of this Theorem was obtained in [7] by studying the finite subquotients of the Galois modules $H^i_{et}(X_{\bar{\mathbb{Q}}}, \mathbb{Q}_5)$ with $1 \leqslant i \leqslant 3$. These Galois modules are unramified outside 5 and their local behaviour at 5 is described by the Fontaine–Laffaille theory [19] of p-adic torsion crystalline representations with HT weights from $[0, p-2]$. The approach in [7] is essentially similar to the approach from [23] but Fontaine considers etale cohomology with coefficients in $\mathbb{Q}_7$. (Of course, these results would be not possible without the great achievements of Fontaine's theory of p-adic periods.)

Part (b) was proved by the author in [7]. The proof requires the study of the Galois module $H^4_{et}(X_{\bar{\mathbb{Q}}}, \mathbb{Q}_5)$, where the tools of the Fontaine–Laffaille theory are not sufficient. For this reason, we developed in [6] a modification of the Fontaine–Laffaille theory for crystalline representations with HT weights from $[0, p-1]$. Note that our modification of Breuil's theory works also in the context of crystalline representations and can be applied to reprove part b) of Theorem 1.2 (and similar results for varieties over $\mathbb{Q}(i)$, $\mathbb{Q}(\sqrt{-3})$ and $\mathbb{Q}(\sqrt{5})$ from [7]). The appropriate comments will be given in due course below.

The constructions in [11] are very technical and we just sketch and discuss their basic steps. Most of them can be illustrated by earlier results related to the Shafarevich Conjecture, cf. Section 1.

In Sections 2–4 we work with a local field $K = \mathrm{Frac}\, W(k)$, where $W(k)$ is the ring of Witt vectors with coefficients in an algebracally closed field k of characteristic p, $p > 2$. Let $\bar{K}$ be an algebraic closure of K and $\Gamma_K = \mathrm{Gal}(\bar{K}/K)$. In Section 2 we outline the construction of the functor $\mathcal{V}^*$ from an appropriate category of filtered modules to the category of $\mathbb{F}_p[\Gamma_K]$-modules. This construction is based on the introduction of a modulo p "truncated" version of Fontaine's ring of p-adic semi-stable periods.

We associate to $\mathcal{V}^*$ the functor $\mathcal{CV}^*$ with values in the category of co-filtered $\mathbb{F}_p[\Gamma_K]$-modules and prove that this functor is fully faithful. In Section 3 we obtain the ramification estimates for the Galois modules H from the image of $\mathcal{V}^*$: if $v > 2 - 1/p$ then the higher ramification subgroups $\Gamma_K^{(v)}$ act trivially on H. We also obtain the ramification estimate for the Galois modules which are associated with the modulo p subquotients of crystalline representations with HT weights from $[0, p)$ and prove that both estimates are sharp. The methods we use here are close to the methods from [8, 9, 10]; one can use also our constructions to show that the estimates from [24] are sharp if $e = n = 1$. In Section 4 we explain the construction of our modification of Breuil's functor $\mathcal{V}^{ft}$. In fact, it is very close to the construction of the modification of the Fontaine–Laffaille functor from [6] but it can be developed in a simpler way due to advantages of Breuil's theory. One of the main features of this construction is that on the level of modulo p subquotients, $\mathcal{V}^{ft}$ essentially coincides with the functor $\mathcal{V}^*$ from Subsection 2. This gives the ramification estimates for modulo p subquotients of semi-stable and crystalline representations with HT weights from $[0, p)$. Finally, in Section 5 we outline the proofs of Theorems 1.1 and 1.2.

1. The Shafarevich Conjecture

Conjecture. (I. R. Shafarevich, 1962) *There are no projective algebraic curves over $\mathbb{Q}$ of genus $g \geqslant 1$ with everywhere good reduction, [29].*

The case $g = 1$ was considered by Shafarevich himself. He has just listed explicitly 22 elliptic curves over $\mathbb{Q}$ with good reduction outside 2 and verified that all these curves have bad reduction at 2. Later his PhD student (Volynsky) studied the case of curves of genus 2. This approach resulted in enormous calculations and was not published. In both cases the approach was based on the study of canonical equations for these curves. It became clear later that one should study the problem in a more general setting.

Conjecture. *There are no abelian varieties A over $\mathbb{Q}$ of dimension $g \geq 1$ with everywhere good reduction.*

This statement is easier to approach. The existence of such abelian variety would have provided examples of non-trivial p-divisible groups over $\mathbb{Z}$ (for all prime numbers p). The question about the existence of such p-divisible groups was asked by J.Tate in [31]. In this way the conjecture was proved in [21, 3] in 1985. The main features of used methods will be described below.

1.1. Small values of g

In [1, 2] it was proved that any 2-divisible group over $\mathbb{Z}$ of height $h \leqslant 6$ is isogeneous to the trivial 2-divisible group. This gave the cases $g = 2$ and $g = 3$ of the Shafarevich Conjecture. The method can be explained as follows.

Suppose G is a f.f.g.s. (finite flat group scheme) over $\mathbb{Z}$ such that $2\mathrm{id}_G = 0$. Then

(a) if the order $|G| = 2$ then G is either etale $(\mathbb{Z}/2)_{\mathbb{Z}} = \mathrm{Spec}(\mathbb{Z} \oplus \mathbb{Z})$ or multiplicative $\mu_2 = \mathrm{Spec}\ \mathbb{Z}[x]/(x^2 - 1)$ f.f.g.s. over $\mathbb{Z}$, [31];

(b) if $|G| = 4$ and $G = \mathrm{Spec}\ A(G)$ is not a product of f.f.g.s. of order 2 then there is a short exact sequence of f.f.g.s.

$$0 \longrightarrow \mu_2 \longrightarrow G \longrightarrow (\mathbb{Z}/2)_{\mathbb{Z}} \longrightarrow 0$$

and $A(G) = A(\mu_2) \oplus \mathbb{Z}[i]$, [1]. In particular, $A(G)_{\mathbb{Q}} \neq \mathbb{Q} \oplus K$, where $[K : \mathbb{Q}] = 3$. (Use that $A(G)_{\mathbb{Q}}$ is etale over $\mathbb{Q}$ and there are no cube field extensions $K/\mathbb{Q}$ unramified outside 2.)

(c) there are similar short exact sequences for f.f.g.s. G over $\mathbb{Z}$ of order 2^n with $n = 3, 4, 5, 6$,

$$0 \longrightarrow \mu_2^a \longrightarrow G \longrightarrow (\mathbb{Z}/2)_{\mathbb{Z}}^b \longrightarrow 0,$$

where $a + b = n$, [2]. This statement is highly non-trivial because the Galois group of the field-of-definition $\mathbb{Q}(G)$ of $\bar{\mathbb{Q}}$-points of f.f.g.s. of order 2^n is not generally soluble if $n \geqslant 4$. On the one hand, we used the Tate formula for the discriminant of $A(G)$ from [31], $v_2(D(A(G)) = d2^n$, where $d = \dim(G \otimes \mathbb{F}_2)$ (it implies that $v_2(D(A(G))) \leqslant 192$ because we can assume that $d \leqslant 3$ by switching, if necessary, from G to its Cartier dual). On the other hand, we used the Odlyzko lower bounds for the minimal discriminants of algebraic number fields, cf. [30, 18, 25];

(d) in the special pre-abelian category of f.f.g.s. G over $\mathbb{Z}$ such that $2\mathrm{id}_G = 0$, one has

$$\mathrm{Ext}(\mu_2, (\mathbb{Z}/2)_{\mathbb{Z}}) = \mathrm{Ext}((\mathbb{Z}/2)_{\mathbb{Z}}, (\mathbb{Z}/2)_{\mathbb{Z}}) = \mathrm{Ext}(\mu_2, \mu_2) = 0.$$

Therefore, the above exact sequences for G and devissage in the pre-abelian category of finite flat 2-group schemes over $\mathbb{Z}$ give the following exact sequence of 2-divisible groups over $\mathbb{Z}$

$$0 \longrightarrow \{\mu_{2^n}\}_{n\geq 1}^a \longrightarrow \mathcal{G} \longrightarrow (\mathbb{Q}_2/\mathbb{Z}_2)^b \longrightarrow 0, \tag{1.1}$$

where $\mathcal{G}$ is of height $a + b \leqslant 6$ (for more details about devissage in pre-abelian categories cf. Appendix, especially Theorem A.1);

(e) such 2-divisible group $\mathcal{G}$ never comes from a non-trivial abelian scheme A over $\mathbb{Z}$. Otherwise, looking at dimensions we obtain $b \neq 0$, but the exact sequence of 2-divisible groups from (d) splits over $\mathbb{F}_2$ and, therefore, A has infinitely many $\mathbb{F}_2$-points. The contradiction.

The above method does not work in higher dimensions.

Indeed, suppose A is an abelian scheme over $\mathbb{Z}$ and $G = \text{Ker}(2\text{id}_A)$ is a group scheme of points of order $\leqslant 2$ on A. Then $|G| = 2^{2g}$, $\dim(G \otimes \mathbb{F}_2) = g$ and Tate's formula gives $v_2(D(A(G))^{1/2g}) = g$. Note that $A(G) \otimes \mathbb{Q}$ is the product of algebraic number fields (because $G \otimes \mathbb{Q}$ is etale) and these fields are unramified outside 2 (because $G \otimes \mathbb{Z}_l$ is etale if $l \neq 2$). Therefore, the normalized discriminant of $A(G)$ equals 2^g and tends to infinity if $g \to \infty$.

On the other hand, if $\mathbb{Q}(G)$ is the field-of-definition of $\bar{\mathbb{Q}}$-points of G, then $\text{Gal}(\mathbb{Q}(G)/\mathbb{Q}) \subset \text{SL}(2g, \mathbb{F}_2)$ is not generally soluble if $g \geqslant 2$, and the only global idea we can use in this situation is related to lower bounds of minimal discriminants of algebraic number fields. The best known bounds are the Odlyzko estimates and they are given by the tables of real numbers $\{d_N \mid N \in \mathbb{N}\}$ such that if $[K : \mathbb{Q}] = N$ then $|D(K/\mathbb{Q})|^{1/N} \geq d_N$. For large N, $d_N \approx d_\infty \approx 22.3$; under *GRH* there are better estimates $\{d_N^* \mid N \in \mathbb{N}\}$ in this case $d_\infty^* \approx 44.76$, [30, 18, 25].

Unfortunately, an analogue of Odlyzko estimates under additional assumption that $K/\mathbb{Q}$ is ramified only over 2, does not exist. Nonetheless, $A(G)$ is considerably smaller than its integral closure and Tate's formula can be replaced by a much better upper estimate for the 2-adic valuation of the normalized discriminant of $\mathbb{Q}(G)$. The evidence for its existence is illustrated in the next section.

1.2. The Shafarevich Conjecture, the ordinary case

Suppose our abelian variety A has good ordinary reduction at 2. Then:

(a) $G := \text{Ker}(2\text{id}_A)$ is a f.f.g.s. over $\mathbb{Z}$ of order 2^{2g};

(b) there is a short exact sequence of f.f.g.s. over $\mathbb{Z}_2$

$$0 \longrightarrow H^{mult} \longrightarrow G \otimes_{\mathbb{Z}} \mathbb{Z}_2 \longrightarrow H^{et} \longrightarrow 0,$$

where H^{mult} is multiplicative and H^{et} is etale group schemes over $\mathbb{Z}_2$ of order 2^g;

(c) because $H^{et} \otimes W(\bar{\mathbb{F}}_2) = \prod_j (\mathbb{Z}/2)_{W(\bar{\mathbb{F}}_2)}$ and $H^{mult} \otimes W(\bar{\mathbb{F}}_2) = \prod_i \mu_{2,W(\bar{F}_2)}$, we have

$$G \otimes W(\bar{\mathbb{F}}_2) = \sum_{i,j} G_{ij} \in \oplus_{i,j} \text{Ext}((\mathbb{Z}/2)_{W(\bar{\mathbb{F}}_2)}, \mu_{2,W(\bar{\mathbb{F}}_2)}),$$

where for all i, j, there are short exact sequences of f.f.g.s.

$$0 \longrightarrow \mu_{2,W(\bar{\mathbb{F}}_2)} \longrightarrow G_{ij} \longrightarrow (\mathbb{Z}/2)_{W(\bar{\mathbb{F}}_2)} \longrightarrow 0;$$

(d) the field-of-definition of geometric points of G_{ij} over the maximal unramified extension $\mathbb{Q}_{2,ur}$ of $\mathbb{Q}_2$, is $\mathbb{Q}_{2,ur}(\sqrt{v_{ij}})$, where all v_{ij} are principal units in $\mathbb{Q}_{2,ur}$, cf. Appendix by J.Tate in [26]. Therefore, for all $v > 1$, the higher ramification subgroups $\Gamma_{\mathbb{Q}_2}^{(v)}$ of $\Gamma_{\mathbb{Q}_2} = \mathrm{Gal}(\bar{\mathbb{Q}}_2/\mathbb{Q}_2)$ act trivially on the field-of-definition $\mathbb{Q}_2(G)$ of all $\bar{\mathbb{Q}}_2$-points of G;

(e) the triviality of $\Gamma_{\mathbb{Q}_2}^{(v)}$-action, where $v > 1$, implies the inequality $|D(\mathbb{Q}(G)/\mathbb{Q})|^{1/[\mathbb{Q}(G):\mathbb{Q}]} < 2^2$ (e.g. use Prop 9.4 of Ch. 1, [12]). But the Odlyzko estimate $d_4 < 4$ and we obtain $[\mathbb{Q}(G) : \mathbb{Q}] < 4$. Therefore, $\mathbb{Q}(G) \subset \mathbb{Q}(i)$, we can use devissage to obtain the exact sequence (1.1) for $a = b = g$ and finish the proof similarly to the case of small g.

In the above discussion, the prime number 2 can be replaced by arbitrary prime number p. If $A \otimes \mathbb{F}_p$ is ordinary and $G = \mathrm{Ker}(p\,\mathrm{id}_A)$ then for $v > 1/(p-1)$, the ramification subgroups $\Gamma_{\mathbb{Q}_p}^{(v)}$ act trivially on $\mathbb{Q}_p(G)$ and using the Odlyzko estimates we can see that for $3 \leqslant p \leqslant 17$, $\mathbb{Q}(G) \subset \mathbb{Q}(\sqrt[p]{1})$. This implies that G is the product of constant etale and multiplicative f.f.g.s. over $\mathbb{Z}$, the corresponding p-divisible group of A will be just the product of several copies of trivial etale $(\mathbb{Q}_p/\mathbb{Z}_p)_{\mathbb{Z}}$ and multiplicative $\{\mu_{2^n,\mathbb{Z}}\}_{n\geqslant 1}$ p-divisible groups over $\mathbb{Z}$ and, therefore, such abelian variety does not exist.

The above case of the Shafarevich Conjecture was not published but gave a right direction towards the proof of the general case.

1.3. The Shafarevich Conjecture, the general case

In this case the same ramification estimates are proved in the general situation [21, 5]: if G is a finite flat group scheme over $W(k)$, where k is a perfect field of characteristic p, $p\,\mathrm{id}_G = 0$ and Frac $W(k) = K$ then the higher ramification subgroups $\Gamma_K^{(v)}$ act trivially on the field-of-definition of $\bar{K}$-points of G for all $v > 1/(p-1)$.

Essentially, Fontaine found ramification estimates for any finite flat p-group schemes over the valuation ring O_L of complete discrete valuation field $L \supset \mathbb{Q}_p$. His method uses the rigidity properties of p-divisible groups defined over valuation rings and a very elegant interpretation of Krasner's Lemma. The methods in [3, 5] are much more computational and use Fontaine's theory of f.f.g.s. over Witt vectors, [20]. In Section 3 we present an alternative proof of ramification estimates. It works also equally well for the subquotients of crystalline and semi-stable p-adic representations.

In our approach from [3, 5] we treated systematically also the case $p = 2$. Here the category of f.f.g.s. over $W(k)$ is not abelian contrary to the case $p \neq 2$, but one can still proceed with the devissage. This gave us not only the bigger list of algebraic number fields where the Shafarevich conjecture about the non-existence of abelian varieties with everywhere good reduction holds. Our main idea [4] of removing the restriction to unipotent objects in Fontaine's classification of 2-group schemes in [20] gave later a right approach to the constructions of modifications of the Fontaine–Laffaille [6] and Breuil [11] functors. These modifications allow us to obtain the ramification estimates for all modulo p subquotients of representations with HT weights from $[0, p)$. They also provide us with the nullity of some groups of extensions in the category of Galois modules appearing as such subquotients. As a matter of fact, these two key ingredients resulted finally in proving Theorem 1.1 and part (*b*) of Theorem 1.2.

2. The functor $\mathcal{CV}^*$

Let $\mathcal{W}_1 = k[[u]]$, where u is an indeterminate. Denote by σ the automorphism of k induced by the p-th power map on k and agree to use the same symbol for the continuous extension of σ to $\mathcal{W}_1$ such that $\sigma(u) = u^p$. Denote by $N : \mathcal{W}_1 \longrightarrow \mathcal{W}_1$ the unique continuous k-differentiation such that $N(u) = -u$.

2.1. Categories of filtered modules

Introduce the following categories:

- the category $\widetilde{\underline{\mathcal{L}}}{}_0^*$ – its objects are $\mathcal{L} = (L, F(L), \varphi)$, where L and $F(L)$ are $\mathcal{W}_1$-modules, $L \supset F(L)$ and $\varphi : F(L) \longrightarrow L$ is a σ-linear morphism of $\mathcal{W}_1$-modules; the morphisms are $\mathcal{W}_1$-linear maps of filtered modules which commute with the corresponding σ-linear maps φ;
- the category $\widetilde{\underline{\mathcal{L}}}{}^*$ – its objects are $\mathcal{L} = (L, F(L), \varphi, N)$, where $(L, F(L), \varphi) \in \widetilde{\underline{\mathcal{L}}}{}_0^*$ and $N : L \longrightarrow L/u^pL$ is such that for $w \in \mathcal{W}_1$ and $l \in L$, $N(wl) = N(w)l + wN(l)$ (we use the same notation l for the image of l in L/u^pL); the morphisms are the morphisms from $\widetilde{\underline{\mathcal{L}}}{}_0^*$ which commute with the corresponding differentiations N;
- the category $\underline{\mathcal{L}}_0^*$ is a full subcategory of $\widetilde{\underline{\mathcal{L}}}{}_0^*$ consisting of $\mathcal{L} = (L, F(L), \varphi)$ such that the module L is free of finite rank, $u^{p-1}L \subset F(L)$ and the natural embedding $\varphi(F(L)) \subset L$ induces the identification

$$\varphi(F(L)) \otimes_{\sigma(\mathcal{W}_1)} \mathcal{W}_1 = L;$$

- the category $\underline{\mathcal{L}}^*$ is a full subcategory of $\widetilde{\underline{\mathcal{L}}}{}^*$ consisting of $\mathcal{L} = (L, F(L), \varphi, N)$ such that $(L, F(L), \varphi) \in \underline{\mathcal{L}}_0^*$, for any $l \in F(L)$, one has $uN(l) \in F(L) \bmod u^pL$ and $N(\varphi(l)) = \varphi(uN(l))$ (we use the same notation φ for the morphism $\varphi \bmod u^pL$).

The above categories are analogs of the categories of filtered modules from [14], Subsection 2.1.2, but we work with the category of $\mathcal{W}_1$-modules. (Breuil uses modules over the appropriate divided power envelope of $W(k)[[u]]$.) Note that in the context of $\mathcal{W}_1$-modules the monodromy operator N can't be defined as a map with values in L. In [11], Subsection 1.1, we proved that N can be defined as a map from L to $L/u^{2p}L$ and it appears as a unique lift of its reduction $N_1 = N \bmod u^p L$. (We used the existence of such a lift when proving in [11] that the category $\underline{\mathcal{L}}^*$ is pre-abelian; we also need this property when defining the functor $\mathcal{V}^*$ in Subsection 2.3 below.) In this chapter we use the notation N for this $(\bmod\, u^p)$-map N_1;

- the category $\underline{\mathcal{L}}^*_{cr}$ is a full subcategory in $\underline{\mathcal{L}}^*$ consisting of the objects $(L, F(L), \varphi, N)$ such that $N(\varphi(F(L))) = 0$.

For obvious reasons, $(L, F(L), \varphi, N) \in \underline{\mathcal{L}}^*_{cr}$ is completely determined by $(L, F(L), \varphi) \in \underline{\mathcal{L}}^*_0$. Note that the category $\underline{\mathcal{L}}^*_{cr}$ is very closely related to the category of Fontaine–Laffaille modules, cf. [11], Subsection 1.3.

According to above definitions the objects of the categories $\underline{\mathcal{L}}^*_0$, $\underline{\mathcal{L}}^*$ and $\underline{\mathcal{L}}^*_{cr}$ are filtered free $\mathcal{W}_1$-modules with additional structures. The category of filtered free $\mathcal{W}_1$-modules is a typical example of a special pre-abelian category, i.e. it is an additive category with kernels and cokernels and nicely behaving bifunctor Ext, cf. Appendix. In Subsection 1.1 of [11] we verified that $\underline{\mathcal{L}}^*_0$, $\underline{\mathcal{L}}^*$ and $\underline{\mathcal{L}}^*_{cr}$ inherit the property of being special pre-abelian.

There are the concepts of etale, unipotent, connected and multiplicative objects in our categories defined in the following way; for more details cf. Subsection 1.2 of [11].

Suppose $\mathcal{L} = (L, F(L), \varphi, N) \in \underline{\mathcal{L}}^*$.

Introduce a σ-linear map $\phi : L \longrightarrow L$ via $\phi : l \mapsto \varphi(u^{p-1}l)$. The module $\mathcal{L}$ is etale (resp., connected) if $\phi \bmod u$ is invertible (resp., nilpotent) on L/uL. Denote by $\underline{\mathcal{L}}^{*et}$ (resp, $\underline{\mathcal{L}}^{*c}$) the full subcategory of $\underline{\mathcal{L}}^*$ consisting of etale (resp. connected) objects. Then any $\mathcal{L} \in \underline{\mathcal{L}}^*$ contains a unique maximal etale subobject $(\mathcal{L}^{et}, i^{et})$ and a unique maximal connected quotient object $(\mathcal{L}^c, j^c)$ and the sequence

$$0 \longrightarrow \mathcal{L}^{et} \xrightarrow{i^{et}} \mathcal{L} \xrightarrow{j^c} \mathcal{L}^c \longrightarrow 0$$

is short exact.

Note that $\varphi(F(L))$ is a $\sigma(\mathcal{W}_1)$-module and $L = \varphi(F(L)) \otimes_{\sigma(\mathcal{W}_1)} \mathcal{W}_1$. If $l \in L$ and for $0 \leqslant i < p$, the elements $l^{(i)} \in F(L)$ are such that $l = \sum_{0 \leqslant i < p} \varphi(l^{(i)}) \otimes u^i$, set $V(l) = l^{(0)}$. Then $V \bmod u$ is a σ^{-1}-linear endomorphism of the k-vector space L/uL.

The module $\mathcal{L}$ is multiplicative (resp., unipotent) if V mod u is invertible (resp., nilpotent) on L/uL. Denote by $\underline{\mathcal{L}}^{*m}$ (resp, $\underline{\mathcal{L}}^{*u}$) the full subcategory of $\underline{\mathcal{L}}^*$ consisting of multiplicative (resp. unipotent) objects. Then any $\mathcal{L} \in \underline{\mathcal{L}}^*$ contains a unique maximal multiplicative quotient object $(\mathcal{L}^m, j^m)$ and a unique maximal unipotent subobject $(\mathcal{L}^u, i^u)$ and the sequence

$$0 \longrightarrow \mathcal{L}^u \xrightarrow{i^u} \mathcal{L} \xrightarrow{j^m} \mathcal{L}^m \longrightarrow 0$$

is short exact.

Note that $\underline{\mathcal{L}}^{*c}$ and $\underline{\mathcal{L}}^{*u}$ are abelian categories: it follows easily from the description of simple objects of $\underline{\mathcal{L}}^*$ in Subsection 1.4 of [11].

2.2. The object $\mathcal{R}^0_{st} \in \widetilde{\underline{\mathcal{L}}}^*$

Let $R = \varprojlim_n (\bar{O}/p)_n$ be Fontaine's ring; it has a natural structure of k-algebra via the map $k \longrightarrow R$ given by $\alpha \mapsto \varprojlim_n ([\sigma^{-n}\alpha] \bmod p)$, where $[\gamma] \in W(k) \subset \bar{O}$ denotes the Teichmüller representative of $\gamma \in k$. Let m_R be the maximal ideal of R.

Choose $x_0 = (x_0^{(n)} \bmod p)_{n\geqslant 0} \in R$ and $\varepsilon = (\varepsilon^{(n)} \bmod p)_{n\geqslant 0}$ such that for all $n \geqslant 0$, $x_0^{(n+1)p} = x_0^{(n)}$ and $\varepsilon^{(n+1)p} = \varepsilon^{(n)}$ with $x_0^{(0)} = -p$, $\varepsilon^{(0)} = 1$ but $\varepsilon^{(1)} \neq 1$. Denote by v_R the valuation on R such that $v_R(x_0) = 1$.

Let Y be an indeterminate.

Consider the divided power envelope $R\langle Y\rangle$ of $R[Y]$ with respect to the ideal (Y). If for $j \geqslant 0$, $\gamma_j(Y)$ is the j-th divided power of Y then $R\langle Y\rangle = \oplus_{j\geqslant 0} R\gamma_j(Y)$. Denote by R_{st} the completion $\prod_{j\geqslant 0} R\gamma_j(Y)$ of $R\langle Y\rangle$ and set, $\mathrm{Fil}^p R_{st} = \prod_{j\geqslant p} R\gamma_j(Y)$. Define the σ-linear morphism of the R-algebra R_{st} by the correspondence $Y \mapsto x_0^p Y$; it will be denoted below by the same symbol σ.

Introduce a $\mathcal{W}_1$-module structure on R_{st} by the k-algebra morphism $\mathcal{W}_1 \longrightarrow R_{st}$ such that $u \mapsto \iota(u) := x_0 \exp(-Y) = x_0 \sum_{j\geqslant 0} (-1)^j \gamma_j(Y)$.

Set $F(R_{st}) = \sum_{0\leqslant i<p} x_0^{p-1-i} R\gamma_i(Y) + \mathrm{Fil}^p R_{st}$.

Define the continuous σ-linear morphism of R-modules $\varphi : F(R_{st}) \longrightarrow R_{st}$ by setting for $0 \leqslant i < p$, $\varphi(x_0^{p-1-i}\gamma_i(Y)) = \gamma_i(Y)(1 - (i/2)x_0^p Y)$, and for $i \geqslant p$, $\varphi(\gamma_i(Y)) = 0$.

Let N be a unique R-differentiation of R_{st} such that $N(Y) = 1$.

Note that $(R_{st}, F(R_{st}), \varphi, N)$ is not an object of $\widetilde{\underline{\mathcal{L}}}^*$, e.g. φ is not a σ-linear morphism of $\mathcal{W}_1$-modules. Nevertheless, all appropriate compatibilities between above introduced additional structures on R_{st} hold modulo $x_0^{2p} R_{st}$, cf. Proposition 2.1 in [11], and we can introduce

$$\mathcal{R}_{st}^0 = (R_{st}^0, F(R_{st}^0), \varphi, N) \in \widetilde{\underline{\mathcal{L}}}^*,$$

where $R_{st}^0 = R_{st} \operatorname{mod} x_0^p \mathrm{m}_R$ and $F(R_{st}^0) = F(R_{st}) \operatorname{mod} x_0^p \mathrm{m}_R$ with the appropriate induced maps φ and N.

In our theory $\mathcal{R}_{st}^0$ plays a role of the ring $\hat{A}_{st}$ from the theory of p-adic semi-stable representations [14], Subsection 3.1.1. In particular, R_{st}^0 can be provided with continuous Galois action as follows. For any $\tau \in \Gamma_K$, let $k(\tau) \in \mathbb{Z}$ be such that $\tau(x_0) = \varepsilon^{k(\tau)} x_0$ and let $\widetilde{\log}(1+X) = X - X^2/2 + \dots - X^{p-1}/(p-1)$ be the truncated logarithm. Define a map $\tau : R_{st} \longrightarrow R_{st}$ by extending the natural action of τ on R and setting for all $j \geqslant 0$,

$$\tau(\gamma_j(Y)) := \sum_{0 \leqslant i \leqslant \min\{j, p-1\}} \gamma_{j-i}(Y) \gamma_i(\widetilde{\log} \varepsilon).$$

Then the correspondences $\gamma_j(Y) \mapsto \tau(\gamma_j(Y))$ induce a Γ_K-action on the $\mathcal{W}_1$-algebra R_{st}^0 which extends the natural Γ_K-action on R and respects the structure of $\mathcal{R}_{st}^0$ as an object of the category $\widetilde{\underline{\mathcal{L}}}^*$, cf. Proposition 2.2 in [11].

2.3. The functor $\mathcal{V}^*$

For any $\mathcal{L} = (L, F(L), \varphi, N) \in \underline{\mathcal{L}}^*$, consider the Γ_K-module $\mathcal{V}^*(\mathcal{L}) = \operatorname{Hom}_{\widetilde{\underline{\mathcal{L}}}^*}(\mathcal{L}, \mathcal{R}_{st}^0)$. Note that in this definition we need N to be defined slightly better than just modulo $u^p L$ (we work modulo $x_0^p \mathrm{m}_R$ rather than modulo $x_0^p R$) but such lift exists and unique, cf. Subsection 2.1. The Galois module $\mathcal{V}^*(\mathcal{L})$ can be studied via the following method from [15], Subsection 2.3.

Let $\mathcal{R}^0 = (R^0, F(R^0), \varphi) \in \widetilde{\underline{\mathcal{L}}}_0^*$, where $R^0 = R/x_0^p \mathrm{m}_R$, $F(R^0) = x_0^{p-1} R^0$, the $\mathcal{W}_1$-module structure on R^0 is given via $u \mapsto x_0$ and ϕ is induced by the map $r \mapsto r^p / x_0^{p(p-1)}$, $r \in x_0^{p-1} R$.

If $f \in \mathcal{V}^*(\mathcal{L})$ and $i \geqslant 0$, introduce k-linear morphisms $f_i : L \longrightarrow R^0$ such that for any $l \in L$, $f(l) = \sum_{i \geqslant 0} f_i(l) \gamma_i(Y)$. The correspondence $f \mapsto f_0$ gives the homomorphism of abelian groups $\mathrm{pr}_0 : \mathcal{V}^*(\mathcal{L}) \longrightarrow \mathcal{V}_0^*(\mathcal{L}) := \operatorname{Hom}_{\widetilde{\underline{\mathcal{L}}}_0^*}(\mathcal{L}, \mathcal{R}^0)$. Then, cf. Subsection 2.2 of [11],

- pr_0 is an isomorphism of abelian groups;
- if $\mathrm{rk}_{\mathcal{W}_1} L = s$ then $|\mathcal{V}_0^*(\mathcal{L})| = p^s$.

Therefore, $\mathcal{V}^*$ is an exact functor from $\underline{\mathcal{L}}^*$ to the category of finite $\mathbb{F}_p[\Gamma_K]$-modules.

Introduce the ideal $\widetilde{J} = \sum_{0 \leqslant i < p} x_0^{p-i} \mathrm{m}_R \gamma_i(Y) + \mathrm{Fil}^p R_{st}^0$ in R_{st}^0. Then $F(R_{st}^0) \supset \widetilde{J}$ and $\varphi|_{\widetilde{J}}$ is nilpotent. Therefore, we can introduce $\widetilde{\mathcal{R}}_{st}^0 = (R_{st}^0/\widetilde{J}, F(R_{st}^0)/\widetilde{J}, \varphi \operatorname{mod} \widetilde{J}) \in \widetilde{\underline{\mathcal{L}}}_0^*$, there is a natural projection $\mathcal{R}_{st}^0 \longrightarrow \widetilde{\mathcal{R}}_{st}^0$ in $\widetilde{\underline{\mathcal{L}}}_0^*$ and for any $\mathcal{L} \in \underline{\mathcal{L}}_0^*$, $\operatorname{Hom}_{\widetilde{\underline{\mathcal{L}}}_0^*}(\mathcal{L}, \mathcal{R}_{st}^0) = \operatorname{Hom}_{\widetilde{\underline{\mathcal{L}}}_0^*}(\mathcal{L}, \widetilde{\mathcal{R}}_{st}^0)$. This implies the following description of the Γ_K-modules $\mathcal{V}^*(\mathcal{L})$, $\mathcal{L} \in \underline{\mathcal{L}}^*$,

$$\mathcal{V}^*(\mathcal{L}) = \left\{ \sum_{0\leqslant i<p} N^{*i}(f_0)\gamma_i(Y) \bmod \widetilde{J} \mid f_0 \in \mathcal{V}_0^*(\mathcal{L}) \right\}. \tag{1.2}$$

Note that for $i \geqslant 1$, it is sufficient to know the maps $N^{*i}(f_0)$ modulo x_0^p and this requires just the $(\bmod u^p)$-version of N, cf. discussion in Subsection 2.1. For future applications also notice the following two special cases of above general description (1.2).

(a) Let $\Gamma_{K,1} = \mathrm{Gal}(\bar{K}/K(\sqrt[p]{p}) \subset \Gamma_K$. Then this group acts trivially on $Y \bmod \widetilde{J}$ and $x_0 \bmod x_0^p \mathrm{m}_R$. Therefore, for any $\mathcal{L} \in \underline{\mathcal{L}}^*$, the map $\mathrm{pr}_0 : \mathcal{V}^*(\mathcal{L}) \longrightarrow \mathcal{V}_0^*(\mathcal{L})$ is an isomorphism of $\Gamma_{K,1}$-modules.
(b) Suppose $\mathcal{L} = (L, F(L), \varphi, N) \in \underline{\mathcal{L}}_{cr}^*$. Then there is a $\mathcal{W}_1$-basis $l_1, \dots, l_s \in \varphi(F(L))$ and integers $0 \leqslant a_1, \dots, a_s \leqslant p-1$ such that $l_1' = u^{a_1} l_1, \dots, l_s' = u^{a_s} l_s$ is a $\mathcal{W}_1$-basis of $F(L)$, cf. Subsection 1.4 of [11]. Then there is a matrix $A \in \mathrm{GL}_s(k)$ such that

$$(\varphi(l_1'), \dots, \varphi(l_s')) = (l_1, \dots, l_s)A \bmod u^p L,$$

and or $f \in \mathcal{V}^*(\mathcal{L})$, $f_0 = \mathrm{pr}_0(f)$ and all i, we have

- $f(l_i) \equiv f_0(l_i) \bmod \widetilde{J}$;
- $x_0^{a_i} f_0(l_i) \equiv f(u^{a_i} l) \bmod \widetilde{J}$.

Let $b_i = p-1-a_i$, where $1 \leqslant i \leqslant s$. Then the Galois module $\mathcal{V}^*(\mathcal{L})$ is isomorphic to the Galois module of all $(r_1, \dots, r_s) \bmod x_0^p \mathrm{m}_R \in R^s \bmod x_0^p \mathrm{m}_R$ such that

$$(r_1^p / x_0^{pb_1}, \dots, r_s^p / x_0^{pb_s}) \equiv (r_1, \dots, r_s)(\sigma A) \bmod x_0^p \mathrm{m}_R.$$

2.4. The category $\underline{\mathrm{CM}\Gamma}_K$ and the functor $\mathcal{CV}^*$

Let $\underline{\mathrm{M}\Gamma}_K$ be the category of continuous $\mathbb{Z}_p[\Gamma_K]$-modules. The objects of the category $\underline{\mathrm{CM}\Gamma}_K$ are the triples $\mathcal{H} = (H, H^0, j)$, where $H, H^0 \in \underline{\mathrm{M}\Gamma}_K$ are finite, Γ_K acts trivially on H^0 and $j : H \longrightarrow H^0$ is an epimorphic map in $\underline{\mathrm{M}\Gamma}_K$. If $\mathcal{H}_1 = (H_1, H_1^0, j_1) \in \underline{\mathrm{CM}\Gamma}_K$ then $\mathrm{Hom}_{\underline{\mathrm{CM}\Gamma}_K}(\mathcal{H}_1, \mathcal{H})$ consists of the couples (f, f^0), where $f : H_1 \longrightarrow H$ and $f^0 : H_1^0 \longrightarrow H^0$ are morphisms in $\underline{\mathrm{M}\Gamma}_K$ such that $jf = f^0 j_1$.

The category $\underline{\mathrm{CM}\Gamma}_K$ is special pre-abelian and its objects have a natural group structure.

Definition. Suppose $\mathcal{L} \in \underline{\mathcal{L}}^*$ and $i^{et} : \mathcal{L}^{et} \longrightarrow \mathcal{L}$ is the maximal etale subobject. Then $\mathcal{CV}^* : \underline{\mathcal{L}}^* \longrightarrow \underline{\mathrm{CM}\Gamma}_K$ is the functor such that $\mathcal{CV}^*(\mathcal{L}) = (\mathcal{V}^*(\mathcal{L}), \mathcal{V}^*(\mathcal{L}^{et}), \mathcal{V}^*(i^{et}))$.

The simple objects in $\underline{\mathrm{CM}\Gamma}_K$ are of the form either $(H, 0, 0)$, where H is a simple $\mathbb{Z}_p[\Gamma_K]$-module, or $(\mathbb{F}_p, \mathbb{F}_p, \mathrm{id})$, where $\mathbb{F}_p$ is provided with the trivial Γ_K-action. The functor $\mathcal{CV}^*$ establishes a bijection of the families of simple objects in $\underline{\mathcal{L}}^*$ and $\underline{\mathrm{CM}\Gamma}_K$, cf. Proposition 2.8 of [11].

In particular, let $\mathcal{L}_0 = (\mathcal{W}_1, u^{p-1}\mathcal{W}_1, \varphi) \in \underline{\mathcal{L}}^*_{cr}$ be such that $\varphi(u^{p-1}) = 1$, and $\mathcal{L}_{p-1} = (\mathcal{W}_1, \mathcal{W}_1, \varphi) \in \underline{\mathcal{L}}^*_{cr}$ be such that $\varphi(1) = 1$. Then $\mathcal{CV}^*(\underline{\mathcal{L}}_0) = \mathcal{F}_0 := (\mathbb{F}_p, \mathbb{F}_p, \mathrm{id})$ and $\mathcal{CV}^*(\mathcal{L}_{p-1}) = \mathcal{F}_{p-1} := (\mathbb{F}_p, 0, 0)$.

The functor $\mathcal{CV}^*$ is fully faithful, cf. Proposition 2.13 in [11].

By devissage the proof of this result is reduced to the fact that for any two simple objects $\mathcal{L}', \mathcal{L}'' \in \underline{\mathcal{L}}^*, \mathcal{CV}^*$ induces an injective map from $\mathrm{Ext}_{\underline{\mathcal{L}}^*}(\mathcal{L}', \mathcal{L}'')$ to $\mathrm{Ext}_{\underline{\mathrm{CM}\Gamma}_K}(\mathcal{CV}^*(\mathcal{L}''), \mathcal{CV}^*(\mathcal{L}'))$. The first group was explicitly described in Subsection 1.5 of [11] and the corresponding objects of $\underline{\mathrm{CM}\Gamma}_K$ were studied in Subsections 2.5-2.8 of [11] by the use of (1.2).

Example. One can verify that (remind that $p > 2$)

$$\mathrm{Ext}_{\underline{\mathcal{L}}^*}(\mathcal{L}_{p-1}, \mathcal{L}_0) = \mathrm{Ext}_{\underline{\mathcal{L}}^*_{cr}}(\mathcal{L}_{p-1}, \mathcal{L}_0) \simeq k.$$

Explicitly this isomorphism is described via $\mathcal{L}[\gamma] \mapsto \gamma$, where for $\gamma \in k$, $\mathcal{L}[\gamma] = (L, F(L), \varphi, N) \in \underline{\mathcal{L}}^*_{cr}$ is such that $L = \mathcal{W}_1 l_0 \oplus \mathcal{W}_1 l_1$, $F(L) = \mathcal{W}_1(u^{p-1}l_0) + \mathcal{W}_1(l_1 + \gamma l_0)$, $\varphi(u^{p-1}l_0) = l_0$ and $\varphi(l_1 + \gamma l_0) = l_1$.

Then $\mathcal{CV}^* : \mathrm{Ext}_{\underline{\mathcal{L}}^*}(\mathcal{L}_{p-1}, \mathcal{L}_0) \longrightarrow \mathrm{Ext}_{\underline{\mathrm{CM}\Gamma}_F}(\mathcal{F}_0, \mathcal{F}_{p-1})$ is injective.

Indeed, for any $\gamma \in k$, $\mathcal{CV}^*(\mathcal{L}[\gamma]) = (V[\gamma], \mathbb{F}_p, j) \in \underline{\mathrm{CM}\Gamma}_K$, where the Galois module $V[\gamma]$ is identified with the module of all vectors $\bar{r} = (r_0, r_1)\mathrm{mod} x_0^p \mathrm{m}_R \in R^2 \,\mathrm{mod}\, x_0^p \mathrm{m}_R$ such that $r_0^p \equiv r_0 \,\mathrm{mod} x_0^p \mathrm{m}_R$ and $(r_1/x_0^p) - (r_1/x_0^p)^p \equiv \gamma^p r_0^p / x_0^{p^2} \,\mathrm{mod}\, \mathrm{m}_R$.

Then $V[\gamma]$ can be included into the short exact sequence of Γ_K-modules $0 \longrightarrow \mathbb{F}_p h^1 \longrightarrow V[\gamma] \xrightarrow{j} \mathbb{F}_p j(h^0) \longrightarrow 0$, where $h^0, h^1 \in V[\gamma]$ are such that $h^0 = (1, \alpha) \,\mathrm{mod}\, x_0^p \mathrm{m}_R$, $h^1 = (0, x_0^p) \,\mathrm{mod}\, x_0^p \mathrm{m}_R$. Here $\alpha \in R$ is such that $\alpha - \alpha^p = \gamma^p / x_0^{p^2}$. So, $V[\gamma]$ can be described as an element of $\mathrm{Ext}_{\underline{\mathrm{M}\Gamma}_K}(\mathbb{F}_p j(h^0), \mathbb{F}_p h^1)$ via the cocycle $\Theta_\gamma \in \mathrm{Hom}(\Gamma_K, \mathbb{F}_p)$ such that $\Theta_\gamma(\tau) = (\tau\alpha - \alpha) \,\mathrm{mod}\, \mathrm{m}_R$. Clearly, $\Theta_\gamma = 0$ iff $\gamma = 0$.

3. Ramification estimates

3.1. Ramification estimates

For any rational number $v \geqslant 0$, denote by $\Gamma_K^{(v)}$ the higher ramification subgroup of Γ_K in upper numbering, [28]. In this section we prove the following theorem.

Theorem 1.3. *(a) If $\mathcal{L} \in \underline{\mathcal{L}}^*$ and $v > 2 - \frac{1}{p}$ then $\Gamma_K^{(v)}$ acts trivially on $\mathcal{V}^*(\mathcal{L})$.*
(b) If $\mathcal{L} \in \underline{\mathcal{L}}^*_{cr}$ and $v > 1$ then $\Gamma_K^{(v)}$ acts trivially on $\mathcal{V}^*(\mathcal{L})$.
(c) The above ramification estimates are sharp.

The proof of part (a) was only outlined in Subsection 2.9 of [11]. In Subsections 3.3–3.5 we shall give a proof based on our characteristic p approach from [8, 9, 10]. One can also apply the methods from [24].

3.2. Review of ramification theory

The following brief sketch of ramification theory of complete discrete valuation fields with perfect residue field is based on the papers [17, 32, 33].

Let E be a complete discrete valuation field with perfect residue field k_E and the maximal ideal m_E. Let E_{sep} be a separable closure of E. Denote by v_E a unique extension of the normalized valuation on E to E_{sep}.

Let $\mathcal{I}_E$ be the group of all continuous automorphisms of E_{sep} which are compatible with v_E and induce the identity map on the residue field of E_{sep}. If F is a finite extension of E in E_{sep} then we always assume that $F_{sep} = E_{sep}$ and, therefore, we have a natural identification $\mathcal{I}_E = \mathcal{I}_F$.

Note that $\Gamma_E = \mathrm{Gal}(E_{sep}/E) \supset \{\iota \in \mathcal{I}_E \mid \iota|_E = \mathrm{id}\}$ and if E is unramified over $\mathbb{Q}_p$ then $\mathcal{I}_E$ is identified with the inertia subgroup of Γ_E. If the characteristic of E is p then $\mathcal{I}_E$ is considerably bigger: it contains the subgroup $\mathrm{Aut}^0_E E_{sep} = \{\iota \in \mathcal{I}_E \mid \iota(E) = E\}$ which is mapped onto the group of "analytic" automorphisms $\mathrm{Aut}^0 E$ of E via $\iota \mapsto \iota|_E$.

Denote by $\mathcal{I}_{F/E}$ the set of all continuous embeddings of F into E_{sep} which induce the identity map on E and k_F. For $v \geqslant 0$, let

$$\mathcal{I}_{F/E}^{(v)} = \{\iota \in \mathcal{I}_{F/E} \mid v_F(\iota(a) - a) \geqslant 1 + v \quad \forall a \in \mathrm{m}_F\}.$$

If $\iota_1, \iota_2 \in \mathcal{I}_{F/E}$ then they are *v-equivalent* iff for any $a \in \mathrm{m}_F$, it holds $v_F(\iota_1(a) - \iota_2(a)) \geqslant 1 + v$. The number of v-equivalent classes in $\mathcal{I}_{F/E}$ we shall denote by $(\mathcal{I}_{F/E} : \mathcal{I}_{F/E}^{(v)})$. Then the Herbrand function can be defined as $\varphi_{F/E}(x) = \int_0^x (\mathcal{I}_{F/E} : \mathcal{I}_{F/E}^{(v)})^{-1} dv$, $x \geqslant 0$. It has the following properties:

- $\varphi_{F/E}$ is a piece-wise linear function with finitely many edges;
- if $L \supset F \supset E$ is a tower of finite field extensions then for any $x \geqslant 0$, $\varphi_{L/E}(x) = \varphi_{F/E}(\varphi_{L/F}(x))$.

We define the ramification filtration $\{\mathcal{I}_E^{(v)}\}_{v \geqslant 0}$ on $\mathcal{I}_E$ as follows:

Definition. The subset $\mathcal{I}_E^{(v)}$ of $\mathcal{I}_E$ consists of $\iota \in \mathcal{I}_E$ such that for any finite extension F of E in E_{sep} and $a \in \mathrm{m}_F$, $v_F(\iota(a) - a) \geqslant 1 + \varphi_{F/E}^{-1}(v)$.

Remark. (a) If $\varphi_{F/E}(v_1) = v$ then $\mathcal{I}_E^{(v)} = \mathcal{I}_F^{(v_1)}$ (with respect to the natural identification $\mathcal{I}_E = \mathcal{I}_F$); (b) $\Gamma_E^{(v)} = \Gamma_E \cap \mathcal{I}_E^{(v)}$ is just the usual higher ramification subgroup of Γ_E with upper number v.

The ramification theory is perfectly compatible with the field-of-norms functor of Fontaine–Wintenberger, [32]. Suppose $\widetilde{E}/E$ is an infinite strictly APF-extension in E_{sep}. Then one can define the Herbrand function $\widetilde{\varphi} = \varphi_{\widetilde{E}/E}$ as the limit of Herbrand functions of all finite extensions of E in $\widetilde{E}$. In this situation the field-of-norms functor $\mathcal{X}$ gives a complete discrete valuation field $\mathcal{E} = \mathcal{X}(E)$ of characteristic p, its separable closure $\mathcal{E}_{sep} = \mathcal{X}(E_{sep})$ and the embedding $\mathcal{X} : \mathcal{I}_E \longrightarrow \mathcal{I}_{\mathcal{E}}$.

With the above notation, the compatibility of the field-of-norms functor $\mathcal{X}$ with the ramification filtration means that for any $v \geqslant 0$,

$$\mathcal{X}(\mathcal{I}_E^{(\widetilde{\varphi}(v))}) = \mathcal{X}(\mathcal{I}_E) \cap \mathcal{I}_{\mathcal{E}}^{(v)}.$$

We apply this general theory in the following situation.

Fix an algebraic closure $\bar{K}$ of K and set for $n \geqslant 0$, $K_n = K(\sqrt[p^n]{p}) \subset \bar{K}$. Then $\widetilde{K} = \bigcup_n K_n$ is a strictly APF-extension and by [32],

- $\mathcal{K} = \mathcal{X}(\widetilde{K}) = k((x_0)) \subset \mathrm{Frac}\ R$;
- $\mathcal{X}(\bar{K}) = \mathcal{K}_{sep}$ is a separable closure of $\mathcal{K}$ in $\mathrm{Frac}\ R$;
- $\mathcal{X}$ transforms the action of Γ_K on $\bar{K}$ to the natural action of Γ_K on $\mathrm{Frac}\ R$ and $\Gamma_K \simeq \mathcal{X}(\Gamma_K) \subset \mathcal{I}_{\mathcal{K}}$ (remind that the residue field k of K is assumed to be algebraically closed).

Note that for the derivative of the Herbrand function $\varphi_{\widetilde{K}/K}$ it holds

$$\varphi'_{\widetilde{K}/K}(x) = \begin{cases} 1, & \text{if } 0 < x < p/(p-1) \\ 1/p, & \text{if } p(p-1) < x < p^2/(p-1). \end{cases}$$

Therefore,

$$\mathcal{X}(\Gamma_K^{(2-1/p)}) = \mathcal{X}(\Gamma_K) \cap \mathcal{I}_{\mathcal{K}}^{(p-1)}, \quad \mathcal{X}(\Gamma_K^{(1)}) = \mathcal{X}(\Gamma_K) \cap \mathcal{I}_{\mathcal{K}}^{(1)}. \tag{1.3}$$

3.3. Proof of part (a) of Theorem 1.3

Consider a filtered module $\mathcal{L} = (L, F(L), \varphi, N) \in \underline{\mathcal{L}}^*$. Then its structure can be specified as follows.

Choose a $\mathcal{W}_1$-basis $f_1, \ldots, f_s$ of $F(L)$, let $l_i = \varphi(f_i)$ for $1 \leqslant i \leqslant s$, and set $\bar{f} = (f_1, \ldots, f_s)$ and $\bar{l} = (l_1, \ldots, l_s)$. Let $C \in M_s(\mathcal{W}_1)$ be such that $\bar{f} = \bar{l}C$. Note that $\mathcal{W}_1$ is identified with a subring of R by $u \mapsto x_0$. Therefore, C can be considered as $(s \times s)$-matrix with coefficients in $k[[x_0]] \subset R$. Note, C divides the scalar matrix $x_0^{p-1} I_s$.

Let $H = \mathcal{V}^*(\mathcal{L})$. Because $\Gamma_{K,1} = \mathrm{Gal}(\bar{K}/K(\sqrt[p]{p})) \supset \Gamma_K^{(2-1/p)}$, we can assume that $H = \mathcal{V}_0^*(\mathcal{L})$.

Lemma 1.4. *There is a natural identification of $\mathbb{F}_p[\Gamma_{K,1}]$-modules*

$$H = \{\bar{r} \in R^s \bmod x_0 \mathrm{m}_R \mid \sigma(\bar{r})C \equiv x_0^{p-1}\bar{r} \bmod x_0^p \mathrm{m}_R\}.$$

Proof of Lemma. Indeed, if $h \in H$, then $h(\bar{l}) = \bar{r}^* \bmod x_0^p \mathrm{m}_R$, where $\bar{r}^* \in R^s$ is such that

$$\frac{\sigma(\bar{r}^*)\sigma(C)}{x_0^{p(p-1)}} \equiv \bar{r}^* \bmod x_0^p \mathrm{m}_R.$$

Then $\sigma^{-1}(\bar{r}^*) = \bar{r}$ satisfies the congruence $\sigma(\bar{r})C \equiv x_0^{p-1}\bar{r} \bmod x_0^p \mathrm{m}_R$. It remains to verify that $h \mapsto \bar{r}$ gives the required identification. □

On the other hand, for any $\tau \in \mathcal{I}_{\mathcal{K}}^{(v)}$ with $v > p-1$, it holds $\tau(C) \equiv C \bmod x_0^p \mathrm{m}_R$ and this implies (use Lemma 1.4) for any $h \in H$, that $\tau(h) \in H$. Therefore, our proposition will be proved if we show that for any such τ and any $h \in H$, $\tau(h) = h$.

From the left-continuity of the ramification filtration $\{\mathcal{I}_{\mathcal{K}}^{(v)} \mid v \geqslant 0\}$ it follows the existence of a minimal $v^* = v^*(H)$ such that for any $v > v^*$ and $\tau \in \mathcal{I}_{\mathcal{K}}^{(v)}$, $\tau|_H = \mathrm{id}$.

If $v^* \leqslant p-1$ there is nothing to prove.

Otherwise, choose $r^* \in (p-1, v^*)$ such that $v_p(r^*) = 0$. Such r^* can be always written in the form $r^* = m/(q-1)$, where $m \in \mathbb{N}$ is prime to p and q is an integral power of p. For the following Lemma cf. [8], Subsection 1.5.

Lemma 1.5. *With above chosen r^* and q there is a field extension $\mathcal{K}' = k((x_0'))$ of $\mathcal{K} = k((x_0))$ such that*

a) $[\mathcal{K}' : \mathcal{K}] = q$;

b) $\varphi'_{\mathcal{K}'/\mathcal{K}}(x) = \begin{cases} 1, & \text{if } 0 < x < r^* \\ 1/p, & \text{if } x > r^* \end{cases}$

c) $x_0 \equiv x_0'^{q}(1 - x_0'^{r^*(q-1)}) \bmod x_0'^{q+2r^*(q-1)}$.

Note that for above chosen r^*, the appropriate m and q are not defined uniquely, e.g. for any $a \in \mathbb{N}$, it holds also that $r^* = m_a/(q^a - 1)$, where $m_a = m(1 + q + \cdots + q^{a-1})$. Therefore, we can assume additionally that q is large enough to provide us with the following inequality

$$r^*(1 - 1/q) > p - 1. \tag{1.4}$$

Choose a field isomorphism $\kappa : \mathcal{K} \longrightarrow \mathcal{K}'$ such that $\kappa(x_0) = x_0'$ and $\kappa|_k = \sigma^{-q}$. Note that by Lemma 1.5 (c) and assumption (1.4), for any

$\gamma \in k[[x_0]]$, $\kappa(\gamma)^{q_*} \equiv \gamma \bmod x_0^p \mathrm{m}_R$. The isomorphism κ can be extended to an isomorphism of separable closures of these fields in R. Therefore, we have the bijection $\kappa^* : \mathcal{I}_{\mathcal{K}} \longrightarrow \mathcal{I}_{\mathcal{K}'}$ such that for any $v \geqslant 0$, $\kappa^*(\mathcal{I}_{\mathcal{K}}^{(v)}) = \mathcal{I}_{\mathcal{K}'}^{(v)}$. In particular, if

$$h' \in H' = \{\bar{r}' \in R^s \bmod x_0' \mathrm{m}_R \mid \sigma(\bar{r}')\kappa(C) \equiv x_0'^{p-1}\bar{r}' \bmod x_0'^p \mathrm{m}_R\}$$

then for any $v > v^*$ and $\tau \in \mathcal{I}_{\mathcal{K}'}^{(v)}$, $\tau(h') = h'$.

On the other hand, from Lemma 1.5 (c) it follows that

$$H = \{\bar{r} \in R^s \bmod x_0 \mathrm{m}_R \mid \sigma(\bar{r})\sigma^q(\kappa(C)) \equiv x_0^{p-1}\bar{r} \bmod x_0^p \mathrm{m}_R\}.$$

Therefore, the map $\bar{r}' \mapsto \sigma^q(\bar{r}')$ establishes a Galois equivariant bijection of H' and H. Because, $\mathcal{I}_{\mathcal{K}'}^{(v^*)} = \mathcal{I}_{\mathcal{K}}^{(\varphi_{\mathcal{K}'/\mathcal{K}}(v^*))}$, this implies that for any $\tau \in \mathcal{I}_{\mathcal{K}}^{(v)}$ with $v > \varphi_{\mathcal{K}'/\mathcal{K}}(v^*)$, $\tau|_H = \mathrm{id}$. But $\varphi_{\mathcal{K}'/\mathcal{K}}(v^*) < v^*$. The contradiction.

3.4.

Prove that the ramification estimate from Theorem 1.3(a) is sharp.

Introduce $\mathcal{L} = (L, F(L), \varphi, N) \in \widetilde{\underline{\mathcal{L}}}^*$ such that:

- $L = \sum_{p>i\geqslant 0} \mathcal{W}_1 l_i$;
- $F(L) = \sum_{p>i\geqslant 0} \mathcal{W}_1 f_i$, where for $p > i \geqslant 1$, $f_i = u^i(l_{p-1} + \cdots + l_i)$ and $f_0 = u\sum_{p-1\geqslant i\geqslant 2}(i-1)l_i + l_1 + l_0$;
- for all i, $\varphi(f_i) = l_i$;
- $N(l_{p-1}) = 0$, if $p - 1 > i \geqslant 1$ then $N(l_i) = l_{i+1} \bmod u^p L$, and $N(l_0) = -l_2 \bmod u^p L$.

A direct verification shows that $\mathcal{L} \in \underline{\mathcal{L}}^*$. In particular:

- for $p > i \geqslant 2$, $\varphi(uN(f_i)) = \varphi(f_{i+1}) = l_{i+1} = N(l_i) = N(\varphi(f_i))$;
- $N(f_0) = -u(l_{p-1} + \cdots + l_2)$ and, therefore, $\varphi(uN(f_0)) = \varphi(-f_2) = -l_2 = N(l_0) = N(\varphi(f_0))$.

By Lemma 1.4, the $\Gamma_{K,1}$-module $\mathcal{V}_0^*(\mathcal{L}) = H$ is identified with the $\Gamma_{K,1}$-module of all $\bar{r} = (r_{p-1}, \ldots, r_1, r_0) \in R^p \bmod x_0 \mathrm{m}_R$ such that

$$r_{p-1}^p \equiv r_{p-1} \bmod x_0 \mathrm{m}_R$$

$$r_{p-1}^p + r_{p-2}^p \equiv x_0 r_{p-2} \bmod x_0^2 \mathrm{m}_R$$

$$\cdots$$

$$r_{p-1}^p + \cdots + r_1^p \equiv x_0^{p-2} r_1 \bmod x_0^{p-1} \mathrm{m}_R$$

$$r_{p-1}^p(p-2)x_0 + \cdots + r_2^p x_0 + r_1^p + r_0^p \equiv x_0^{p-1} r_0 \bmod x_0^p \mathrm{m}_R$$

Let $v^* \geqslant 0$ be the minimal such that for all $v > v^*$, $\mathcal{I}_{\mathcal{K}}^{(v)}$ acts trivially on H. We must prove that $v^* = p-1$.

Suppose $v^* < p-1$.

Choose $r^* \in (v^*, p-1)$ such that $v_p(r^*) = 0$. As earlier, we can assume that $r^* = m/(q-1)$, where $m \in \mathbb{N}$, q is a power of p and $r^*(1-1/q) > p-2$.

Apply Lemma 1.5 and consider the appropriate fields isomorphism $\kappa : \mathcal{K} \longrightarrow \mathcal{K}'$. If $\kappa(x_0) = x_0'$ then for $i \geqslant 2$, $x_0'^{iq} \equiv x_0^i \operatorname{mod} x_0^p \mathrm{m}_R$, $x_0'^q \equiv x_0 + x_0'^{q+r^*(q-1)} \operatorname{mod} x_0^p \mathrm{m}_R$ and $x_0'^{q+r^*(q-1)} \in x_0^{p-1}\mathrm{m}_R$.

This implies that for $p-1 \geqslant i \geqslant 1$, $r_i \equiv \sigma^q(r_i') \operatorname{mod} x_0 \mathrm{m}_R$ and $r_i^p \equiv \sigma^q(r_i')^p \operatorname{mod} x_0^p \mathrm{m}_R$. Therefore,

$$\bar{r} \equiv \sigma^q(\bar{r}') + (0, \dots, 0, Y_0) \operatorname{mod} x_0 \mathrm{m}_R,$$

where $x_0^{p-1}Y_0 - Y_0^p \equiv (r_{p-1}^p(p-2) + \cdots + r_2^p)x_0'^{q+r^*(q-1)}$

$$\equiv r_{p-1}x_0'^{q+r^*(q-1)} \operatorname{mod} x_0^p \mathrm{m}_R.$$

Note that this relation can be rewritten in the following form

$$(Y_0/x_0) - (Y_0/x_0)^p \equiv r_{p-1}x_0'^{-a} \operatorname{mod} \mathrm{m}_R,$$

where $\varphi_{\mathcal{K}'/\mathcal{K}}(a) = p-1$. Therefore, $\mathcal{I}_{\mathcal{K}'}^{(a)} = \mathcal{I}_{\mathcal{K}}^{(p-1)}$ acts non-trivially on $Y_0 \operatorname{mod} x_0 \mathrm{m}_R$.

On the other hand, we assumed that $v^* < p-1$ and, therefore, for $\tau \in \mathcal{I}_{\mathcal{K}'}^{(p-1)}$, we have $\tau(\bar{r}') = \bar{r}'$. But $a > p-1$ and this implies $\mathcal{I}_{\mathcal{K}}^{(p-1)} = \mathcal{I}_{\mathcal{K}'}^{(a)} \subset \mathcal{I}_{\mathcal{K}'}^{(p-1)}$. Therefore, any $\tau \in \mathcal{I}_{\mathcal{K}}^{(p-1)}$ acts trivially on $(\bar{r} - \sigma^q(\bar{r}')) \operatorname{mod} x_0 \mathrm{m}_R = (0, \dots, Y_0) \operatorname{mod} x_0 \mathrm{m}_R$. The contradiction.

3.5. Proof of estimate (b) of Theorem 1.3

Suppose $\mathcal{L} \in \mathcal{L}_{cr}^*$ is given in the notation of Subsection 2.3 (b). By (1.3) we must prove that $\mathcal{I}_{\mathcal{K}}^{(v)}$ acts trivially on $H = \mathcal{V}_0^*(\mathcal{L})$ if $v > 1$.

Let v^* be the maximal such that $\mathcal{I}_{\mathcal{K}}^{(v^*)}$ acts non-trivially on H.

Assume that $v^* > 1$. Choose $r^* \in (1, v^*)$ such that $r^* = m/(q-1)$, where $m \in \mathbb{N}$, q is a power of p and $r^*(1-1/q) > 1$. Consider the appropriate field isomorphism $\kappa : \mathcal{K} \longrightarrow \mathcal{K}'$ and its extension to $\mathcal{K}_{sep} = \mathcal{K}'_{sep}$. Let $x_0' = \kappa(x_0)$ and $\bar{r}' = (r_1', \dots, r_s') = \kappa(\bar{r})$. If

$$H' = \left\{ \bar{r}' \operatorname{mod} x_0'^p \mathrm{m}_R \mid \left(\frac{r_1'^p}{x_0'^{pb_1}}, \dots, \frac{r_s'^p}{x_0'^{pb_s}} \right) = \bar{r}'(\sigma^{1-q}A) \operatorname{mod} x_0'^p \mathrm{m}_R \right\}$$

then for any $v > v^*$, $\mathcal{I}_{\mathcal{K}'}^{(v)}$ acts trivially on H'.

Note that the assumption $r^*(1-1/q) > 1$ implies that $x_0'^{r^*(q-1)} \in x_0\mathrm{m}_R$ and for all $1 \leqslant i \leqslant s$,

$$\frac{r_i^p}{x_0^{pb_i}} \equiv \frac{(\sigma^q r_i')^p}{x_0'^{qpb_i}} \bmod x_0^p \mathrm{m}_R.$$

Therefore, $\bar{r}' \mapsto \sigma^q \bar{r}'$ induces $\mathrm{Aut}^0_{\mathcal{K}}\mathcal{K}_{sep}$-equivariant isomorphism of H' and H.

If $v' \geqslant 0$ is such that $\varphi_{\mathcal{K}'/\mathcal{K}}(v') = v^*$ then $v' > v^*$ and $\mathcal{I}_{\mathcal{K}'}^{(v')} = \mathcal{I}_{\mathcal{K}}^{(v^*)}$ acts trivially on H' but not on H. The contradiction.

3.6.

The example in Subsection 2.4 shows that for all $\gamma \neq 0$, $\mathcal{I}_{\mathcal{K}}^{(1)}$ acts non-trivially on $h^0 \in V[\gamma] = \mathcal{V}^*(\mathcal{L}[\gamma])$. Therefore, the estimate from (b) is sharp.

4. A construction of modification of Breuil's functor

Generalize slightly the initial data from Section 2 as follows.

Let $\mathcal{W} = W(k)[[u]]$, where u is an indeterminate. Denote by σ the automorphism of $W(k)$ induced by the p-th power map on k and agree to use the same symbol for the continuous extension of σ to $\mathcal{W}$ such that $\sigma(u) = u^p$. Denote by $N : \mathcal{W} \longrightarrow \mathcal{W}$ the unique continuous $W(k)$-differentiation such that $N(u) = -u$. We denote by S the divided power envelope of $\mathcal{W}$ with respect to the ideal $(u+p)$.

4.1. Breuil's functor

We work with Breuil's theory of semi-stable p-adic representations of $\Gamma_K = \mathrm{Gal}(\bar{K}/K)$, [14]–[16]. This theory allows us to construct Γ_K-invariant lattices in semi-stable $\mathbb{Q}_p[\Gamma_K]$-modules with Hodge–Tate weights from $[0, p)$. The construction is done via Breuil's functor $\mathcal{S}_{p-1} \longrightarrow \underline{\mathrm{M}\Gamma}_K$, where $\mathcal{S}_{p-1}$ is a suitable category of free S-modules M with filtration by a submodule $F(M)$ and additional structures involving σ-linear morphisms $\varphi : F(M) \longrightarrow M$ and differentiations $N : M \longrightarrow M$, [16], Subsection 2.2. The objects of $\mathcal{S}_{p-1}$ satisfy the properties similar to those from the definition of the category $\underline{\mathcal{L}}^*$ from Subsection 2.1. Breuil's functor appears in the form $\mathcal{M} \mapsto \mathrm{Hom}_{F,\varphi,N}(M, \hat{A}_{st})$, where $\hat{A}_{st}$ is the ring of semi-stable p-adic periods [14], Subsection 3.1. Note that $\hat{A}_{st}$ is provided with the appropriate S-module structure, filtration, morphisms φ and N, and Γ_K-module structure. The notation $\mathrm{Hom}_{F,\varphi,N}$ means the set of all S-linear homomorphisms compatible with filtrations and the morphisms φ and N. Breuil's theory allows also to construct crystalline representations of Γ_K with HT weights from $[0, p)$ by the use of the appropriate subcategory $\mathcal{S}_{p-1}^{cr}$ of $\mathcal{S}_{p-1}$. (The objects of $\mathcal{S}_{p-1}^{cr}$ come from the Fontaine–Laffaille modules with filtration of length p.)

Similarly to the Fontaine–Laffaille theory the Breuil theory perfectly describes all Γ_K-invariant lattices of semi-stable representations with HT weights from $[0, p-2]$ but does not give generally all such lattices for representations with weights from $[0, p)$.

4.2. Modification of Breuil's functor

In Subsection 4 of [11] we constructed a modification of Breuil's functor which allows us to construct all Galois invariant lattices and study all subquotients modulo p of semi-stable representations with weights from $[0, p)$. We shall give below a brief explanation of our construction from Subsection 4 of [11] together with a modelled example.

4.2.1.

As a first step, we prove that Breuil's category of filtered S-modules $\underline{\mathcal{S}}_{p-1}$ can be replaced by a similar category $\underline{\mathcal{L}}^f$ of free filtered $\mathcal{W}$-modules $(M, F(M))$ with σ-linear maps $\varphi : F(M) \longrightarrow M$ and differentiations $N : M \longrightarrow M \otimes_{\mathcal{W}} S$. Then we define a torsion analogue $\underline{\mathcal{L}}^t$ of the category $\underline{\mathcal{L}}^f$. As a result, we can use Breuil's functor in the form $\mathcal{V}^t : \mathcal{M} \mapsto \mathrm{Hom}_{F,\varphi,N}(\mathcal{M}, A_{st,\infty})$, where $\mathcal{M} \in \underline{\mathcal{L}}^t$ and $A_{st,\infty}$ is a torsion analogue of Fontaine's ring of semi-stable periods, [14], Subsection 3.1. Note that $\underline{\mathcal{L}}^t$ contains the full subcategory $\underline{\mathcal{L}}^{ft}$ whose objects are subquotients of objects of $\underline{\mathcal{L}}^f$ and this subcategory is strictly smaller than $\underline{\mathcal{L}}^t$. This is very special feature of "semi-stable" theory: if we start with the subcategory $\mathcal{S}_{p-1}^{cr}$ then the appropriate categories $\mathcal{L}_{cr}^t$ and $\mathcal{L}_{cr}^{ft}$ coincide. Denote the restriction of $\mathcal{V}^t$ to $\underline{\mathcal{L}}^{ft}$ by $\mathcal{V}^{ft}$.

Following general formalism we prove that $\underline{\mathcal{L}}^t$ is special pre-abelian and there is a concept of p-divisible object in $\underline{\mathcal{L}}^t$ (just mimic Tate's definition of p-divisible groups in the pre-abelian category of group schemes). Such p-divisible objects will be called p-divisible groups if there is no risk of confusion. Then the objects of $\underline{\mathcal{L}}^f$ can be recovered as "Tate's modules" associated with p-divisible groups in $\underline{\mathcal{L}}^t$. In particular, a p-divisible group in $\underline{\mathcal{L}}^t$ is inductive limit of objects from $\underline{\mathcal{L}}^{ft}$. As we have just noted, there is no similar problem for the appropriate subcategory $\underline{\mathcal{L}}_{cr}^t$ of "crystalline" filtered modules in $\underline{\mathcal{L}}^t$: any such module comes as a subquotient of a "crystalline" module from $\underline{\mathcal{L}}_{cr}^f \subset \underline{\mathcal{L}}^f$.

4.2.2.
If $\mathcal{M}=(M,F(M))\in\underline{\mathcal{L}}^t$ then it is called multiplicative if $M=F(M)$ and etale if $F(M)=u^{p-1}M$. As usually, any $\mathcal{M}\in\underline{\mathcal{L}}^t$ has a unique maximal etale subobject $\iota^{et}:\mathcal{M}^{et}\longrightarrow\mathcal{M}$ and a unique maximal multiplicative quotient object $\iota^m:\mathcal{M}\longrightarrow\mathcal{M}^m$. We call $\mathcal{M}\in\underline{\mathcal{L}}^t$ unipotent if $\mathcal{M}^m=0$.

By compairing $\mathcal{V}^t$ and the functor $\mathcal{V}^*$ from Subsection 2.3 we deduce that $\mathcal{V}^t$ is fully faithful on the subcategory $\underline{\mathcal{L}}^{t,u}$ of unipotent objects of $\underline{\mathcal{L}}^t$. Quite oppositely, if $\mathcal{M}\in\underline{\mathcal{L}}^t$, $p\mathcal{M}=0$ and

$$0\longrightarrow\mathcal{M}^u\longrightarrow\mathcal{M}\longrightarrow\mathcal{M}^m\longrightarrow 0$$

is the standard short exact sequence with unipotent $\mathcal{M}^u$ and multiplicative $\mathcal{M}^m$, then the corresponding exact sequence

$$0\longrightarrow\mathcal{V}^t(\mathcal{M}^m)\longrightarrow\mathcal{V}^t(\mathcal{M})\longrightarrow\mathcal{V}^t(\mathcal{M}^u)\longrightarrow 0$$

has a functorial splitting in $\underline{\mathrm{M}\Gamma}_K$. Denote the appropriate splitting maps by $\Theta:\mathcal{V}^t(\mathcal{M}^u)\longrightarrow\mathcal{V}^t(\mathcal{M})$ and $\widetilde{\Theta}:\mathcal{V}^t(\mathcal{M})\longrightarrow\mathcal{V}^t(\mathcal{M}^m)$.

Let us illustrate this splitting phenomenon by the following explicit calculations. Let A_{cr} be Fontaine's crystalline ring. It is the p-adic closure of the DP-envelope of $W(R)$ with respect to the ideal $([x_0]+p)$. Let $F(A_{cr})$ be the $(p-1)$-st divided power of $([x_0]+p)$ and $\psi:A_{cr}\longrightarrow A_{cr}$ be the map induced by $\sigma:R\longrightarrow R$. Set $\varphi=\psi/p^{p-1}$. Notice that A_{cr} is provided with the natural continuous Γ_K-action. Then $A_{cr,1}=A_{cr}/pA_{cr}$ is provided with induced filtration $F(A_{cr,1})$, morphism φ and Γ_K-action, and we obtain $\mathcal{A}_{cr,1}=(A_{cr,1},F(A_{cr,1}),\varphi)\in\widetilde{\underline{\mathcal{L}}}_0^*$ by defining the $\mathcal{W}_1$-module structure on $A_{cr,1}$ via $u\mapsto[x_0]$.

If A_{st} is Fontaine's ring of semi-stable periods then A_{st} is obtained from A_{cr} in the same way as R^0_{st} was obtained from R^0 in Subsections 2.2 and 2.3. Therefore, if $\mathcal{L}\in\underline{\mathcal{L}}^t$ and $p\mathcal{L}=0$ then we can illustrate the splitting phenomenon by studying the abstract module $\mathcal{V}_0(\mathcal{L})=\mathrm{Hom}_{\widetilde{\underline{\mathcal{L}}}_0^*}(\mathcal{L},\mathcal{A}_{cr,1})$. Even more, we can treat $\underline{\mathcal{L}}^*_{cr}$ as a full subcategory of $\underline{\mathcal{L}}^t$ and then for $\mathcal{L}\in\underline{\mathcal{L}}^*_{cr}$, $\mathcal{V}_0(\mathcal{L})=\mathcal{V}^t(\mathcal{L})$ even as Galois modules.

From the definition of A_{cr} it follows that $A_{cr,1}=(R/x_0^p)[T_1,T_2,\dots]$, where for all $i\geqslant 1$, T_i comes from the divided powers $\gamma_{p^i}([x_0]+p)$ and $T_i^p=0$. Let J be the ideal in $A_{cr,1}$ generated by T_1^2 and T_i with $i\geqslant 2$. Then $\widetilde{A}_{cr,1}=A_{cr,1}/J=(R/x_0^p)T_1\oplus(R/x_0^p)$ is provided with the induced filtration $F(\widetilde{A}_{cr,1})=(R/x_0^p)T_1\oplus(x_0^{p-1}R/x_0^p)$ and σ-linear $\varphi:F(\widetilde{A}_{cr,1})\longrightarrow\widetilde{A}_{cr,1}$ such that $\varphi(x_0^{p-1})=1-T_1$ and $\varphi(T_1)=1$. This gives the object $\widetilde{\mathcal{A}}_{cr,1}$ in $\widetilde{\underline{\mathcal{L}}}_0^*$ together with the natural projection $j_{cr,1}:\mathcal{A}_{cr,1}\longrightarrow\widetilde{\mathcal{A}}_{cr,1}$. Using that $\varphi(J)=0$ we obtain that $j_{cr,1*}$ induces the identification $\mathcal{V}_0(\mathcal{L})=\mathrm{Hom}_{\widetilde{\underline{\mathcal{L}}}_0^*}(\mathcal{L},\widetilde{\mathcal{A}}_{cr,1})$.

Consider $\mathcal{L}[\gamma] \in \underline{\mathcal{L}}^*_{cr}$ from Subsection 2.4.

Then we have the following standard exact sequence

$$0 \longrightarrow \mathcal{L}_0 \longrightarrow \mathcal{L}[\gamma] \to \mathcal{L}_{p-1} \to 0$$

where $\mathcal{L}_0 = (\mathcal{W}_1 l_0, u^{p-1}\mathcal{W}_1 l_0, \varphi)$ is a simple etale subobject in $\mathcal{L}[\gamma]$ and $\mathcal{L}_{p-1} = (\mathcal{W}_1 \bar{l}_1, \mathcal{W}_1 \bar{l}_1, \varphi)$ is its simple multiplicative quotient object.

Consider the corresponding short exact sequence of $\mathbb{F}_p[\Gamma_K]$-modules

$$0 \longrightarrow H_{p-1} \longrightarrow H \longrightarrow H_0 \longrightarrow 0$$

where $H_{p-1} = \mathcal{V}_0(\mathcal{L}_{p-1})$, $H = \mathcal{V}_0(\mathcal{L}[\gamma])$, $H_0 = \mathcal{V}_0(\mathcal{L}_0)$. Note that $H_0 = H_{p-1} \simeq \mathbb{F}_p$ are trivial Γ_K-modules.

Lemma 1.6. *For any $\alpha \in R$, there is a unique $r^*(\alpha) \in R \bmod x_0^p$ such that $r^*(\alpha)^p / x_0^{p(p-1)} - r^*(\alpha) \equiv \alpha \bmod x_0^p R$.*

Proof. It follows from the congruence $A^p - A = \alpha x_0^{-p} \bmod R$, where $A = r^*(\alpha) x_0^{-p} \in \operatorname{Frac} R$. □

Now direct calculations show that:

- $H_0 = \operatorname{Hom}_{\underline{\widetilde{\mathcal{L}}}_0^*}(\mathcal{L}_0, \widetilde{\mathcal{A}}_{cr,1})$ consists of $h : l_0 \mapsto r_{-1}T_1 + r_0$ such that $r_0 = -r_{-1} = f$, where f_0 runs over all elements of $\mathbb{F}_p \subset R/x_0^p$;
- $H_{p-1} = \operatorname{Hom}_{\underline{\widetilde{\mathcal{L}}}_0^*}(\mathcal{L}_{p-1}, \widetilde{\mathcal{A}}_{cr,1})$ consists of $h : l_1 \mapsto r_{-1}T_1 + r_0$ such that $r_0 = r^*(f_1)$ and $r_{-1} = -(r^*(f_1) + f_1)$, where f_1 runs over $\mathbb{F}_p$;
- $H = \operatorname{Hom}_{\underline{\widetilde{\mathcal{L}}}_0^*}(\mathcal{L}[\gamma], \widetilde{\mathcal{A}}_{cr,1})$ consists of
$$h : (l_0, l_1) \mapsto (-f_0 T_1 + f_0, r_{-1}T_1 + r_0),$$
where $r_0 = -(r_{-1} + f_1) = r^*(f_1) - r^*(\gamma^p f_0 / x_0^{p(p-1)})$ and $f_0, f_1 \in \mathbb{F}_p$;
- the splitting Θ is defined via the submodule $\Theta(H_0)$ of H consisting of $h \in H$ such that $f_1 = 0$.

Let $A^0_{cr,1} = \{r_{-1}T_1 + r_0 \mid r_{-1} = -r_0\} \subset \widetilde{A}_{cr,1}$ with the induced filtration and the morphism φ, and denote by $\mathcal{A}^0_{cr,1}$ the appropriate object of $\underline{\widetilde{\mathcal{L}}}_0^*$. Then

$$\Theta(H_0) = \operatorname{Hom}_{\underline{\widetilde{\mathcal{L}}}_0^*}(\mathcal{L}[\gamma], \mathcal{A}^0_{cr,1}). \tag{1.5}$$

Remark. Relation (1.5) determines the splitting Θ for any $\mathcal{L} \in \underline{\mathcal{L}}^*_{cr}$.

4.2.3.

For $\mathcal{M} \in \underline{\mathcal{L}}^{ft}$, let $\mathcal{M}' \in \underline{\mathcal{L}}^{ft}$ be such that $p\mathcal{M}' = \mathcal{M}$ and let

$$C_p = \operatorname{Coker} p|_{\mathcal{M}'} : \mathcal{M}' \longrightarrow {}_p\mathcal{M}', \quad K_p = \operatorname{Ker} p|_{\mathcal{M}'} : \mathcal{M}'_p \longrightarrow \mathcal{M}'.$$

Consider the following sequence of objects and maps in $\underline{\mathrm{M}\Gamma}_K$

$$\mathcal{V}^{ft}({}_p\mathcal{M}'^u) \xrightarrow{\Theta} \mathcal{V}^{ft}({}_p\mathcal{M}') \xrightarrow{\mathcal{V}^{ft}(C_p)} \mathcal{V}^{ft}(\mathcal{M}') \xrightarrow{\mathcal{V}^{ft}(K_p)}$$
$$\mathcal{V}^{ft}(\mathcal{M}'_p) \xrightarrow{\widetilde{\Theta}} \mathcal{V}^{ft}(\mathcal{M}'^m_p).$$

Then:

- $\widetilde{\Theta} \circ V^{ft}(K_p) \circ V^{ft}(C_p) \circ \Theta = 0$;
- Γ_K-module $\widetilde{\mathcal{V}}^{ft}(\mathcal{M}) := \mathrm{Ker}(\widetilde{\Theta} \circ \mathcal{V}^{ft}(K_p))/\mathrm{Im}\,(\mathcal{V}^{ft}(C_p) \circ \Theta)$ does not depend on a choice of $\mathcal{M}'$;
- $\widetilde{\mathcal{V}}^{ft}(\mathcal{M}^u) = \mathcal{V}^{ft}(\mathcal{M}^u)$, $\widetilde{\mathcal{V}}^{ft}(\mathcal{M}^m) = \mathcal{V}^{ft}(\mathcal{M}^m)$;
- there is a canonical epimorphism $\widetilde{\mathcal{V}}^{ft}(i^{et}) : \widetilde{\mathcal{V}}^{ft}(\mathcal{M}) \longrightarrow \widetilde{\mathcal{V}}^{ft}(\mathcal{M}^{et})$.

Definition. The modification of Breuil's functor $\widetilde{\mathcal{CV}}^{ft} : \underline{\mathcal{L}}^{ft} \longrightarrow \underline{\mathrm{CM}\Gamma}_K$ is induced by the correspondence $\mathcal{M} \mapsto (\widetilde{\mathcal{V}}^{ft}(\mathcal{M}), \widetilde{\mathcal{V}}^{ft}(\mathcal{M}^{et}), \widetilde{\mathcal{V}}^{ft}(i^{et}))$.

Let us illustrate the above definition by the following explicit construction. Having in mind that A_{cr} is related to the DP-envelope of $W(R)$ we can describe explicitly $A_{cr,2} = A_{cr}/p^2A_{cr}$ and similarly to the case of $A_{cr,1}$ introduce an appropriate simpler object $\widetilde{A}_{cr,2}$ as follows:

- the elements of $\widetilde{A}_{cr,2}$ are written in the form
$$[r_{-1}]T_1 + [r_0] + p[r_1],$$
where $r_{-1}, r_1 \in R/x_0^p$ and $r_0 \in R/x_0^{2p}$; the operations are induced by those on the Teichmuller representatives of r_{-1}, r_0, r_1 via the relations $pT = [x_0]^p$ and $p^2 = 0$;
- the $\mathcal{W}$-module structure on $\widetilde{A}_{cr,2}$ is induced by the $W(k)$-algebra morphism $\mathcal{W} \longrightarrow W(R)$ such that $u \mapsto [x_0] + p$;
- $F(\widetilde{A}_{cr,2})$ is generated over $W(R)$ by $f_1 = T_1$ and $f_2 = [x_0]^{p-1} - p[x_0]^{p-2}$;
- $\varphi : F(\widetilde{A}_{cr,2}) \longrightarrow \widetilde{A}_{cr,2}$ is uniquely determined by $\varphi(f_1) = 1 + [x_0]^p$ and $\varphi(f_2) = -T_1 + 1$.

Note that $p\widetilde{\mathcal{A}}_{cr,2} = \widetilde{\mathcal{A}}_{cr,1}$.

Consider again $\mathcal{L}[\gamma] \in \underline{\mathcal{L}}^*_{cr}$ from Subsection 2.4. Choose $\mathcal{L}' \in \underline{\mathcal{L}}^{ft}$ such that $p\mathcal{L}' = \mathcal{L}[\gamma]$. For simplicity assume that the corresponding $\mathcal{W}$-module is $(\mathcal{W}/p^2)l_0 \oplus (\mathcal{W}/p^2)l_1$. In this case $C_p : \mathcal{L}' \longrightarrow {}_p\mathcal{L}'$ is just the natural projection $\mathcal{L}' \longrightarrow \mathcal{L}'/p\mathcal{L}' = \mathcal{L}$ and $K_p : \mathcal{L}'_p \longrightarrow \mathcal{L}'$ is just the natural embedding $\mathcal{L} = p\mathcal{L}' \longrightarrow \mathcal{L}'$. Therefore,

- $\mathrm{Ker}(\widetilde{\Theta}\circ\mathcal{V}_0(K_p))$ appears as the kernel of the map

$$\mathcal{V}_0(\mathcal{L}')\longrightarrow\mathcal{V}_0(\mathcal{L}')/p\mathcal{V}_0(\mathcal{L}')=\mathcal{V}_0(\mathcal{L})=H\xrightarrow{\widetilde{\Theta}}H_1$$

and equals $\mathrm{Hom}_{F,\varphi}(\mathcal{L}',\mathcal{A}^0_{cr,2})\subset\mathrm{Hom}_{F,\varphi}(\mathcal{L}',\widetilde{\mathcal{A}}_{cr,2})=\mathcal{V}_0(\mathcal{L}')$.
- $\mathrm{Im}(\mathcal{V}_0(C_p)\circ\Theta)$ appears as the image of the map

$$H_0\xrightarrow{\Theta}H=p\mathcal{V}_0(\mathcal{L}')\subset\mathcal{V}_0(\mathcal{L}')$$

and equals $\mathrm{Hom}_{F,\varphi}(\mathcal{L}',p\mathcal{A}^0_{cr,2})\subset\mathrm{Hom}_{F,\varphi}(\mathcal{L}',\widetilde{\mathcal{A}}_{cr,2})$.
- $\widetilde{\mathcal{V}}^{ft}(\mathcal{L})=\mathrm{Hom}_{\widetilde{\underline{\mathcal{L}}}_0^*}(\mathcal{L},\mathcal{A}^0_{cr,2}/p\mathcal{A}^0_{cr,2})$.

Note that the correspondence

$$[r_0 \bmod x_0^p]T_1+[r_0]+p[r_1]\mapsto(r_0+x_0^pr_1)\bmod x_0^p\mathrm{m}_R$$

determines an epimorphic map $\mathcal{A}^0_{cr,2}/p\mathcal{A}^0_{cr,2}\longrightarrow\mathcal{R}^0$ in the category $\widetilde{\underline{\mathcal{L}}}_0^*$ and this map induces isomorphism of Γ_K-modules $\widetilde{\mathcal{V}}^{ft}(\mathcal{L}[\gamma])$ and $\mathcal{V}^*(\mathcal{L}[\gamma])$.

4.3. Properties of modified functor

The following property was our main target.

Theorem 1.7. *$\widetilde{\mathcal{CV}}^{ft}$ is fully faithful.*

Proof. By devissage it will be sufficient to verify this statement on the level of the subcategories of killed by p objects. The corresponding restriction of $\widetilde{\mathcal{CV}}^{ft}$ is equivalent then to the functor $\mathcal{CV}^*$ from Section 2 (cf. also the example in Subsection 4.2.3) but the functor $\widetilde{\mathcal{CV}}^*$ is fully faithful, cf. Subsection 2.4. □

Suppose V is a finite dimensional vector space over $\mathbb{Q}_p$ with continuous Γ_K-action and H a Γ_K-invariant lattice in V.

Corollary 1.8. *If V is semi-stable (resp., crystalline) with HT weights from $[0,p)$ then the higher ramification subgroups $\Gamma_K^{(v)}$ act trivially on H/pH for all $v>2-1/p$ (resp., $v>1$).*

Proof. As it was noted in Subsections 4.1–4.2, Breuil's functor allows us to obtain a Galois invariant lattice H_0 in V in the form $H_0=\varprojlim_n\mathcal{V}^t(\mathcal{M}_n)$, where $\{\mathcal{M}_n\}_{n\geqslant0}$ is a p-divisible group in the category $\underline{\mathcal{L}}^{ft}$. Then $H_1=\varprojlim_n\widetilde{\mathcal{V}}^{ft}(\mathcal{M}_n)$ is again a Galois invariant lattice in V. (Use that the p-divisible group $\{\widetilde{\mathcal{V}}^{ft}\}_{n\geqslant0}$ is isogeneous to $\{\mathcal{V}^t(\mathcal{M}_n)\}_{n\geqslant0}$, cf. Subsection 4.2.3.) We can assume that $H_1 \supset H \supset pH \supset p^m H_1$ with some $m \in \mathbb{N}$. Then by Theorem 1.7 there is a subquotient modulo p, $\mathcal{M} \in \underline{\mathcal{L}}^{ft}$ of $\mathcal{M}_m$, such that $\widetilde{\mathcal{CV}}^{ft}(\mathcal{M}) = (H/pH, \widetilde{\mathcal{V}}^{ft}(\mathcal{M}^{et}), \widetilde{\mathcal{V}}^{ft}(i^{et}))$. Therefore, the Γ_K-module H/pH belongs to the image of the functor $\mathcal{V}^*$ and we can apply Theorem 1.3(a). If V is crystalline then $\mathcal{M} \in \underline{\mathcal{L}}^{ft}_{cr}$ and our assertion follows from Theorem 1.3(b). □

Remark. If in the above Corollary V has HT weights from $[0, A]$, where $2 \leqslant A \leqslant p-2$, then $\Gamma_K^{(v)}$ acts trivially on H/pH for all $v > 1 + A/(p-1) - 1/p$ in the semi-stable case and for $v > A/(p-1)$ in the crystalline case. These ramification estimates have been proved in [24, 23, 7] and can be also obtained via methods from Section 3.

Let $\underline{\mathrm{CM}\Gamma}^{st}_{1,K}$, resp., $\underline{\mathrm{CM}\Gamma}^{cr}_{1,K}$, be the full subcategory in $\underline{\mathrm{CM}\Gamma}_K$ consisting of $\widetilde{\mathcal{CV}}^*(\mathcal{L})$ where $\mathcal{L}$ runs over the family of all objects of the category $\underline{\mathcal{L}}^*$, resp., $\underline{\mathcal{L}}^*_{cr}$.

Consider the simple objects $\mathcal{F}_j \in \underline{\mathrm{CM}\Gamma}^{cr}_{1,K} \subset \underline{\mathrm{CM}\Gamma}^{st}_{1,K}$ such that for $j = 0$, $\mathcal{F}_0 = (\mathbb{F}_p, \mathbb{F}_p, \mathrm{id})$ and for $1 \leqslant j \leqslant p-1$, $\mathcal{F}_j = (\mathbb{F}_p(j), 0, 0)$, where $\mathbb{F}_p(j)$ is the j-th Tate twist. (The objects $\mathcal{F}_0$ and $\mathcal{F}_{p-1}$ have already appeared in Subsection 2.4.)

Corollary 1.9. *(a) If $j_1 \geqslant j_2$ then* $\mathrm{Ext}_{\underline{\mathrm{CM}\Gamma}^{cr}_{1,K}}(\mathcal{F}_{j_1}, \mathcal{F}_{j_2}) = 0$.
(b) If $j_1 = 0$ or $j_2 = p-1$ then $\mathrm{Ext}_{\underline{\mathrm{CM}\Gamma}^{st}_{1,K}}(\mathcal{F}_{j_1}, \mathcal{F}_{j_2}) = 0$.

Proof. This follows from the appropriate statements in $\underline{\mathcal{L}}^*_{cr}$ and $\underline{\mathcal{L}}^*$. As a matter of fact, the cases $j_1 = 0$ or $j_2 = p-1$ are just the existence of a maximal etale subobject and a maximal multiplicative quotient. In the case (a), the appropriate statements in $\underline{\mathcal{L}}^*_{cr}$ are just easy exercises or one can use very general approach from Subsection 1.5 of [11]. □

Remark. An analogue of property (a) for $\underline{\mathrm{CM}\Gamma}^{st}_{1,K}$ is false because there are appropriate non-trivial extensions in the category $\underline{\mathcal{L}}^*$. Nevertheless, there is a chance to have such analogue in the smaller category $\underline{\mathrm{CM}\Gamma}^{0,st}_{1,K}$ of the objects $\widetilde{\mathcal{CV}}^{ft}(\mathcal{L})$ such that $\mathcal{L} \in \underline{\mathcal{L}}^{ft}[1] := \{\mathcal{L} \in \underline{\mathcal{L}}^{ft} \mid p\mathcal{L} = 0\}$. A partial evidence for this is given in Subsection 5.5 below.

5. Generalization of the Shafarevich Conjecture

As earlier, p is a fixed prime number, $p > 2$. Suppose $k = \bar{\mathbb{F}}_p$, $K = W(k)[1/p]$, $\underline{\mathrm{M}\Gamma}_K$ and $\underline{\mathrm{M}\Gamma}_{\mathbb{Q}}$ are the categories of $\mathbb{Z}_p$-modules with continuous action of Γ_K and, resp., $\Gamma_{\mathbb{Q}}$. Choose an extension of the p-adic valuation to $\bar{\mathbb{Q}}$ and use it to identify Γ_K with a subgroup of $\Gamma_{\mathbb{Q}}$.

5.1. The category $\underline{\mathrm{M}\Gamma}_{\mathbb{Q}}^{p,cr}$

The objects of the category $\underline{\mathrm{M}\Gamma}_{\mathbb{Q}}^{p,cr}$ are the pairs $H_{\mathbb{Q}} = (H, \widetilde{H}_{cr})$ such that

- $H \in \underline{\mathrm{M}\Gamma}_{\mathbb{Q}}$ is unramified outside of p;
- $\widetilde{H}_{cr} = (H_{cr}, H^0, j) \in \underline{\mathrm{CM}\Gamma}_K^{cr}$ – the full subcategory in $\underline{\mathrm{CM}\Gamma}_K$ of the objects of the form $\widetilde{CV}^{ft}(\mathcal{L})$, where $\mathcal{L} \in \underline{\mathcal{L}}_{cr}^{ft}$;
- $H|_{\Gamma_K} = H_{cr}$;
- morphisms in $\underline{\mathrm{M}\Gamma}_{\mathbb{Q}}^{p,cr}$ are compatible morphisms of Galois modules.

Clearly, $\underline{\mathrm{M}\Gamma}_{\mathbb{Q}}^{p,cr}$ is a special pre-abelian category.

Let $\underline{\mathrm{M}\Gamma}_{\mathbb{Q}}^{p,cr}[1]$ be the subcategory in $\underline{\mathrm{M}\Gamma}_{\mathbb{Q}}^{p,cr}$ consisting of all $H_{\mathbb{Q}} = (H, \widetilde{H}_{cr})$ such that $pH = 0$. Denote by $\mathbb{Q}_{cr}(p)$ the field-of-definition of all $H \in \underline{\mathrm{M}\Gamma}_{\mathbb{Q}}$ such that $H_{\mathbb{Q}} = (H, \widetilde{H}_{cr}) \in \underline{\mathrm{M}\Gamma}_{\mathbb{Q}}^{p,cr}[1]$. In other words, $\tau \in \mathrm{Gal}(\bar{\mathbb{Q}}/\mathbb{Q}_{cr}(p))$ iff τ acts trivially on the first components H of all $H_{\mathbb{Q}} \in \underline{\mathrm{M}\Gamma}_{\mathbb{Q}}^{p,cr}[1]$.

Let $\mathcal{H}_{\mathbb{Q}} = \{(H^{(n)}, \widetilde{H}_{cr}^{(n)})\}_{n\geqslant 0}$ be a p-divisible group in $\underline{\mathrm{M}\Gamma}_{\mathbb{Q}}^{p,cr}$. Then $\mathcal{H} = \{H^{(n)}\}_{n\geqslant 0}$ is a p-divisible group in $\underline{\mathrm{M}\Gamma}_{\mathbb{Q}}$.

Proposition 1.10. *If $\mathbb{Q}_{cr}(p)$ is totally ramified at p then there are p-divisible groups $\mathcal{H} = \mathcal{H}_0 \supset \mathcal{H}_1 \supset \cdots \supset \mathcal{H}_{p-1} \supset \mathcal{H}_p = 0$ in $\underline{\mathrm{M}\Gamma}_{\mathbb{Q}}$ such that for all $0 \leqslant i < p$, $\mathcal{H}_i/\mathcal{H}_{i+1}$ is the product of several copies of the Tate twist $\mathbb{Q}_p/\mathbb{Z}_p(i)$ of the trivial p-divisible group $\mathbb{Q}_p/\mathbb{Z}_p$.*

Proof. We have $\mathrm{Gal}(\mathbb{Q}_{cr}(p)/\mathbb{Q}) = \mathrm{Gal}(\mathbb{Q}_{cr}(p)K/K)$. Therefore, we can apply local results about Galois modules from $\underline{\mathrm{CM}\Gamma}_{1,K}^{cr}$ to the objects of the category $\underline{\mathrm{M}\Gamma}_{\mathbb{Q}}^{p,cr}[1]$. In particular, the tamely ramified part of $\mathrm{Gal}(\mathbb{Q}_{cr}(p)K/K)$ comes from $\mathrm{Gal}(\mathbb{Q}(\zeta_p)/\mathbb{Q})$ where ζ_p is a primitive p-th root of unity. (Indeed, it is a quotient of prime to p order of the Galois group of the maximal abelian extension of $\mathbb{Q}$ unramified outside of p.) Therefore, any simple subquotient of $(H^{(1)}, \widetilde{H}_{cr}^{(1)}) \in \underline{\mathrm{M}\Gamma}_{\mathbb{Q}}^{p,cr}$ comes from simple subquotients $\mathcal{F}_j$, $0 \leqslant j < p$, of $\widetilde{H}_{cr}^{(1)}$ (cf. Subsection 4.3 for the definition of $\mathcal{F}_j$). It remains to apply Corollary 1.9 and Theorem 1.A.15 from Appendix. □

5.2. The category $\underline{\mathrm{M}\Gamma}_{\mathbb{Q}}^{p,st}$

The objects of the category $\underline{\mathrm{M}\Gamma}_{\mathbb{Q}}^{p,st}$ are the pairs $H_{\mathbb{Q}} = (H, \widetilde{H}_{st})$ such that

- $H \in \underline{\mathrm{M}\Gamma}_{\mathbb{Q}}$ is unramified outside of p;
- $\widetilde{H}_{st} = (H_{st}, H^0, j) \in \underline{\mathrm{CM}\Gamma}_K^{0,st}$ – the full subcategory in $\underline{\mathrm{CM}\Gamma}_K$ consisting of $\widetilde{CV}^{ft}(\mathcal{L})$ such that $\mathcal{L} \in \underline{\mathcal{L}}^{ft}$;

- $H|_{\Gamma_K} = H_{st}$;
- morphisms in $\underline{\mathrm{M}\Gamma}_{\mathbb{Q}}^{p,st}$ are compatible morphisms of Galois modules.

Clearly, $\underline{\mathrm{M}\Gamma}_{\mathbb{Q}}^{p,st}$ is a special pre-abelian category.

Let $\underline{\mathrm{M}\Gamma}_{\mathbb{Q}}^{p,st}[1]$ be the subcategory in $\underline{\mathrm{M}\Gamma}_{\mathbb{Q}}^{p,st}$ consisting of all $H_{\mathbb{Q}} = (H, \widetilde{H}_{st})$ such that $pH = 0$. Denote by $\mathbb{Q}_{st}(p)$ the field-of-definition of all $H \in \underline{\mathrm{M}\Gamma}_{\mathbb{Q}}$ such that $H_{\mathbb{Q}} = (H, \widetilde{H}_{st}) \in \underline{\mathrm{M}\Gamma}_{\mathbb{Q}}^{p,st}[1]$.

Let $\mathcal{H}_{\mathbb{Q}} = \{(H^{(n)}, \widetilde{H}_{cr}^{(n)})\}_{n\geqslant 0}$ be a p-divisible group in $\underline{\mathrm{M}\Gamma}_{\mathbb{Q}}^{p,cr}$. Similarly to Proposition 1.10 we obtain the following property.

Proposition 1.11. *If $\mathbb{Q}_{st}(p)$ is totally ramified at p then there are p-divisible groups $\mathcal{H} = \mathcal{H}_0 \supset \mathcal{H}_1 \supset \mathcal{H}_2$ in $\underline{\mathrm{M}\Gamma}_{\mathbb{Q}}$ such that $\mathcal{H}_0/\mathcal{H}_1$ is the product of several copies of $\mathbb{Q}_p/\mathbb{Z}_p$, $\mathcal{H}_2$ is the product of several copies of $(\mathbb{Q}_p/\mathbb{Z}_p)(p-1)$ and all simple subquotients in $\mathcal{H}_1/\mathcal{H}_2$ come from objects $\mathcal{F}_j = (\mathbb{F}_p(j), 0, 0)$ with $1 \leqslant j \leqslant p-2$.*

5.3. General criterion

Suppose $X/\mathbb{Q}$ is a projective variety, p is a prime number and $N \in \mathbb{N}$. Then $V = H_{et}^N(X_{\bar{\mathbb{Q}}}, \mathbb{Q}_p)$ is a finite dimensional $\mathbb{Q}_p$-vector space with continuous $\Gamma_{\mathbb{Q}}$-action.

Proposition 1.12. *Suppose there is a filtration of $\mathbb{Q}_p[\Gamma_{\mathbb{Q}}]$-modules*

$$V = V_0 \supset V_1 \supset \cdots \supset V_N \supset V_{N+1} = 0$$

such that for all i, $V_i/V_{i+1} \simeq \mathbb{Q}_p(i)^{s_i}$, where $s_i \geqslant 0$ and $\mathbb{Q}_p(i)$ is the Tate twist. Then $h^{a,b}(X_{\mathbb{C}}) = 0$ if $a+b = N$ and $a \neq b$.

Proof. The relation between the etale and de Rham cohomology of X implies that $h^{a,b}(X_{\mathbb{C}}) = s_a$. Choose a prime $l \neq p$ such that the variety X has good reduction modulo l. Then the corresponding Frobenius σ_l acts on V with eigenvalues λ such that $|\lambda| = l^{N/2}$. But for any i, σ_l acts on $\mathbb{Q}_p(i)$ via the multiplication by l^i. Therefore, $h^{a,b}(Y_{\mathbb{C}})$ with $a+b = N$ can be different from 0 only if $l^a = l^{N/2}$. □

5.4. Crystalline case

Suppose X has everywhere good reduction. Consider $\mathbb{Q}_5[\Gamma_{\mathbb{Q}}]$-module $V = H_{et}^4(X_{\bar{Q}}, \mathbb{Q}_5)$. All subquotients of V come from appropriate filtered modules associated with de Rham cohomology of X via Breuil's functor. Therefore, any finite subquotient H of V appears as the first component of an appropriate object $H_{\mathbb{Q}} = (H, \widetilde{H}_{cr})$ of the category $\underline{\mathrm{M}\Gamma}_{\mathbb{Q}}^{5,cr}$. In particular, if T is a Galois invariant lattice in V then the 5-divisible group $\{T/5^n\}_{n\geqslant 0}$ appears as the first component of the appropriate 5-divisible group in $\underline{\mathrm{M}\Gamma}_{\mathbb{Q}}^{5,cr}$. Therefore, part (b) of Theorem 1.2 is implied by the following Proposition. (This Proposition was stated without proof at the end of [7].)

Lemma 1.13. *Modulo GRH (Generalised Riemann Hypothesis) the field* $\mathbb{Q}_{cr}(5)$ *is totally ramified at 5.*

Proof. Because the higher ramification subgroup $\Gamma_K^{(1)}$ acts trivially on $\mathbb{Q}_{cr}(5)$ the normalized discriminant of any subfield L in $\mathbb{Q}_{cr}(5)$, which is finite over $\mathbb{Q}$, is less than $25 < d^*_{340}$, cf. [25]. Here for $N \in \mathbb{N}$, d^*_N is the Odlyzko estimate for the normalized discriminant of algebraic number fields of given degree N under GRH. Therefore, $[\mathbb{Q}_{cr}(5) : \mathbb{Q}] < 340$.

The maximal abelian extension of $\mathbb{Q}$ in $\mathbb{Q}_{cr}(5)$ equals $\mathbb{Q}(\zeta_{25})$, where ζ_{25} is a 25-th primitive root of unity. Let L_2 be the maximal abelian extension of L_1 inside $\mathbb{Q}_{cr}(5)$. The class number $h(L_1) = 1$ implies that L_2 is totally ramified at 5 over $\mathbb{Q}$ and L_2/L_1 is a 5-extension. Because the total degree is less than 340 we have $[L_2 : L_1] \leqslant 5$ and one can see that $L_2 = L_1(\sqrt[5]{2+\sqrt{5}})$. Then the maximal upper ramification number of this field extension is 1 and the maximal lower number is 8, therefore, the normalized discriminant of L_2 equals $21.6288\ldots < d^*_{160}$, cf. [25]. This implies that $h(L_2) = 1$, the maximal abelian extension L_3 of L_2 inside $\mathbb{Q}_{cr}(5)$ is totally ramified at 5 and L_3/L_2 is a 5-extension. But $[L_3 : L_2] \leqslant 3$ implies that $L_3 = L_2$. □

Remark. (a) For part (a) of Theorem 1.2 take $V = H^3(X_{\bar{\mathbb{Q}}}, \mathbb{Q}_5)$. Then the corresponding ramification estimates are better, cf. Subsection 4.3. As a result, one can use unconditional Odlyzko estimates to find that all modulo 5 subquotients of V are defined over $\mathbb{Q}(\zeta_5, \sqrt[5]{\zeta_5 + \zeta_5^{-1}})$, cf. [7], Subsection 7.5.1, where this field was denoted by $\mathbb{Q}(5, 3)$.

(b) Unconditional Odlyzko estimates are still sufficient to prove that $h^2(X_{\mathbb{C}}) = h^{1,1}(X_{\mathbb{C}})$, when X has everywhere good reduction and is defined over $\mathbb{Q}(\sqrt{-3})$, $\mathbb{Q}(\sqrt{-1})$ or $\mathbb{Q}(\sqrt{5})$, cf. Section 7 of [7].

5.5. Semi-stable case

Suppose X has semi-stable reduction modulo 3 and good reduction modulo all primes $l \neq 3$. Consider $V = H^2_{et}(X_{\bar{\mathbb{Q}}}, \mathbb{Q}_3)$ and proceed similarly to Subsection 5.4. We need the following lemma.

Lemma 1.14. $\mathbb{Q}_{st}(3)$ *is totally ramified at 3.*

Proof. For a complete proof cf. [11], Lemma 5.2. Note that the upper estimate for the normalized discriminant of $\mathbb{Q}_{st}(3)/\mathbb{Q}$ is $3^{3-1/3} < d_{238}$, therefore $[\mathbb{Q}_{st}(3) : \mathbb{Q}] < 238$. Because $K_1 = \mathbb{Q}(\sqrt[3]{3}, \exp(2\pi i/9))$ is contained in $\mathbb{Q}_{st}(3)$, the group $\mathrm{Gal}(\mathbb{Q}_{st}(3)/\mathbb{Q})$ is soluble. Then the proof that $\mathbb{Q}_{st}(3) = K_1$ requires calculations with fundamental units inside K_1. Namely, we need that:

- the class number of K_1 is 1;
- for any unit $u \in K_1^* \setminus K_1^{*3}$, $\sqrt[3]{u} \notin \mathbb{Q}_{st}(3)$.

Both properties were verified via the computing package SAGE, cf. www.sagemath.org and Appendix B of [11]. □

Note that under the condition that $\mathbb{Q}_{st}(p)$ is totally ramified at p we can identify $\underline{\mathrm{M}\Gamma}_{\mathbb{Q}}^{p,st}$ with full subcategories in $\underline{\mathrm{M}\Gamma}_{\mathbb{Q}}$ and in $\underline{\mathrm{CM}\Gamma}_K^{0,st}$.

Apply Proposition 1.11. Theorem 1.1 will be proved if we show that the 3-divisible group $\mathcal{H}_1/\mathcal{H}_2$ is the product of 3-divisible groups $(\mathbb{Q}_3/\mathbb{Z}_3)(1)$.

This would follow from $\mathrm{Ext}_{\underline{\mathrm{M}\Gamma}_{\mathbb{Q}}^{3,st}[1]}(\mathcal{F}_1, \mathcal{F}_1) = 0$. The most natural way is to verify this in the category $\underline{\mathrm{CM}\Gamma}_{1,K}^{0,st}$ or, equivalently, in the category $\underline{\mathcal{L}}^{ft}[1] = \{\mathcal{L} \in \underline{\mathcal{L}}^{ft} \mid p\mathcal{L} = 0\}$. Here we come again to the principal difference between the theories of torsion crystalline and torsion semi-stable Galois modules. In the semi-stable case we can efficiently work only with the Galois modules obtained from the objects of $\underline{\mathcal{L}}^t[1] = \{\mathcal{L} \in \underline{\mathcal{L}}^t \mid p\mathcal{L} = 0\}$ (which can be identified with $\underline{\mathcal{L}}^*$), and this category is strictly bigger than $\underline{\mathcal{L}}^{ft}[1]$. (In the crystalline case $\underline{\mathcal{L}}_{cr}^t = \underline{\mathcal{L}}_{cr}^{ft}$, cf. Subsection 4.2.1.)

If $\mathcal{L}_1 \in \underline{\mathcal{L}}^t$ is such that $\mathcal{V}^t(\mathcal{L}_1) = \mathcal{F}_1$, then $\mathrm{Ext}_{\underline{\mathcal{L}}^t[1]}(\mathcal{L}_1, \mathcal{L}_1) \neq 0$ and the question about $\mathrm{Ext}_{\underline{\mathcal{L}}^{ft}[1]}(\mathcal{L}_1, \mathcal{L}_1) = 0$ is still open. We did not resolve this issue in [11] but treated it in the following simpler situation. Remind that all points of the Galois modules coming from $\underline{\mathrm{M}\Gamma}_{\mathbb{Q}}^{3,st}$ are defined over the relatively small field $K\mathbb{Q}_{st}(3)$. This allows us to replace the category $\underline{\mathcal{L}}^t[1]$ by a smaller one $\underline{\mathcal{L}}_{\mathbb{Q}}^t[1]$, where $\mathrm{Ext}_{\underline{\mathcal{L}}_{\mathbb{Q}}^t[1]}(\mathcal{L}_1, \mathcal{L}_1) = \mathbb{Z}/3\mathbb{Z}$ is generated by one object $\mathcal{L}_{11}$ (we use in [11], Subsection 5.4, slightly different notation). Then we prove that any object of $\underline{\mathcal{L}}_{\mathbb{Q}}^t[1]$, which has only subquotients of the form $\mathcal{L}_1$, is isomorphic to the product of several copies of $\mathcal{L}_1$ and $\mathcal{L}_{11}$.

Now come back to our 3-divisible group $\mathcal{H}_1/\mathcal{H}_2$ viewed as a 3-divisible group in the category $\underline{\mathrm{CM}\Gamma}_K^{0,st}$. If any subquotient of this group contains $H_{11} = \mathcal{CV}^*(\mathcal{L}_{11})$ then we apply the devissage from Appendix A to deduce the existence of a 3-divisible group $\mathcal{H} = \{H^{(n)}\}_{n\geqslant 0}$ in $\underline{\mathrm{CM}\Gamma}_K^{0,st}$ such that $H^{(1)} = H_{11}$. The height of this 3-divisible group is 2 and it determines a 2-dimensional semi-stable $\mathbb{Q}_3[\Gamma_K]$-module. But the existence of such 2-dimensional representation contradicts Theorem 6.1.1.2 from [14].

Therefore, all $\mathcal{L} \in \underline{\mathcal{L}}^{ft}[1]$ with simple subquotients isomorphic to $\mathcal{L}_1$ are just the products of copies of $\mathcal{L}_1$. In particular, $\mathrm{Ext}_{\underline{\mathcal{L}}^{ft}[1]}(\mathcal{L}_1, \mathcal{L}_1) = 0$ and $\mathcal{H}_2/\mathcal{H}_1$ is the product of copies of the trivial 3-divisible group $(\mathbb{Q}_3/\mathbb{Z}_3)(1)$.

Remark. The situation from Theorem 1.1 is quite exceptional. For semi-stable $\mathbb{Q}_p[\Gamma_K]$-modules with $p > 3$ and $2 \leqslant N < p$, the appropriate estimates for normalized discriminants of the fields-of-definition of their subquotients modulo p are bigger than the appropriate Odlyzko estimates (even under GRH). In addition, any explicit calculations with elements of algebraic number fields of degree bigger than, say, 100 are already very difficult (if possible).

Appendix A. Formalism of pre-abelian categories

This is an expository version of Appendix A of [11].

A pre-abelian category $\mathcal{C}$ is an additive category such that any of its morphisms has kernel and cokernel. A morphism u of $\mathcal{C}$ is STRICT if the canonical map $\mathrm{Coim}\, u := \mathrm{Coker}\,\mathrm{Ker}\, u \longrightarrow \mathrm{Im}\, u := \mathrm{Ker}\,\mathrm{Coker}\, u$ is isomorphism. Then $0 \longrightarrow A \xrightarrow{u} B \xrightarrow{v} C \longrightarrow 0$ is a short exact sequence in $\mathcal{C}$ if u is strict monomorphism, v is strict epimorphism and $\mathrm{Coker} u = \mathrm{Ker}\, v$.

A pre-abelian category $\mathcal{C}$ is SPECIAL if the group of classes of equivalent short exact sequences is functorial in both arguments and there are standard 6-terms Hom –Ext exact sequences.

A typical example of special pre-abelian category is the category $\underline{\mathrm{FMod}}_R$ of free modules over a ring R with filtration $F(H) \subset H$. The morphisms $\mathrm{Hom}_{\underline{\mathrm{FMod}}_R}((H, F(H)), (H_1, F(H_1))$ are morphisms $f : H \longrightarrow H_1$ of R-modules such that $f(F(H)) \subset F(H_1)$. The morphism f is a strict monomorphism iff $H_1/f(H)$ has no R-torsion and f is a strict epimorphism iff $f(H) = H_1$ and $f(F(H)) = F(H_1)$.

Denote by $\mathcal{C}(1)$ the full subcategory of killed by p (i.e. such that $p\mathrm{id}_A = 0$) objects A of a special pre-abelian category $\mathcal{C}$.

Mimicing Tate's definition [31] introduce the concept of a p-divisible object (or just p-divisible group if there is no risk of confusion) in $\mathcal{C}$.

Suppose $C = \{C^{(n)}\}_{n \geqslant 1}$ is a p-divisible group in $\mathcal{C}$. The following result provides us with very convenient devissage technique in $\mathcal{C}$. (For a complete proof of these statements see Theorems A.1 and A.2 of [11].)

Theorem 1.A.15. *(a) Suppose*

$$0 \longrightarrow D_1 \longrightarrow C^{(1)} \longrightarrow D_2 \longrightarrow 0$$

is a short exact sequence in $\mathcal{C}(1)$ *and* $\mathrm{Ext}_{\mathcal{C}(1)}(D_1, D_2) = 0$. *Then there is a short exact sequence of p-divisible groups*

$$0 \longrightarrow C_1 \longrightarrow C \longrightarrow C_2 \longrightarrow 0$$

such that $C_1^{(1)} = D_1$ *and* $C_2^{(1)} = D_2$.

(b) *Suppose* $\mathrm{Ext}_{\mathcal{C}(1)}(C^{(1)}, C^{(1)}) = 0$ *then any* p*-divisible group* D *in* $\mathcal{C}$ *such that* $D^{(1)} \simeq C^{(1)}$ *is isomorphic to* C.

References

[1] V. A. ABRASHKIN, Good reduction of two-dimensional Abelian varieties (Russian) *Izv. Akad. Nauk SSSR Ser. Mat.*, **40** (1976), no. 2, 262–272

[2] V. A. ABRASHKIN, p-divisible groups over $\mathbb{Z}$ (Russian) *Izv. Akad. Nauk SSSR Ser. Mat.*, **41** (1977), no. 5, 987–1007

[3] V. A. ABRASHKIN, Group schemes of period p over the ring of Witt vectors (Russian) *Dokl. Akad. Nauk SSSR*, **283** (1985), no. 6, 1289–1294

[4] V. A. ABRASHKIN, Honda systems of group schemes of period p (Russian) *Izv. Akad. Nauk SSSR Ser. Mat.*, **51** (1987), no. 3, 451–484, 688; translation in *Math. USSR-Izv.* **30** (1988), no. 3, 419–453

[5] V. A. ABRASHKIN, Galois modules of group schemes of period p over the ring of Witt vectors (Russian) *Izv. Akad. Nauk SSSR Ser. Mat.*, **51** (1987), no. 4, 691–736; translation in *Math. USSR-Izv.* **31** (1988), no. 1, 1–46

[6] V. A. ABRASHKIN, Modification of the Fontaine-Laffaille functor (Russian), *Izv. Akad. Nauk SSSR Ser Mat*, **53** (1989), 451–497; translation in *Math. USSR-Izv.*, **34** (1990), 57–97

[7] V. A. ABRASHKIN, Modular representations of the Galois group of a local field and a generalization of a conjecture of Shafarevich (Russian) *Izv. Akad. Nauk SSSR Ser. Mat.* **53** (1989), no. 6, 1135–1182; translation in *Math. USSR-Izv*. **35** (1990), no. 3, 469–518

[8] V. A. ABRASHKIN, A ramification filtration of the Galois group of a local field. III. (Russian) *Izv. Ross. Akad. Nauk Ser. Mat.* **62** (1998), no. 5, 3–48; translation in *Izv. Math.* **62** (1998), no. 5, 857–900

[9] V. ABRASHKIN, Galois modules arising from Faltings's strict modules. *Compos. Math.* **142** (2006), no. 4, 867–888

[10] V. ABRASHKIN, Characteristic p analogue of modules with finite crystalline height, *Pure Appl. Math.* Q., **5** (2009), 469–494

[11] V. ABRASHKIN, Projective varieties with bad reduction at 3 only, *Doc. Math.* **18** (2013), 547–619.

[12] *Algebraic number theory.* Proceedings of an instructional conference organized by the LMS. Edited by J. W. S. Cassels and A. Fröhlich Academic Press, London; Thompson Book Co., Inc., Washington, D.C. 1967

[13] A. BRUMER, K. KRAMER Non-existence of certain semistable abelian varieties, *Manuscripta Math.*, **106** (2001), 291–304

[14] CH. BREUIL, Construction de représentations p-adiques semi-stable, *Ann. Sci. École Norm. Sup., 4 Ser.*, **31** (1998), 281–327

[15] CH. BREUIL, Représentations semi-stables et modules fortement divisibles, *Invent. Math.*, **136** (1999), 89–122

[16] CH. BREUIL, Integral p-adic Hodge theory, *Adv. Stud. Pure Math.* **36** (2002), 51–80

[17] P. DELIGNE, *Les corps locaux de caractristique p, limites de corps locaux de caractristique 0.* (French) Representations of reductive groups over a local field, 119–157, Travaux en Cours, Hermann, Paris, 1984

[18] DIA Y DIAZ, *Tables minorant la racine n-ième du discriminant d'un corps de degré n* Publications Mathematiques d'Orsay, 80.06, Université de Paris-Sud, Départment de Mathématique, Bat. 425, 91405, Orsay, France

[19] J.-M. FONTAINE, G. LAFFAILLE, Construction de représentations p-adiques, *Ann. Sci. École Norm. Sup., 4 Ser.* **15** (1982), 547–608

[20] J.-M. FONTAINE, Groupes finis commutatifs sur les vecteurs de Witt, *C. R. Acad. Sci. Paris Sér. A-B* **280** (1975), A1423A1425

[21] J.-M. FONTAINE, Il n'y a pas de variété abélienne sur $\mathbb{Z}$, *Invent. Math.*, **81** (1985), 515–538

[22] J.-M. FONTAINE, W. MESSING, p-adic periods and p-adic étale cohomology. In: *Current Trends in Arithmetical Algebraic Geometry*, K. Ribet, Editor, Contemporary Math. 67, Amer. Math. Soc., Providence (1987), 179–207

[23] J.-M. FONTAINE, Schémas propres et lisses sur $\mathbb{Z}$. *Proceedings of the Indo-French Conference on Geometry* (Bombay, 1989), 43–56, Hindustan Book Agency, Delhi, 1993

[24] SH. HATTORI, On a ramification bound of torsion semi-stable representations over a local field, *J. Number theory*, **129** (2009), 2474–2503

[25] J. MARTINET, *Petits discriminants des corps de nombres*, London Math. Soc - Lect. note ser., 56 (1982), 151–193

[26] W. MESSING *The crystals associated to Barsotti-Tate groups: with applications to abelian schemes*, Lecture notes in Mathematics. **264**, Springer-Verlag, 1972

[27] R.SCHOOF, Abelian varieties over $\mathbb{Q}$ with bad reduction in one prime only. *Comp. Math.*, **141** (2005), 847–868.

[28] J.-P.SERRE, *Local Fields* Berlin, New York: Springer-Verlag, 1980

[29] I. R. SHAFAREVICH, Algebraic number fields. *Proc. Internat. Congr. Math.* (Stockholm, 1962), Inst. Mittag-Leffler, Djursholm, (1963), 163–176

[30] G. POITOU, Minorations de discriminants (d'apr A. M. Odlyzko). Séminaire Bourbaki, Vol. 1975/76 28me anne, Exp. No. 479, pp. 136–153. *Lecture Notes in Math.*, 567, Springer, Berlin, 1977 Berlin, New York: Springer-Verlag, 1980

[31] J. TATE, p-divisible groups, *Proc. Conf. Local Fields* (Dreibergen, 1966), Springer-Verlag, Berlin and New-York, 1967, pp. 158–183

[32] J.-P. WINTENBERGER, Le corps des normes de certaines extensions infinies des corps locaux; application *Ann. Sci. Ec. Norm. Super., IV. Ser*, **16** (1983), 59–89

[33] J.-P. WINTENBERGER, Extensions de Lie et groupes d'automorphismes des corps locaux de caractristique p. (French) *C. R. Acad. Sci. Paris Sér.* **A-B 288** (1979), no. 9, A477–A479

2

Irreducible modular representations of the Borel subgroup of $GL_2(\mathbf{Q}_p)$

Laurent Berger and Mathieu Vienney

Abstract

Let E be a finite extension of $\mathbf{F}_p$. Using Fontaine's theory of (φ, Γ)-modules, Colmez has shown how to attach to any irreducible E-linear representation of $\mathrm{Gal}(\overline{\mathbf{Q}}_p/\mathbf{Q}_p)$ an infinite dimensional smooth irreducible E-linear representation of $B_2(\mathbf{Q}_p)$ that has a central character. We prove that every such representation of $B_2(\mathbf{Q}_p)$ arises in this way.

Our proof extends to algebraically closed fields E of characteristic p. In this case, infinite dimensional smooth irreducible E-linear representations of $B_2(\mathbf{Q}_p)$ having a central character arise in a similar way from irreducible E-linear representations of the Weil group of $\mathbf{Q}_p$.

Contents

Introduction *page* 33
Notation 34
1 (φ, Γ)-modules and (ψ, Γ)-modules 34
2 Construction of Galois representations 36
3 Topological representations of profinite groups 41
4 Colmez's functor 44
5 Representations of $B_2(\mathbf{Q}_p)$ 46
References 50

1991 *Mathematics Subject Classification*. 11F; 11S; 20C; 20G; 22E; 46A.
Key words and phrases. Modular representation; Borel subgroup; Galois representation; (φ, Γ)-module; (ψ, Γ)-module; p-adic Langlands correspondence; linear topology; non-commutative polynomial.

This research is partially supported by the ANR grant ThéHopaD (Théorie de Hodge p-adique et Développements) ANR-11-BS01-005.

Automorphic Forms and Galois Representations, ed. Fred Diamond, Payman L. Kassaei and Minhyong Kim. Published by Cambridge University Press.

Introduction

This chapter is inspired by the p-adic local Langlands correpondence for $\mathrm{GL}_2(\mathbf{Q}_p)$, which is a bijection between some 2-dimensional representations of $\mathrm{Gal}(\overline{\mathbf{Q}}_p/\mathbf{Q}_p)$ and some representations of $\mathrm{GL}_2(\mathbf{Q}_p)$. Colmez observed that this bijection, whose existence had been conjectured by Breuil, can be constructed using Fontaine's theory of (φ, Γ)-modules, in order to obtain representations of $\mathrm{B}_2(\mathbf{Q}_p)$, the upper triangular Borel subgroup of $\mathrm{GL}_2(\mathbf{Q}_p)$, from 2-dimensional representations of $\mathrm{Gal}(\overline{\mathbf{Q}}_p/\mathbf{Q}_p)$. In this chapter, we determine completely which class of representations of $\mathrm{B}_2(\mathbf{Q}_p)$ can be constructed by applying Colmez's method to irreducible mod p representations of $\mathrm{Gal}(\overline{\mathbf{Q}}_p/\mathbf{Q}_p)$ of any dimension.

Let E be a finite extension of $\mathbf{F}_p$. If V is a finite dimensional E-linear representation of $\mathrm{Gal}(\overline{\mathbf{Q}}_p/\mathbf{Q}_p)$ and if χ is a smooth character of $\mathbf{Q}_p^\times$, then Colmez's functor "$\varprojlim_\psi D^\natural(\cdot)$" allows us to construct a smooth representation $\Omega_\chi(V) = (\varprojlim_\psi D^\natural(V))^*$ of the group $\mathrm{B}_2(\mathbf{Q}_p)$, having χ as central character. Our first result is the following (see theorem 4.2 and remark 4.3 of [Vie12b]).

Theorem A. *If E is a finite field, and if Π is an infinite dimensional smooth irreducible E-linear representation of $\mathrm{B}_2(\mathbf{Q}_p)$ having a central character χ, then there exists an irreducible E-linear representation V of $\mathrm{Gal}(\overline{\mathbf{Q}}_p/\mathbf{Q}_p)$ such that $\Pi = \Omega_\chi(V)$.*

Our proof extends to representations with coefficients in an algebraically closed field E of characteristic p. The theory of (φ, Γ)-modules is then less satisfactory, but one can still carry out Colmez's construction and prove an analogue of Theorem A.

Theorem A′. *If E is an algebraically closed field of characteristic p, and if Π is an infinite dimensional smooth irreducible E-linear representation of $\mathrm{B}_2(\mathbf{Q}_p)$ having a central character χ, then there exists an irreducible E-linear representation V of the Weil group of $\mathbf{Q}_p$ such that $\Pi = \Omega_\chi(V)$.*

This extension of Theorem A depends on the following result, which (following a suggestion of Colmez) extends Fontaine's theory of (φ, Γ)-modules to algebraically closed coefficient fields of characteristic p.

Theorem B. *If E is an algebraically closed field of characteristic p, then there is a natural bijection between the set of irreducible E-linear representations of the Weil group of $\mathbf{Q}_p$ and the set of irreducible (φ, Γ)-modules over $E((X))$.*

This bijection, which is compatible with the usual theory of (φ, Γ)-modules, does not seem to extend to reducible objects if E is not an algebraic extension of $\mathbf{F}_p$.

In order to prove Theorems A and A′, we need to "invert" Colmez's construction $V \mapsto (\varprojlim_\psi D^\natural(V))^*$. This was done in some cases by Colmez (see §IV of [Col10b] as well as §4 of Emerton's [Eme08]) and in much greater generality by Schneider and Vignéras (see [SV11]). Our method is similar. The finiteness result that we need in order to conclude is provided by Emerton (see [Eme08]).

Note that if E is a field of characteristic different from p, then determining the smooth irreducible representations of $\mathrm{B}_2(\mathbf{Q}_p)$ is a much easier problem (see for instance §8 of [BH06] for the case $E = \mathbf{C}$). Likewise, it is a simple exercise to determine the finite dimensional smooth irreducible E-linear representations of $\mathrm{B}_2(\mathbf{Q}_p)$.

Notation

The letter E stands for a field of characteristic p. Throughout this chapter, E is given the discrete topology. We let $\mathrm{B} = \mathrm{B}_2(\mathbf{Q}_p)$ and write $\mathcal{G}_{\mathbf{Q}_p}$ for $\mathrm{Gal}(\overline{\mathbf{Q}}_p/\mathbf{Q}_p)$. We define a map $n : \mathcal{G}_{\mathbf{Q}_p} \to \widehat{\mathbf{Z}}$ as follows: if $g \in \mathcal{G}_{\mathbf{Q}_p}$, then the image of g in $\mathrm{Gal}(\overline{\mathbf{F}}_p/\mathbf{F}_p)$ is $\mathrm{Frob}_p^{n(g)}$ where $\mathrm{Frob}_p = [x \mapsto x^p]$. The Weil group of $\mathbf{Q}_p$ is $\mathcal{W}_{\mathbf{Q}_p} = \{g \in \mathcal{G}_{\mathbf{Q}_p}$ such that $n(g) \in \mathbf{Z}\}$ and $\mathcal{I}_{\mathbf{Q}_p}$ denotes the inertia subgroup of $\mathcal{G}_{\mathbf{Q}_p}$.

In order to retain the spirit of the lectures given at the LMS Durham Symposium, we explain the idea of the proofs of some of the technical results that are taken from other papers, in order for this chapter to be more easily readable by newcomers to the subject.

1. (φ, Γ)-modules and (ψ, Γ)-modules

In this section, we recall the definition of (φ, Γ)-modules and (ψ, Γ)-modules and we explain how these objects are related to each other.

The ring $E[\![X]\!]$ is given the X-adic topology, for which it is complete, and the field $E(\!(X)\!) = \cup_{n \geqslant 0} X^{-n} E[\![X]\!]$ is given the inductive limit topology when necessary.

The rings $E[\![X]\!]$ and $E(\!(X)\!)$ are equipped with a continuous Frobenius map φ given by $(\varphi f)(X) = f(X^p)$. Let Γ stand for the group $\mathbf{Z}_p^\times$, the element of Γ corresponding to $a \in \mathbf{Z}_p^\times$ being denoted by $[a]$. The rings $E[\![X]\!]$ and $E(\!(X)\!)$ are also equipped with an action of Γ, given by $([a]f)(X) = f((1+X)^a - 1)$. This action is continuous and commutes with φ.

Definition 2.1. A (φ, Γ)-module is an $E(\!(X)\!)$-vector space D of dimension d, equipped with a semilinear Frobenius map $\varphi : \mathrm{D} \to \mathrm{D}$ whose matrix in some basis belongs to $\mathrm{GL}_d(E(\!(X)\!))$, and a continuous semilinear action of Γ that commutes with φ.

Example 2.2. If $\delta : \mathbf{Q}_p^\times \to E^\times$ is a continuous character, then we define $E(\!(X)\!)(\delta)$ as the (φ, Γ)-module of dimension 1 having e_δ as a basis, where $\varphi(e_\delta) = \delta(p)e_\delta$ and $[a]e_\delta = \delta(a)e_\delta$. Every (φ, Γ)-module of dimension 1 is then isomorphic to $E(\!(X)\!)(\delta)$ for a well-defined character $\delta : \mathbf{Q}_p^\times \to E^\times$.

If $\alpha(X) \in E(\!(X)\!)$, then we can write $\alpha(X) = \sum_{j=0}^{p-1}(1 + X)^j\alpha_j(X^p)$ in a unique way, and we define a map $\psi : E(\!(X)\!) \to E(\!(X)\!)$ by the formula $\psi(\alpha)(X) = \alpha_0(X)$. A direct computation shows that if $0 \leqslant r \leqslant p - 1$ then $\psi(X^{pm+r}) = (-1)^r X^m$.

If D is a (φ, Γ)-module over $E(\!(X)\!)$ and if $y \in \mathrm{D}$, then we can write as above $y = \sum_{j=0}^{p-1}(1+X)^j\varphi(y_j)$, and we set $\psi(y) = y_0$. The operator ψ thus defined commutes with the action of Γ and satisfies $\psi(\alpha(X)\varphi(y)) = \psi(\alpha)(X)y$ and $\psi(\alpha(X^p)y) = \alpha(X)\psi(y)$ (in particular, it is a left inverse of φ).

Definition 2.3. A (ψ, Γ)-module is an $E[\![X]\!]$-module M of finite type, equipped with an E-linear map $\psi : \mathrm{M} \to \mathrm{M}$ such that $\psi(f(X^p)y) = f(X)\psi(y)$, and a continuous semilinear action of Γ that commutes with ψ. We say that

(1) M is surjective if $\psi : \mathrm{M} \to \mathrm{M}$ is surjective;
(2) M is non-degenerate if $\ker(\psi : \mathrm{M} \to \mathrm{M})$ does not contain an $E[\![X]\!]$-submodule (in other words: if $y \in \mathrm{M}$ satisfies $\psi(f(X)y) = 0$ for all $f(X) \in E[\![X]\!]$, then $y = 0$);
(3) M is irreducible if it has no non-trivial sub-(ψ, Γ)-module.

Note that an irreducible (ψ, Γ)-module is surjective and non-degenerate. It is also torsion-free unless it is finite-dimensional over E.

Theorem 2.4. *If* D *is a* (φ, Γ)*-module, then* D *contains a surjective sub-*(ψ, Γ)*-module* M *such that* $\mathrm{D} = E(\!(X)\!) \otimes_{E[\![X]\!]} \mathrm{M}$. *In addition,*

(1) *if* D *is irreducible of dimension* $\geqslant 2$, *then* M *is uniquely determined;*
(2) *if* D *is of dimension* 1, *and we write* $\mathrm{D} = E(\!(X)\!)(\delta)$, *then either* $\mathrm{M} = E[\![X]\!] \cdot e_\delta$ *or* $\mathrm{M} = X^{-1}E[\![X]\!] \cdot e_\delta$.

Proof. This is proved in §II.4 and §II.5 of [Col10a] if E is a finite field, and more generally in §4.3 of [Vie12a]. Note that if D is of dimension 1, then the existence of M and the fact that either $\mathrm{M} = E[\![X]\!] \cdot e_\delta$ or $\mathrm{M} = X^{-1}E[\![X]\!] \cdot e_\delta$

are both simple exercises. In general, Colmez constructs both a smallest and a largest such sub-(ψ, Γ)-module, denoted by $\mathrm{D}^{\natural}$ and $\mathrm{D}^{\sharp}$ respectively. He then proves (see corollary II.5.21 of [Col10a] and theorem 4.3.50 of [Vie12a]) that if D is irreducible of dimension $\geqslant 2$, then $\mathrm{D}^{\natural} = \mathrm{D}^{\sharp}$. □

Definition 2.5. We denote by M(D) the surjective (ψ, Γ)-module attached to an irreducible (φ, Γ)-module D (if D is of dimension 1, then we take $\mathrm{M(D)} = E[\![X]\!] \cdot e_\delta$), so that our M(D) is Colmez's $\mathrm{D}^{\natural}$.

Theorem 2.6. *If M is a surjective (ψ, Γ)-module that is non-degenerate and free over $E[\![X]\!]$, then there exists a compatible (φ, Γ)-module structure on $\mathrm{D} = E(\!(X)\!) \otimes_{E[\![X]\!]} \mathrm{M}$.*

Proof. Let $\mathrm{D} = E(\!(X)\!) \otimes_{E[\![X]\!]} \mathrm{M}$ and let $\widetilde{\mathrm{D}}$ be D but with the $E(\!(X)\!)$-vector space structure given by $f(X) \cdot y = f(X^p) y$ so that $\widetilde{\mathrm{D}}$ is an $E(\!(X)\!)$-vector space of dimension pd. Let $\psi_j : \mathrm{D} \to \mathrm{D}$ be the map $y \mapsto \psi((1+X)^{-j} y)$, so that $\psi_j : \widetilde{\mathrm{D}} \to \mathrm{D}$ is a surjective linear map. Its kernel is therefore of dimension $pd - d$ and $\mathrm{N} = \cap_{i=1}^{p-1} \ker \psi_j$ is an $E(\!(X)\!)$-vector space of dimension at least $pd - (p-1)d = d$. The non-degeneracy of M implies that $\psi : \mathrm{N} \to \mathrm{D}$ is injective, so that $\dim \mathrm{N} = d$ and $\psi : \mathrm{N} \to \mathrm{D}$ is bijective.

Let $\varphi : \mathrm{D} \to \mathrm{N} \subset \mathrm{D}$ denote its inverse. It is easily checked that φ and Γ give rise to a (φ, Γ)-module structure on D, compatible with the (ψ, Γ)-module structure on M. □

We finish this section with a technical result on regularization by Frobenius. Let R be a ring equipped with an automorphism φ, which is extended to $R[\![X]\!]$ by $\varphi(X) = X^p$.

Lemma 2.7. *If $P \in \mathrm{GL}_d(R[\![X]\!])$, then there exists a matrix $M \in \mathrm{GL}_d(R[\![X]\!])$ such that $M^{-1} P \varphi(M) = P(0) \in \mathrm{GL}_d(R)$.*

Proof. This is a standard result, which is proved by successive approximation: if there exists a matrix $M_i \in \mathrm{GL}_d(R[\![X]\!])$ such that $M_i^{-1} P \varphi(M_i) = P(0) + P_i X^i + \mathrm{O}(X^{i+1})$ with $P_i \in \mathrm{M}_d(R)$ and if $Q_i = P_i P(0)^{-1}$, then

$$(1 + X^i Q_i)^{-1} M_i^{-1} \cdot P \cdot \varphi(M_i(1 + X^i Q_i)) = P(0) + \mathrm{O}(X^{i+1}),$$

so that one can set $M_{i+1} = M_i \cdot (1 + X^i Q_i)$ and take $M = \lim_{i \to +\infty} M_i$. □

2. Construction of Galois representations

In this section, we recall Fontaine's equivalence between (φ, Γ)-modules and representations of $\mathcal{G}_{\mathbf{Q}_p}$ over finite fields. After that, we explain how to extend this equivalence to irreducible representations of $\mathcal{W}_{\mathbf{Q}_p}$ over algebraically closed fields.

Let $\mathbf{E}_{\mathbf{Q}_p} = \mathbf{F}_p(\!(X)\!)$ and recall that if K is a finite Galois extension of $\mathbf{Q}_p$, then there exists a finite extension $\mathbf{E}_K$ of $\mathbf{E}_{\mathbf{Q}_p}$ attached to it by the theory of the field of norms (see [Win83] and A3 of [Fon90]), and that $\mathcal{G}_{\mathbf{Q}_p}$ acts on $\mathbf{E}_K$. For example, $\mathcal{G}_{\mathbf{Q}_p}$ acts on $\mathbf{E}_{\mathbf{Q}_p}$ by $g(f(X)) = f([\chi_{\mathrm{cycl}}(g)](X))$. We have $\mathbf{E}_{\mathbf{Q}_p}^{\mathrm{sep}} = \cup_{K/\mathbf{Q}_p}\mathbf{E}_K$ and if $\mathcal{H}_{\mathbf{Q}_p}$ denotes the kernel of $\chi_{\mathrm{cycl}} : \mathcal{G}_{\mathbf{Q}_p} \to \mathbf{Z}_p^\times$, then the map $\mathcal{H}_{\mathbf{Q}_p} \to \mathrm{Gal}(\mathbf{E}_{\mathbf{Q}_p}^{\mathrm{sep}}/\mathbf{E}_{\mathbf{Q}_p})$ is an isomorphism.

If E is a finite field and if D is a (φ, Γ)-module over $E(\!(X)\!)$, then $V(\mathrm{D}) = (\mathbf{E}_{\mathbf{Q}_p}^{\mathrm{sep}} \otimes_{\mathbf{F}_p(\!(X)\!)} \mathrm{D})^{\varphi=1}$ is an E-vector space and the group $\mathcal{G}_{\mathbf{Q}_p}$ acts on $V(\mathrm{D})$ by the formula $g(\alpha \otimes y) = g(\alpha) \otimes [\chi_{\mathrm{cycl}}(g)](y)$. This way, we get a functor from the category of (φ, Γ)-modules over $E(\!(X)\!)$ to the category of E-linear representations of $\mathcal{G}_{\mathbf{Q}_p}$. The following theorem is proved in §1.2 of [Fon90].

Theorem 2.8. *If* D *is a* (φ, Γ)*-module over* $E(\!(X)\!)$*, then* $V(\mathrm{D})$ *is an* E*-vector space of dimension* $\dim(\mathrm{D})$*, and the functor* $\mathrm{D} \mapsto V(\mathrm{D})$ *gives rise to an equivalence of categories between the category of* (φ, Γ)*-modules over* $E(\!(X)\!)$ *and the category of* E*-linear representations of* $\mathcal{G}_{\mathbf{Q}_p}$.

Proof. We give a sketch of Fontaine's proof. Assume first that $E = \mathbf{F}_p$ and let D be a (φ, Γ)-module over $\mathbf{F}_p(\!(X)\!)$. If we choose a basis of D and if $\mathrm{Mat}(\varphi) = (p_{ij})_{1 \leqslant i,j \leqslant \dim(\mathrm{D})}$ in that basis, then the algebra

$$A = \mathbf{F}_p(\!(X)\!)[X_1, \ldots, X_{\dim(\mathrm{D})}]/(X_j^p - \sum_i p_{ij} X_i)_{1 \leqslant j \leqslant \dim(\mathrm{D})}$$

is an étale $\mathbf{F}_p(\!(X)\!)$-algebra of rank $p^{\dim(\mathrm{D})}$ and

$$V(\mathrm{D}) = \mathrm{Hom}_{\mathbf{F}_p(\!(X)\!)-\mathrm{algebra}}(A, \mathbf{F}_p(\!(X)\!)^{\mathrm{sep}})$$

so that $V(\mathrm{D})$ is an $\mathbf{F}_p$-vector space of dimension $\dim(\mathrm{D})$.

Given the isomorphism $\mathcal{H}_{\mathbf{Q}_p} \simeq \mathrm{Gal}(\mathbf{F}_p(\!(X)\!)^{\mathrm{sep}}/\mathbf{F}_p(\!(X)\!))$, Hilbert's theorem 90 tells us that $\mathrm{H}^1_{\mathrm{discrete}}(\mathcal{H}_{\mathbf{Q}_p}, \mathrm{GL}_d(\mathbf{F}_p(\!(X)\!)^{\mathrm{sep}})) = \{1\}$ if $d \geqslant 1$. If V is an $\mathbf{F}_p$-linear representation of $\mathcal{H}_{\mathbf{Q}_p}$ then $\mathbf{F}_p(\!(X)\!)^{\mathrm{sep}} \otimes_{\mathbf{F}_p} V \simeq (\mathbf{F}_p(\!(X)\!)^{\mathrm{sep}})^{\dim(V)}$ as representations of $\mathcal{H}_{\mathbf{Q}_p}$ so that the $\mathbf{F}_p(\!(X)\!)$-vector space $\mathrm{D}(V) = (\mathbf{F}_p(\!(X)\!)^{\mathrm{sep}} \otimes_{\mathbf{F}_p} V)^{\mathcal{H}_{\mathbf{Q}_p}}$ is of dimension $\dim(V)$ and $V = (\mathbf{F}_p(\!(X)\!)^{\mathrm{sep}} \otimes_{\mathbf{F}_p(\!(X)\!)} \mathrm{D}(V))^{\varphi=1}$.

It is then easy to check that the functors $V \mapsto \mathrm{D}(V)$ and $\mathrm{D} \mapsto V(\mathrm{D})$ are inverse of each other. Finally, if $E \neq \mathbf{F}_p$ then one can consider an E-linear representation as an $\mathbf{F}_p$-linear representation with an E-linear structure and likewise for (φ, Γ)-modules, so that the equivalence carries over. □

For example, if δ is a character of $\mathbf{Q}_p^\times$, then the representation arising from $E(\!(X)\!)(\delta)$ is the character of $\mathcal{G}_{\mathbf{Q}_p}$ corresponding to δ by local class field theory.

If E is not a finite extension of $\mathbf{F}_p$, then Theorem 2.8 above may well fail. Suppose for instance that $E = \mathbf{F}_p(t)$ and that $\mathrm{D} = E(\!(X)\!)(\delta)$ where $\delta(p) = t$ and $\delta|_{\mathbf{Z}_p^\times} = 1$. This (φ, Γ)-module "should" correspond to the unramified character of $\mathcal{G}_{\mathbf{Q}_p}$ sending Frob_p to t^{-1}, but there is no such character because the map $n \mapsto t^{-n}$ does not extend to $\widehat{\mathbf{Z}}$. There is however such a character of the Weil group $\mathcal{W}_{\mathbf{Q}_p}$ of $\mathbf{Q}_p$ and in the rest of this section, we construct a bijection between the set of irreducible representations of $\mathcal{W}_{\mathbf{Q}_p}$ and the set of irreducible (φ, Γ)-modules over $E(\!(X)\!)$, for any algebraically closed field E.

Assume for the rest of this section that E is an algebraically closed field of characteristic p. We first explain how to attach an irreducible (φ, Γ)-module over $E(\!(X)\!)$ to an irreducible E-linear representation of $\mathcal{W}_{\mathbf{Q}_p}$. If $\lambda \in E^\times$, let $\mu_\lambda : \mathcal{W}_{\mathbf{Q}_p} \to E^\times$ denote the character defined by $g \mapsto \lambda^{-n(g)}$. Take $n \geqslant 1$ and let $\omega_n : \mathcal{I}_{\mathbf{Q}_p} \to \overline{\mathbf{F}}_p^\times$ be one of Serre's fundamental characters of level n (see [Ser72]). If $h \in \mathbf{Z}$ is not divisible by any of the $(p^n - 1)/(p^d - 1)$ for $d < n$ dividing n (we then say that h is primitive), then let $\mathrm{ind}(\omega_n^h)$ be the unique irreducible representation of $\mathcal{G}_{\mathbf{Q}_p}$ whose restriction to $\mathcal{I}_{\mathbf{Q}_p}$ is $\oplus_{i=0}^{n-1} \omega_n^{p^i h}$ and whose determinant is ω^h.

The representation $\mathrm{ind}(\omega_n^h)$ is actually defined over $\mathbf{F}_p$, as we now show. Let $W = \{\alpha \in \overline{\mathbf{F}}_p$ such that $\alpha^{p^n} = (-1)^{n-1}\alpha\}$ so that W is a $\mathbf{F}_{p^n}$-vector space of dimension 1 and hence a $\mathbf{F}_p$-vector space of dimension n. Choose $\pi_n \in \overline{\mathbf{Q}}_p$ such that $\pi_n^{p^n-1} = -p$. By composing the map $\mathrm{Gal}(\mathbf{Q}_p^{\mathrm{nr}}(\pi_n)/\mathbf{Q}_p) \xrightarrow{\sim} \mathbf{F}_{p^n}^\times \rtimes \widehat{\mathbf{Z}}$ with the map $\mathbf{F}_{p^n}^\times \rtimes \widehat{\mathbf{Z}} \to \mathrm{End}_{\mathbf{F}_p}(W)$ given by $(x, 0) \mapsto m_x^h$ (where m_x is the multiplication by x map) and by $(1, 1) \mapsto (\alpha \mapsto \alpha^p)$, we make W into an n-dimensional $\mathbf{F}_p$-linear representation of $\mathcal{G}_{\mathbf{Q}_p}$. We leave it as an exercise to check that $W = \mathrm{ind}(\omega_n^h)$.

Proposition 2.9. *If V is an irreducible n-dimensional E-linear representation of $\mathcal{W}_{\mathbf{Q}_p}$, then there exists $h \in \mathbf{Z}$ and $\lambda \in E^\times$ such that*

$$V = (E \otimes_{\mathbf{F}_p} \mathrm{ind}(\omega_n^h)) \otimes \mu_\lambda.$$

Proof. The proof is the same as in §2.1 of [Ber10a]: by §1.6 of [Ser72], $V|_{\mathcal{I}_{\mathbf{Q}_p}}$ splits as a direct sum of n tame characters and since V is irreducible, these characters are transitively permuted by Frobenius, so that they are of level n. Therefore, there exists a primitive h such that $V = \oplus_{i=0}^{n-1} V_i$ where $\mathcal{I}_{\mathbf{Q}_p}$ acts on V_i by $\omega_n^{p^i h}$. Since ω_n extends to $\mathrm{Gal}(\overline{\mathbf{Q}}_p/\mathbf{Q}_{p^n})$, each V_i is stable under the Weil group of $\mathrm{Gal}(\overline{\mathbf{Q}}_p/\mathbf{Q}_{p^n})$, which then acts on V_i by $\omega_n^{p^i h}\chi_i$ where χ_i is an unramified character. The lemma then follows from Frobenius reciprocity. □

Definition 2.10. To $V = (E \otimes_{\mathbf{F}_p} \operatorname{ind}(\omega_n^h)) \otimes \mu_\lambda$, we then attach the (φ, Γ)-module $\mathrm{D}(V)$ having a basis $e_0, \ldots, e_{n-1}$ in which $[a](e_j) = (aX/[a](X))^{hp^j(p-1)/(p^n-1)} e_j$ if $a \in \mathbf{Z}_p^\times$ and $\varphi(e_j) = e_{j+1}$ for $0 \leqslant j \leqslant n-2$ and $\varphi(e_{n-1}) = (-1)^{n-1}\lambda^n X^{-h(p-1)} e_0$.

Different choices of h and λ can give rise to the same representation V, but we can check that the (φ, Γ)-module $\mathrm{D}(V)$ thus defined depends only on V. Indeed, if $\lambda \in \overline{\mathbf{F}}_p$, then $(E \otimes_{\mathbf{F}_p} \operatorname{ind}(\omega_n^h)) \otimes \mu_\lambda$ extends to $\mathcal{G}_{\mathbf{Q}_p}$ and the (φ, Γ)-module above is the extension of scalars of the one given by Fontaine's construction, by the results of §2.1 of [Ber10a].

We now explain how to attach an irreducible representation of $\mathcal{W}_{\mathbf{Q}_p}$ to an irreducible (φ, Γ)-module over $E(\!(X)\!)$. Let F be a field that is complete for a discrete valuation $\operatorname{val}(\cdot)$ and endowed with an automorphism φ, such that $\operatorname{val}(\varphi(y)) = p \cdot \operatorname{val}(y)$ (in the sequel, we'll have $F = E(\!(Y)\!)$ where $Y^n = X$ and $\operatorname{val} = \operatorname{val}_X$). Let $F\{\varphi\}$ denote the non-commutative ring of polynomials in φ with coefficients in F. If $P(\varphi) = a_0 + a_1\varphi + \cdots + a_n\varphi^n \in F\{\varphi\}$, then the Newton polygon $\mathrm{NP}(P)$ of P is the convex polygon whose support consists of the points $([k], \operatorname{val}(a_k))$ where $[k] = (p^k - 1)/(p-1)$. The slopes of $\mathrm{NP}(P)$ are the opposites of the slopes of the segments of the polygon. If $P(\varphi) = a_0 + a_1\varphi + \cdots + a_n\varphi^n \in F\{\varphi\}$, and if $y \in F$, then $P(\varphi)y = a_0 y + a_1\varphi(y)\varphi + \cdots + a_n\varphi^n(y)\varphi^n$.

Lemma 2.11. *If $P(\varphi) \in F\{\varphi\}$ is isoclinic of slope s, and if $y \in F$ satisfies $\operatorname{val}(y) = r$, then $P(\varphi)y$ is isoclinic of slope $s - (p-1)r$.*

Proof. We have $\operatorname{val}(\varphi^k(y)a_k) = p^k \operatorname{val}(y) + \operatorname{val}(a_k)$ so that

$$\frac{\operatorname{val}(\varphi^k(y)a_k) - \operatorname{val}(\varphi^j(y)a_j)}{[k]-[j]} = \frac{\operatorname{val}(a_k) - \operatorname{val}(a_j)}{[k]-[j]} + (p-1)r.$$

□

Proposition 2.12. *If $P(\varphi) \in F\{\varphi\}$ is irreducible, then it is isoclinic.*

Proof. See §2.4 of [Ked08] as well as §3.2 of [Vie12a]. We give a sketch of the proof. Let $F\{\varphi^{\pm 1}\}$ be the space of polynomials in φ and φ^{-1}. Since $\varphi : F \to F$ is not necessarily invertible, $F\{\varphi^{\pm 1}\}$ is not a ring, but it is a left $F\{\varphi\}$-module. If $r \in \mathbf{R}$ and $P \in F\{\varphi^{\pm 1}\}$, let $\operatorname{val}_r(P) = \min_{i \in \mathbf{Z}}(\operatorname{val}(a_i) + r[i])$. Using successive approximations, we can show that if $R \in F\{\varphi^{\pm 1}\}$ and $r \in \mathbf{R}$ are such that $\operatorname{val}_r(R-1) > 0$, then there exists $P \in F\{\varphi\}$ and $Q \in F\{\varphi^{-1}\}$ such that $R = PQ$. Using this factorization result, we can now prove that if $P \in F\{\varphi\}$ and $\mathrm{NP}(P)$ has a breakpoint, then P can be factored in $F\{\varphi\}$. □

Note that in general, if $P = P_1P_2$, then the set of slopes of $\mathrm{NP}(P)$ is not the union of the sets of slopes of $\mathrm{NP}(P_1)$ and $\mathrm{NP}(P_2)$.

We denote by val_X the X-adic valuation on $\mathbf{E}_K$, by $\mathbf{E}_K^+$ the ring of integers of $\mathbf{E}_K$ for val_X and by k_K the residue field of $\mathbf{E}_K$ (it is the residue field of $K(\mu_{p^\infty})$).

Proposition 2.13. *If* D *is an irreducible* (φ, Γ)*-module over* $E(\!(X)\!)$*, then there exists a finite extension* K *of* $\mathbf{Q}_p$*, such that* $\mathbf{E}_K \otimes_{\mathbf{E}_{\mathbf{Q}_p}} \mathrm{D}$ *has a basis in which* $\mathrm{Mat}(\varphi)$ *belongs to* $\mathrm{GL}_d(k_K \otimes_{\mathbf{F}_p} E)$*.*

The $k_K \otimes_{\mathbf{F}_p} E$*-module generated by this basis depends only on* D*, and in particular it is stable under the action of* $\mathcal{G}_{\mathbf{Q}_p}$ *given by* $g(\alpha \otimes y) = g(\alpha) \otimes [\chi_{\mathrm{cycl}}(g)](y)$*.*

Proof. Let us first show that the $k_K \otimes_{\mathbf{F}_p} E$-module generated by such a basis is unique. If $M \in \mathrm{M}_d(\mathbf{E}_K \otimes_{\mathbf{E}_{\mathbf{Q}_p}} E(\!(X)\!))$, then let $\mathrm{val}_X(M)$ be the minimum of the valuations of the entries of M.

If $\mathbf{E}_K \otimes_{\mathbf{E}_{\mathbf{Q}_p}} \mathrm{D}$ admits two bases in which $\mathrm{Mat}(\varphi) \in \mathrm{GL}_d(k_K \otimes_{\mathbf{F}_p} E)$, then let P_1 and P_2 be the two matrices of φ and let $B \in \mathrm{GL}_d(\mathbf{E}_K \otimes_{\mathbf{E}_{\mathbf{Q}_p}} E(\!(X)\!))$ be the change of basis matrix. We then have $P_2 = B^{-1}P_1\varphi(B)$ so that $\varphi(B) = P_1^{-1}BP_2$. This implies that $\mathrm{val}_X(\varphi(B)) = \mathrm{val}_X(B)$ so that $\mathrm{val}_X(B) = 0$, and hence $B \in \mathrm{M}_d(\mathbf{E}_K^+ \otimes_{\mathbf{E}_{\mathbf{Q}_p}^+} E[\![X]\!])$. The same argument applied to B^{-1} shows that $B \in \mathrm{GL}_d(\mathbf{E}_K^+ \otimes_{\mathbf{E}_{\mathbf{Q}_p}^+} E[\![X]\!])$. If we write $B = B_0 + C$ where $B_0 \in \mathrm{GL}_d(k_K \otimes_{\mathbf{F}_p} E)$ and $\mathrm{val}_X(C) > 0$, then the formula $\varphi(B) = P_1^{-1}BP_2$ implies likewise that $\mathrm{val}_X(C) = +\infty$ so that $C = 0$. The $k_K \otimes_{\mathbf{F}_p} E$-module generated by these two bases is therefore the same.

We now show the existence of such a basis. We can assume that D is irreducible as a φ-module; indeed, if M is an irreducible sub-φ-module of D, then we can write $\mathrm{D} = \sum_{i=1}^n \gamma_i(\mathrm{M})$ with $\gamma_i \in \Gamma$. We can assume that n is minimal, so that the sum is direct and the existence result for D follows from the result for each of the φ-modules $\gamma_i(\mathrm{M})$.

If $m \in \mathrm{D}$ is non-zero, then it generates D as an $E(\!(X)\!)\{\varphi\}$-module since D is assumed to be irreducible. Let $P(\varphi)$ be a non-zero polynomial of degree $\dim \mathrm{D}$ such that $P(\varphi)(m) = 0$. If $P(\varphi)$ were reducible, then this would correspond to a non-trivial sub-φ-module of D so that $P(\varphi)$ is irreducible and by Proposition 2.12, $P(\varphi)$ is isoclinic. If s is the slope of $P(\varphi)$, then there exists a finite extension K of $\mathbf{Q}_p$ and an element $y \in \mathbf{E}_K$ of valuation $s/(p-1)$. Lemma 2.11 shows that if we replace m by ym, then the resulting polynomial $Q(\varphi)$ is isoclinic of slope 0. This implies that there exists a basis of $\mathbf{E}_K \otimes_{\mathbf{E}_{\mathbf{Q}_p}} \mathrm{D}$ in which $\mathrm{Mat}(\varphi) \in \mathrm{GL}_d((k_K \otimes_{\mathbf{F}_p} E)[\![Y]\!])$. Lemma 2.7 now implies that there exists a basis of $\mathbf{E}_K \otimes_{\mathbf{E}_{\mathbf{Q}_p}} \mathrm{D}$ in which $\mathrm{Mat}(\varphi) \in \mathrm{GL}_d(k_K \otimes_{\mathbf{F}_p} E)$, which is the sought-after result. □

Let D be an irreducible (φ, Γ)-module over $E(\!(X)\!)$, and let K be as above. Since E is algebraically closed, we have $k_K \otimes_{\mathbf{F}_p} E = E^n$ with $n = [k_K : \mathbf{F}_p]$. We denote by $\pi_k : E^n \to E$ the projection on the k-th factor. Let $V_K(\mathrm{D})$ be the E^n-module generated by the basis afforded by Proposition 2.13. This module is stable under $\mathcal{G}_{\mathbf{Q}_p}$ which acts by k_K-semilinear automorphisms. We define an action of $\mathcal{W}_{\mathbf{Q}_p}$ on $V_K(\mathrm{D})$ by $\rho(g)(y) = \varphi^{-n(g)}(g(y))$. This action is now E^n-linear, and commutes with φ. In particular, $V_K(\mathrm{D}) = \pi_1 V_K(\mathrm{D}) \oplus \cdots \oplus \pi_n V_K(\mathrm{D})$ and $\varphi(\pi_k V_K(\mathrm{D})) = \pi_{k+1} V_K(\mathrm{D})$ (with $\pi_{n+1} = \pi_1$) so that all the representations $\pi_k V_K(\mathrm{D})$ are isomorphic. We let $V(\mathrm{D}) = \pi_1 V_K(\mathrm{D})$.

Proposition 2.14. *The representation $V(\mathrm{D})$ defined above is irreducible.*

Proof. Note that φ^n gives rise to an endomorphism of $V(\mathrm{D})$. Since E is algebraically closed, φ^n has an eigenvalue λ, and the space $V(\mathrm{D})^{\varphi^n=\lambda}$ is stable under $\mathcal{W}_{\mathbf{Q}_p}$, so that it contains an irreducible sub-representation W of $\mathcal{W}_{\mathbf{Q}_p}$.

The $k_K \otimes_{\mathbf{F}_p} E$-module $M = W \oplus \varphi(W) \oplus \cdots \oplus \varphi^{n-1}(W)$ is then a subspace of $V_K(\mathrm{D})$, which is stable under $\mathcal{W}_{\mathbf{Q}_p}$ and φ, so that it is also stable under $\mathcal{G}_{\mathbf{Q}_p}$ and φ. The space $\mathbf{E}_K \otimes_{k_K} M$ is then a sub-(φ, Γ)-module of $\mathbf{E}_K \otimes_{\mathbf{E}_{\mathbf{Q}_p}} \mathrm{D}$ that is stable under φ and $\mathcal{G}_{\mathbf{Q}_p}$. By Galois descent (see for instance proposition 2.2.1 of [BC08]), $\mathbf{E}_K \otimes_{k_K} M$ comes by extension of scalars from a sub-(φ, Γ)-module of D. If D is irreducible, then $M = \{0\}$ or $M = V_K(\mathrm{D})$ and hence $V(\mathrm{D})$ is irreducible. □

Theorem 2.15. *The two constructions $V \mapsto \mathrm{D}(V)$ and $\mathrm{D} \mapsto V(\mathrm{D})$ defined above are inverse of each other and give rise to dimension preserving bijections between the set of irreducible E-linear representations of $\mathcal{W}_{\mathbf{Q}_p}$ and the set of irreducible (φ, Γ)-modules over $E(\!(X)\!)$.*

Proof. The fact that dimensions are preserved is clear from the constructions. The fact that the two constructions are inverse of each other is a tedious but straightforward exercise. □

3. Topological representations of profinite groups

In this section, we first gather some results about topological E-vector spaces and duality, which generalize Pontryagin's theorems to certain E-vector spaces (see §II.6 of [Lef42]). After that, we look at continuous representations of certain topological groups.

Recall that E is a field that is taken with the discrete topology. A topological E-vector space V is said to be linearly topologized if V is separated (Hausdorff) and if $\{0\}$ has a basis of neighborhoods that are all vector spaces. For example, the discrete topology on V is a linear topology. We denote by $\mathrm{Vec}_{\mathrm{disc}}(E)$ the category whose objects are the E-linear vector spaces with the discrete topology, with continuous linear maps as morphisms.

We say that an affine subspace W of a linearly topologized E-vector space V is linearly compact if every family $\{W_i\}_{i\in I}$ of closed affine subspaces of W having the finite intersection property has a non-empty intersection. Linearly compact affine spaces generally enjoy the same properties as compact topological spaces (see (27) of §II.6 of [Lef42]). For example, a linearly compact subspace of V is closed in V, its image under a continuous linear map is linearly compact, and a product of linearly compact spaces is linearly compact. A finite dimensional discrete E-vector space is linearly compact. If V is linearly compact and if W is a closed subspace of V, then W is open in V if and only if it is of finite codimension.

We say that an E-vector space is of profinite dimension if it is an inverse limit of finite dimensional discrete E-vector spaces. For example, $E[\![X]\!]$ with the X-adic topology is of profinite dimension. Such a space is then linearly compact and conversely, by (32) of §II.6 of [Lef42], linearly compact spaces are profinite dimensional. We denote by $\mathrm{Vec}_{\mathrm{comp}}(E)$ the category whose objects are the linearly compact E-vector spaces, with continuous linear maps as morphisms.

If V is a topological vector space, we denote by V^* its continuous dual. This space is given a linear topology by choosing as a basis of neighborhoods of $\{0\}$ the set $\{E^\perp\}_E$ where E runs through all linearly compact subspaces of V, and $E^\perp = \{f \in V^* \text{ such that } f(v) = 0 \text{ for all } v \in E\}$.

Theorem 2.16. *The duality functor $V \mapsto V^*$ gives rise to equivalences of categories $\mathrm{Vec}_{\mathrm{disc}}(E) \to \mathrm{Vec}_{\mathrm{comp}}(E)$ and $\mathrm{Vec}_{\mathrm{comp}}(E) \to \mathrm{Vec}_{\mathrm{disc}}(E)$.*

Moreover, the natural map $V \to (V^*)^*$ is an isomorphism.

Proof. See (29) in §II.6 of [Lef42]. □

We now turn to group representations. Let G be a topological group and let $\mathrm{Vec}^G_{\mathrm{disc}}(E)$ and $\mathrm{Vec}^G_{\mathrm{comp}}(E)$ denote the categories of continuous E-linear representations of G on either discrete or linearly compact spaces. If V is a representation of G, then V^* is a representation of G, with the usual action given by $(gf)(v) = f(g^{-1}v)$.

Proposition 2.17. *If $V \in \mathrm{Vec}^G_{\mathrm{disc}}(E)$ or $V \in \mathrm{Vec}^G_{\mathrm{comp}}(E)$ is topologically irreducible, then so is its dual V^*.*

Proof. If W is a closed subspace of V^* stable under G, then let $W^\perp = \{v \in V$ such that $f(v) = 0$ for all $f \in W\}$. The natural map $W^\perp \to (V^*/W)^*$ is an isomorphism by Theorem 2.16. Moreover, $W^\perp$ is a closed subspace of V, that is also stable under G, so that either $W^\perp = \{0\}$ and $W = V^*$ or $W^\perp = V$ and $W = \{0\}$. □

Assume now that G is a topologically finitely generated profinite group (in this chapter, we only need the case $G = \mathbf{Z}_p$). Denote by $V(G)$ the sub-E-vector space of V generated by the elements $(g-1)v$ where $g \in G$ and $v \in V$.

Proposition 2.18. *If $V \in \mathrm{Vec}^G_{\mathrm{comp}}(E)$, then $V(G)$ is a closed subspace of V.*

Proof. Let $g_1, \ldots, g_n$ be elements generating a dense subgroup G' of G. The subspace $(g_i - 1)V$ is the image of a linearly compact subspace by a continuous linear map and is hence linearly compact. This implies that $V(G') = \sum_{i=1}^n (g_i - 1)V$ is linearly compact and therefore closed in V.

If $v \in V$, then the image of G' under the map $g \mapsto (g-1)v$ is contained in $V(G')$ and, since G' is dense in G and $V(G')$ is closed in V, the image of G is also contained in $V(G')$ so that $V(G) = V(G')$ and $V(G)$ is closed in V. □

Note that the same is trivially true if $V \in \mathrm{Vec}^G_{\mathrm{disc}}(E)$. We set $V_G = V/V(G)$.

Proposition 2.19. *If $V \in \mathrm{Vec}^G_{\mathrm{disc}}(E)$ or $V \in \mathrm{Vec}^G_{\mathrm{comp}}(E)$, then $(V^G)^* = (V^*)_G$.*

Proof. If $f \in V^*$, then $f(gv) = f(v)$ for all $g \in G$ and $v \in V$ if and only if f is zero on $V(G)$. This implies that $(V^*)^G = (V_G)^*$. Replacing V by V^* in this formula and dualizing gives us the proposition. □

Let $E[\![G]\!] = \varprojlim_N E[G/N]$ denote the completed group algebra of G, where N runs through the set of open normal subgroups of G.

Proposition 2.20. *If $V \in \mathrm{Vec}^G_{\mathrm{comp}}(E)$ or $V \in \mathrm{Vec}^G_{\mathrm{disc}}(E)$, then V is an $E[\![G]\!]$-module.*

Proof. If $V \in \mathrm{Vec}^G_{\mathrm{disc}}(E)$, then this is immediate, so assume that $V \in \mathrm{Vec}^G_{\mathrm{comp}}(E)$. The space V is a projective limit of finite dimensional E-vector spaces. We first show that if $V \in \mathrm{Vec}^G_{\mathrm{comp}}(E)$, then V is a projective limit of finite dimensional E-linear representations of G. It is enough to prove that if W is an open subspace of V, then it contains an open subspace stable under G. By continuity, for each $g \in G$, there exists an open neighborhood H_g of g in G and an open subspace W_g of V such that $H_g \cdot W_g \subset W$. By compacity of

G, there exists $g_1, \dots, g_n \in G$ such that $G = H_{g_1} \cup \dots \cup H_{g_n}$ and if we set $X = W_{g_1} \cap \dots \cap W_{g_n}$, then X is an open subspace of W and $G \cdot X \subset W$. The vector space generated by $G \cdot X$ is then open in W and stable under G.

Since V is a projective limit of finite dimensional E-linear representations of G by the above, and since each of them is an $E[\![G]\!]$-module, then so is V. □

We now assume that $G = \mathbf{Z}_p$ so that a choice of a topological generator of $\mathbf{Z}_p$ gives rise to an isomorphism $E[\![G]\!] = E[\![X]\!]$. The following result is a variant of Nakayama's lemma.

Theorem 2.21. *If $V \in \mathrm{Vec}_{\mathrm{comp}}(E)$ is a topological $E[\![X]\!]$-module, then V is finitely generated over $E[\![X]\!]$ if and only if V/XV is a finite dimensional E-vector space.*

Proof. The fact that if V is finitely generated over $E[\![X]\!]$, then V/XV is a finite dimensional E-vector space is immediate, so let us prove the converse.

Let $v_1, \dots, v_n$ be elements of V that generate V/XV over E, and let W be the $E[\![X]\!]$-module generated by $v_1, \dots, v_n$. The E-vector space W is linearly compact, and therefore so is V/W. In addition, $(V/W)/X = \{0\}$. It is therefore enough to show that if $V \in \mathrm{Vec}_{\mathrm{comp}}(E)$ is a topological $E[\![X]\!]$-module such that $V/XV = \{0\}$, then $V = \{0\}$.

Let U be an open subspace of V. By continuity, there exists an open subspace W of U and $k_0 \geqslant 1$ such that $X^k W \subset U$ if $k \geqslant k_0$. Since W is open, it is of finite codimension in V and there exists $v_1, \dots, v_n \in V$ such that $V = W + Ev_1 + \dots + Ev_n$. For each i, there exists k_i such that $X^k v_i \in U$ if $k \geqslant k_i$. If $k \geqslant \max(k_0, \dots, k_n)$, then $X^k V \subset U$. But $X^k V = V$ so that the only open subspace of V is V itself and hence $V = \{0\}$. □

4. Colmez's functor

In this section, we recall Colmez's construction of representations of $\mathrm{B} = \mathrm{B}_2(\mathbf{Q}_p)$ starting from Galois representations (see §III of [Col10a]).

If M is a (ψ, Γ)-module, then we denote by $\varprojlim_{\psi} \mathrm{M}$ the set of sequences $\{m_n\}_{n \in \mathbf{Z}}$ where $m_n \in \mathrm{M}$ and $\psi(m_{n+1}) = m_n$ for all $n \in \mathbf{Z}$. Let $\chi : \mathbf{Q}_p^\times \to E^\times$ be a smooth character. We endow $\varprojlim_{\psi} \mathrm{M}$ with an action of B in the following way

$$\begin{aligned}
\left(\left(\begin{smallmatrix} z & 0 \\ 0 & z \end{smallmatrix}\right) \cdot y\right)_i &= \chi(z)^{-1} y_i; \\
\left(\left(\begin{smallmatrix} 1 & 0 \\ 0 & p \end{smallmatrix}\right) \cdot y\right)_i &= y_{i-1} = \psi(y_i); \\
\left(\left(\begin{smallmatrix} 1 & 0 \\ 0 & a \end{smallmatrix}\right) \cdot y\right)_i &= [a^{-1}](y_i); \\
\left(\left(\begin{smallmatrix} 1 & z \\ 0 & 1 \end{smallmatrix}\right) \cdot y\right)_i &= \psi^j((1+X)^{p^{i+j}z} y_{i+j}), \text{ for } i + j \geqslant -\mathrm{val}(z).
\end{aligned}$$

It is straightforward to check that these formulas give rise to an action of B, and make $\varprojlim_\psi \mathrm{M}$ into a profinite dimensional topological representation, M itself being separated and complete for the X-adic topology (warning: the normalization for the central character is the one chosen in §1.2 of [Ber10b] and it differs from the one in §2.2 of [Ber10a]). Note that if M_1 and M_2 are two (ψ, Γ)-modules, and there is a map $\mathrm{M}_1 \to \mathrm{M}_2$, then there is a map $\varprojlim_\psi \mathrm{M}_1 \to \varprojlim_\psi \mathrm{M}_2$.

Proposition 2.22. *If Σ is a closed subspace of $\varprojlim_\psi \mathrm{M}$ stable under* B, *then there exists a surjective sub-(ψ, Γ)-module* N *of* M *such that $\Sigma = \varprojlim_\psi \mathrm{N}$.*

Proof. This is lemma III.3.6 of [Col10a]. We recall the idea of the proof: if N_k is the set of $m \in \mathrm{M}$ such that there exists $x \in \Sigma$ with $m = x_k$, then Colmez shows that N_k is a (ψ, Γ)-module that is independent of k and that we can take $\mathrm{N} = \mathrm{N}_k$. □

Theorem 2.23. *If Σ is an infinite dimensional topologically irreducible subrepresentation of $\varprojlim_\psi \mathrm{M}$ for some (ψ, Γ)-module* M, *then there exists a (ψ, Γ)-module* N *that is irreducible and free over $E[\![X]\!]$, such that $\Sigma = \varprojlim_\psi \mathrm{N}$.*

Proof. Let $\mathrm{M}_{\mathrm{tor}}$ denote the torsion submodule of M. We then have an exact sequence $\varprojlim_\psi \mathrm{M}_{\mathrm{tor}} \to \varprojlim_\psi \mathrm{M} \to \varprojlim_\psi \mathrm{M}/\mathrm{M}_{\mathrm{tor}}$. If the image of Σ in $\varprojlim_\psi \mathrm{M}/\mathrm{M}_{\mathrm{tor}}$ is non-zero, then we have reduced to the case where M is torsion-free.

Otherwise, Σ injects in $\varprojlim_\psi \mathrm{M}_{\mathrm{tor}}$ and $\mathrm{M}_{\mathrm{tor}}$ is a finite dimensional E-vector space. Proposition 2.22 shows that $\Sigma = \varprojlim_\psi \mathrm{N}$ where N is a finite dimensional E-vector space. Since $\psi : \mathrm{N} \to \mathrm{N}$ is surjective, it is injective, and then $\varprojlim_\psi \mathrm{N} = \mathrm{N}$ so that Σ itself is a finite dimensional E-vector space.

We can therefore assume that M is torsion free. Let M be such that Σ injects in $\varprojlim_\psi \mathrm{M}$, with M torsion free, surjective and of minimal rank. If N is a sub-(ψ, Γ)-module of M, then the same argument as above shows that Σ injects in either $\varprojlim_\psi \mathrm{N}$ or $\varprojlim_\psi \mathrm{M}/\mathrm{N}$. This implies that the rank of N is equal to the rank of M, so there exists $n \geqslant 0$ such that $X^n\mathrm{M} \subset \mathrm{N}$. Repeatedly applying ψ shows that $X\mathrm{M} \subset \mathrm{N}$. Since M/X is a finite dimensional E-vector space, there is therefore a smallest M such that Σ injects in $\varprojlim_\psi \mathrm{M}$, and this M is then irreducible. □

If V is an irreducible representation of either $\mathcal{G}_{\mathbf{Q}_p}$ (when E is a finite field) or $\mathcal{W}_{\mathbf{Q}_p}$ (when E is an algebraically closed field), then by the results of §2, we can attach to it a (φ, Γ)-module $\mathrm{D}(V)$ and then by Definition 2.5 an irreducible

(ψ, Γ)-module $\mathrm{M}(V) = \mathrm{M}(\mathrm{D}(V))$. Let χ be a smooth character of $\mathbf{Q}_p^\times$. The space $\varprojlim_\psi \mathrm{M}(V)$ is of profinite dimension and gives rise to a continuous representation of B, which is topologically irreducible by Proposition 2.22. Its dual $\Omega_\chi(V) = (\varprojlim_\psi \mathrm{M}(V))^*$ is therefore a smooth irreducible representation of B, with central character χ. We finish by recalling a result of [Ber10b] to the effect that $\Omega_\chi(V)$ determines χ and V.

Proposition 2.24. *If V_1 and V_2 are irreducible and $\Omega_{\chi_1}(V_1)$ is isomorphic to $\Omega_{\chi_2}(V_2)$ as representations of* B, *then $\chi_1 = \chi_2$ and $V_1 = V_2$.*

Proof. This is proposition 1.2.3 of [Ber10b] in the case that E is a finite field, and the proof is similar if E is algebraically closed. We recall the main ideas: since χ is the central character of $\Omega_\chi(V)$, it is immediate that $\chi_1 = \chi_2$ so we need to show that if there is an equivariant map $f : \varprojlim_\psi \mathrm{M}(V_1) \to \varprojlim_\psi \mathrm{M}(V_2)$, then $V_1 = V_2$. Let $\mathrm{pr}_k : \varprojlim_\psi \mathrm{M} \to \mathrm{M}$ denote the map $\{m_n\}_{n \in \mathbf{Z}} \mapsto m_k$. If $n \geqslant 0$, let K_n be the set of elements m of $\varprojlim_\psi \mathrm{M}(V_1)$ such that $\mathrm{pr}_k(m) = 0$ for $k \leqslant n$. The module K_n is a closed sub-$E[\![X]\!]$-module of $\varprojlim_\psi \mathrm{M}(V_1)$ that is stable under ψ and Γ, and $\psi(K_n) = K_{n+1}$. This implies that $\mathrm{pr}_0 \circ f(K_n)$ is a sub-(ψ, Γ)-module of $\mathrm{M}(V_2)$. Since $\mathrm{M}(V_2)$ is irreducible, we have either $\mathrm{pr}_0 \circ f(K_n) = \{0\}$ or $\mathrm{pr}_0 \circ f(K_n) = \mathrm{M}(V_2)$. In addition, $\psi(\mathrm{pr}_0 \circ f(K_n)) = \mathrm{pr}_0 \circ f(K_{n+1})$ and $\mathrm{pr}_0 \circ f(K_n) = \{0\}$ for $n \gg 0$ by continuity, so that $\mathrm{pr}_0 \circ f(K_n) = \{0\}$ for all $n \geqslant 0$. This implies that $\mathrm{pr}_0 \circ f(m)$ depends only on m_0.

The map $m_0 \mapsto \mathrm{pr}_0 \circ f(m)$ from $\mathrm{M}(V_1)$ to $\mathrm{M}(V_2)$ is therefore a well-defined map of (ψ, Γ)-modules, which is non-zero because f is an isomorphism. By proposition II.3.4 of [Col10a], it extends to a map $\mathrm{D}(V_1) \to \mathrm{D}(V_2)$ so that $\mathrm{D}(V_1) \simeq \mathrm{D}(V_2)$ and $V_1 \simeq V_2$. □

5. Representations of $\mathbf{B}_2(\mathbf{Q}_p)$

In this section, we prove that every infinite dimensional smooth irreducible representation of B having a central character is of the form $\Omega_\chi(V)$ for some V and χ. We start by studying representations of B. Let $\mathrm{Z} = \{a \cdot \mathrm{Id}, a \in \mathbf{Q}_p^\times\}$ be the center of B, and let

$$\mathrm{K} = \mathrm{B}_2(\mathbf{Z}_p) = \begin{pmatrix} \mathbf{Z}_p^\times & \mathbf{Z}_p \\ 0 & \mathbf{Z}_p^\times \end{pmatrix}.$$

If $\beta \in \mathbf{Q}_p$ and $\delta \in \mathbf{Z}$, let

$$g_{\beta,\delta} = \begin{pmatrix} 1 & \beta \\ 0 & p^\delta \end{pmatrix}.$$

Let $A = \cup_{n\geqslant 1}\{\alpha_n p^{-n} + \cdots + \alpha_1 p^{-1}$ where $0 \leqslant \alpha_j \leqslant p-1\}$ so that A is a system of representatives of $\mathbf{Q}_p/\mathbf{Z}_p$. The following is lemma 1.2.1 of [Ber10a].

Lemma 2.25. *We have* $\mathrm{B} = \coprod_{\beta\in A,\delta\in\mathbf{Z}} g_{\beta,\delta}\cdot \mathrm{KZ}$.

If σ_1 and σ_2 are two smooth characters $\sigma_i : \mathbf{Q}_p^\times \to E^\times$, then let $\sigma = \sigma_1\otimes\sigma_2 : \mathrm{B}\to E^\times$ be the character $\sigma : \left(\begin{smallmatrix} a & b \\ 0 & d\end{smallmatrix}\right) \mapsto \sigma_1(a)\sigma_2(d)$ and let $\mathrm{ind}_{\mathrm{KZ}}^{\mathrm{B}}\sigma$ be the set of functions $f : \mathrm{B}\to E$ satisfying $f(kg) = \sigma(k)f(g)$ if $k\in\mathrm{KZ}$ and such that f has compact support modulo Z. If $g\in\mathrm{B}$, denote by $[g]$ the function $[g] : \mathrm{B}\to E$ defined by $[g](h) = \sigma(hg)$ if $h\in\mathrm{KZ}g^{-1}$ and $[g](h) = 0$ otherwise. Every element of $\mathrm{ind}_{\mathrm{KZ}}^{\mathrm{B}}\sigma$ is a finite linear combination of some functions $[g]$. We make $\mathrm{ind}_{\mathrm{KZ}}^{\mathrm{B}}\sigma$ into a representation of B in the usual way: if $g\in\mathrm{B}$, then $(gf)(h) = f(hg)$. In particular, we have $g[h] = [gh]$ in addition to the formula $[gk] = \sigma(k)[g]$ for $k\in\mathrm{KZ}$.

Theorem 2.26. *If* Π *is a smooth irreducible representation of* B *having a central character, then there exists* $\sigma = \sigma_1\otimes\sigma_2$ *such that* Π *is a quotient of* $\mathrm{ind}_{\mathrm{KZ}}^{\mathrm{B}}\sigma$.

Proof. This is theorem 1.2.3 of [Ber10a]; we recall the proof here. The group I_1 defined by

$$\mathrm{I}_1 = \begin{pmatrix} 1+p\mathbf{Z}_p & \mathbf{Z}_p \\ 0 & 1+p\mathbf{Z}_p\end{pmatrix}$$

is a pro-p-group and hence $\Pi^{\mathrm{I}_1}\neq 0$. Furthermore, I_1 is a normal subgroup of K so that Π^{I_1} is a representation of $\mathrm{K}/\mathrm{I}_1 = \mathbf{F}_p^\times\times\mathbf{F}_p^\times$. Since that group is a finite group of order prime to p, we have $\Pi^{\mathrm{I}_1} = \oplus_\eta \Pi^{\mathrm{K}=\eta}$ where η runs over the characters of $\mathbf{F}_p^\times\times\mathbf{F}_p^\times$ and since Z acts through a character by hypothesis, there exists a character σ of KZ and $v\in\Pi$ such that $k\cdot v = \sigma(k)v$ for $k\in\mathrm{KZ}$. By Frobenius reciprocity, we get a non-trivial map $\mathrm{ind}_{\mathrm{KZ}}^{\mathrm{B}}\sigma\to\Pi$ and this map is surjective since Π is irreducible. □

Note that if μ is a character of $\mathbf{Q}_p^\times$ that is trivial on $\mathbf{Z}_p^\times$, then $\mathrm{ind}_{\mathrm{KZ}}^{\mathrm{B}}\sigma_1\mu\otimes\sigma_2\mu^{-1} = \mathrm{ind}_{\mathrm{KZ}}^{\mathrm{B}}\sigma_1\otimes\sigma_2$. We can therefore assume that $\sigma_2(p) = 1$, which we now do.

Write $\sigma = \sigma_1\otimes\sigma_2$. By Lemma 2.25, each $f\in\mathrm{ind}_{\mathrm{KZ}}^{\mathrm{B}}\sigma$ can be written in the form $f = \sum_{\beta\in A,\delta\in\mathbf{Z}}\alpha(\beta,\delta)[g_{\beta,\delta}]$.

Definition 2.27. Let $s : \mathrm{ind}_{\mathrm{KZ}}^{\mathrm{B}}\sigma\to E$ be the map

$$s : \sum_{\beta\in A,\delta\in\mathbf{Z}}\alpha(\beta,\delta)[g_{\beta,\delta}] \mapsto \sum_{\beta\in A,\delta\in\mathbf{Z}}\alpha(\beta,\delta).$$

Note that if $\sigma = 1 \otimes 1$, then $\mathrm{ind}_{\mathrm{KZ}}^{\mathrm{B}}\sigma$ is the set of functions with finite support on the set of the vertices of the Bruhat–Tits tree, and s is then the "sum of the values" function. The following lemma results from a straightforward calculation (recall that $\sigma_2(p) = 1$).

Lemma 2.28. *The map* $s : \mathrm{ind}_{\mathrm{KZ}}^{\mathrm{B}}\sigma \to E(\sigma)$ *is* B*-equivariant.*

Let B^+ and B^- denote the monoids

$$\mathrm{B}^+ = \left\{ \begin{pmatrix} p^{\mathbf{Z}_{\geqslant 0}}\mathbf{Z}_p^\times & \mathbf{Z}_p \\ 0 & \mathbf{Z}_p^\times \end{pmatrix} \right\} \subset \mathrm{B}, \qquad \mathrm{B}^- = \left\{ \begin{pmatrix} \mathbf{Z}_p^\times & \mathbf{Z}_p \\ 0 & p^{\mathbf{Z}_{\geqslant 0}}\mathbf{Z}_p^\times \end{pmatrix} \right\} \subset \mathrm{B},$$

and let $(\mathrm{ind}_{\mathrm{KZ}}^{\mathrm{B}}\sigma)^+$ denote the set of elements of $\mathrm{ind}_{\mathrm{KZ}}^{\mathrm{B}}\sigma$ with support in B^+. Since

$$\begin{pmatrix} p^n a & b \\ 0 & d \end{pmatrix} = \begin{pmatrix} 1 & p^{-n}bd^{-1} \\ 0 & p^{-n} \end{pmatrix} \begin{pmatrix} a & 0 \\ 0 & d \end{pmatrix} \begin{pmatrix} p^n & 0 \\ 0 & p^n \end{pmatrix},$$

$(\mathrm{ind}_{\mathrm{KZ}}^{\mathrm{B}}\sigma)^+$ is the set of $f = \sum \alpha(\beta,\delta)[g_{\beta,\delta}]$ with $\delta \leqslant 0$ and $\beta \in p^{-\delta}\mathbf{Z}_p/\mathbf{Z}_p$.

Lemma 2.29. *If* $y = \sum_{\beta \in A, \delta \in \mathbf{Z}} \alpha(\beta,\delta)[g_{\beta,\delta}] \in (\mathrm{ind}_{\mathrm{KZ}}^{\mathrm{B}}\sigma)^+$, *then* $y \in \left(\left(\begin{smallmatrix} 1 & 1 \\ 0 & 1 \end{smallmatrix}\right) - \mathrm{Id}\right) \cdot (\mathrm{ind}_{\mathrm{KZ}}^{\mathrm{B}}\sigma)^+$ *if and only if* $\sum_{\beta \in A} \alpha(\beta,\delta) = 0$ *for all* $\delta \leqslant 0$.

Proof. We have $\left(\begin{smallmatrix} 1 & 1 \\ 0 & 1 \end{smallmatrix}\right)[g_{\beta,\delta}] = [g_{\beta+p^{-\delta},\delta}]$ so that

$$X \cdot \sum_{\beta \in A, \delta \in \mathbf{Z}} \alpha(\beta,\delta)[g_{\beta,\delta}] = \sum_{\beta \in A, \delta \in \mathbf{Z}} (\alpha(\beta - p^{-\delta},\delta) - \alpha(\beta,\delta))[g_{\beta,\delta}].$$

Since $\beta \in p^{-\delta}\mathbf{Z}_p/\mathbf{Z}_p$, the lemma follows from the fact that the image of the map $(x_i)_i \mapsto (x_{i-1} - x_i)_i$ from $E^{\mathbf{Z}/p^\delta\mathbf{Z}}$ to itself is the set of sequences $(x_i)_i$ with $\sum_i x_i = 0$. □

Write $F = \left(\begin{smallmatrix} p & 0 \\ 0 & 1 \end{smallmatrix}\right)$ and $X = \left(\begin{smallmatrix} 1 & 1 \\ 0 & 1 \end{smallmatrix}\right) - \mathrm{Id}$ so that $A = E[\![X]\!]$ is the completed group ring of $\left(\begin{smallmatrix} 1 & \mathbf{Z}_p \\ 0 & 1 \end{smallmatrix}\right)$ and let $A\{F\}$ be the non-commutative ring of polynomials in F with coefficients in A, where $FX = X^pF$. If $\Pi = \mathrm{ind}_{\mathrm{KZ}}^{\mathrm{B}}\sigma/R$ is a quotient of $\mathrm{ind}_{\mathrm{KZ}}^{\mathrm{B}}\sigma$, let Π^+ denote the image of $(\mathrm{ind}_{\mathrm{KZ}}^{\mathrm{B}}\sigma)^+$ in Π. The space Π^+ is then a left $A\{F\}$-module, as well as a torsion A-module (since Π is smooth). Recall (see §3 of [Eme08]) that an admissible A-module is an A-module M that is torsion and such that $M^{X=0}$ is finite dimensional.

Proposition 2.30. *If M is a finitely generated left $A\{F\}$-module that is torsion over A, then M is admissible as an A-module if and only if the quotient M/XM is finite dimensional over E.*

Proof. This is proposition 3.5 of [Eme08]. □

Lemma 2.31. *The map* $(\mathrm{ind}_{\mathrm{KZ}}^{\mathrm{B}}\sigma)^+ \to E[F]$ *given by*

$$\sum_{\beta \in A, \delta \leqslant 0} \alpha(\beta, \delta)[g_{\beta,\delta}] \mapsto \sum_{n \geqslant 0} \left(\sum_{\beta \in A} \alpha(\beta, -n) \right) F^n$$

(which arises from "retracting the building to the apartment") gives rise to an isomorphism of $A\{F\}$*-modules* $(\mathrm{ind}_{\mathrm{KZ}}^{\mathrm{B}}\sigma)^+/X = E[F]$.

Proof. It is straightforward to check that the given map $(\mathrm{ind}_{\mathrm{KZ}}^{\mathrm{B}}\sigma)^+ \to E[F]$ is a surjective map of $A\{F\}$-modules. Its kernel is $X \cdot (\mathrm{ind}_{\mathrm{KZ}}^{\mathrm{B}}\sigma)^+$ by Lemma 2.29. □

Lemma 2.32. *The* $A\{F\}$*-module* $(\mathrm{ind}_{\mathrm{KZ}}^{\mathrm{B}}\sigma)^+$ *is generated by* [Id].

Proof. The fact that if $n \geqslant 0$, $a, d \in \mathbf{Z}_p^\times$ and $b \in \mathbf{Z}_p$, then $[\begin{pmatrix} p^n a & b \\ 0 & d \end{pmatrix}]$ belongs to the $A\{F\}$-module generated by [Id] follows from the formula

$$\begin{pmatrix} p^n a & b \\ 0 & d \end{pmatrix} = \begin{pmatrix} 1 & (b - p^n a)d^{-1} \\ 0 & 1 \end{pmatrix} \begin{pmatrix} p^n & 0 \\ 0 & 1 \end{pmatrix} \begin{pmatrix} a & 0 \\ 0 & d \end{pmatrix}.$$

□

Theorem 2.33. *If* Π *has no quotient isomorphic to* $E(\sigma)$*, then the* A*-module* Π^+ *is admissible.*

Proof. By Proposition 2.30 above (Emerton's theorem), it is enough to show that Π^+ is finitely generated over $A\{F\}$ and that $\Pi^+/X\Pi^+$ is a finite dimensional E-vector space. The finite generation follows from the fact that Π^+ is a quotient of $(\mathrm{ind}_{\mathrm{KZ}}^{\mathrm{B}}\sigma)^+$, which is generated by one element over $A\{F\}$ by Lemma 2.32.

Let $R^+ = (\mathrm{ind}_{\mathrm{KZ}}^{\mathrm{B}}\sigma)^+ \cap R$. We have an exact sequence of $A\{F\}$-modules $R^+/X \to (\mathrm{ind}_{\mathrm{KZ}}^{\mathrm{B}}\sigma)^+/X \to \Pi^+/X \to 0$. By Lemma 2.31, we have an isomorphism of $A\{F\}$-modules $(\mathrm{ind}_{\mathrm{KZ}}^{\mathrm{B}}\sigma)^+/X = E[F]$. Since any non-trivial quotient of $E[F]$ is finite dimensional over E, it is enough to show that R^+ has non-trivial image in $(\mathrm{ind}_{\mathrm{KZ}}^{\mathrm{B}}\sigma)^+/X$. If this was not the case, then we would have $R^+ \subset X \cdot (\mathrm{ind}_{\mathrm{KZ}}^{\mathrm{B}}\sigma)^+$. Lemma 2.29 shows that $X \cdot (\mathrm{ind}_{\mathrm{KZ}}^{\mathrm{B}}\sigma)^+ \subset \ker(s)$ where s is the map of Definition 2.27. If $y \in R$, then $F^n y \in R^+$ for $n \gg 0$ so that $R \subset \ker(s)$ and therefore by Lemma 2.28, there is a surjective map $\Pi \to E(\sigma)$. □

Proof of Theorems A and A′. Let Π be an infinite dimensional smooth irreducible representation of B having a central character. By Theorem 2.26, we can write $\Pi = \mathrm{ind}_{\mathrm{KZ}}^{\mathrm{B}}\sigma/R$ and by Theorem 2.33, Π^+ is an admissible $E[\![X]\!]$-module. Its dual $\mathrm{M} = (\Pi^+)^*$ is therefore a linearly compact topological

E-vector space, and an $E[\![X]\!]$-module by Proposition 2.20. In addition, the space of coinvariants $\mathrm{M}/X\mathrm{M} = \mathrm{M}_{\mathbf{Z}_p}$ is finite dimensional by Proposition 2.19. By Theorem 2.21 (Nakayama's lemma), M is finitely generated over $E[\![X]\!]$.

Since Π^+ is a representation of $\mathrm{B}^+\mathrm{Z}$, its dual M is a representation of $\mathrm{B}^-\mathrm{Z}$. We define a (ψ, Γ)-module structure on M as follows: we know that it is a finitely generated module over $E[\![X]\!]$ and we set $\psi(m) = \begin{pmatrix} 1 & 0 \\ 0 & p \end{pmatrix} m$ and $[a](m) = \begin{pmatrix} 1 & 0 \\ 0 & a^{-1} \end{pmatrix} m$ if $a \in \mathbf{Z}_p^\times$.

If $f : \Pi \to E$ is an element of Π^*, let f_n denote the restriction of $\begin{pmatrix} 1 & 0 \\ 0 & p^n \end{pmatrix} f$ to Π^+. The map $f \mapsto \{f_n\}_{n \in \mathbf{Z}}$ gives rise to an equivariant map $\Pi^* \to \varprojlim_\psi \mathrm{M}$. Since Π^* is irreducible by Proposition 2.17, Theorem 2.23 applied to $\Sigma = \Pi^*$ gives us a free irreducible (ψ, Γ)-module N such that $\Pi^* = \varprojlim_\psi \mathrm{N}$. Theorem 2.6 now says that $\mathrm{N} = \mathrm{M}(\mathrm{D})$ for some irreducible (φ, Γ)-module D so that $\Pi^* = \varprojlim_\psi \mathrm{M}(\mathrm{D})$. Theorem 2.16 finally says that $\Pi = (\varprojlim_\psi \mathrm{M}(\mathrm{D}))^*$ which proves Theorems A and A′ by the bijections constructed in §2 (Theorem 2.8 if E is a finite field and Theorem 2.15 if E is algebraically closed). □

Remark 2.34. Theorem A′ and Proposition 2.9 imply that if E is algebraically closed, and Π is an infinite dimensional smooth irreducible representation of B having a central character, then there exists an infinite dimensional smooth irreducible $\mathbf{F}_p$-linear representation Π_0 of B having a central character, and a smooth character $\mu : \mathrm{B} \to E^\times$, such that $\Pi = (E \otimes_{\mathbf{F}_p} \Pi_0) \otimes \mu$.

In particular, we can apply the same methods as in [Ber12] in order to prove that in fact, every smooth irreducible representation of B over an algebraically closed field necessarily has a central character.

References

[BC08] L. Berger & P. Colmez – "Familles de représentations de de Rham et monodromie p-adique", *Astérisque* (2008), no. 319, 303–337, Représentations p-adiques de groupes p-adiques. I. Représentations galoisiennes et (φ, Γ)-modules.

[Ber10a] L. Berger – "On some modular representations of the Borel subgroup of $\mathrm{GL}_2(\mathbf{Q}_p)$", *Compos. Math.* **146** (2010), no. 1, 58–80.

[Ber10b] L. Berger – "Représentations modulaires de $\mathrm{GL}_2(\mathbf{Q}_p)$ et représentations galoisiennes de dimension 2", *Astérisque* (2010), no. 330, 263–279.

[Ber12] L. Berger – "Central characters for smooth irreducible modular representations of $\mathrm{GL}_2(\mathbf{Q}_p)$", *Rend. Semin. Mat. Univ. Padova* **128** (2012), 1–6.

[BH06] C. J. Bushnell & G. Henniart – *The local Langlands conjecture for* GL(2), Grundlehren der Mathematischen Wissenschaften, vol. 335, Springer-Verlag, Berlin, 2006.

[Col10a] P. COLMEZ – “(φ, Γ)-modules et représentations du mirabolique de $GL_2(\mathbf{Q}_p)$”, *Astérisque* (2010), no. 330, 61–153.

[Col10b] P. COLMEZ – “Représentations de $GL_2(\mathbf{Q}_p)$ et (φ, Γ)-modules”, *Astérisque* (2010), no. 330, 281–509.

[Eme08] M. EMERTON – “On a class of coherent rings, with applications to the smooth representation theory of $GL_2(\mathbf{Q}_p)$ in characteristic p”, preprint, 2008.

[Fon90] J.-M. FONTAINE – “Représentations p-adiques des corps locaux. I”, in *The Grothendieck Festschrift, Vol. II*, Progr. Math., vol. 87, Birkhäuser Boston, Boston, MA, 1990, pp. 249–309.

[Ked08] K. S. KEDLAYA – “Slope filtrations for relative Frobenius”, *Astérisque* (2008), no. 319, 259–301, Représentations p-adiques de groupes p-adiques. I. Représentations galoisiennes et (φ, Γ)-modules.

[Lef42] S. LEFSCHETZ – *Algebraic topology*, American Mathematical Society Colloquium Publications, v. 27, American Mathematical Society, New York, 1942.

[Ser72] J.-P. SERRE – “Propriétés galoisiennes des points d’ordre fini des courbes elliptiques”, *Invent. Math.* **15** (1972), no. 4, 259–331.

[SV11] P. SCHNEIDER & M.-F. VIGNERAS – “A functor from smooth o-torsion representations to (φ, Γ)-modules”, in *On certain L-functions*, Clay Math. Proc., vol. 13, Amer. Math. Soc., Providence, RI, 2011, pp. 525–601.

[Vie12a] M. VIENNEY – “Construction de (φ, Γ)-modules en caractéristique p”, PhD, UMPA, ENS de Lyon, 2012.

[Vie12b] M. VIENNEY – “Représentations modulo p d’un sous-groupe de Borel de $GL_2(\mathbf{Q}_p)$”, *C. R. Math. Acad. Sci. Paris* **350** (2012), no. 13–14, 651–654.

[Win83] J.-P. WINTENBERGER – “Le corps des normes de certaines extensions infinies de corps locaux; applications”, *Ann. Sci. École Norm. Sup. (4)* **16** (1983), no. 1, 59–89.

3

p-adic L-functions and Euler systems: a tale in two trilogies

Massimo Bertolini, Francesc Castella, Henri Darmon, Samit Dasgupta, Kartik Prasanna, and Victor Rotger

Abstract

This chapter surveys six different special value formulae for p-adic L-functions, stressing their common features and their eventual arithmetic applications via Kolyvagin's theory of "Euler systems", in the spirit of Coates–Wiles and Kato–Perrin-Riou.

Contents

Introduction *page* 52
1 Classical examples 54
1.1 Circular units 54
1.2 Elliptic units 61
1.3 Heegner points 70
2 Euler systems of Garrett–Rankin–Selberg type 74
2.1 Beilinson–Kato elements 76
2.2 Beilinson–Flach elements 83
2.3 Gross–Kudla–Schoen cycles 88
Conclusion 94
References 98

Introduction

This chapter surveys six different special value formulae for p-adic L-functions, stressing their common features and their eventual arithmetic

Automorphic Forms and Galois Representations, ed. Fred Diamond, Payman L. Kassaei and Minhyong Kim. Published by Cambridge University Press.

applications via Kolyvagin's theory of "Euler systems", in the spirit of Coates–Wiles and Kato–Perrin-Riou. The most classical instances are:

(1) Leopoldt's formula for the value at $s = 1$ of the Kubota-Leopoldt p-adic L-function in terms of p-adic logarithms of *circular units*;
(2) Katz's p-adic Kronecker limit formula for values of the two variable p-adic L-function of a quadratic imaginary field at finite order characters in terms of p-adic logarithms of associated *elliptic units*.

They are reviewed in Sections 1.1 and 1.2 respectively. Section 1.3 describes the more recent formula of [BDP] and explains why it is a direct generalisation of the formulae of Leopoldt and Katz in the setting where special units are replaced by *Heegner points*. The three parallel treatments in Section 1 suggest that both elliptic and circular units might be viewed as *degenerate cases* of the Euler system of Heegner points, obtained by successively replacing the cusp forms and ordinary CM points that arise in the latter setting by Eisenstein series and cusps.

The second part of this survey attempts to view the Euler system introduced by Kato [Kato] in a similar way, as the "most degenerate instance" of a broader class of examples. Referred to as "Euler systems of Garrett–Rankin–Selberg type" because of the role played by the formulae of Rankin–Selberg and Garrett in relating them to special values of L-functions, these examples consist of:

(1) Kato's original Euler system of (p-adic families of) Beilinson elements in the second K-group of modular curves, whose global objects are indexed by pairs $\{u_1, u_2\}$ of modular units. Their connection to L-values follows from Rankin's method applied to a cusp form and the pair of weight two Eisenstein series corresponding to the logarithmic derivatives of u_1 and u_2.
(2) The Euler system of *Beilinson–Flach* elements in the first K-group of a product of two modular curves.
(3) The Euler system of *generalised Gross–Kudla–Schoen* diagonal cycles, whose connection with L-values arises from the formula of Garrett for the central critical value of the convolution L-series attached to a triple of newforms.

The global cohomology classes in (3) are indexed by triples (f, g, h) of cusp forms and take values in the tensor product of the three p-adic representations attached to f, g and h. Example (1) (resp. (2)) can in some sense be viewed as a degenerate instance of (3) in which g and h (resp. h only) are replaced by Eisenstein series.

Ever since the seminal work of Kolyvagin [Ko], there have been many proposals for axiomatising and classifying the Euler systems that should arise in nature (cf. [Ru], [PR3], [Cz1], [Cz2], [MR], ...) with the goal of understanding them more conceptually and systematising the process whereby arithmetic information is coaxed from their behaviour. The present survey is less ambitious, focussing instead on six settings where the associated global objects have been constructed unconditionally, attempting to organise them coherently, and suggesting that they arise from two rather than six fundamentally distinct classes of examples. Like the ten plagues of Egypt in the Jewish Passover Haggadah, Euler Systems can surely be counted in many ways. The authors believe (and certainly hope) that their "two trilogies" are but the first instalments of a richer story in which higher-dimensional cycles on Shimura varieties and p-adic families of automorphic forms are destined to play an important role.

Acknowledgements: The authors are grateful to Minhyong Kim for encouraging them to write this survey, and to Benedict Gross and the anonymous referee for some enlightening feedback.

1. Classical examples

1.1. Circular units

Let $\chi : (\mathbb{Z}/N\mathbb{Z})^\times \longrightarrow \mathbb{C}^\times$ be a primitive, non-trivial even Dirichlet character of conductor N. The values of the Dirichlet L-function $L(s, \chi)$ at the negative odd integers belong to the field $\mathbb{Q}_\chi \subset \bar{\mathbb{Q}}$ generated by the values of χ. This can be seen by realising $L(1-k, \chi)$ for even $k \geq 2$ as the constant term of the holomorphic Eisenstein series

$$E_{k,\chi}(q) := L(1-k, \chi)+2\sum_{n=1}^{\infty} \sigma_{k-1,\chi}(n)q^n, \qquad \sigma_{k-1,\chi}(n) := \sum_{d|n} \chi(d)d^{k-1} \tag{3.1}$$

of weight k, level N and character χ, and invoking the q-expansion principle to argue that the constant term in (3.1) inherits the rationality properties of the coefficients $\sigma_{k-1,\chi}(n)$.

If p is any prime (possibly dividing N), the *ordinary p-stabilisation*

$$E^{(p)}_{k,\chi}(q) := E_{k,\chi}(q) - \chi(p)p^{k-1}E_{k,\chi}(q^p) \tag{3.2}$$

has Fourier expansion given by

$$E^{(p)}_{k,\chi}(q) = L_p(1-k, \chi) + 2\sum_{n=1}^{\infty} \sigma^{(p)}_{k-1,\chi}(n)q^n, \tag{3.3}$$

where

$$L_p(1-k,\chi) = (1-\chi(p)p^{k-1})L(1-k,\chi), \qquad \sigma^{(p)}_{k-1,\chi}(n) = \sum_{p\nmid d|n} \chi(d)d^{k-1}. \tag{3.4}$$

For each $n \geq 1$, the function on $\mathbb{Z}$ sending k to the n-th Fourier coefficient $\sigma^{(p)}_{k-1,\chi}(n)$ extends to a p-adic analytic function of $k \in (\mathbb{Z}/(p-1)\mathbb{Z}) \times \mathbb{Z}_p$. The article [Se] explains why the constant term of (3.3) inherits the same property. The resulting extension to $\mathbb{Z}/(p-1)\mathbb{Z} \times \mathbb{Z}_p$ of $L_p(s,\chi)$, defined *a priori* as a function on the negative odd integers, is the *Kubota–Leopoldt p-adic L-function* attached to χ. The elegant construction of $L_p(s,\chi)$ arising from this circle of ideas (and its subsequent extension to totally real fields) was one of the original motivations for the theory of p-adic modular forms initiated in [Se].

The collection of eigenforms in (3.3) is a prototypical example of a *p-adic family of modular forms*, whose specialisations at even integers $k \leq 0$, while not classical, continue to admit a geometric interpretation as *p-adic modular forms*[1] of weight k and level N_0, the prime-to-p part of N. When $k = 0$, these are just rigid analytic functions on the *ordinary locus* $\mathcal{A} \subset X_1(N_0)(\mathbb{C}_p)$ obtained by deleting from $X_1(N_0)(\mathbb{C}_p)$ all the residue discs attached to supersingular elliptic curves in characteristic p. In particular, the special value $L_p(1,\chi)$ can be interpreted as the value at the cusp ∞ of such a rigid analytic function, namely the weight 0 Eisenstein series

$$E^{(p)}_{0,\chi}(q) = L_p(1,\chi) + 2\sum_{n=1}^{\infty}\left(\sum_{p\nmid d|n}\chi(d)d^{-1}\right)q^n. \tag{3.5}$$

An independent expression for this function can be derived in terms of the Siegel units $g_a \in \mathcal{O}^\times_{Y_1(N)}$ attached to a fixed choice of primitive N-th root of unity ζ and a parameter $1 \leq a \leq N-1$, whose q-expansions are given by

$$g_a(q) = q^{1/12}(1-\zeta^a)\prod_{n>0}(1-q^n\zeta^a)(1-q^n\zeta^{-a}). \tag{3.6}$$

More precisely, let Φ be the *canonical lift of Frobenius* on $\mathcal{A}$ which sends the point corresponding to the pair $(E,t) \in \mathcal{A}$ to the pair $(E/C, t+C)$ where $C \subset E(\mathbb{C}_p)$ is the *canonical subgroup* of order p in E. The rigid analytic function

$$g_a^{(p)} := \Phi^*(g_a)g_a^{-p} = g_{pa}(q^p)g_a(q)^{-p} \tag{3.7}$$

maps the ordinary locus $\mathcal{A}$ to the residue disc of 1 in $\mathbb{C}_p$, and therefore its p-adic logarithm $\log_p g_a^{(p)}$ is a rigid analytic function on $\mathcal{A}$, with q-expansion given by

[1] They are even overconvergent, but this stronger property will not be exploited here.

$$\log_p g_a^{(p)} = \log_p \left(\frac{1-\zeta^{ap}}{(1-\zeta^a)^p} \right) + p \sum_{n=1}^{\infty} \left(\sum_{p \nmid d | n} \frac{\zeta^{ad} + \zeta^{-ad}}{d} \right) q^n.$$

Letting

$$\mathfrak{g}(\chi) = \sum_{a=1}^{N-1} \chi(a) \zeta^a \tag{3.8}$$

denote the Gauss sum attached to χ, a direct computation shows that the rigid analytic function on $\mathcal{A}$ given by

$$h_\chi^{(p)} := \frac{1}{p\mathfrak{g}(\chi^{-1})} \times \sum_{a=1}^{N-1} \chi^{-1}(a) \log_p g_a^{(p)} \tag{3.9}$$

has q-expansion equal to

$$\begin{aligned} h_\chi^{(p)}(q) = &- \frac{(1-\chi(p)p^{-1})}{\mathfrak{g}(\chi^{-1})} \sum_{a=1}^{N-1} \chi^{-1}(a) \log_p(1-\zeta^a) \\ &+ 2 \sum_{n=1}^{\infty} \left(\sum_{p \nmid d | n} \chi(d) d^{-1} \right) q^n. \end{aligned} \tag{3.10}$$

Theorem 3.1 (Leopoldt). *Let χ be a non-trivial even primitive Dirichlet character of conductor N. Then*

$$L_p(1, \chi) = - \frac{(1-\chi(p)p^{-1})}{\mathfrak{g}(\chi^{-1})} \sum_{a=1}^{N-1} \chi^{-1}(a) \log_p(1-\zeta^a).$$

Proof. Comparing q-expansions in (3.5) and (3.10) shows that the difference $E_{0,\chi}^{(p)} - h_\chi^{(p)}$ is constant on the residue disc of a cusp, and hence on all of $\mathcal{A}$ since it is rigid analytic on this domain. In fact,

$$E_{0,\chi}^{(p)} = h_\chi^{(p)}, \tag{3.11}$$

since both these p-adic modular functions have nebentype character $\chi \neq 1$. Leopoldt's formula follows by equating the constant terms in the q-expansions in (3.5) and (3.10). For more details on this "modular" proof of Leopoldt's formula, see [Katz, §10.2]. □

Remark: Recall the customary notations in which $E_k(\psi, \chi)$ denotes the Eisenstein series attached to a pair (ψ, χ) of Dirichlet characters, having the q-expansion

$$E_k(\psi, \chi) = \delta_{\psi=1} L(1-k, \psi\chi) + 2\sum_{n=1}^{\infty} \left(\sum_{d|n} \psi(n/d)\chi(d)d^{k-1} \right) q^n,$$

and let χ_p be the Dirichlet character of modulus Np which agrees with χ on $(\mathbb{Z}/Np\mathbb{Z})^\times$, so that $E_k(1,\chi) = E_{k,\chi}$ and $E_k(1,\chi_p) = E_{k,\chi}^{(p)}$. The rigid analytic function $E_{0,\chi}^{(p)} = h_\chi^{(p)}$ which is a key actor in the proof of Leopoldt's formula above is a *Coleman primitive* of the weight two Eisenstein series $E_2(\chi_p, 1)$, a non-ordinary modular form of *critical slope* since its p-th Fourier coefficient is equal to p. The pattern whereby special values of p-adic L-series outside the range of classical interpolation arise as values of Coleman primitives of p-adic modular forms at distinguished points of the modular curve (namely cusps, or ordinary CM points) will recur in Sections 1.2 and 1.3.

The expressions of the form $(1-\zeta^a)$ (when N is composite) and $\frac{1-\zeta^a}{1-\zeta^b}$ (when N is prime) that occur in Leopoldt's formula are called *circular units*. These explicit units play an important role in the arithmetic of the cyclotomic field $\mathbb{Q}(\zeta)$. Letting F_χ denote the field cut out by χ viewed as a Galois character, and $\mathbb{Z}_\chi$ the ring generated by its values, the expression

$$u_\chi := \prod_{a=1}^{N-1} (1-\zeta^a)^{\chi^{-1}(a)} \in (\mathcal{O}_{F_\chi}^\times \otimes \mathbb{Z}_\chi)^\chi$$

is a distinguished unit in F_χ (or rather, a $\mathbb{Z}_\chi$-linear combination of such) which lies in the χ-eigenspace for the natural action of the absolute Galois group $G_\mathbb{Q}$ of $\mathbb{Q}$ (in which the second factor $\mathbb{Z}_\chi$ in the tensor product is fixed by this group).

A notable feature of the unit u_χ is that it is essentially a "universal norm" over the tower of cyclotomic fields whose n-th layer is $F_{\chi,n} = F_\chi(\mu_{p^n})$. More precisely, after fixing a sequence $(\zeta = \zeta_N, \zeta_{Np}, \zeta_{Np^2}, \ldots, \zeta_{Np^n}, \ldots)$ of primitive Np^n-th roots of unity which are compatible under the p-power maps, and setting

$$u_{\chi,n} = \prod_{a=1}^{N-1} (1-\zeta_{Np^n}^a)^{\chi^{-1}(a)} \quad \in (\mathcal{O}_{F_{\chi,n}}^\times \otimes \mathbb{Z}_\chi)^\chi,$$

we find that

$$\mathrm{Norm}_{F_{\chi,n}}^{F_{\chi,n+1}}(u_{\chi,n+1}) = \begin{cases} u_{\chi,n} & \text{if } n \geq 1, \\ u_\chi \otimes (1-\chi^{-1}(p)) & \text{if } n = 0. \end{cases}$$

After viewing χ as a $\mathbb{C}_p$-valued character, let $\mathbb{Z}_{p,\chi}$ be the ring generated over $\mathbb{Z}_p$ by the values of χ (endowed with the trivial $G_\mathbb{Q}$-action) and let

$\mathbb{Z}_{p,\chi}(\chi)$ be the free module of rank one over $\mathbb{Z}_{p,\chi}$ on which $G_{\mathbb{Q}}$ acts via the character χ. More generally, denote by $\mathbb{Z}_{p,\chi}(m)(\chi)$ the m-th Tate twist of $\mathbb{Z}_{p,\chi}(\chi)$, on which $G_{\mathbb{Q}}$ acts via the m-th power of the cyclotomic character times χ. The symbols $\mathbb{Q}_{p,\chi}$, $\mathbb{Q}_{p,\chi}(\chi)$, and $\mathbb{Q}_{p,\chi}(m)(\chi)$ are likewise given the obvious meaning. The images

$$\kappa_{\chi,n} := \delta u_{\chi,n} \in H^1(F_{\chi,n}, \mathbb{Z}_{p,\chi}(1))^{\chi} = H^1(F_n, \mathbb{Z}_{p,\chi}(1)(\chi^{-1})),$$

(where $F_n := \mathbb{Q}(\mu_{p^n})$) under the connecting homomorphism $\delta : (F_{\chi,n}^{\times} \otimes \mathbb{Z}_{\chi})^{\chi} \longrightarrow H^1(F_{\chi,n}, \mathbb{Z}_{p,\chi}(1))^{\chi}$ of Kummer theory can be patched together in a canonical element $\kappa_{\chi,\infty} := (\kappa_{\chi,n})_{n\geq 0}$ belonging to

$$\begin{aligned} \varprojlim_n H^1(F_n, \mathbb{Z}_{p,\chi}(1)(\chi^{-1})) &= \varprojlim_n H^1(\mathbb{Q}, \mathbb{Z}_p[G_n] \otimes_{\mathbb{Z}_p} \mathbb{Z}_{p,\chi}(1)(\chi^{-1})) \quad (3.12)\\ &= H^1(\mathbb{Q}, \Lambda_{\mathrm{cyc}} \otimes_{\mathbb{Z}_p} \mathbb{Z}_{p,\chi}(1)(\chi^{-1})), \end{aligned}$$

where

- $\mathbb{Z}_p[G_n]$ is the group ring of $G_n := \mathrm{Gal}(F_n/\mathbb{Q}) = (\mathbb{Z}/p^n\mathbb{Z})^{\times}$, equipped with the tautological action of $G_{\mathbb{Q}}$ in which $\sigma \in G_{\mathbb{Q}}$ acts via multiplication by its image in G_n, and the identification (3.12) follows from Shapiro's lemma;
- $\Lambda_{\mathrm{cyc}} = \varprojlim_n \mathbb{Z}_p[G_n] = \mathbb{Z}_p[[\mathbb{Z}_p^{\times}]]$ is the completed group ring of $\mathbb{Z}_p^{\times}$ equipped with the similar "tautological" action of $G_{\mathbb{Q}}$.

The Galois module Λ_{cyc} can be viewed as a p-adic interpolation of the Tate twists $\mathbb{Z}_p(k)$ for all $k \in \mathbb{Z}$. More precisely, given $k \in \mathbb{Z}$ and a Dirichlet character ξ of p-power conductor, let $\nu_{k,\xi} : \Lambda \longrightarrow \mathbb{Z}_{p,\xi}$ be the ring homomorphism sending the group-like element $a \in \mathbb{Z}_p^{\times}$ to $a^{k-1}\xi^{-1}(a)$. It induces a $G_{\mathbb{Q}}$-equivariant specialisation map

$$\nu_{k,\xi} : \Lambda_{\mathrm{cyc}} \longrightarrow \mathbb{Q}_{p,\xi}(k-1)(\xi^{-1}),$$

giving rise to a collection of global cohomology classes

$$\kappa_{k,\chi\xi} := \nu_{k,\xi}(\kappa_{\chi,\infty}) \quad \in \quad H^1(\mathbb{Q}, \mathbb{Q}_{p,\chi,\xi}(k)((\chi\xi)^{-1})),$$

(where $\mathbb{Q}_{\chi,\xi} := \mathbb{Q}_{p,\chi} \otimes \mathbb{Q}_{p,\xi}$). These classes can be viewed as the "arithmetic specialisations" of the p-adic family $\kappa_{\chi,\infty}$ of cohomology classes.

Given any Dirichlet character η with $\eta(p) \neq 1$, let $F_{p,\eta}$ be the finite extension of $\mathbb{Q}_p$ cut out by the corresponding Galois character and denote by

$G_\eta = \mathrm{Gal}(F_{p,\eta}/\mathbb{Q}_p)$ its Galois group. Restriction to $F_{p,\eta}$ composed with δ^{-1} leads to the identifications

$$\begin{aligned} H^1(\mathbb{Q}_p, \mathbb{Q}_{p,\eta}(1)(\eta)) &= H^1(F_{p,\eta}, \mathbb{Q}_{p,\eta}(1)(\eta))^{G_\eta} \\ &= \left(H^1(F_{p,\eta}, \mathbb{Q}_p(1)) \otimes_{\mathbb{Q}_p} \mathbb{Q}_{p,\eta}(\eta)\right)^{G_\eta} \\ &= \left(\mathcal{O}_{F_{p,\eta}}^\times \otimes_{\mathbb{Q}_p} \mathbb{Q}_{p,\eta}(\eta)\right)^{G_\eta}. \end{aligned}$$

Applying the p-adic logarithm $\log_p : \mathcal{O}_{F_{p,\eta}}^\times \to F_{p,\eta}$ to this last module leads to the map

$$\begin{aligned} \log_\eta : H^1(\mathbb{Q}_p, \mathbb{Q}_{p,\eta}(1)(\eta)) &\longrightarrow (F_{p,\eta} \otimes_{\mathbb{Q}_p} \mathbb{Q}_{p,\eta}(\eta))^{G_\eta} \\ &= (\mathbb{C}_p \otimes_{\mathbb{Q}_{p,\eta}} \mathbb{Q}_{p,\eta}(\eta))^{G_{\mathbb{Q}_p}}, \end{aligned} \quad (3.13)$$

where the Tate–Sen isomorphism $F_{p,\eta} = \mathbb{C}_p^{\mathrm{Gal}(\bar{\mathbb{Q}}_p/F_{p,\eta})}$ has been used to make the last identification.

The module $\mathbb{D}_{\mathbb{C}_p}(\mathbb{Q}_{p,\eta}(\eta)) := (\mathbb{C}_p \otimes_{\mathbb{Q}_{p,\eta}} \mathbb{Q}_{p,\eta}(\eta))^{G_{\mathbb{Q}_p}}$ appearing as the target in (3.13) is the Dieudonné module (with $\mathbb{C}_p$ as "period ring") attached to the Artin representation $\mathbb{Q}_{p,\eta}(\eta)$. It is a one-dimensional $\mathbb{Q}_{p,\eta}$-vector space generated by the "Gauss sum"

$$\mathfrak{g}(\eta) = \sum_{a=1}^{m-1} \zeta_m^a \otimes \eta(a)$$

which can thus be viewed as a "p-adic period" attached to the Galois representation $\mathbb{Q}_{p,\eta}(\eta)$. Theorem 3.1 can be re-phrased as the following relationship between the classes $\kappa_{1,\chi\xi}$ and the values of the Kubota–Leopoldt L-function at $s = 1$, twisted eventually by *finite order characters*:

$$L_p(1, \chi\xi) = -\frac{(1 - \chi\xi(p)p^{-1})}{(1 - (\chi\xi)^{-1}(p))} \times \frac{\log_{\chi\xi}(\kappa_{1,\chi\xi})}{\mathfrak{g}((\chi\xi)^{-1})}, \quad (3.14)$$

(after extending the map $\log_{\chi\xi}$ by linearity.) Note in particular that the global classes $\kappa_{1,\chi\xi}$ determine the Kubota–Leopoldt L-function completely, since an element of the Iwasawa algebra has finitely many zeros.

For all $k \geq 1$ and characters η of conductor prime to p (with $\eta(p) \neq 1$ when $k = 1$), Bloch and Kato have defined a generalisation of the map $\log_\eta$ of (3.13) for the representation $\mathbb{Q}_{p,\eta}(k)(\eta)$, in which $\mathbb{C}_p$ is replaced by the larger Fontaine period ring B_{dR}:

$$\log_{k,\eta} : H^1(\mathbb{Q}_p, \mathbb{Q}_{p,\eta}(k)(\eta)) \longrightarrow \mathbb{D}_{\mathrm{dR}}(\mathbb{Q}_{p,\eta}(k)(\eta))$$
$$:= (B_{\mathrm{dR}} \otimes_{\mathbb{Q}_p} \mathbb{Q}_{p,\eta}(k)(\eta))^{G_{\mathbb{Q}_p}}. \qquad (3.15)$$

(See for instance [Kato], [Cz1, §2.6]; we have implicitly used the fact that for $k \geq 1$, all extensions of $\mathbb{Q}_p$ by $\mathbb{Q}_p(k)$ are crystalline, and that $\mathrm{Fil}^0\mathbb{D}_{\mathrm{dR}}(\mathbb{Q}_{p,\eta}(k)(\eta)) = 0$.) The target of the logarithm map is a one-dimensional $\mathbb{Q}_{p,\eta}$-vector space with a canonical generator $t^{-k}\mathfrak{g}(\eta)$, where $t \in B_{\mathrm{dR}}$ is Fontaine's p-adic analogue of $2\pi i$ on which $G_{\mathbb{Q}_p}$ acts as $\sigma t = \chi_{\mathrm{cyc}}(\sigma)t$. The Bloch–Kato logarithm $\log_{k,\eta}$ with $k = 1$ is related to the map $\log_\eta$ of (3.13) by the formula

$$\log_\eta = t \log_{1,\eta},$$

and (3.14), specialised to $\xi = 1$, admits the following extension for all $k \geq 1$ (cf. [PR4, 3.2.3]):

$$L_p(k, \chi) = \frac{(1 - \chi(p)p^{-k})}{(1 - \chi^{-1}(p)p^{k-1})} \times \frac{(-t)^k}{(k-1)!\mathfrak{g}(\chi^{-1})} \times \log_{k,\chi}(\kappa_{k,\chi}). \qquad (3.16)$$

When $k \leq 0$, (and $\eta(p) \neq 1$ when $k = 0$) the source and target of the logarithm map are both zero, and (3.16) does not extend to the negative integers. An interpretation of $L_p(k, \chi)$ can be given in terms of the *dual exponential map* of Bloch–Kato,

$$\exp^*_{k,\eta} : H^1(\mathbb{Q}_p, \mathbb{Q}_{p,\eta}(k)(\eta)) \longrightarrow \mathbb{D}_{\mathrm{dR}}(\mathbb{Q}_{p,\eta}(k)(\eta)), \qquad (3.17)$$

obtained by dualising the map $\exp_{1-k,\eta^{-1}} := \log^{-1}_{1-k,\eta^{-1}}$ and combining local Tate duality with the natural duality between $\mathbb{D}_{\mathrm{dR}}(\mathbb{Q}_{p,\eta}(1-k)(\eta^{-1}))$ and $\mathbb{D}_{\mathrm{dR}}(\mathbb{Q}_{p,\eta}(k)(\eta))$. The kernel of the map $\exp^*_{k,\eta}$ consists precisely of the extensions of $\mathbb{Q}_p$ by $\mathbb{Q}_{p,\eta}(k)(\eta)$ which are crystalline, and $\exp^*_{k,\eta}$ is an isomorphism of one-dimensional $\mathbb{Q}_{p,\eta}$-vector spaces for all $k \leq 0$. The following theorem for $k \leq 0$ [PR4, 3.2.2] is one of the simplest instances of so-called *reciprocity laws* relating L-values to distinguished global cohomology classes with values in the associated p-adic representation:

$$L_p(k, \chi) = -\frac{(1 - \chi(p)p^{-k})}{(1 - \chi^{-1}(p)p^{k-1})} \times \frac{(-k)!t^k}{\mathfrak{g}(\chi^{-1})} \times \exp^*_{k,\chi}(\kappa_{k,\chi}). \qquad (3.18)$$

Equations (3.16) and (3.18) give a satisfying intepretation of $L_p(k, \chi)$ at *all* integers $k \in \mathbb{Z}$ in terms of the global classes $\kappa_{k,\chi}$. In particular the global classes $\kappa_{k,\chi}$ are non-trivial, and in fact non-crystalline, whenever $L_p(k, \chi) \neq 0$ and $k < 0$. Since k is then in the region of classical interpolation defining $L_p(s, \chi)$, the non-vanishing of $L_p(k, \chi)$ is directly related to the behaviour of the corresponding classical L-function. In this region of classical interpolation, the complex L-value attached to the p-adic representation $\mathbb{Q}_{p,\chi}(k)(\eta)$ is thus given new meaning, as the *obstruction* to the global class $\kappa_{k,\chi} \in H^1(\mathbb{Q}, \mathbb{Q}_{p,\chi}(k)(\chi))$ being cristalline at p.

1.2. Elliptic units

In addition to the cusps, modular curves are endowed with a second distinguished class of algebraic points: the *CM points* attached to the moduli of elliptic curves with complex multiplication by an order in a quadratic imaginary field K. The values of modular units at such points give rise to units in abelian extensions of K, the so-called *elliptic units*, which play the same role for abelian extensions of K as circular units in the study of cyclotomic fields. The resulting p-adic families of global cohomology classes are the main ingredient in the seminal work of Coates and Wiles [CW] on the arithmetic of elliptic curves with complex multiplication.

The Eisenstein series $E_{k,\chi}$ of (3.1), viewed as a function of a variable τ in the complex upper half-plane $\mathcal{H}$ by setting $q = e^{2\pi i\tau}$, is given by the well-known formula

$$E_{k,\chi}(\tau) := N^k \mathfrak{g}(\bar{\chi})^{-1} \frac{(k-1)!}{(2\pi i)^k} \sum_{(m,n)\in N\mathbb{Z}\times\mathbb{Z}}{}' \frac{\bar{\chi}(n)}{(m\tau+n)^k}, \tag{3.19}$$

where $\mathfrak{g}(\bar{\chi})$ is the Gauss sum defined in (3.8), and the superscript $'$ indicates that the sum is to be taken over the non-zero lattice vectors in $N\mathbb{Z} \times \mathbb{Z}$.

Assume for simplicity that K has class number one, trivial unit group $\mathcal{O}_K^\times = \pm 1$, and odd discriminant $D < 0$. Assume also that there is an integral ideal – and hence, an element $\mathfrak{n}$ of $\mathcal{O}_K$ – satisfying

$$\mathcal{O}_K/\mathfrak{n} = \mathbb{Z}/N\mathbb{Z}. \tag{3.20}$$

One then says that the Eisenstein series $E_{k,\chi}$ satisfies the *Heegner hypothesis* relative to K. Under this hypothesis, the even character χ gives rise to a finite order character $\chi_{\mathfrak{n}}$ of conductor $\mathfrak{n}$ on the ideals of K by the rule

$$\chi_{\mathfrak{n}}((\alpha)) := \bar{\chi}(\alpha \bmod \mathfrak{n}). \tag{3.21}$$

After writing

$$\tau_{\mathfrak{n}} = \frac{b+\sqrt{D}}{2N}, \qquad \text{where} \quad \mathfrak{n} = \mathbb{Z}N + \mathbb{Z}\frac{b+\sqrt{D}}{2}, \tag{3.22}$$

a direct calculation using (3.19) shows that

$$E_{k,\chi}(\tau_{\mathfrak{n}}) = N^k \mathfrak{g}(\bar{\chi})^{-1} \frac{(k-1)!}{(2\pi i)^k} L(K, \chi_{\mathfrak{n}}, k, 0), \qquad (3.23)$$

where for all $k_1, k_2 \in \mathbb{Z}$ with $k_1 + k_2 > 2$,

$$L(K, \chi_{\mathfrak{n}}, k_1, k_2) := \sum_{\alpha \in \mathcal{O}_K}' \chi_{\mathfrak{n}}(\alpha) \alpha^{-k_1} \bar{\alpha}^{-k_2}.$$

Note that $L(K, \chi_{\mathfrak{n}}, s) := \frac{1}{2} L(K, \chi_{\mathfrak{n}}, s, s)$, viewed as a function of a complex variable s, is the usual Hecke L-function attached to the finite order character $\chi_{\mathfrak{n}}$. The relation (3.23) between values of $E_{k,\chi}$ at CM points and L-series of quadratic imaginary fields is the direct counterpart of (3.1) relating the value at the cusps with Dirichlet L-series. For the purpose of p-adic interpolation it will be convenient, at least initially, to consider the values $L(K, \chi_{\mathfrak{n}}, k, 0)$ rather than the values $L(K, \chi_{\mathfrak{n}}, k, k)$.

Under the moduli interpretation of $\Gamma_1(N)\backslash\mathcal{H} = Y_1(N)(\mathbb{C})$, the point $\tau \in \mathcal{H}$ corresponds to the pair $(\mathbb{C}/\mathbb{Z} \oplus \mathbb{Z}\tau, 1/N)$ consisting of an elliptic curve over $\mathbb{C}$ with a marked point of order N. The point $\tau_{\mathfrak{n}}$ corresponds in this way to the pair $(\mathbb{C}/\bar{\mathfrak{n}}^{-1}, 1/N)$. Since $\bar{\mathfrak{n}}^{-1}$ is a fractional ideal of $\mathcal{O}_K$, the elliptic curve $A = \mathbb{C}/\bar{\mathfrak{n}}^{-1}$ has complex multiplication by $\mathcal{O}_K$ and hence has a model defined over K and even over $\mathcal{O}_K$, while $1/N$ represents an $\mathfrak{n}$-torsion point of A, denoted $t_{\mathfrak{n}}$, and hence is defined over the ray class field $K_{\mathfrak{n}}$ of K of conductor $\mathfrak{n}$. In particular the point $P_{\mathfrak{n}}$ of $X_1(N)$ attached to the pair $(A, t_{\mathfrak{n}})$ is rational over $K_{\mathfrak{n}}$.

Choose a Néron differential $\omega_A \in \Omega^1(A/\mathcal{O}_K)$ and let $\Omega_K \cdot \mathcal{O}_K \subset \mathbb{C}$ be the associated period lattice. This determines the complex number Ω_K uniquely up to sign, once ω_A has been chosen. Following Katz, an algebraic modular form of weight k on $\Gamma_1(N)$ can be viewed as a function f on the isomorphism classes of triples (E, t, ω_E) consisting of an elliptic curve (E, t) with $\Gamma_1(N)$-level structure and a choice of regular differential ω_E on E, satisfying a weight $-k$-homogeneity condition under scaling of ω_E. The convention relating both points of view is that $f(\mathbb{C}/\mathbb{Z} + \mathbb{Z}\tau, 1/N, dz) := f(\tau)$ for $\tau \in \mathcal{H}$, where dz is the standard differential on $\mathbb{C}/\Lambda$ whose period lattice is equal to Λ. In particular,

$$E_{k,\chi}(A, t_{\mathfrak{n}}, \omega_A) = E_{k,\chi}(\mathbb{C}/\mathbb{Z} + \mathbb{Z}\tau_{\mathfrak{n}}, 1/N, \Omega_K \bar{\mathfrak{n}} dz) = \frac{E_{k,\chi}(\tau_{\mathfrak{n}})}{(\Omega_K \bar{\mathfrak{n}})^k}, \qquad (3.24)$$

where by an abuse of notation $\bar{\mathfrak{n}}$ is identified with one of its generators.

Because $E_{k,\chi}$ is defined over the field $\mathbb{Q}_\chi$ generated by the values of χ and the triple $(A, t_{\mathfrak{n}}, \omega_A)$ is defined over $K_{\mathfrak{n}}$, the quantities in (3.24) are algebraic, and in fact belong to the compositum $K_{\mathfrak{n},\chi}$ of $K_{\mathfrak{n}}$ and $\mathbb{Q}_\chi$. Therefore by (3.23), the normalised L-value

$$\frac{E_{k,\chi}(\tau_{\mathfrak{n}})}{(\Omega_K \bar{\mathfrak{n}})^k} = \mathfrak{n}^k \mathfrak{g}(\bar{\chi})^{-1} \frac{(k-1)!}{(2\pi i \cdot \Omega_K)^k} L(K, \chi_{\mathfrak{n}}, k, 0) \tag{3.25}$$

also belongs to $K_{\mathfrak{n},\chi}$.

Fix a prime $p \nmid N$, and let $X_{01}(p, N)$ be the modular curve attached to $\Gamma_0(p) \cap \Gamma_1(N)$. Recall the affinoid region $\mathcal{A}$ of Section 1.1 obtained by deleting the supersingular residue discs from $X_1(N)$. The natural algebraic projection $X_{01}(p, N) \longrightarrow X_1(N)$ admits a rigid analytic section $s : \mathcal{A} \longrightarrow X_{01}(p, N)(\mathbb{C}_p)$ over this ordinary locus, which sends the point attached to an ordinary pair (E, t) with $\Gamma_1(N)$-level structure to the point attached to the triple (E, C, t) where C is the *canonical subgroup* of E of order p. The ordinary p-stabilisation $E_{k,\chi}^{(p)}$ of $E_{k,\chi}$ defined in (3.2) can thus be viewed either as a classical modular form on $X_{01}(p, N)$ or as a p-adic modular form on $X_1(N)$ by pulling back to $\mathcal{A}$ via s.

Assume further that $p = \mathfrak{p}\bar{\mathfrak{p}}$ splits in K, let $K \hookrightarrow \mathbb{Q}_p$ be the embedding of K into its completion at $\mathfrak{p}$, and extend this to an embedding $\iota_p : \bar{\mathbb{Q}} \hookrightarrow \mathbb{C}_p$. The elliptic curve $A/\mathbb{C}_p$ deduced from A via ι_p is ordinary at p, and its canonical subgroup of order p is equal to the group scheme of its $\mathfrak{p}$-division points. In particular, this group is defined over K. It follows that the image $s(P_{\mathfrak{n}})$ of $P_{\mathfrak{n}} = (A, t_{\mathfrak{n}})$ belongs to $X_{01}(p, N)(K_{\mathfrak{n}})$. A direct calculation shows that it is represented by the pair $(\mathbb{C}/\mathbb{Z} + \mathbb{Z}\tau_{\mathfrak{p}\mathfrak{n}}, \frac{\bar{\mathfrak{p}}_{\mathfrak{n}}^{-1}}{pN})$, where $\tau_{\mathfrak{p}\mathfrak{n}} \in \Gamma_{01}(p, N)\backslash\mathcal{H}$ is defined by the same equation as (3.22) but with $\mathfrak{n}$ replaced by $\mathfrak{p}\mathfrak{n}$:

$$\tau_{\mathfrak{p}\mathfrak{n}} = \frac{b + \sqrt{D}}{2Np}, \qquad \text{where} \quad \mathfrak{p}\mathfrak{n} = \mathbb{Z}pN + \mathbb{Z}\frac{b + \sqrt{D}}{2}, \tag{3.26}$$

and $\bar{\mathfrak{p}}_{\mathfrak{n}}^{-1}$ is any element of $\mathcal{O}_K/\mathfrak{n}\mathcal{O}_K$ which is congruent to $\bar{\mathfrak{p}}^{-1}$. The value of the p-adic modular form $E_{k,\chi}^{(p)}$ on the ordinary triple $(A, t_{\mathfrak{n}}, \omega_A)$ can therefore be calculated as

$$\begin{aligned} E_{k,\chi}^{(p)}(A, t_{\mathfrak{n}}, \omega_A) &= E_{k,\chi}^{(p)}\left(\mathbb{C}/\mathbb{Z} + \mathbb{Z}\tau_{\mathfrak{p}\mathfrak{n}}, \frac{\bar{\mathfrak{p}}_{\mathfrak{n}}^{-1}}{pN}, \bar{\mathfrak{p}}\bar{\mathfrak{n}}\Omega_K dz\right) \\ &= \frac{\chi_{\mathfrak{n}}(\bar{\mathfrak{p}}) E_{k,\chi}^{(p)}(\tau_{\mathfrak{p}\mathfrak{n}})}{(\bar{\mathfrak{p}}\bar{\mathfrak{n}}\Omega_K)^k} \\ &= (1 - \chi_{\mathfrak{n}}^{-1}(\mathfrak{p})\mathfrak{p}^k/p) \times \mathfrak{n}^k \mathfrak{g}(\bar{\chi})^{-1} \\ &\quad \frac{(k-1)!}{(2\pi i \cdot \Omega_K)^k} L(K, \chi_{\mathfrak{n}}, k, 0), \end{aligned} \tag{3.27}$$

where the second and third occurrence of $E_{k,\chi}^{(p)}$ are treated as classical modular forms of level pN, and (3.27) follows from a direct calculation based on (3.19) and (3.26). The fact that all the coefficients in the Fourier expansion (3.3) of $E_{k,\chi}^{(p)}$ extend to p-adic analytic functions of k on weight space $W = \mathbb{Z}/(p-1)\mathbb{Z} \times \mathbb{Z}_p$ and that the pair $(A, t_{\mathfrak{n}})$ belongs to $\mathcal{A}$ suggests that the right-hand side of (3.27), normalised by a suitable p-adic period, should admit a similar prolongation. More precisely, the formal completion $\hat{A}$ of $A_{\mathcal{O}_{\mathbb{C}_p}}$ along its identity section is isomorphic to the formal multiplicative group $\hat{\mathbf{G}}_m$, and upon choosing an isomorphism $\iota : \hat{A} \longrightarrow \hat{\mathbf{G}}_m$ we may define a p-adic period $\Omega_p \in \mathbb{C}_p^\times$ by the rule

$$\omega_A = \Omega_p \cdot \omega_{\mathrm{can}},$$

where $\omega_{\mathrm{can}} := \iota^* \frac{dt}{t}$, with $\frac{dt}{t}$ the canonical differential on $\hat{\mathbf{G}}_m$, plays the role of the complex differential dz in the p-adic setting. The function $L_p(K, \chi_{\mathfrak{n}}, k)$ of k defined by

$$L_p(K, \chi_n, k) = E_{k,\chi}^{(p)}(A, t_{\mathfrak{n}}, \omega_{\mathrm{can}}) = \Omega_p^k \cdot E_{k,\chi}^{(p)}(A, t_{\mathfrak{n}}, \omega_A), \qquad (3.28)$$

extends to a p-adic analytic function of $k \in W$ and is equal (up to the p-adic period Ω_p^k) to the right-hand side of (3.27) for all integers $k \geq 2$. It is called the *Katz one-variable p-adic L-function* attached to K and to the character $\chi_{\mathfrak{n}}$.

Recall the Siegel units g_a and $g_a^{(p)}$ defined in Equations (3.6) and (3.7) of Section 1.1. Evaluating these functions at the pair $(A, t_{\mathfrak{n}})$ gives rise to the *elliptic units*

$$u_{a,\mathfrak{n}} := g_a(A, t_{\mathfrak{n}}) = g_a(\tau_{\mathfrak{n}}), \qquad u_{a,\mathfrak{n}}^{(p)} := g_a^{(p)}(A, t_{\mathfrak{n}}) = g_a^{(p)}(\tau_{\mathfrak{p}\mathfrak{n}}) = u_{a,\mathfrak{n}}^{\sigma_{\mathfrak{p}}-p}$$

in $\mathcal{O}_{K_{\mathfrak{n}}(\mu_N)}^\times$, where $\sigma_{\mathfrak{p}} \in \mathrm{Gal}(K_{\mathfrak{n}}(\mu_N)/K)$ denotes the Frobenius element at $\mathfrak{p}$.

The following result of Katz (as well as its proof) is the direct counterpart of Leopoldt's formula (Theorem 3.1) in which cusps are replaced by CM points and circular units by elliptic units.

Theorem 3.2 (Katz). *Let χ be a non-trivial even primitive Dirichlet character of conductor N and let K be a quadratic imaginary field equipped with an ideal $\mathfrak{n}$ satisfying $\mathcal{O}_K/\mathfrak{n} = \mathbb{Z}/N\mathbb{Z}$. Let $\chi_{\mathfrak{n}}$ be the ideal character of K associated to the pair $(\chi, \mathfrak{n})$ as in* (3.21). *Then*

$$L_p(K, \chi_{\mathfrak{n}}, 0) = -\frac{(1 - \chi_{\mathfrak{n}}(\mathfrak{p})p^{-1})}{\mathfrak{g}(\bar{\chi})} \times \sum_{a=1}^{N-1} \chi^{-1}(a) \log_p u_{a,\mathfrak{n}}.$$

Proof. Setting $k = 0$ in (3.28) gives

$$L_p(K, \chi_{\mathfrak{n}}, 0) = E_{0,\chi}^{(p)}(A, t_{\mathfrak{n}}),$$

where $E_{0,\chi}^{(p)}(A, t_{\mathfrak{n}})$ refers to $E_{0,\chi}^{(p)}(A, t_{\mathfrak{n}}, \omega)$ for *any* choice of regular differential ω on A, since $E_{0,\chi}$ is of weight zero and this value is therefore independent of ω. But by Equations (3.11) and (3.9) in the proof of Theorem 3.1,

$$
\begin{aligned}
E_{0,\chi}^{(p)}(A,t_{\mathfrak{n}}) = h_{\chi}^{(p)}(A,t_{\mathfrak{n}}) &= \frac{1}{p\mathfrak{g}(\bar{\chi})} \times \sum_{a=1}^{N-1} \chi^{-1}(a) \log_p g_a^{(p)}(A,t_{\mathfrak{n}}) \\
&= \frac{1}{p\mathfrak{g}(\bar{\chi})} \times \sum_{a=1}^{N-1} \chi^{-1}(a) \log_p u_{a,\mathfrak{n}}^{\sigma_{\mathfrak{p}}-p} \\
&= \frac{\chi_{\mathfrak{n}}(\mathfrak{p}) - p}{p\mathfrak{g}(\bar{\chi})} \times \sum_{a=1}^{N-1} \chi^{-1}(a) \log_p u_{a,\mathfrak{n}}.
\end{aligned}
$$

The theorem follows. □

The calculations above can be extended by introducing the Katz two-variable p-adic L-function $L_p(K,\chi_{\mathfrak{n}},k_1,k_2)$ which interpolates the values of $L(K,\chi_{\mathfrak{n}},k_1,k_2)$ as k_1 and k_2 both vary over weight space. These more general L-values are related to the values at CM points of the non-holomorphic Eisenstein series

$$
E_{k_1,k_2,\chi}(\tau) = N^{k_1+k_2}\mathfrak{g}(\bar{\chi})^{-1}\frac{(k_1-1)!}{(2\pi i)^{k_1}}(\tau-\bar{\tau})^{k_2} \sum_{(m,n)\in N\mathbb{Z}\times\mathbb{Z}}{}' \frac{\bar{\chi}(n)}{(m\tau+n)^{k_1}(m\bar{\tau}+n)^{k_2}}
$$

by the formula generalising (3.23)

$$
E_{k_1,k_2,\chi}(\tau_{\mathfrak{n}}) = N^{k_1}\mathfrak{g}(\bar{\chi})^{-1}\frac{(k_1-1)!}{(2\pi i)^{k_1}}\sqrt{D}^{-k_2} L(K,\chi_{\mathfrak{n}},k_1,k_2). \tag{3.29}
$$

The function $E_{k_1,k_2,\chi}$ is a real analytic function on $\mathcal{H}$ which transforms like a modular form of weight k_1-k_2 and character χ under the action of $\Gamma_0(N)$. Although it is non-holomorphic in general, it can sometimes be expressed as the image of holomorphic modular forms under iterates of the *Shimura–Maass* derivative operator

$$
\delta_k = \frac{1}{2\pi i}\left(\frac{d}{d\tau} + \frac{k}{\tau-\bar{\tau}}\right) \tag{3.30}
$$

sending real analytic modular forms of weight k to real analytic modular forms of weight $k+2$. More precisely, after setting $\delta_k^r = \delta_{k+2r-2}\circ\cdots\circ\delta_{k+2}\circ\delta_k$, a direct calculation reveals that

$$
\delta_k^r E_{k,\chi} = E_{k+r,-r,\chi}.
$$

A *nearly holomorphic modular form of weight* k *on* $\Gamma_1(N)$ is a linear combination

$$
f = \sum_{i=1}^{t} \delta_{k-2j_i}^{j_i} f_i, \qquad f_i \in M_{k-2j_i}(\Gamma_1(N)),
$$

where the f_i are classical modular forms of weight $k - 2j_i$ on $\Gamma_1(N)$. If the Fourier expansions of all the f_i at all the cusps have Fourier coefficients in a field L, then the nearly holomorphic modular form f is said to be *defined over* L.

Shimura proved that nearly holomorphic modular forms of weight k which are defined over $\bar{\mathbb{Q}}$ take algebraic values at CM triples like $(A, t_{\mathfrak{n}}, \omega_A)$. More precisely, if f is defined over $L \subset \bar{\mathbb{Q}}$, then

$$f(A, t_{\mathfrak{n}}, \omega_A) := \frac{f(\tau_{\mathfrak{n}})}{(\bar{\mathfrak{n}}\Omega_K)^k} \text{ belongs to } LK_{\mathfrak{n}}. \tag{3.31}$$

A conceptual explanation for this striking algebraicity result rests on the relationship between nearly holomorphic modular forms of weight k and global sections of an algebraic vector bundle arising from the relative de Rham cohomology of the universal elliptic curve over $Y_1(N)$, and on the resulting interpretation of the Shimura–Maass derivative in terms of the Gauss–Manin connection on this vector bundle. See for instance Section 1.5 of [BDP] or Section 2.4 of [DR1] for a brief account of this circle of ideas, and Section 10.1 of [Hi2] for a more elementary treatment.

Specialising (3.31) to the setting where f is the Eisenstein series $E_{k+r,-r,\chi}$ of weight $k + 2r$ with $k \geq 2$ and $r \geq 0$, and invoking (3.29) leads to the conclusion that the special values

$$E_{k+r,-r,\chi}(A, t_{\mathfrak{n}}, \omega_A) = \mathfrak{n}^{k+r}\bar{\mathfrak{n}}^{-r}\,\mathfrak{g}(\bar{\chi})^{-1}\frac{(k+r-1)!}{(2\pi i)^{k+r}\Omega_K^{k+2r}}\sqrt{D}^{-r}$$
$$L(K, \chi_{\mathfrak{n}}, k+r, -r) \tag{3.32}$$

belong to the compositum $K_{\mathfrak{n},\chi}$ of $K_{\mathfrak{n}}$ and $\mathbb{Q}_\chi$, just as in (3.25), for all $k \geq 2$ and $r \geq 0$.

In light of this algebraicity result, it is natural to attempt to interpolate these values p-adically as a function of both k and r in weight space. This p-adic interpolation rests on the fact that the Shimura–Maass derivative admits a counterpart in the realm of p-adic modular forms: the Atkin–Serre operator d which raises the weight by two and acts as the differential operator $d = q\frac{d}{dq}$ on q-expansions. If f is a classical modular form of weight k with rational Fourier coefficients, the nearly holomorphic modular form $\delta_k^r f$ and the p-adic modular form $d^r f$ are objects of a very different nature, but nonetheless their values *agree* on ordinary CM triples (where it makes sense to compare them) so that in particular

$$\delta_k^r f(A, t_{\mathfrak{n}}, \omega_A) = d^r f(A, t_{\mathfrak{n}}, \omega_A), \qquad \text{for all } r \geq 0. \tag{3.33}$$

The reason, which is explained for instance in Section 1.5 of [BDP], is that d admits *the same algebraic description* as δ_k in terms of the Gauss–Manin connection on a relative de Rham cohomology sheaf, with the sole difference that the (non-holomorphically varying) Hodge decomposition on the complex de Rham cohomology of the fibers is replaced in the p-adic setting by the *Frobenius decomposition* of the de Rham cohomology of the universal (ordinary) elliptic curve over $\mathcal{A}$. But the functorial action of the endomorphism algebra on algebraic de Rham cohomology causes these two decompositions to agree for ordinary CM elliptic curves.

Equation (3.33) means that, as far as values at ordinary CM triples are concerned, the p-adic modular form $d^r E_{k,\chi}$ is a perfect substitute for $E_{k+r,-r,\chi} = \delta_k^r E_{k,\chi}$. The Fourier expansion of this p-adic avatar is given, for $r \geq 1$, by

$$d^r E_{k,\chi} = \sum_{n=1}^{\infty} n^r \sigma_{k-1,\chi}(n) q^n. \tag{3.34}$$

The coefficients of q^n when $p|n$ do not extend to a p-adic analytic function of $(k, r) \in W^2$, and this difficulty persists after replacing $E_{k,\chi}$ by its ordinary p-stabilisation $E_{k,\chi}^{(p)}$. One is therefore led to consider instead the so-called *p-depletion* of $E_{k,\chi}$, defined by

$$E_{k,\chi}^{[p]}(\tau) = E_{k,\chi}(\tau) - (1 + \chi(p)p^{k-1})E_{k,\chi}(p\tau) + \chi(p)p^{k-1}E_{k,\chi}(p^2\tau), \tag{3.35}$$

so that $E_{k+r,-r,\chi}^{[p]} := \delta_k^r E_{k,\chi}^{[p]}$ – a nearly holomorphic modular form of level Np^2 – is given by

$$\begin{aligned} E_{k+r,-r,\chi}(\tau) - (p^r + \chi(p)p^{k+r-1})E_{k+r,-r,\chi}(p\tau) \\ + \chi(p)p^{k+2r-1}E_{k+r,-r,\chi}(p^2\tau), \end{aligned} \tag{3.36}$$

while its p-adic avatar $d^r E_{k,\chi}^{[p]}$ – a p-adic modular form of level N – has Fourier expansion

$$d^r E_{k,\chi}^{[p]} = \sum_{p \nmid n}^{\infty} n^r \sigma_{k-1,\chi}(n) q^n.$$

The coefficients in this expansion do extend to p-adic analytic functions of k and r on weight space $W = \mathbb{Z}/(p-1)\mathbb{Z} \times \mathbb{Z}_p$. Just as in the construction of $L_p(K, \chi_{\mathfrak{n}}, k)$, this suggests that the function $L_p(K, \chi_{\mathfrak{n}}, k_1, k_2)$ defined (for $k \geq 2$ and $r \geq 0$) by the rule

$$\begin{aligned}\frac{L_p(K,\chi_{\mathfrak{n}},k+r,-r)}{\Omega_p^{k+2r}} &:= d^r E^{[p]}_{k,\chi}(A,t_{\mathfrak{n}},\omega_A)\\ &= E^{[p]}_{k+r,-r,\chi}\left(\mathbb{C}/\mathbb{Z}+\mathbb{Z}\tau_{\mathfrak{p}^2\mathfrak{n}},\frac{\bar{\mathfrak{p}}_{\mathfrak{n}}^{-2}}{p^2N},\bar{\mathfrak{p}}^2\bar{\mathfrak{n}}\Omega_K dz\right)\\ &= (1-\chi_{\mathfrak{n}}^{-1}(\mathfrak{p})\mathfrak{p}^{k+r}\bar{\mathfrak{p}}^{-r}/p)\times(1-\chi_{\mathfrak{n}}(\bar{\mathfrak{p}})\mathfrak{p}^r\bar{\mathfrak{p}}^{-k-r})\times \qquad (3.37)\\ &\quad \mathfrak{n}^{k+r}\bar{\mathfrak{n}}^{-r}\mathfrak{g}(\bar{\chi})^{-1}\frac{(k+r-1)!}{(2\pi i)^{k+r}\Omega_K^{k+2r}}\sqrt{D}^{-r}L(K,\chi_{\mathfrak{n}},k+r,-r)\end{aligned}$$

extends to an analytic function of $(k,r)\in W^2$. The function $L_p(K,\chi_{\mathfrak{n}},k_1,k_2)$ is called the Katz *two-variable p-adic L-function* attached to the character $\chi_{\mathfrak{n}}$. Note that the restriction of $L_p(K,\chi_{\mathfrak{n}},k_1,k_2)$ to the line $k_2=0$ is related to the Katz one-variable p-adic L-function by the rule

$$L_p(K,\chi_{\mathfrak{n}},k,0)=(1-\chi_{\mathfrak{n}}(\bar{\mathfrak{p}})\bar{\mathfrak{p}}^{-k})L_p(K,\chi_{\mathfrak{n}},k).$$

The ratio of the two sides (which can be seen to be a p-adic analytic function of k, since $\bar{\mathfrak{p}}$ belongs to $\mathcal{O}_{K_{\mathfrak{p}}}^\times$) reflects the difference between working with the ordinary p-stabilisation $E^{(p)}_{k,\chi}$ and the p-depletion $E^{[p]}_{k,\chi}$.

The following variant of Katz's Theorem 3.2 expresses the special value $L_p(K,\chi_{\mathfrak{n}}^{-1},1,1)$ in terms of the elliptic units $u^*_{a,\mathfrak{n}}:=g_a(w(A,t_{\mathfrak{n}}))$, where w is the *Atkin–Lehner involution* such that $g_a(w(A,t_{\mathfrak{n}}))=g_a(-1/N\tau_{\mathfrak{n}})$.

Theorem 3.3 (Katz). *With notation as in Theorem 3.2,*

$$L_p(K,\chi_{\mathfrak{n}}^{-1},1,1)=(1-\chi_{\mathfrak{n}}^{-1}(\bar{\mathfrak{p}}))(1-\chi_{\mathfrak{n}}(\mathfrak{p})/p)\times\sum_{a=1}^{N-1}\chi^{-1}(a)\log_p u^*_{a,\mathfrak{n}}.$$

Proof. By setting $k=2$ and $r=-1$ in Equation (3.37) (and replacing $\chi_{\mathfrak{n}}$ by $\chi_{\mathfrak{n}}^{-1}$), we obtain

$$L_p(K,\chi_{\mathfrak{n}}^{-1},1,1)=d^{-1}E^{[p]}_{2,\chi^{-1}}(A,t_{\mathfrak{n}},\omega_A).$$

The p-adic modular form of weight 0 which appears on the right-hand side of this identity has the same Fourier expansion as the p-depletion of the Eisenstein series $E_0(\chi^{-1},1)$ introduced in the remark following Theorem 3.1 of Section 1. Combined with [Hi1, Lemma 5.3], we thus see that

$$L_p(K,\chi_{\mathfrak{n}}^{-1},1,1)=\mathfrak{g}(\bar{\chi})wE^{[p]}_{0,\chi}(A,t_{\mathfrak{n}},\omega_A),$$

where w is the *Atkin–Lehner involution* at N introduced above. The result then follows from an almost identical calculation to the proof of Theorem 3.2. □

In stark analogy with the case of circular units, the expressions

$$u_{\chi_n} := \prod_{a=1}^{N-1} u_{a,n}^{\chi^{-1}(a)} = \prod_{\sigma \in \mathrm{Gal}(K_n/K)} (\sigma u_{1,n})^{\chi_n(\sigma)}$$

are (formal $\mathbb{Q}_\chi$-linear combination of) special units in K_n lying in the χ_n^{-1}-eigenspace for the natural action of G_K.

These units arise as the "bottom layer" of a norm-coherent family of elliptic units over the two-variable $\mathbb{Z}_p$-extension K_∞ of K. The same construction as in Equation (3.12) of Section 1.1 leads to a global cohomology class

$$\kappa_{\chi_n,\infty} \in H^1(K, \Lambda_K(\chi_n)),$$

where $\Lambda_K = \mathbb{Z}_p[\![\mathrm{Gal}(K_\infty/K)]\!]$ is the *two-variable* Iwasawa algebra attached to K, equipped with its tautological G_K-action. The Galois module $\Lambda_K(\chi_n)$ gives a p-adic interpolation of the Hecke characters of the form $\chi_n\phi$ where ϕ is a Hecke character of p-power conductor.

In particular, if ψ is a Hecke character of infinity type (k_1, k_2) arising as a specialisation of $\Lambda_K(\chi_n)$, the global class

$$\kappa_\psi \in H^1(K, \mathbb{Q}_p(\psi))$$

obtained by specialising $\kappa_{\chi_n,\infty}$ at ψ, although it arises from elliptic units, encodes arithmetic information about a Galois representation V_ψ of K attached to a Hecke character of possible infinite order.

The reciprocity law of Coates and Wiles expresses the Katz two-variable p-adic L-function $L_p(K, \chi_n, k_1, k_2)$ as the image under a "big exponential map" of the global class $\kappa_{n,\infty}$. By an analogue of (3.18) of Section 1.1, the special value $L(\psi^{-1}, 0)$ can then be interpreted as the obstruction to the class κ_ψ being cristalline at $\mathfrak{p}$, whenever ψ lies in the range of classical interpolation for the Katz p-adic L-function.

A classical result of Deuring asserts that the p-adic Tate module of a CM elliptic curve, viewed as a representation of G_K, is always of the form V_ψ for a suitable Hecke character ψ of K of infinity type $(1, 0)$. The global class κ_ψ for such a ψ acquires a special interest in relation to the Birch and Swinnerton-Dyer conjecture for the elliptic curve A_ψ attached to it. This connection between $L(A_\psi, 1)$ and the singular parts of global classes κ_ψ was used by Coates–Wiles to give what historically was the first broad piece of convincing supporting evidence for the conjecture of Birch and Swinnerton-Dyer, notably the implication

$$L(A, 1) \neq 0 \implies A(K) \otimes \mathbb{Q} = \{0\},$$

for any elliptic curve $A/\mathbb{Q}$ with complex multiplication by K.

1.3. Heegner points

This section replaces the Eisenstein series $E_{k,\chi}$ of Section 1.2 by a *cusp form* of weight k. The argument used in the proof of Theorem 3.3 then leads naturally to the p-adic Gross–Zagier formula of [BDP] expressing the special values of certain *p-adic Rankin L-series* attached to f and K in terms of the p-adic Abel–Jacobi images of so-called *generalised Heegner cycles*.

For simplicity, let $f \in S_k(N)$ be a normalised cuspidal eigenform of *even weight* k on $\Gamma_1(N)$ with rational Fourier coefficients and trivial nebentypus character. (When $k = 2$, the form f is therefore associated to an elliptic curve $E/\mathbb{Q}$.) Let K be a quadratic imaginary field satisfying all the hypotheses of Section 1.2, including the Heegner assumption (3.20). We will also assume that $p = \mathfrak{p}\bar{\mathfrak{p}}$ is, as before, a rational prime which splits in K and does not divide N.

By analogy with the construction of the Katz two-variable p-adic L-function, it is natural to consider the quantities

$$\delta_k^r f(A, t_{\mathfrak{n}}, \omega_A) = d^r f(A, t_{\mathfrak{n}}, \omega_A), \tag{3.38}$$

which belong to $K_{\mathfrak{n}}$ for all $r \geq 0$. The role of the relatively elementary formula (3.32) of Section 1.2 relating such quantities to L-values when f is an Eisenstein series is played in this context by a seminal formula of Waldspurger, whose importance for the arithmetic study of generalised Heegner cycles and points would be hard to overstate, even though its proof lies beyond the scope of this survey.

Waldspurger's formula relates (3.38) to the L-function of f twisted (over K) by certain unramified Hecke characters of K. If ϕ is such a character (viewed as a mutiplicative function on fractional ideals of K in the usual way) then the L-series $L(f, K, \phi, s)$ of f/K twisted by ϕ is defined (for $s \in \mathbb{C}$ in some right-half plane) by the Euler product

$$L(f, K, \phi, s) = \prod_{\mathfrak{l}} \left[(1 - \alpha_{\mathbf{N}\mathfrak{l}}(f) \cdot \phi(\mathfrak{l})\mathbf{N}\mathfrak{l}^{-s})(1 - \beta_{\mathbf{N}\mathfrak{l}}(f) \cdot \phi(\mathfrak{l})\mathbf{N}\mathfrak{l}^{-s})\right]^{-1}$$

taken over the prime ideals $\mathfrak{l}$ in $\mathcal{O}_K$, where $\alpha_\ell(f)$ and $\beta_\ell(f)$ are the roots of the Hecke polynomial $x^2 - a_\ell(f)x + \ell^{k-1}$ for f at ℓ, and we set $\alpha_{\mathbf{N}\mathfrak{l}} := \alpha_\ell(f)^t$ and $\beta_{\mathbf{N}\mathfrak{l}} := \beta_\ell(f)^t$ if $\mathbf{N}\mathfrak{l} = \ell^t$. Rankin's method can be used to show that $L(f, K, \phi, s)$ admits an analytic continuation to the entire complex plane.

If k_1 and k_2 are integers with the same parity, let ϕ_{k_1,k_2} be the unramified Hecke character of K of infinity type (k_1, k_2) defined on fractional ideals by the rule

$$\phi_{k_1,k_2}((\alpha)) = \alpha^{k_1}\bar{\alpha}^{k_2}, \tag{3.39}$$

and set

$$L(f, K, k_1, k_2) := L(f, K, \phi_{k_1,k_2}^{-1}, 0).$$

Because

$$L(f, K, k_1 + s, k_2 + s) = L(f, K, \phi_{k_1,k_2}^{-1}, s),$$

we may view $L(f, K, k_1, k_2)$ as a function on pairs $(k_1, k_2) \in \mathbb{C}^2$ with $k_1 - k_2 \in 2\mathbb{Z}$. The functional equation relates the values $L(f, K, k_1, k_2)$ and $L(f, K, k - k_2, k - k_1)$ and hence corresponds to the reflection about the line $k_1 + k_2 = k$ in the (k_1, k_2)-plane, preserving the perpendicular lines $k_1 - k_2 = k + 2r$ for any $r \in \mathbb{Z}$. The restriction of $L(f, K, k_1, k_2)$ to such a line therefore admits a functional equation whose sign depends on r in an interesting way. More precisely, it turns out that assumption (3.20) forces this sign to be -1 when $1 - k \le r \le -1$, and to be $+1$ for other integer values of r. In particular, the central critical value $L(f, K, k + r, -r)$ vanishes for reasons of sign when $1 - k \le r \le -1$, but is expected to be non-zero for infinitely many $r \ge 0$. The quantities $L(f, K, k + r, -r)$ are precisely the special values that arise in the formula of Waldspurger, which asserts (cf. [BDP, Thm. 5.4]) that

$$\left(\delta_k^r f(A, t_{\mathfrak{n}}, \omega_A)\right)^2 = 1/2 \cdot (2\pi/\sqrt{D})^{k+2r-1} r!(k+r-1)! \cdot \frac{L(f, K, k + r, -r)}{(2\pi i \cdot \bar{\pi}\Omega_K)^{2(k+2r)}}. \tag{3.40}$$

Note the square that appears on the left-hand side. It is an enlightening exercise to recover Equation (3.32) of Section 1.2 (up to an elementary constant) by setting $f = E_{k,\chi}$ and replacing $\phi_{k+r,-r}$ by $\chi_{\mathfrak{n}}\phi_{k+r,-r}$ in Equation (3.40). This suggests that Waldspurger's formula should be viewed as the natural extension of (3.32) to the setting where Eisenstein series are replaced by cusp forms. An important difference with the setting of Section 1.2 is that the quantities $\delta_k^r f(A, t_{\mathfrak{n}}, \omega_A)$ only encode the *central critical values* $L(f, K, k + r, -r)$. In the setting of Eisenstein series the analogous value $L(E_{k,\chi}, K, (\chi_{\mathfrak{n}}\phi_{k+r,-r})^{-1}, 0)$ is also central critical, but breaks up as a product

$$L(E_{k,\chi}, K, (\chi_{\mathfrak{n}}\phi_{k+r,-r})^{-1}, 0) = L(K, (\chi_{\mathfrak{n}}\phi_{k+r,-r})^{-1}, 0) \times L(K, (\chi_{\mathfrak{n}}^{-1}\phi_{1-k-r,1+r})^{-1}, 0)$$

whose two factors, which are values of L-functions of Hecke characters, are interchanged by the functional equation and are non-self dual in general.

In order to interpolate the values $d^r f(A, t_{\mathfrak{n}}, \omega_A)$ p-adically, we replace, exactly as in (3.35), the modular form f by its p-depletion

$$f^{[p]}(\tau) := f(\tau) - a_p(f) f(p\tau) + p^{k-1} f(p^2\tau). \tag{3.41}$$

A direct calculation using (3.41) shows that for all $r \geq 0$,

$$d^r f^{[p]}(A, t_{\mathfrak{n}}, \omega_A) = (1 - a_p \mathfrak{p}^r \bar{\mathfrak{p}}^{-k-r} + \mathfrak{p}^{k+2r-1} \bar{\mathfrak{p}}^{-k-2r-1}) d^r f(A, t_{\mathfrak{n}}, \omega_A). \tag{3.42}$$

Since the collection of modular forms indexed by the parameter r

$$d^r f^{[p]}(q) = \sum_{(n,p)=1} n^r a_n(f) q^n \tag{3.43}$$

is a p-adic family of modular forms in the sense defined in Section 1.1, it follows, just as in Section 1.2, that the product of (3.42) by the p-adic period Ω_p^{k+2r} extends to a p-adic analytic function of $r \in W = (\mathbb{Z}/(p-1)\mathbb{Z}) \times \mathbb{Z}_p$. In light of (3.38) and (3.40), the quantity defined by

$$L_p(f, K, k+r, -r) := \Omega_p^{2(k+2r)} \times d^r f^{[p]}(A, t_{\mathfrak{n}}, \omega_A)^2 \tag{3.44}$$

is referred to as the *anticyclotomic p-adic L-function* attached to f and K.

The value $L_p(f, K, k+r, -r)$ of this p-adic L-function for $r < 0$ is defined by p-adic continuity rather than by the direct interpolation of classical L-values. For $r \in [1-k, -1]$, the quantity $L_p(f, K, k+r, -r)$ should be regarded as a genuinely p-adic avatar of the special value $L(f/K, \phi_{k+r,-r}^{-1}, 0)$ – or rather, of the first derivative $L'(f/K, \phi_{k+r,-r}^{-1}, 0)$ since the value vanishes for parity reasons. The main result of [BDP] relates it to the image under the p-adic Abel–Jacobi map of the *generalised Heegner cycles* introduced in [BDP].

We will now state the main result of [BDP] in the illustrative special case where $k = 2$ and f is attached to an elliptic curve E of conductor N. Let $P_K \in J_0(N)(K)$ be the class of the degree 0 divisor $(A, t_{\mathfrak{n}}) - (\infty)$ in the Jacobian variety $J_0(N)$ of $X_0(N)$, let $P_{f,K}$ denote its image in $E(K)$ under the modular parametrisation

$$\varphi_E : J_0(N) \longrightarrow E$$

arising from the modular form f, and let ω_E be the regular differential on E satisfying $\varphi_E^*(\omega_E) = \omega_f := (2\pi i) f(\tau) d\tau$.

Theorem 3.4. *Let $f \in S_2(N)$ be a normalised cuspidal eigenform of level $\Gamma_0(N)$ with N prime to p and let K be a quadratic imaginary field equipped with an integral ideal $\mathfrak{n}$ satisfying $\mathcal{O}_K/\mathfrak{n} = \mathbb{Z}/N\mathbb{Z}$. Then*

$$L_p(f, K, 1, 1) = \left(\frac{1 - a_p(f) + p}{p}\right)^2 \log_p(P_{K,f})^2,$$

where $\log_p$ is the formal group logarithm on E associated with the regular differential ω_E.

Sketch of proof. By (3.44) with $k = 2$ and $r = -1$,

$$L_p(f, K, 1, 1) = d^{-1} f^{[p]}(A, t_{\mathfrak{n}}, \omega_A)^2. \qquad (3.45)$$

As in the proof of Theorems 3.1, 3.2 and 3.3, the result will be obtained by interpreting the p-adic modular function $F^{[p]} := d^{-1} f^{[p]}$ as the rigid analytic primitive of the differential $\omega_{f^{[p]}}$ on $\mathcal{A}$ which vanishes at the cusp ∞, and relating this rigid analytic function to the *Coleman primitive* of ω_f.

Recall the lift of Frobenius Φ on the ordinary locus $\mathcal{A} \subset X_0(N)(\mathbb{C}_p)$. A direct computation shows that

$$\Phi\omega_f = p\omega_{Vf} \qquad (3.46)$$

as sections of $\Omega^1_{X_0(N)}$ over $\mathcal{A}$, where V is the operator on p-adic modular forms acting as $f(q) \mapsto f(q^p)$ on q-expansions.

Let F denote the Coleman primitive of ω_f on $\mathcal{A}$. It is a *locally analytic* function on $\mathcal{A}$, which is well-defined up to a constant and satisfies $dF = \omega_f$ on $\mathcal{A}$. By definition of the Coleman primitive,

$$F(A, t_{\mathfrak{n}}) = \log_{\omega_f}(P_K) = \log_p(P_{K,f}),$$

where $\log_{\omega_f}$ is the p-adic logarithm on $J_0(N)$ attached to the differential ω_f, and we have used the fact that $\varphi_E(P_K) = P_{K,f}$ and $\varphi_E^*(\omega_E) = \omega_f$ to derive the second equality. The result now follows from noting that

$$F^{[p]}(A, t_{\mathfrak{n}}) = (1 - a_p(f)/p + 1/p)F(A, t_{\mathfrak{n}}).$$

□

Like circular units and elliptic units, the Heegner point $P_{K,f} \in E(K)$ arises naturally as a *universal norm* of a compatible system of points defined over the so-called *anticyclotomic* $\mathbb{Z}_p$*-extension* of K. This extension, denoted K_∞^-, is contained in the two-variable $\mathbb{Z}_p$-extension K_∞ of Section 1.2 and is the largest subextension which is Galois over $\mathbb{Q}$ and for which $\mathrm{Gal}(K/\mathbb{Q})$ acts as -1 on $\mathrm{Gal}(K_\infty^-/K)$ via conjugation. After letting $\Lambda_K^- = \mathbb{Z}_p[[\mathrm{Gal}(K_\infty^-/K)]]$ be the Iwasawa algebra attached to this extension, equipped with its tautological action of G_K, this norm-compatible collection of Heegner points can be parlayed into the construction of a global cohomology class

$$\kappa_{f,K,\infty} \in H^1(K, V_p(E) \otimes \Lambda_K^-),$$

where $V_p(E)$ is the Galois representation attached to the p-adic Tate module of E.

The module $\Lambda_K^{-}\otimes_{\mathbb{Z}_p} V_p(E)$ is a deformation of $V_p(E)$ which p-adically interpolates the twists of $V_p(E)$ by the *anticyclotomic* Hecke characters $\phi_{r,-r}$, and hence the class $\kappa_{f,K,\infty}$ admits specialisations

$$\kappa_{f,K,r} \in H^1(K, V_p(E) \otimes \mathbb{Q}_p(\phi_{r,-r})).$$

When $r = 0$ this class arises from the image of Heegner points under the Kummer map, and hence is cristalline at p. Theorem 3.4 asserts that its p-adic logarithm is related to the values of the anti-cyclotomic p-adic L-function $L_p(f, K, 1+r, 1-r)$ at the point $(1, 1)$ that lies outside the range of classical interpolation.

In contrast, the classes $\kappa_{f,K,r}$ need not be cristalline at p when $r > 0$, and the formalism described in the previous sections suggests that the image of $\kappa_{f,K,r}$ under the dual exponential map at p should be related to the value $L_p(f, K, 1+r, 1-r)$, a simple non-zero multiple of $L(f, K, \phi_{r,-r}^{-1}, 1)$ since $(1+r, 1-r)$ lies in the range of classical interpolation defining the anticyclotomic p-adic L-function. One thus expects, just as in Sections 1.1 and 1.2, a direct relation between the images of the "higher weight" specialisations $\kappa_{f,K,r}$ under the Bloch–Kato dual exponential map and the central critical values $L(f, \phi_{-r,r}, 1)$. Current ongoing work of the second author aims to exploit the classes $\kappa_{f,K,r}$ and Kolyvagin's "method of Euler systems", as summarised in [Kato, §13], for example, to derive new cases of the Bloch–Kato conjecture of the form

$$L(f, \phi_{r,-r}, 1) \neq 0 \quad \Longrightarrow \quad \mathrm{Sel}_K(V_p(E)(\phi_{r,-r})) = \{0\}, \qquad (r > 0). \tag{3.47}$$

Using Hida families and a p-adic deformation along them of the Euler system of Heegner points due to Howard [How], one can more generally hope to establish the analogue of (3.47) for p-ordinary cuspidal eigenforms of even weight $k \geq 2$. This program is being carried out in [Cas2] based in part on a suitable extension of the circle of ideas described in this section and on [Cas1].

2. Euler systems of Garrett–Rankin–Selberg type

The three formulae described in Chapter 1 relate the p-adic logarithms of circular units, elliptic units and Heegner points to values of associated p-adic L-functions at points that lie outside their range of classical interpolation. The construction of the p-adic L-function in all three cases rests on formulae for critical L-values in terms of the *values* of modular forms at distinguished points of modular curves, namely cusps or CM points.

There is a different but equally useful class of special value formulae arising from the Rankin–Selberg method and the work of Garrett [Gar]. The prototypical such formula is concerned with a triple (f, g, h) of eigenforms of weights k, ℓ, and m respectively with $k = \ell + m + 2r$ and $r \geq 0$. It involves the Petersson scalar product

$$I(f, g, h) := \langle f, g \times \delta_m^r h \rangle,$$

and relates the square of this quantity to the *central critical value* $L(f \otimes g \otimes h, \frac{k+\ell+m-2}{2})$ of the convolution L-function attached to f, g and h. An over-arching theme of Section 2 is that the quantity $I(f, g, h)$ can be p-adically interpolated as f, g and h are made to vary over a suitable set of classical specialisations of Hida families. In particular, when f, g and h are of weight two, forcing r to tend p-adically to -1 in weight space, the resulting p-adic limit of $I(f, g, h)$, denoted $I_p(f, g, h)$ – a p-adic L-value – acquires an interpretation as the Bloch–Kato p-adic logarithm of a global cohomology class arising from a suitable geometric construction, thereby motivating the study of the following Euler systems:

(1) When g and h are Eisenstein series, the invariant $I_p(f, g, h)$ is related in Section 2.1 to the *p-adic regulator*

$$\mathbf{reg}_p\{u_g, u_h\}(\eta_f),$$

were u_g and u_h are the modular units whose logarithmic derivatives are equal to g and h respectively, $\{u_g, u_h\} \in K_2(X_1(N))$ is the *Beilinson element* in the second K-group of $X_1(N)$ formed essentially by taking the cup-product of these two units, and η_f is a suitable class in $H^1_{\mathrm{dR}}(X_1(N))$ attached to f. The p-adic regulator has a counterpart in p-adic étale cohomology and the images of the Beilinson elements under this map lead to a system of global cohomology classes which underlie Kato's study of the Mazur–Swinnerton-Dyer p-adic L-function attached to classical modular forms via the theory of Euler systems.

(2) When only h is an Eisenstein series and f and g are cuspidal, the invariant $I_p(f, g, h)$ is again related, in Section 2.2, to a p-adic regulator of the form

$$\mathbf{reg}_p(\Delta_{u_h})(\eta_f \wedge \omega_g),$$

where Δ_{u_h} is a *Beilinson–Flach element* in $K_1(X_1(N) \times X_1(N))$ attached to the modular unit u_h viewed as a function on a diagonally embedded copy of $X_1(N) \subset X_1(N)^2$. The Euler system of *Beilinson–Flach elements* is obtained by replacing the p-adic regulator by its p-adic étale counterpart; some of its possible arithmetic applications are discussed in Section 2.2.

(3) When f, g and h are all cusp forms, the invariant $I_p(f, g, h)$ is related in Section 2.3 to

$$\mathrm{AJ}_p(\Delta)(\eta_f \wedge \omega_g \wedge \omega_h),$$

where

$$\mathrm{AJ}_p : \mathrm{CH}^2(X_1(N)^3)_0 \longrightarrow \mathrm{Fil}^2(H^3_{\mathrm{dR}}(X_1(N)^3))^\vee$$

is the p-adic Abel-Jacobi map, and Δ is the *Gross–Kudla–Schoen* cycle obtained by a simple modification of the diagonal cycle in $X_1(N)^3$. The resulting Euler system of *Gross–Kudla–Schoen diagonal cycles* and some of its eventual arithmetic applications are described in Section 2.3.

2.1. Beilinson–Kato elements

Let $f \in S_2(N)$ be a cuspidal eigenform on $\Gamma_0(N)$ (not necessarily new of level N), and let p be an odd prime not dividing N. Assume that p is ordinary for f, relative to a fixed embedding of $\bar{\mathbb{Q}}$ into $\bar{\mathbb{Q}}_p$. Denote by $L_p(f, s)$ the Mazur–Swinnerton-Dyer p-adic L-function attached to f. This p-adic L-function is defined as in [MTT] via the p-adic interpolation of the complex central critical values $L(f, \xi, 1)$, with ξ varying among the Dirichlet characters of p-power conductor.

One goal of this section is to explain the connection between the value of $L_p(f, s)$ at the point $s = 2$, lying outside the range of classical interpolation for $L_p(f, s)$, and the image of so-called Beilinson–Kato elements by the p-adic syntomic regulator on K_2 of the modular curve of level N. The resulting formula is a p-adic analogue of Beilinson's theorem [Bei2] relating $L(f, 2)$ to the complex regulator of the Beilinson–Kato elements considered above.

Write Y for the open modular curve $Y_1(N)$ over $\mathbb{Q}$, and X for its canonical compactification $X_1(N)$; furthermore, denote by $\bar{Y}$ and $\bar{X}$ the extension to $\bar{\mathbb{Q}}$ of Y and X, respectively. Let F be a field of characteristic 0, and let $\mathrm{Eis}_2(\Gamma_1(N), F)$ denote the F-vector space of weight 2 Eisenstein series on $\Gamma_1(N)$ with Fourier coefficients in F. There is a surjective homomorphism

$$\mathcal{O}^\times_{\bar{Y}} \otimes F \xrightarrow{\mathrm{dlog}} \mathrm{Eis}_2(\Gamma_1(N), F), \tag{3.48}$$

sending a modular unit u (or rather, a multiplicative F-linear combination of such) to the Eisenstein series $\frac{1}{2\pi i}\frac{u'(z)}{u(z)}$. Given $u_1, u_2 \in \mathcal{O}^\times_{\bar{Y}}$, write

$$\{u_1, u_2\} \in K_2(\bar{Y}) \otimes \mathbb{Q}$$

for their Steinberg symbol in the second K-group of $\bar{Y}$.

Before stating the main theorem of this section, we recall the definition of the p-adic regulator, following [Bes3] and [BD1], to which we refer for more detailed explanations. This definition builds on the techniques of p-adic integration that played a crucial role in the preceding sections.

As in Section 1.1, let $\mathcal{A}$ be the ordinary locus of Y (viewed here as a rigid analytic curve over $\mathbb{C}_p$), obtained by removing both the supersingular and the cuspidal residue discs. It is equipped with a system of *wide open neighbourhoods* $\mathcal{W}_\epsilon$ in the terminology of Coleman, as described for example in [DR1] and [BD1]. Let Φ be the canonical lift of Frobenius on the collection of $\mathcal{W}_\epsilon$, and let $\Phi \times \Phi$ be the corresponding lift of Frobenius on $\mathcal{W}_\epsilon \times \mathcal{W}_\epsilon$. Choose a polynomial $P(t) \in \mathbb{Q}[t]$ satisfying

- $P(\Phi \times \Phi)$ annihilates the class of $\frac{du_1}{u_1} \otimes \frac{du_2}{u_2}$ in $H^2_{\mathrm{rig}}(\mathcal{W}_\epsilon \times \mathcal{W}_\epsilon)$,
- $P(\Phi)$ acts invertibly on $H^1_{\mathrm{rig}}(\mathcal{W}_\epsilon)$.

Such a P exists because the eigenvalues of Φ on $H^1_{\mathrm{rig}}(\mathcal{W}_\epsilon)$ have complex absolute value $\sqrt{p}$ and p, while $\Phi \times \Phi$ acts on $\frac{du_1}{u_1} \otimes \frac{du_2}{u_2}$ with eigenvalues of modulus p^2. The first condition on P implies the existence of a rigid analytic one-form ρ_P in $\Omega^1(\mathcal{W}_\epsilon^2)$ such that

$$d\rho_P = P(\Phi \times \Phi)\left(\frac{du_1}{u_1} \otimes \frac{du_2}{u_2}\right). \tag{3.49}$$

The form ρ_P is well-defined up to closed rigid one-forms on $\mathcal{W}_\epsilon \times \mathcal{W}_\epsilon$. Fix a base point $x \in \mathcal{W}_\epsilon$, and set

$$\tilde{\xi}_{P,x} := (\delta^* - h_x^* - v_x^*)\rho_P \in \Omega^1(\mathcal{W}_\epsilon),$$

where $\delta(w) := (w, w)$, $h_x(w) := (w, x)$ and $v_x(w) := (x, w)$ are the diagonal, horizontal and vertical inclusions of $\mathcal{W}_\epsilon$ in $\mathcal{W}_\epsilon^2$, respectively. As explained in [BD1], the image $\xi_{P,x}$ of $\tilde{\xi}_{P,x}$ in $H^1_{\mathrm{rig}}(\mathcal{W}_\epsilon)$ does not depend on the choice of a form ρ_P satisfying Equation (3.49). Moreover, setting $\xi_x := P(\Phi)^{-1}\xi_{P,x} \in H^1_{\mathrm{rig}}(\mathcal{W}_\epsilon)$ in view of the second condition on P, one shows that the class ξ_x is independent of the choice of P. Write

$$\mathrm{spl}_X : H^1_{\mathrm{rig}}(\mathcal{W}_\epsilon) \longrightarrow H^1_{\mathrm{dR}}(X)$$

for the canonical Frobenius equivariant splitting of the exact sequence induced by the natural inclusion of $H^1_{\mathrm{dR}}(X)$ into $H^1_{\mathrm{rig}}(\mathcal{W}_\epsilon)$. As is shown in [BD1], the image $\xi := \mathrm{spl}_X(\xi_x)$ of ξ_x under spl_X does not depend on the choice of the base point x. The p-adic regulator of $\{u_1, u_2\}$ is defined as

$$\mathbf{reg}_p\{u_1, u_2\} := \xi \in H^1_{\mathrm{dR}}(X). \tag{3.50}$$

By Poincaré duality, the p-adic regulator $\mathbf{reg}_p\{u_1, u_2\}$ can and will be identified with a linear functional on $H^1_{\mathrm{dR}}(X)$.

We are now ready to state the main theorem of this section. Let χ be a primitive, even Dirichlet character of conductor N. Recall the Eisenstein series $E_{2,\chi}$ appearing in Equation (3.1) of Section 1.1, and let u_χ be a modular unit satisfying

$$\mathrm{dlog}(u_\chi) = E_{2,\chi}. \tag{3.51}$$

Normalise the Petersson scalar product on real analytic modular forms of weight k and character ψ on $\Gamma_0(N)$ by setting

$$\langle f_1, f_2\rangle_{k,N} := \int_{\Gamma_0(N)\backslash\mathcal{H}} y^k \overline{f_1(z)} f_2(z) \frac{dxdy}{y^2}. \tag{3.52}$$

Write $\alpha_p(f)$, resp. $\beta_p(f)$ for the unit, resp. non-unit root of the Frobenius polynomial $x^2 - a_p(f)x + p$ associated to f. Consider the unit root subspace

$$H^1_{\mathrm{dR}}(X)^{f,\mathrm{ur}} \subset H^1_{\mathrm{dR}}(X)^f$$

of the f-isotypic part of $H^1_{\mathrm{dR}}(X)$, on which Frobenius acts as multiplication by $\alpha_p(f)$. We attach to f a canonical element η_f^{ur} of $H^1_{\mathrm{dR}}(X)^{f,\mathrm{ur}}$ in the following way (cf. [BD1], Sections 2.5 and 3.1 for more details). First, we define an anti-holomorphic differential

$$\eta_f^{\mathrm{ah}} := \langle f, f\rangle_{2,N}^{-1} \cdot \bar{f}(z) d\bar{z}. \tag{3.53}$$

The differential η_f^{ah} gives rise to a class in $H^1_{\mathrm{dR}}(X_{\mathbb{C}})$, whose natural image in $H^1(X_{\mathbb{C}}, \mathcal{O}_X)$ is in fact defined over $\bar{\mathbb{Q}}$ via our fixed embedding of $\bar{\mathbb{Q}}$ into $\mathbb{C}$. Using now the embedding of $\bar{\mathbb{Q}}$ into $\mathbb{C}_p$ we obtain a class η_f in $H^1(X, \mathcal{O}_X)$, and a lift η_f^{ur} of η_f to $H^1_{\mathrm{dR}}(X)^{f,\mathrm{ur}}$.

Theorem 3.5. *The equality*

$$L_p(f, 2) \cdot \frac{L(f, \chi, 1)}{\Omega_f^+} = (2iN^{-2}\mathfrak{g}(\chi))(1 - \beta_p(f)p^{-2})(1 - \beta_p(f)) \cdot \mathbf{reg}_p\{u_{\chi^{-1}}, u_\chi\}(\eta_f^{\mathrm{ur}})$$

holds, where Ω_f^+ is a real period attached to f as in Section 2.3 of [BD1], *and $\mathfrak{g}(\chi)$ is the Gauss sum defined in Equation* (3.8).

Remark 3.6. A version of Theorem 3.5 has been obtained by Brunault [Br], as a consequence of Kato's reciprocity law [Kato]. A different proof, proposed by

Bannai–Kings [BK] and Niklas [Nik], relies on the Eisenstein measures introduced by Panchiskine. The approach sketched here is based on the methods of [BD1], depending crucially on Hida's p-adic deformation of f.

The key steps in this approach consist in:

- The p-adic approximation of the value $L_p(f, 2)$ (lying outside the range of classical interpolation of $L_p(f, s)$) by means of values in the range of classical interpolation of the so-called Mazur–Kitagawa p-adic L-function.
- The description of the Mazur–Kitagawa p-adic L-function as a factor of a p-adic Rankin L-series $L_p(\mathbf{f}, \mathbf{E}_\chi)$ associated to the convolution of the Hida families $\mathbf{f}$ and $\mathbf{E}_\chi$ interpolating in weight 2 the ordinary forms f and $E_{2,\chi}$, respectively. (This factorisation follows from a corresponding factorisation of complex special values.)
- The explicit evaluation of $L_p(\mathbf{f}, \mathbf{E}_\chi)$ at the weights $(2, 2)$ (lying outside the range of classical interpolation for this p-adic L-function), which yields an expression directly related to the p-adic regulator $\mathbf{reg}_p\{u_{\chi^{-1}}, u_\chi\}(\eta_f^{\mathrm{ur}})$ described above.

Write $U_\mathbf{f} \subset (\mathbb{Z}/(p-1)\mathbb{Z}) \times \mathbb{Z}_p$ for the weight space attached to $\mathbf{f}$, and denote by $f_k \in S_k(N)$ the classical eigenform whose ordinary p-stabilisation is equal to the weight k specialisation of $\mathbf{f}$, for all k in the space of classical weights $U_{\mathbf{f},\mathrm{cl}} := U_\mathbf{f} \cap \mathbb{Z}^{\geq 2}$. (In particular, $f_2 = f$.) Let $L_p(f_k, \rho, s)$ be the Mazur–Swinnerton-Dyer p-adic L-function [MTT] associated to f_k and to a Dirichlet character ρ (equal to $\mathbf{1}$ or to χ in our study). Thus $L_p(f_k, \rho, s)$ interpolates the special values $L_p(f_k \otimes \xi\rho, j)$, for $1 \leq j \leq k-1$ and ξ in the set of Dirichlet characters of p-power conductor. As k varies, the p-adic L-functions $L_p(f_k, \rho, s)$ can be patched together to yield the Mazur–Kitagawa two-variable p-adic L-function $L_p(\mathbf{f}, \rho)(k, s)$, defined on the domain $U_\mathbf{f} \times \mathbb{Z}_p$. For $k \in U_{\mathbf{f},\mathrm{cl}}$, one has the identity

$$L_p(\mathbf{f}, \rho)(k, s) = \lambda(k) \cdot L_p(f_k, \rho, s),$$

where $\lambda(k)$ is a p-adic period equal to 1 at $k = 2$ and non-vanishing in a neighbourhood of $k = 2$. Note that $L_p(f, 2)$ can be described as the p-adic limit, as $(k, \ell) \in U_{\mathbf{f},\mathrm{cl}} \times [1, k/2]$ tends to $(2, 2)$, of the values $L_p(\mathbf{f}, \mathbf{1})(k, k/2 + \ell - 1)$ occurring in the range of classical interpolation for $L_p(\mathbf{f}, \mathbf{1})$.

Let $\mathbf{E}_\chi$ be the Hida family of Eisenstein series whose weight $\ell \in \mathbb{Z}^{\geq 2}$ specialisation is equal to the ordinary p-stabilisation of the Eisenstein series $E_{\ell,\chi}$. We recall the definition [BD1] of the two-variable Rankin p-adic L-function $L_p(\mathbf{f}, \mathbf{E}_\chi)(k, \ell)$, where (k, ℓ) belongs to $U_\mathbf{f} \times \mathbb{Z}_p$. It is defined via the p-adic interpolation of the critical values

$$L(f_k \otimes E_{\ell,\chi}, k/2+\ell-1), \qquad \ell \in [1, k/2].$$

Set $t := k/2 - \ell$ (so that $t \geq 0$). Let δ_ℓ be the Shimura–Maass differential operator of Equation (3.30), mapping the space $M_\ell^{\mathrm{nh}}(N, \phi)$ of weight ℓ nearly holomorphic modular forms on $\Gamma_0(N)$ and character ϕ to $M_{\ell+2}^{\mathrm{nh}}(N, \phi)$, and let δ_ℓ^t denote its t-fold iterate $\delta_{\ell+2t-2} \circ \ldots \circ \delta_\ell$. Theorem 2 of [Sh1] yields the special value formula

$$L(f_k \otimes E_{\ell,\chi}, k/2+\ell-1) = B_{N,k,\ell,\chi} \cdot \langle f_k(z), (\delta_\ell^t E_{\ell,\chi^{-1}}) \cdot E_{\ell,\chi} \rangle_{k,N}, \quad (3.54)$$

where $B_{N,k,\ell,\chi}$ is an explicit non-zero algebraic constant depending on N, k, ℓ and χ. Set $\Xi_{k,\ell} := (\delta_\ell^t E_{\ell,\chi^{-1}}) \cdot E_{\ell,\chi}$, and denote by $\Xi_{k,\ell}^{\mathrm{hol}}$ its holomorphic projection. Hence

$$\langle f_k(z), \Xi_{k,\ell} \rangle_{k,N} = \langle f_k(z), \Xi_{k,\ell}^{\mathrm{hol}} \rangle_{k,N},$$

and (3.54) implies that the ratio

$$\frac{L(f_k \otimes E_{\ell,\chi}, k/2+\ell-1)}{\langle f_k, f_k \rangle_{k,N}}$$

is algebraic. (More precisely, it belongs to the extension of $\mathbb{Q}$ generated by the Fourier coefficients of f_k and the values of χ.) Viewing $\Xi_{k,\ell}^{\mathrm{hol}}$ as a p-adic modular form, the calculations of Section 3.1 of [BD1] – see in particular equation (46) – show that the ratios

$$\frac{\langle f_k(z), \Xi_{k,\ell}^{\mathrm{hol}} \rangle_{k,N}}{\langle f_k, f_k \rangle_{k,N}},$$

normalised by multiplying them by suitable Euler factors, are interpolated p-adically by a p-adic L-function $L_p(\mathbf{f}, \mathbf{E}_\chi)(k, \ell)$. As a by-product of this construction, one obtains the formula (cf. equation (50) of [BD1])

$$L_p(\mathbf{f}, \mathbf{E}_\chi)(2, 2) = \frac{1}{1 - \beta_p(f)^2 p^{-1}} \langle \eta_f^{\mathrm{ur}}, e_{\mathrm{ord}}(d^{-1} E_{2,\chi^{-1}}^{[p]} \cdot E_{2,\chi}) \rangle_Y, \quad (3.55)$$

where e_{ord} denotes Hida's ordinary projector, $E_{2,\chi^{-1}}^{[p]}$ is the p-depletion of $E_{2,\chi^{-1}}$ defined in equation (3.35), $d = q\frac{d}{dq}$ is the Atkin–Serre derivative operator, and $\langle\ ,\ \rangle_Y$ is the natural Poincaré pairing on Y. The right-hand side of (3.55) is equal to $\mathbf{reg}_p\{u_{\chi^{-1}}, u_\chi\}(\eta_f^{\mathrm{ur}})$ up to an Euler factor (whose precise form is given in equation (60) of [BD1]).

Remark 3.7. Note that the rigid-analytic function $d^{-1} E_{2,\chi^{-1}}^{[p]}$ is the Coleman primitive of $E_{2,\chi^{-1}}^{[p]}$, so that Equation (3.55) expresses values of p-adic L-functions in terms of the theory of p-adic integration. This feature is in common with the formulae presented in the previous sections, with the notable difference that in this case (and in the cases described in the following sections) it is a so-called "iterated Coleman integral" that makes its appearance.

As a final step, we remark that the factorisation of complex L-functions

$$L(f_k \otimes E_{\ell,\chi}, k/2+\ell-1) = L(f_k, k/2+\ell-1) \cdot L(f_k, \chi, k/2)$$

implies directly the factorisation of p-adic L-functions

$$L_p(\mathbf{f}, \mathbf{E}_\chi)(k,\ell) = \eta(k) \cdot L_p(\mathbf{f}, \mathbf{1})(k, k/2+\ell-1) \cdot L_p(\mathbf{f}, \chi)(k, k/2), \quad (3.56)$$

where $\eta(k)$ is a p-adic analytic function whose exact value at 2 is determined in Theorem 3.4 of [BD1]. Theorem 3.5 follows by combining Equations (3.55) and (3.56), given the interpretation of the right-hand side of (3.55) as a p-adic regulator. This concludes our outline of the proof of Theorem 3.5.

We now turn to a brief discussion of the theory of Euler systems, which in the current context aims to relate the values of $L_p(f,s)$ at integer points to a collection of classes in various continuous Galois cohomology groups associated to f. Let F be a p-adic field containing the values of χ, and let $\delta_{\chi^\pm}$ be the image of $u_{\chi^\pm}$ in $H^1_{\mathrm{et}}(Y, F(1))$ arising from Kummer theory. Define the (p-adic) étale regulator

$$\mathbf{reg}_{\mathrm{et}}\{u_{\chi^{-1}}, u_\chi\} := \delta_{\chi^{-1}} \cup \delta_\chi \in H^2_{\mathrm{et}}(Y, F(2)) = H^1(\mathbb{Q}, H^1_{\mathrm{et}}(\bar{Y}, F(2))),$$

where the last identification is a consequence of the Hochschild–Serre spectral sequence. Consider the isomorphism

$$\log_{Y,2} : H^1(\mathbb{Q}_p, H^1_{\mathrm{et}}(\bar{Y}, F(2))) \longrightarrow D_{\mathrm{dR}}(H^1_{\mathrm{et}}(\bar{Y}, F(2))) = H^1_{\mathrm{dR}}(Y/F), \quad (3.57)$$

where the first map is the Bloch–Kato logarithm (which is an isomorphism in our setting), and the second equality follows from the comparison theorem between étale and de Rham cohomology. The map (3.57) sends the restriction at p of the étale regulator, denoted $\mathbf{reg}_{\mathrm{et},p}$, to the p-adic regulator:

$$\log_{Y,2}(\mathbf{reg}_{\mathrm{et},p}\{u_{\chi^{-1}}, u_\chi\}) = \mathbf{reg}_p\{u_{\chi^{-1}}, u_\chi\} \quad (3.58)$$

(cf. Proposition 9.11 and Corollary 9.10 of [Bes2]). Thus Theorem 3.5 can be rephrased as a relation between the value $L_p(f,2)$ and the Bloch–Kato logarithm of the étale regulator:

$$L_p(f,2) \cdot \frac{L(f,\chi,1)}{\Omega_f^+} = (2iN^{-2}\mathfrak{g}(\chi))(1-\beta_p(f)p^{-2})(1-\beta_p(f)) \quad (3.59)$$
$$\times \log_{Y,2}(\mathbf{reg}_{\mathrm{et},p}\{u_{\chi^{-1}}, u_\chi\})(\eta_f^{\mathrm{ur}}).$$

The identity (3.59) should be viewed as the analogue of Equation (3.14) of Section 1.1. More general versions of (3.59) can be obtained by replacing u_χ

by the modular units corresponding to the Eisenstein series $E_2(\chi_1, \chi_2)$ defined in Section 1.1.

Extend F so that it contains the Fourier coefficients of f, and fix a Galois and Hecke equivariant (for the prime-to-N Hecke operators) projection

$$\pi_f : H^1_{\mathrm{et}}(\bar{Y}, F) \longrightarrow V_f,$$

where V_f is the Galois representation attached to f. Let κ_f be the image in $H^1(\mathbb{Q}, V_f(2))$ of $\mathbf{reg}_{\mathrm{et}}\{u_{\chi^{-1}}, u_\chi\}$ by the natural map induced by π_f. Although κ_f depends on the choice of an auxiliary character χ, this dependency may be eliminated by "stripping off" the scalar $\frac{L(f,\chi,1)}{\Omega_f^+} \in F$, which can assumed to be *non-zero* by a judicious choice of χ.

Kato shows that $\kappa_f = \kappa_{f,0}$ is the bottom element of a norm-compatible system of classes $\kappa_{f,n} \in H^1(\mathbb{Q}(\mu_{p^n}), V_f(2))$, constructed from Eisenstein series of level divisible by p^n (cf. [Kato] and [Br]). The formalism outlined in Section 1.1 identifies the system of classes $(\kappa_{f,n})_{n\geq 0}$ with an element $\kappa_{f,\infty}$ of $H^1(\mathbb{Q}, \Lambda_{\mathrm{cyc}} \otimes V_f(2))$. Setting

$$\kappa_{f,\xi}(k) := \nu_{k,\xi}(\kappa_{f,\infty}) \in H^1(\mathbb{Q}, \mathbb{Q}_{p,\xi} \otimes V_f(1+k)(\xi^{-1}))$$

for the specialisation of $\kappa_{f,\infty}$ under $\nu_{k,\xi}$, it turns out that the image of $\kappa_{f,\xi}(1)$ by the Bloch–Kato logarithm encodes the values $L_p(f, \xi, 2)$, thereby generalising (3.59) (cf. [Kato], (16.6)). In this way, $L_p(f, s)$ can be "read-off" from the class $\kappa_{f,\infty}$. The work [BD2] undertakes the task of obtaining such a description of $L_p(f, s)$ by extending the techniques of [BD1] outlined above.

More generally, Perrin–Riou's description of $L_p(f, s)$ as the image of $\kappa_{f,\infty}$ by a "big" log-map (interpolating the Bloch–Kato logarithms) allows one to recover the values $L_p(f, \xi, 1 + k)$ for all $k \geq 1$ from the logarithmic images of the classes $\kappa_{f,\xi}(k)$ (cf. for example [Br], Theorem 23, [PR2], 3.3.10 and [Cz1]).

Consider now the dual exponential map (in the case $k = 0$)

$$\exp^*_{0,\xi} : H^1(\mathbb{Q}_p, \mathbb{Q}_{p,\xi} \otimes V_f(1)(\xi^{-1})) \longrightarrow D_{\mathrm{dR}}(\mathbb{Q}_{p,\xi} \otimes V_f(1)(\xi^{-1})).$$

Kato's explicit reciprocity law describes $L_p(f, \xi, 1)$ in terms of $\exp^*_{0,\xi}(\kappa_{f,\xi}(0))$ (cf. for example [PR2], §2.1). In light of the interpolation formula for $L_p(f, s)$, this gives the chain of equivalences

$$\exp^*_{0,\xi}(\kappa_{f,\xi}(0)) \neq 0 \quad \Leftrightarrow \quad L_p(f, \xi, 1) \neq 0 \quad \Leftrightarrow \quad L(f, \xi, 1) \neq 0. \tag{3.60}$$

When combined with Kolyvagin's theory of Euler systems, (3.60) implies the following case of the Birch and Swinnerton-Dyer conjecture (where ξ can be assumed to be an arbitrary Dirichlet character):

$$L(f,\xi,1) \neq 0 \quad \Rightarrow \quad \mathrm{Hom}(\mathbb{C}(\xi), E(\bar{\mathbb{Q}}) \otimes \mathbb{C}) = 0. \tag{3.61}$$

2.2. Beilinson–Flach elements

Let $f \in S_2(\Gamma_1(N), \chi_f)$ and $g \in S_2(\Gamma_1(N), \chi_g)$ be two cusp forms. In this section we discuss the p-adic Beilinson formula of [BDR1] for the value of the p-adic Rankin L-series attached to $f \otimes g$ at the non-critical value $s = 2$. Roughly speaking, this formula is achieved by applying the formalism of the previous section with the p-adic family of Eisenstein series $\mathbf{E}_\chi$ replaced by the Hida family $\mathbf{g}$ interpolating the cusp form g.

The characters χ_f and χ_g are taken to have modulus N, so $\chi_f(p) = 0$ for $p \mid N$. We assume that the forms f and g are normalised eigenforms of level N (not necessarily new), including for the operators U_p for $p \mid N$. Recall the imprimitive Rankin L-series associated to f and g:

$$\begin{aligned} L^{\mathrm{imp}}(f \otimes g, s) &:= \\ \prod_p \big[(1-\alpha_p(f)\alpha_p(g)p^{-s})&(1-\alpha_p(f)\beta_p(g)p^{-s}) \\ &\times(1-\beta_p(f)\alpha_p(g)p^{-s})(1-\beta_p(f)\beta_p(g)p^{-s})\big]^{-1}, \end{aligned} \tag{3.62}$$

where $\alpha_p(f), \beta_p(f)$ are the roots of the Hecke polynomial $x^2 - a_p(f)x + \chi_f(p)p$ of f if $p \nmid N$, and $(\alpha_p(f), \beta_p(f)) = (a_p(f), 0)$ if $p \mid N$. We adopt similar notations for g. The Rankin series $L^{\mathrm{imp}}(f \otimes g, s)$ is the imprimitive L-function associated to the tensor product $V_f \otimes V_g$ of the motives attached to f and g, and differs from the full $L(f \otimes g, s)$ only possibly in Euler factors at primes $p \mid N$.

Using the Rankin–Selberg method, Shimura gave an explicit formula for $L^{\mathrm{imp}}(f \otimes g, s)$ in terms of the Petersson inner product (3.52). Let $\chi = (\chi_f\chi_g)^{-1}$, and let $E_\chi(z, s-1)$ be the non-holomorphic Eisenstein series of weight zero and character χ:

$$E_\chi(z, s-1) = \sum_{(m,n)\in N\mathbb{Z}\times Z}{}' \chi^{-1}(n)\, \mathrm{Im}(z)^s |mz+n|^{-2s}.$$

Shimura's formula reads (cf. for example equation (4) in [BDR1], with $k = \ell = 2$)

$$L^{\mathrm{imp}}(f \otimes g, s) = \frac{1}{2}\frac{(4\pi)^s}{\Gamma(s)}\langle f^*(z), E_\chi(z, s-1)g(z)\rangle_{2,N}, \tag{3.63}$$

where $f^*(z) = \overline{f(-\bar{z})} \in S_2(N_f, \chi_f^{-1})$ denotes the cusp form obtained by conjugating the Fourier coefficients of f. Since non-holomorphic Eisenstein series satisfy a functional equation relating s and $1-s$, Shimura's formula leads to a functional equation for $L^{\text{imp}}(f \otimes g, s)$ relating the values at s and $3-s$.

Let X denote the modular curve $X_1(N)$. Beilinson gave a geometric interpretation for $L^{\text{imp}}(f \otimes g, s)$ at the near central point $s=2$ in terms of higher Chow groups on the surface $S = X \times X$. The Rankin L-function $L^{\text{imp}}(f \otimes g, s)$ has no critical points, and in particular Beilinson's formula concerns the non-critical point $s=2$. The higher Chow group $\text{CH}^2(S, 1) \cong K_1(S)$ is defined to be the homology (in the middle) of the Gersten complex

$$K_2(K(S)) \xrightarrow{\partial} \oplus_{Z \subset S} K(Z)^* \xrightarrow{\text{div}} \oplus_{P \in S} \mathbb{Z}.$$

Here Z ranges over irreducible curves in S, and P ranges over closed points in S. The map denoted div sends a rational function to its divisor. The map ∂ sends a symbol $\{f, g\} \in K_2(K(S))$ associated to pair of functions $x, y \in K(S)^*$ to the tame symbol

$$\partial(\{x, y\}) = (u_Z)_{Z \subset S}, \qquad u_Z = (-1)^{\nu_Z(x)\nu_Z(y)} \frac{x^{\nu_Z(y)}}{y^{\nu_Z(x)}}. \tag{3.64}$$

Let $u \in K(X)^*$ be a modular unit as in (3.48), i.e. a rational function on X whose divisor is supported on the cusps. By viewing u as a rational function on the diagonal $\Delta \subset S$, one can define certain distiguished elements $\Delta_u \in \text{CH}^2(S, 1)$ as follows.

Lemma 3.8. *Given a modular unit u on X, there exists an element of the form*

$$\Delta_u = (\Delta, u) + \sum a_i (Z_i, u_i) \in \text{CH}^2(S, 1) \otimes \mathbb{Q}, \tag{3.65}$$

where $a_i \in \mathbb{Q}$ and each $Z_i \subset S$ is a horizontal or vertical divisor, i.e. a curve of the form $X \times P$ or $P \times X$ for a point $P \in X$.

Definition 3.9. An element of the form Δ_u as in (3.65) is called a *Beilinson–Flach element* in $\text{CH}^2(S, 1) \otimes \mathbb{Q}$.

We will be interested in the modular unit u_χ such that $\text{dlog}\, u_\chi = E_{2,\chi}$ as in (3.51) and the associated Beilinson–Flach element Δ_{u_χ}.

There is a complex regulator map

$$\mathbf{reg}_{\mathbb{C}} : \text{CH}^2(S, 1) \to (\text{Fil}^1 H^2_{\text{dR}}(S/\mathbb{C}))^\vee$$

defined by

$$\mathbf{reg}_{\mathbb{C}} \left(\sum (Z_i, u_i) \right) (\omega) = \frac{1}{2\pi i} \int_{Z_i'} \omega \log |u_i|,$$

where ω is a smooth (1,1)-form on S and $Z_i' \subset Z_i$ is the locus on which u_i is regular. We may now state Beilinson's formula

Theorem 3.10 (Beilinson). *We have*

$$\frac{L^{\mathrm{imp}}(f \otimes g, 2)}{\langle f^*, f^* \rangle_{2,N}} = C_\chi \mathbf{reg}_{\mathbb{C}}(\Delta_{u_\chi})(\omega_g \otimes \eta_f^{\mathrm{ah}})$$

where $C_\chi = (8i)\pi^3[\Gamma_0(N) : \Gamma_1(N)(\pm)]^{-1}N^{-2}\mathfrak{g}(\chi^{-1})$, $\omega_g = 2\pi i g(z)dz$ *and*

$$\eta_f^{\mathrm{ah}} = \langle f^*, f^* \rangle_{2,N}^{-1} \bar{f}^*(z)d\bar{z}$$

as in (3.53).

The main theorem of [BDR1] is a p-adic analogue of Beilinson's formula. Before stating this result, it is worth noting that the very definition of the p-adic Rankin L-series associated to $f \otimes g$ is subtle for a reason mentioned before: the classical Rankin L-series $L(f \otimes g, s)$ has no critical values. When f and g are ordinary at a prime $p \nmid N$, Hida showed how to define a p-adic Rankin L-series as follows. Let $\mathbf{f}$ and $\mathbf{g}$ denote the Hida families whose weight two specialisations are f and g, respectively. Then for weights k and ℓ such that $2 \leq \ell \leq s \leq k-1$, the values $L^{\mathrm{imp}}(f_k \otimes g_\ell, s)$ are critical. Hida proved that there exists a p-adic L-function $L_p(\mathbf{f}, \mathbf{g})(k, \ell, s)$ interpolating the values

$$\frac{L^{\mathrm{imp}}(f_k \otimes g_\ell, s)}{(2\pi i)^{\ell+2s-1}\langle f_k, f_k \rangle_{k,N}} \in \overline{\mathbb{Q}} \qquad \text{for } 2 \leq \ell \leq s \leq k-1. \qquad (3.66)$$

Note that the roles of $\mathbf{f}$ and $\mathbf{g}$ are not symmetric in this definition, and we therefore obtain two p-adic Rankin L-series associated to f and g, namely:

$$L_p^f(f \otimes g, s) := L_p(\mathbf{f}, \mathbf{g}, 2, 2, s), \qquad L_p^g(f \otimes g, s) := L_p(\mathbf{g}, \mathbf{f}, 2, 2, s).$$

Just as we saw in Section 2.1 regarding the p-adic regulator attached to $K_2(Y_1(N))$, there is a p-adic (or "syntomic") regulator attached to $K_1(S) \cong \mathrm{CH}^2(S, 1)$. This is a map

$$\mathbf{reg}_p : \mathrm{CH}^2(S, 1) \to \mathrm{Fil}^1 H^2_{\mathrm{dR}}(S/\mathbb{Q}_p)^\vee \qquad (3.67)$$

defined by Besser [Bes4] in terms of Coleman's theory of p-adic integration. As for the p-adic regulator discussed in Section 2.1, the map $\mathbf{reg}_p$ of (3.67) satisfies the property that it factors through the étale regulator via the Bloch–Kato logarithm. The main theorem of [BDR1] is as follows.

Theorem 3.11. *We have*

$$L_p^f(f \otimes g, 2) = \frac{\mathcal{E}(f, g, 2)}{\mathcal{E}(f)\mathcal{E}^*(f)} \mathbf{reg}_p(\Delta_{u_\chi})(\omega_g \otimes \eta_f^{\mathrm{ur}}),$$

where

$$\begin{aligned}\mathcal{E}(f,g,2) &= (1-\beta_p(f)\alpha_p(g)p^{-2})(1-\beta_p(f)\beta_p(g)p^{-2})\\ &\quad\times(1-\beta_p(f)\alpha_p(g)\chi(p)p^{-1})(1-\beta_p(f)\beta_p(g)\chi(p)p^{-1})\\ \mathcal{E}(f) &= 1-\beta_p(f)^2\chi_f^{-1}(p)p^{-2}\\ \mathcal{E}^*(f) &= 1-\beta_p(f)^2\chi_f^{-1}(p)p^{-1}\end{aligned}$$

and η_f^{ur} *is as in Section 2.1 (see the discussion following (3.53)).*

As indicated earlier, the proof of Theorem 3.11 follows along that of Theorem 3.5, with the Hida family $\mathbf{E}_\chi$ replaced by $\mathbf{g}$; we therefore indicate now only a few salient aspects. First, one uses Shimura's generalisation of (3.63) for the critical values of $L(f_k \otimes g_\ell, s)$ – essentially (3.54) with $E_{\ell,\chi}$ replaced by g_ℓ – and interprets the right side of this formula in terms of the Poincaré pairing on algebraic de Rham cohomology. This pairing can be realised in p-adic de Rham cohomology, and one obtains that the algebraic numbers (3.66) vary p-adic analytically, up to the multiplication of appropriate Euler factors. This defines a p-adic L-function $L_p(\mathbf{f},\mathbf{g})(k,\ell,s)$, whose value at $k=\ell=s=2$ is given by the formula

$$L_p(\mathbf{f},\mathbf{g})(2,2,2) = \frac{1}{\mathcal{E}^*(f)}\left\langle \eta_f^{\mathrm{ur}}, e_{\mathrm{ord}}(d^{-1}E_{2,\chi}^{[p]}\cdot g)\right\rangle_X. \tag{3.68}$$

Theorem 3.11 is then deduced by using Besser's work [Bes4], which relates the right-hand side of Equation (3.68) to the p-adic regulator.

We conclude this section with a brief discussion of various works in progress regarding the construction of an Euler system of Beilinson–Flach elements along the p-power level tower of self-products of modular curves $X_1(Np^r) \times X_1(Np^r)$, along the lines that were suggested in [DR1] and [BDR1]. This Euler system, which has been developed further by Lei, Loeffler and Zerbes [LLZ] and in [BDR2], holds the promise of several arithmetic applications.

Firstly, such an Euler system would yield a generalisation of Theorem 3.11 that varies in p-adic families. This generalisation would in particular apply when f and g have level divisible by p, and s is a general arithmetic weight (e.g. $s(x)=x^j\psi(x)$ for a p-power conductor Dirichlet character ψ and integer j). Using this generalisation, the fourth author has indicated a proof of a factorisation of Hida's p-adic Rankin L-function $L_p(f\otimes f,s)$ of the Rankin square into the Coates–Schmidt p-adic L-function of the symmetric square of f and a Kubota–Leopoldt p-adic L-function [Das]. The approach taken in [Das] parallels closely the one in [Gr] to factor the restriction to the cyclotomic line of the Katz p-adic L-function into a product of two Kubota–Leopoldt p-adic L-functions, but with the Katz L-function replaced by $L_p(f\otimes f,s)$, elliptic units by Beilinson–Flach elements, and Katz's p-adic Kronecker limit formula by the p-adic Beilinson formula of [BDR1]. As shown by Citro, a suitable twist of this factorisation formula implies Greenberg's trivial zero conjecture for the p-adic L-function of the adjoint of f [Cit].

The construction of an Euler system of Beilinson–Flach elements is also the subject of independent work in progress by Lei, Loeffler, and Zerbes [LLZ] with the goal of studying the Iwasawa theory of modular forms over quadratic imaginary fields.

Finally, we mention the application of an Euler system of Beilinson–Flach elements toward the rank zero BSD conjecture, which serves as the motivation of [BDR1]; details will appear in [BDR2]. Let $E/\mathbb{Q}$ be an elliptic curve, and let $f \in S_2(\Gamma_0(N))$ denote its associated modular form with rational Fourier coefficients. The image of the Beilinson–Flach element $\Delta_{u_\chi} \in \mathrm{CH}^2(S, 1)$ under the étale regulator yields a class in $H^1(\mathbb{Q}, H^2_{\mathrm{et}}(S_{/\bar{\mathbb{Q}}}, \mathbb{Z}_p(2)))$. By projecting onto the (f, g)-isotypic component of $H^2_{\mathrm{et}}(S_{/\bar{\mathbb{Q}}}, \mathbb{Z}_p(2))$, we obtain a class

$$\kappa_E(g) \in H^1(\mathbb{Q}, V_E \otimes V_g(1))$$

where $V_g = H^1_{\mathrm{et}}(X_{/\bar{\mathbb{Q}}}, \mathbb{Z}_p)^g$ denotes the Galois representation coming from the g-isotypic component of the first étale cohomology of the modular curve X and $V_E \cong V_f(1)$ denotes the (rational) p-adic Tate-module of E.

The interpolation of the Beilinson–Flach elements Δ_{u_χ} into an Euler system allows for the construction of a p-adic family of classes

$$\kappa_E(\mathbf{g}) \in H^1(\mathbb{Q}, \mathbf{V}_{E,\mathbf{g}})$$

where $\mathbf{V}_\mathbf{g}$ is Hida's two-dimensional Λ-adic Galois representation associated to the family $\mathbf{g}$, and

$$\mathbf{V}_{E,\mathbf{g}} := V_E \otimes_{\mathbb{Z}_p} \mathbf{V}_\mathbf{g} \otimes_\Lambda \Lambda_{\mathrm{cyc}},$$

where Λ_{cyc} is as in Section 1.1 and the tensor product $\mathbf{V}_\mathbf{g} \otimes_\Lambda \Lambda_{\mathrm{cyc}}$ is taken with respect to a suitable algebra homomorphism $\Lambda \hookrightarrow \Lambda \otimes_{\mathbb{Z}_p} \Lambda$.

An appropriate generalisation of Theorem 3.11 relates the image of $\kappa_E(\mathbf{g})$ under the Bloch–Kato logarithm to the p-adic L-function $L_p(\mathbf{f}, \mathbf{g})(k, \ell, s)$ when $k = 2$.

Let us now suppose that the specialisation of $\mathbf{g}$ in weight 1 is a classical cusp form, and hence its associated Galois representations $V_{\mathbf{g}_1}$ is an odd 2-dimensional Artin representation (i.e. has finite image), which we denote by ρ. The specialisation of the class $\kappa_E(\mathbf{g})$ in weight 1 need not be crystalline at p, since it is defined as the specialisation of a p-adic family of classes at a non-classical weight, and is not directly defined via a geometric construction.

In fact, one shows that $\kappa_E(\mathbf{g})_1$ is crystalline at p if and only if $L(E, \rho, 1) = 0$. Similarly to (3.61), Kolyvagin's theory of Euler systems can then be used to deduce the following case of the BSD conjecture for E (cf. [BDR2]):

$$L(E, \rho, 1) \neq 0 \Longrightarrow \mathrm{Hom}(\rho, E(\overline{\mathbb{Q}}) \otimes \mathbb{C}) = 0.$$

2.3. Gross–Kudla–Schoen cycles

The setting of this section is obtained from that of Section 2.1 and 2.2 by replacing, in the triple of modular forms one starts with, Eisenstein series with cusp forms. Thus, let $f, g, h \in S_2(N)$ be a triple of normalised *cuspidal* eigenforms of weight 2, level N and nebentypus characters χ_f, χ_g, and χ_h, respectively. Assume $\chi_f \cdot \chi_g \cdot \chi_h$ is the trivial character, so that the tensor product $V_{f,g,h} := V_f \otimes V_g \otimes V_h(2)$ of the compatible system of Galois representations associated by Shimura to f, g and h is self-dual and the Garrett–Rankin L-function $L(f, g, h, s)$ of $V_{f,g,h}$ satisfies a functional equation relating the values s and $4 - s$.

Let $\mathbb{Q}_{f,g,h} = \mathbb{Q}(\{a_n(f), a_n(g), a_n(h)\}_{n\geq 1})$ denote the field generated by the Fourier coefficients of f, g and h. For the sake of simplicity in the exposition, let us also assume that N is square-free, the three eigenforms are new in that level, and $a_\ell(f)a_\ell(g)a_\ell(h) = -1$ for all primes $\ell \mid N$. The results described in this section hold (suitably adapted) in much greater generality (cf. [DR1]) – a fact that is important to bear in mind for the arithmetic applications we shall discuss at the end of this section.

Fix a prime $p \nmid N$ and an embedding $\mathbb{Q}_{f,g,h} \hookrightarrow \mathbb{C}_p$ for which the three newforms are ordinary, and let $\mathbf{f} : \Omega_f \longrightarrow \mathbb{C}_p[[q]]$ denote the Hida family of overconvergent p-adic modular forms passing though f. The space Ω_f is a finite rigid analytic covering of the weight space $W = \mathbb{Z}/(p-1)\mathbb{Z} \times \mathbb{Z}_p$.

Write $\Omega_{f,\mathrm{cl}} \subset \Omega_f$ for the subset of points x whose image in $\mathbb{Z}_p^\times$ is an integer $\kappa(x) \geq 2$ and let $f_x \in S_{\kappa(x)}(N)$ denote the eigenform whose p-stabilisation equals the specialisation of $\mathbf{f}$ at x. Adopt similar notations for g and h.

Single out one of the three eigenforms, say f. Building on Hida's definition of the p-adic Rankin L-function discussed in Section 2.2, Harris and Tilouine [HaTi] defined a p-adic L-function of three variables, denoted

$$\mathscr{L}_p{}^f(\mathbf{f}, \mathbf{g}, \mathbf{h}) : \Omega_f \times \Omega_g \times \Omega_h \longrightarrow \mathbb{C}_p. \tag{3.69}$$

See also [DR1, §4] for a description of (3.69) which corrects the Euler factor in the p-adic interpolation formula arising in [HaTi].

This p-adic L-function interpolates the square-root of the central critical value of the complex L-function $L(f_x, g_y, h_z, s)$, as (x, y, z) ranges over those triples $(x, y, z) \in \Omega_{f,\text{cl}} \times \Omega_{g,\text{cl}} \times \Omega_{h,\text{cl}}$ such that $\kappa(x) \geq \kappa(y) + \kappa(z)$. To construct $\mathscr{L}_p{}^f(\mathbf{f}, \mathbf{g}, \mathbf{h})$, one invokes the work of Garrett [Gar], Harris and Kudla [HaKu], which shows that these central critical values are equal, up to certain explicit periods, to the algebraic number

$$J(f_x, g_y, h_z) := \left(\frac{\langle f_x^*, \delta^t(g_y) \times h_z \rangle_N}{\langle f_x^*, f_x^* \rangle_N}\right)^2 \in \bar{\mathbb{Q}},$$

where recall f_x^* is the eigenform obtained from f_x by complex conjugating its Fourier coefficients, $\delta = \delta_{\kappa(y)}$ denotes the Shimura–Maass operator of (3.30) and $t := \frac{\kappa(x)-\kappa(y)-\kappa(z)}{2} \geq 0$ (cf. also [DR1, Thm. 4.4]). After multiplying $J(f_x, g_y, h_z)$ by a suitable Euler factor at p, these quantities vary continuously and interpolate to a function on $\Omega_f \times \Omega_g \times \Omega_h$ denoted $\mathscr{L}_p{}^f(\mathbf{f}, \mathbf{g}, \mathbf{h})$.

Note that the original triple (f, g, h) corresponds to a point (x_0, y_0, z_0) above the triple of weights $(2, 2, 2)$ which lies *outside* the region of interpolation used to define $\mathscr{L}_p{}^f(\mathbf{f}, \mathbf{g}, \mathbf{h})$. By an abuse of notation, let us simply write $\mathscr{L}_p{}^f(\mathbf{f}, \mathbf{g}, \mathbf{h})(2, 2, 2)$ for the value of this function at the point (x_0, y_0, z_0).

Similarly to Sections 2.1 and 2.2, our goal here is to report on a formula which describes $\mathscr{L}_p{}^f(\mathbf{f}, \mathbf{g}, \mathbf{h})(2, 2, 2)$ as the image of the *Gross–Kudla–Schoen diagonal cycle* Δ on the cube of the modular curve $X = X_1(N)/\mathbb{Q}$ under the p-adic syntomic Abel–Jacobi map. As before, the resulting formula is a p-adic analogue of a complex one: in this case Yuan–Zhang–Zhang's theorem [YZZ] relating the first derivative $L'(f, g, h, 2)$ to the Beilinson–Bloch height of Δ. But it can also be viewed, even more suggestively, as the direct generalisation to the setting of diagonal cycles of the p-adic Beilinson formulae of the two previous sections.

The Gross–Kudla–Schoen cycle is essentially the diagonal $X_{123} = \{(x, x, x), x \in X\}$ in X^3, which has to be modified to make it null-homologous. More precisely, fix the cusp $\infty \in X$ at infinity as base point and, following Gross, Kudla and Schoen [GrKu], [GrSc], define Δ to be the class in the Chow group $\mathrm{CH}^2(X^3)$ of codimension 2 cycles in X^3 up to rational equivalence of the formal sum

$$X_{123} - X_{12} - X_{13} - X_{23} + X_1 + X_2 + X_3,$$

where $X_1 = \{(x, \infty, \infty), x \in X\}$, $X_{12} = \{(x, x, \infty), x \in X\}$ and likewise for the remaining summands. One checks that, for any of the standard cohomology theories (e.g. algebraic de Rham, Betti, p-adic syntomic, p-adic étale), the class of Δ in the cohomology group $H^4(X^3)$ vanishes. Thus Δ belongs to the subgroup $\mathrm{CH}^2(X^3)_0$ of null-homologous cycles in $\mathrm{CH}^2(X^3)$, which is the source of the various Abel-Jacobi maps available, e.g.:

$$\begin{array}{rcll} \mathrm{AJ}_{\mathbb{C}}: & \mathrm{CH}^2(X^3)_0 & \longrightarrow & \mathrm{Fil}^2 H^3_{\mathrm{dR}}(X^3_{/\mathbb{C}})^\vee / H_3(X^3(\mathbb{C}), \mathbb{Z}) \\ \mathrm{AJ}_{\mathrm{syn},p}: & \mathrm{CH}^2(X^3)_0 & \longrightarrow & \mathrm{Fil}^2 H^3_{\mathrm{dR}}(X^3_{/\mathbb{Q}_p})^\vee \\ \mathrm{AJ}_{\mathrm{et},p}: & \mathrm{CH}^2(X^3)_0 & \longrightarrow & H^1(G_{\mathbb{Q}}, H^3_{\mathrm{et}}(X^3_{/\bar{\mathbb{Q}}}, \mathbb{Z}_p(2))). \end{array} \tag{3.70}$$

Since it is an essential ingredient of the formula that we aim to describe here, we recall Besser's description [Bes1] of the image of Δ under the p-adic syntomic Abel–Jacobi map. Specialised to our setting, [Bes1, Theorem 1.2] shows that $\mathrm{AJ}_{\mathrm{syn},p}(\Delta)$ is a $\mathbb{Q}_p$-valued functional

$$\mathrm{AJ}_{\mathrm{syn},p}(\Delta) : \mathrm{Fil}^2 H^3_{\mathrm{dR}}(X^3) \longrightarrow \mathbb{Q}_p \tag{3.71}$$

which can be described purely in terms of Coleman integration. In [Bes1], Besser recasts Coleman's integration theory of 1-forms on curves in a cohomological guise, and exploits this interpretation to provide a generalisation of the notion of Coleman's primitive to forms on higher-dimensional varieties $V/\mathbb{Q}_p$ admitting a smooth model $\mathcal{V}$ over $\mathbb{Z}_p$. The spaces in which Besser's primitives live are called *finite polynomial cohomology groups*, denoted $H^i_{\mathrm{fp}}(\mathcal{V}, n)$ for indices $i, n \geq 0$, and his formalism gives rise to a canonical projection $\mathrm{p} : H^i_{\mathrm{fp}}(\mathcal{V}, n) \to \mathrm{Fil}^n H^i_{\mathrm{dR}}(V)$.

In the case of a curve, like our X equipped with its standard integral model $\mathcal{X}/\mathbb{Z}_p$, taking $i = n = 1$ yields an exact sequence $0 \to \mathbb{Q}_p \to H^1_{\mathrm{fp}}(\mathcal{X}, 1) \xrightarrow{\mathrm{p}} \Omega^1(X) \to 0$ where any pre-image $\tilde{\omega} \in H^1_{\mathrm{fp}}(\mathcal{X}, 1)$ of a regular 1-form ω on X may be identified with a choice of a Coleman primitive of ω. That the kernel of p is $\mathbb{Q}_p$ agrees with the well-known fact that such primitives are well-defined only up to constants.

As it will suffice for our purposes, we are content here to describe the restriction of (3.71) to the Künneth component $\mathrm{Fil}^2(H^1_{\mathrm{dR}}(X)^{\otimes 3})$. Up to permutations of the three variables, the typical element in this space is of the type $\eta_1 \otimes \omega_2 \otimes \omega_3$, where η_1 is a class in $H^1_{\mathrm{dR}}(X)$ and $\omega_2, \omega_3 \in \Omega^1(X)$ are regular differential 1-forms. [Bes1, Theorem 1,2] asserts that

$$\begin{aligned} \mathrm{AJ}_p(\Delta)(\eta_1 \otimes \omega_2 \otimes \omega_3) &= \int_\Delta \eta_1 \otimes \omega_2 \otimes \omega_3 \\ &:= \sum_{\emptyset \neq I \subseteq \{1,2,3\}} \mathrm{sign}(I) \mathrm{tr}_I(\iota_I^*(\tilde{\eta}_1 \cup \tilde{\omega}_2 \cup \tilde{\omega}_3)), \end{aligned} \tag{3.72}$$

where $\tilde{\eta}_1 \in H^1_{\mathrm{fp}}(\mathcal{X}, 0)$, $\tilde{\omega}_2, \tilde{\omega}_3 \in H^1_{\mathrm{fp}}(\mathcal{X}, 1)$ are choices of primitives of η_1, ω_2, ω_3, respectively, and $\tilde{\eta}_1 \cup \tilde{\omega}_2 \cup \tilde{\omega}_3 \in H^3_{\mathrm{fp}}(\mathcal{X}^3, 2)$ is their cup-product. Moreover, in the above formula we set $\mathrm{sign}(I) = (-1)^{|I|+1}$, $\iota_I : X \hookrightarrow X^3$ denotes the natural inclusion which maps X onto the curve $X_I \subset X^3$ and $\mathrm{tr}_I : H^3_{\mathrm{fp}}(\mathcal{X}, 2) \xrightarrow{\sim} \mathbb{Q}_p$ is the canonical trace isomorphism of [Bes1, Prop. 2.5 (4)].

While each of the terms $\mathrm{tr}_I(\iota_I^*(\tilde{\eta}_1 \cup \tilde{\omega}_2 \cup \tilde{\omega}_3))$ does depend on the choice of primitives, one checks that their sum does not.

We are finally in position to state the main formula alluded to at the beginning.

Theorem 3.12. *For each $\phi \in \{f, g, h\}$, let $\omega_\phi \in \Omega^1(X)$ denote the regular 1-form associated to ϕ, and $\eta_f^{\mathrm{ur}} \in H^1_{\mathrm{dR}}(X)^{f,\mathrm{ur}}$ be the unique class in the unit root subspace of the f-isotypical component of $H^1_{\mathrm{dR}}(X)$ such that $\langle \omega_f, \eta_f^{\mathrm{ur}} \rangle = 1$. Let also $\alpha_p(\phi)$, $\beta_p(\phi)$ be the two roots of the Hecke polynomial $x^2 - a_p(\phi)x + p$, labelled in such a way that $\alpha_p(\phi)$ is a p-adic unit. Then the equality*

$$\mathscr{L}_p{}^f(\mathbf{f}, \mathbf{g}, \mathbf{h})(2, 2, 2) = \frac{\mathcal{E}(f, g, h)}{\mathcal{E}_0(f)\mathcal{E}_1(f)} \mathrm{AJ}_p(\Delta)(\eta_f^{\mathrm{ur}} \otimes \omega_g \otimes \omega_h) \tag{3.73}$$

holds, where

$$\mathcal{E}(f, g, h) := \left(1 - \beta_p(f)\alpha_p(g)\alpha_p(h)p^{-2}\right) \times \left(1 - \beta_p(f)\alpha_p(g)\beta_p(h)p^{-2}\right) \times \left(1 - \beta_p(f)\beta_p(g)\alpha_p(h)p^{-2}\right) \times \left(1 - \beta_p(f)\beta_p(g)\beta_p(h)p^{-2}\right),$$

$$\mathcal{E}_0(f) := (1 - \beta_p^2(f)\chi_f^{-1}(p)p^{-1}), \qquad \mathcal{E}_1(f) := (1 - \beta_p^2(f)\chi_f^{-1}(p)p^{-2}).$$

This statement holds in greater generality for eigenforms f, g, h of possibly different primitive levels N_f, N_g, N_h, and different weights k, ℓ, m, provided none of the weights is larger than or equal to the sum of the other two. We refer the reader to [DR1] for the precise formulation; here we limit ourselves to provide an overall description of the proof of Theorem 3.12.

To show the identity (3.73), one first shows that

$$\mathrm{AJ}_p(\Delta)(\eta_f^{\mathrm{ur}} \otimes \omega_g \otimes \omega_h) = \langle \eta_f^{\mathrm{ur}}, P(\Phi)^{-1}\xi \rangle \tag{3.74}$$

where, similarly as in the previous sections, Φ is a lift of Frobenius to the system $\{\mathcal{W}_\epsilon\}_{\epsilon>0}$ of wide open neighbourhoods of the ordinary locus of $X(\mathbb{C}_p)$, $P(t) \in \mathbb{C}_p[t]$ is a polynomial satisfying

- $P(\Phi \times \Phi)$ annihilates the class of $\omega_g \otimes \omega_h$ in $H^2_{\mathrm{rig}}(\mathcal{W}_\epsilon^2)$,
- $P(\Phi)$ acts invertibly on $H^1_{\mathrm{dR}}(X)$,
- $P(\Phi)$ annihilates $H^2_{\mathrm{dR}}(X/\mathbb{Q}_p) \simeq \mathbb{Q}_p(-1)$,

and, given a rigid analytic primitive $\rho \in \Omega^1_{\mathrm{rig}}(\mathcal{W}_\epsilon^2)$ of $P(\Phi \times \Phi)(\omega_g \otimes \omega_h)$ and the choice of the cusp ∞ at infinity as base point, we set

$$\xi = (\delta^* - h_\infty^* - v_\infty^*)\rho \in H^1_{\mathrm{dR}}(X) \subset H^1_{\mathrm{rig}}(\mathcal{W}_\epsilon).$$

This is proved in [DR1, Ch. 3] in two steps: a formal calculation permits us first to relate the image under the Abel–Jacobi map of the diagonal cycle

Δ on X^3 to the diagonal D on the square X^2 of the curve; this leads to a simplification, which allows us to apply Besser's machinery [Bes1] to compute $\mathrm{AJ}_p(\Delta)(\eta_f^{\mathrm{ur}} \otimes \omega_g \otimes \omega_h)$ purely in terms of classes of differential 1-forms on X, yielding (3.74).

In view of this formula, it is then not difficult to show that, up to an explicit Euler factor at p, (3.74) is equal to

$$\langle \eta_f^{\mathrm{ur}}, e_{\mathrm{ord}}(d^{-1}(g^{[p]}) \cdot h) \rangle \tag{3.75}$$

where, as in §2.1, e_{ord} denotes Hida's ordinary projector, $g^{[p]} := \sum_{p\nmid n} a_n(g)q^n$ is the p-depletion of g and $d := qd/dq$ is Serre's derivative operator. That the left-hand side of (3.73) equals (3.75), again up to an explicit p-multiplier, follows from the explicit calculations involved in the very construction of the p-adic L-function $\mathscr{L}_p{}^f(\mathbf{f}, \mathbf{g}, \mathbf{h})$.

We bring (3.75) to the reader's attention not only because it stands as the basic bridge between the two sides of the equality in (3.73), but also because this is a quantity which is amenable to effective numerical approximations, thanks to the work of A. Lauder (see [Lau] for more details).

This can be used in turn to design algorithms for computing p-adic numerical approximations to Chow–Heegner and Stark–Heegner points on elliptic curves over various number fields. We refer the reader to [DR2] and the forthcoming works of the third and sixth author, in collaboration with A. Lauder, for some of these constructions. To illustrate the method, we just mention here that the prototypical construction arises when f is taken to be the eigenform associated to an elliptic curve $E/\mathbb{Q}$ and $g = h$. In this setting Zhang [Zh] introduced a rational point $P_{g,f} \in E(\mathbb{Q})$, whose formal group logarithm can be computed as

$$\log_{\omega_f}(P_{g,f}) = -2\frac{\mathcal{E}_1(g)}{\mathcal{E}(g,g,f)}\langle \eta_g^{\mathrm{ur}}, e_{\mathrm{ord}}(d^{-1}(g^{[p]}) \cdot f) \rangle, \tag{3.76}$$

by invoking Theorem 3.12 and the results of [DRS1].

Motivated by the analogies between the description of $P_{g,f}$ given in [DRS1] by means of Chen's complex iterated integrals and the above formula, we refer to (3.75) as a *p-adic iterated integral.*

We close this section by discussing briefly the Euler system underlying the diagonal cycle Δ, and the arithmetic applications that Theorem 3.12 has in this context.

Assume f has rational Fourier coefficients and trivial nebentypus, and let $E/\mathbb{Q}$ be the (isogeny class of the) elliptic curve associated to it by the Eichler–Shimura construction. If instead of applying the p-adic syntomic Abel–Jacobi map, one considers the image of Δ under the p-adic étale Abel–Jacobi map

$\mathrm{AJ}_{\mathrm{et},p}$ recalled in (3.70), one obtains a global cohomology class with values in $H^3_{\mathrm{et}}(X^3_{/\bar{\mathbb{Q}}}, \mathbb{Z}_p(2))$. After projecting to the (f, g, h)-isotypical component of this Galois module, we obtain an element

$$\kappa_E(g, h) \in H^1(\mathbb{Q}, V_p(E) \otimes V_g \otimes V_h(1)), \tag{3.77}$$

where, for any of the forms $\phi = f, g, h$, $V_\phi := H^1_{\mathrm{et}}(X_{/\bar{\mathbb{Q}}}, \mathbb{Z}_p)^\phi$ and $V_p(E) \simeq V_f(1)$ is the Galois representation associated to the p-adic Tate module of E.

Let now Λ_ϕ be the finite extension of the Iwasawa algebra $\Lambda = \mathbb{Z}_p[[\mathbb{Z}_p^\times]]$ corresponding to the space Ω_ϕ as considered at the beginning of the section for $\phi = f$. Hida constructed a rank two Galois representation $\mathbf{V}_\phi$ over Λ_ϕ, interpolating the Galois representations associated by Deligne to each of the classical specialisations of $\underline{\phi}$.

As in §1.1, let Λ_{cyc} denote the Λ-adic Galois representation whose underlying module is Λ itself, equipped with the Galois action induced by the character

$$\mathrm{Gal}\,(\bar{\mathbb{Q}}/\mathbb{Q}) \longrightarrow \mathrm{Gal}\,(\mathbb{Q}(\mu_{p^\infty})/\mathbb{Q}) \xrightarrow{\chi_{\mathrm{cyc}}} \mathbb{Z}_p^\times \hookrightarrow \Lambda^\times,$$

where χ_{cyc} is the cyclotomic character and the latter inclusion maps an element $z \in \mathbb{Z}_p^\times$ to the corresponding group-like element $[z] \in \Lambda^\times$. Define

$$\mathbf{V}_{f,g,h} := V_p(E) \otimes_{\mathbb{Z}_p} (\mathbf{V}_g \otimes_\Lambda \mathbf{V}_h) \otimes_\Lambda \Lambda_{\mathrm{cyc}}$$

where the tensor products over Λ are taken with respect to algebra homomorphisms

$$\Lambda \hookrightarrow \Lambda \otimes \Lambda \subseteq \Lambda_g \otimes \Lambda_h, \quad \Lambda \hookrightarrow \Lambda \otimes \Lambda \subseteq (\Lambda_g \otimes_\Lambda \Lambda_h) \otimes \Lambda_{\mathrm{cyc}}$$

such that for all classical points (x, y, z) of weights $k, \ell, m \geq 2$ with $m = k+\ell-2$, the specialisation of $\mathbf{V}_{f,g,h}$ at (x, y, z) is isomorphic as a $G_{\mathbb{Q}}$-module to $V_{f_x} \otimes V_{g_y} \otimes V_{h_z}(k + \ell - 2)$.

One of the main results of [DR2] is the construction of a cohomology class

$$\kappa_E(\mathbf{g}, \mathbf{h}) \in H^1(\mathbb{Q}, \mathbf{V}_{f,g,h}) \tag{3.78}$$

which satisfies the following interpolation property. Let $(y, z) \in \Omega_g \times_\Omega \Omega_h$ be a pair of classical points over a weight in Ω of the form $z \mapsto \xi(z)z^2$ for some Dirichlet character ξ of p-power conductor. Then the specialisation of $\kappa_E(\mathbf{g}, \mathbf{h})$ at (y, z) satisfies

$$\nu_{y,z}(\kappa_E(\mathbf{g}, \mathbf{h}) = \mathcal{E}_{y,z} \cdot \kappa_E(g_y, h_z).$$

for some explicit Euler factor $\mathcal{E}_{y,z}$.

The most interesting arithmetical applications of the Λ-adic cohomology class (3.78) arise when we deform it to points (y, z) of weight 1 which are *classical*. Indeed, assume that for such a pair, the specialisations g_y and h_z are the q-expansions of classical eigenforms of weight 1; in this case their associated Galois representations are Artin representations, denoted ρ_y and ρ_z respectively. By specialising $\kappa_E(\mathbf{g}, \mathbf{h})$ to this pair, we obtain a cohomology class $\kappa_E(g_y, h_z) \in H^1(\mathbb{Q}, V_p(E) \otimes \rho_y \otimes \rho_z)$. Note that this class needs not be cristalline at p, since it did not arise from a geometric construction, but rather by deforming p-adically a collection of geometric classes. The goal of [DR2] is to show that

$$\kappa_E(g_y, h_z) \text{ is cristalline at } p \text{ if and only if } L(E, \rho_y \otimes \rho_z, 1) = 0. \quad (3.79)$$

The main arithmetical application of (3.79) is the following instance of the Birch and Swinnerton-Dyer conjecture:

Theorem 3.13. *Let E be an elliptic curve over $\mathbb{Q}$ and let*

$$\rho_1, \rho_2 : G_{\mathbb{Q}} \longrightarrow \mathrm{GL}_2(\mathbb{C})$$

be two continuous odd Galois representations, attached to weight one modular forms g and h respectively. Assume $\det(\rho_1) \cdot \det(\rho_2) = 1$ *and there exists $\sigma \in G_{\mathbb{Q}}$ for which $\rho_1 \otimes \rho_2(\sigma)$ has distinct eigenvalues. If $L(E \otimes \rho_1 \otimes \rho_2, 1) \neq 0$, then*

$$\dim_{\mathbb{C}}(\hom(\rho_1 \otimes \rho_2, E(\bar{\mathbb{Q}}) \otimes \mathbb{C})) = 0.$$

Conclusion

The contents of this survey are summarised in the table below, whose rows correspond to the six Euler systems covered in each section. The first and second column list the Euler system and its associated p-adic L-function, while the third gives the p-adic special value formula relating the two. The last column indicates the (eventual) application of each Euler system to the Birch and Swinnerton-Dyer conjecture. In this column, $\mathrm{BSD}_r(E, \rho)$ refers to the implication

$$\mathrm{ord}_{s=1} L(E, \rho, s) = r \quad \Rightarrow \quad \dim_{\mathbb{C}} \hom_{G_{\mathbb{Q}}}(\rho, E(K_\rho) \otimes \mathbb{C}) = r,$$

where E is an elliptic curve over $\mathbb{Q}$,

$$\rho : G_{\mathbb{Q}} \longrightarrow \mathrm{GL}_n(\mathbb{C})$$

is an Artin representation, and $r \geq 0$ is an integer. The letter A alludes to an elliptic curve with complex multiplication, and the letter E to a general (modular) elliptic curve over $\mathbb{Q}$. Likewise, the symbol ρ_ψ refers to the induction from an imaginary quadratic field K to $\mathbb{Q}$ of a *dihedral* character of K, while χ refers to a one-dimensional representation of $G_{\mathbb{Q}}$ (i.e., a Dirichlet character) and ρ, ρ_1 and ρ_2 refer to general odd irreducible two-dimensional Artin representations of $\mathbb{Q}$, with the sole constraint that $\rho_1 \otimes \rho_2$ has trivial determinant.

Euler system	p-adic L-function	p-adic formula	BSD application
Circular units	Kubota–Leopoldt p-adic L-function	Leopoldt's theorem	None
Elliptic units	Katz two-variable p-adic L-function	Katz's p-adic Kronecker limit formula	Coates–Wiles: $\mathrm{BSD}_0(A, 1)$.
Heegner points	Anticyclotomic p-adic L-function of [BDP]	p-adic Gross–Zagier theorem of [BDP]	Gross–Zagier, Kolyvagin: $\mathrm{BSD}_0(E, \rho_\psi)$ and $\mathrm{BSD}_1(E, \rho_\psi)$.
Beilinson–Kato elements	Mazur–Swinnerton-Dyer–Panciskin	p-adic Beilinson formula: [Kato], [Br], [Nik], [BD1]	Kato: $\mathrm{BSD}_0(E, \chi)$.
Beilinson–Flach elements	Hida's p-adic Rankin L-function	p-adic Beilinson formula: [BDR1]	$\mathrm{BSD}_0(E, \rho)$.
Gross–Kudla–Schoen cycles	Hida–Harris–Tilouine triple product p-adic L-function	p-adic Gross–Kudla formula [DR1]	$\mathrm{BSD}_0(E, \rho_1 \otimes \rho_2)$

Before concluding, the following three remarks are in order:

A. Complex formulae

The authors would be remiss if they failed to mention that the p-adic special value formulae described in this survey all have complex counterparts:

1.1. Dirichlet's class number formula relates the complex logarithm of the absolute value of the circular unit u_χ of Section 1.1 to the special value $L(1, \chi)$, or (equivalently, by the functional equation of $L(s, \chi)$) to the first derivative $L'(0, \chi)$ at $s = 0$.

1.2. The Kronecker limit formula relates the complex logarithm of the absolute value of the elliptic unit u_{χ_n} of Section 1.2 to the special value $L(K, \chi_n, 1)$ or (equivalently, by the functional equation) to the first derivative $L'(K, \chi_n, 0)$. The Kronecker limit formula can also be recast as a simple relation between the *square* of this logarithm and the first derivative at $s = 1$ of the Rankin convolution L-series

$$L(E_{2,\chi} \otimes \theta_{\chi_n}, s) = L(\theta_{\chi_n^{-1}}, s-1) L(\theta_{\chi_n}, s), \tag{3.80}$$

where θ_{χ_n} is the weight one theta series attached to the character χ_n. Note that the two factors on the right-hand side, at $s = 1$, are interchanged under the functional equation for $L(\theta_{\chi_n}, s)$.

1.3. The Gross–Zagier formula of [GZ] relates the Néron–Tate canonical height of the Heegner point $P_{K,f}$ of Section 1.3 to the first derivative at $s = 1$ of the Rankin L-series

$$L(f \otimes \theta_K, s). \tag{3.81}$$

This L-series is obtained from the L-series in (3.80) by replacing the Eisenstein series $E_{2,\chi}$ by the weight two cusp form f (and χ by the trivial character).

2.1. Beilinson's formula for $K_2(X_1(N))$ relates the square of the complex regulator of the Beilinson element $\{u_\chi, u_{\chi^{-1}}\}$ of Section 2.1, evaluated at the class η_f, (attached, here as in Section 2.1, to a form f with trivial nebentypus character) to the first derivative at the central value $s = 2$ of the triple convolution L-series

$$L(f \otimes E_{2,\chi} \otimes E_{2,\chi^{-1}}, s) = L(f, s) L(f, s-2) L(f, \chi, s-1) L(f, \bar{\chi}, s-1) \tag{3.82}$$

(up to a simple elementary fudge factor). When $L(f, \chi, 1) \neq 0$, this triple convolution L-series has a *simple zero* at $s = 2$ arising from the known behaviour of $L(f, s)$ at $s = 0$ and $s = 2$.

2.2. Beilinson's formula for $K_1(X_1(N)^2)$, which is stated in Theorem 3.10, can also be viewed as expressing the *square* of the complex regulator of the element Δ_{u_χ} of Section 2.2 in terms of the first derivative at the central value $s = 2$ of the convolution L-series

$$\begin{aligned} L(f \otimes g \otimes E_{2,\chi}, s) &= L(f \otimes g, s) L(f \otimes g, \chi, s-1) \\ &= L(f \otimes g, s) L(\bar{f} \otimes \bar{g}, s-1), \end{aligned} \tag{3.83}$$

(where the last equality follows from the fact that χ is fixed to be the inverse of the product of the nebentypus characters of f and g). Note that the factors on the right of equations (3.82) and (3.83), evaluated at $s = 2$, are interchanged under the functional equation, up to simple constants and Gamma factors, and that the L-series on the left admits a simple zero at $s = 2$.

2.3. The Gross–Kudla–Yuan–Zhang–Zhang formula: As described in [GrKu] and [YZZ], it expresses the Arakelov height of the (f, g, h)-isotypic component of the Gross–Kudla–Schoen diagonal cycle $\Delta \in \mathrm{CH}^2(X_1(N)^3)_0$ of

Section 2.3 in terms of the first derivative at the central value $s = 2$ of the triple convolution L-series

$$L(f \otimes g \otimes h, s). \tag{3.84}$$

B. Other *p*-adic Gross–Zagier formulae

The p-adic formulae of Gross–Zagier type of [BDP] and of [DR1] alluded to in lines 3 and 6 of the table are not the only, or indeed even the most natural, generalisations of the formulae of Gross–Zagier and Gross–Kudla–Zhang to the p-adic setting. Perrin-Riou's p-adic Gross–Zagier formula described in [PR1], which relates the *first derivative* of a suitable p-adic L-function at a point which lies in its range of classical interpolation to the *p-adic height* of a Heegner point, bears a more visible analogy with its original complex counterpart. (The analogue of Perrin-Riou's formula for diagonal cycles has yet to be worked out in the literature, even though it appears to lie within the scope of the powerful techniques developed by Zhang and his school.) In contrast, the p-adic formulae of [BDP] and [DR1] are the direct generalisation of those of Leopoldt and Katz, and are thus better adapted to certain Euler system arguments.

C. Euler systems and central critical zeros of order one

As will be apparent from the discussion in paragraph A above, all of the Euler systems discussed in this survey are governed by the leading terms of certain L-series at their *central points*, and seem to arise when these L-series admit *simple zeros* at the centre, at least generically. The "degenerate instances" described in Sections 1.2, 2.1, and 2.2 correspond to settings where the relevant L-function breaks up into factors that are not central critical but rather are interchanged under the functional equation, as described in (3.80), (3.82), and (3.83). The order of vanishing of an L-series at a non-central point can be read off from the Gamma-factors appearing in its functional equation, and the examples of L-functions admitting simple zeros at such points are essentially exhausted[2] by the examples treated in Sections 1.1, 1.2, 2.1, and 2.2 (along with the simple zeros of Artin L-functions whose leading terms are conjecturally expressed in terms of Stark units). This remark may explain why the Euler systems alluded to in the first, second, fourth and fifth lines of the table, which ultimately rely on properties of modular units, do not generalise readily to other settings (such as totally real base fields), unlike the Euler systems

[2] The authors are thankful to Benedict Gross for pointing this out to them.

of Heegner points and Gross–Kudla–Schoen cycles which are controlled by "genuine" central critical values.

This survey has taken the view that an Euler system is a collection of *global cohomology classes* which can be *related to L-functions* in a precise way and can be made to *vary in (p-adic) families*. The possibility of p-adic variation is an essential feature because it allows the construction of global classes which do not directly arise, in general, from a geometric construction involving étale Abel–Jacobi images of algebraic cycles or étale regulators of elements in K-theory, but rather from *p-adic limits* of such classes. Frequently, the obstruction to such "p-adic limits of geometric classes" being cristalline at p is encoded in a classical critical L-value, thereby tying this L-value to a global object which can be used to bound the associated Bloch–Kato Selmer group.

One aspect of the picture which has been deliberately *downplayed* is the idea that Euler systems should arise from norm-compatible collections of global elements defined over a varying collection of abelian extensions of a fixed ground field. This feature is clearly present in the first five examples considered in this survey, but not in the sixth, where the only variables of "p-adic deformation" are the *weight variables* arising in Hida theory. Over the years, p-adic families of automorphic forms have been studied for a wide variety of reductive groups. This raises the hope that Gross–Kudla–Schoen diagonal cycles will point the way to further fruitful examples of Euler systems, involving for instance (p-adically varying families of) algebraic cycles on Shimura varieties of unitary or orthogonal type. Examples of this kind would comfort the authors in their belief that Euler systems are far more ubiquitous than would appear from the limited panoply of known instances described in this survey.

References

[BD1] M. Bertolini, H. Darmon, Kato's Euler system and rational points on elliptic curves I: A p-adic Beilinson formula, *Israel J. Math.*, to appear.

[BD2] M. Bertolini, H. Darmon, Kato's Euler system and rational points on elliptic curves II: The explicit reciprocity law, in preparation.

[BDP] M. Bertolini, H. Darmon, K. Prasanna, *Generalised Heegner cycles and p-adic Rankin L-series*, Duke Math. J. **162**, (2013) no. 6, 1033–1148.

[BDR1] M. Bertolini, H. Darmon, V. Rotger, Beilinson-Flach elements and Euler systems I: syntomic regulators and p-adic Rankin L-series, submitted for publication.

[BDR2] M. Bertolini, H. Darmon, V. Rotger, Beilinson-Flach elements and Euler systems II: p-adic families and the Birch and Swinnerton-Dyer conjecture, in preparation.

[Bei1] A.A. Beilinson, Higher regulators and values of L-functions, in *Current problems in mathematics* **24**, 181–238, Akad. Nauk SSSR, Vsesoyuz, Inst. Nauchn. Tekhn. Inform., Moscow, 1984.

[Bei2] A.A. Beilinson, Higher regulators of modular curves. *Applications of algebraic K-theory to algebraic geometry and number theory, Part I, II* (Boulder, Colo., 1983), 1-34, Contemp. Math., 55, Amer. Math. Soc., Providence, RI, 1986.

[Bes1] A. Besser, A generalization of Coleman's p-adic integration theory, *Invent. Math.* **142** (2000), 397–434.

[Bes2] A. Besser, Syntomic regulators and p-adic integration I: rigid syntomic regulators, *Israel J. Math.* **120** (2000), 291–334.

[Bes3] A. Besser, Syntomic regulators and p-adic integration, II. K_2 of curves. Proceedings of the Conference on p-adic Aspects of the Theory of Automorphic Representations (Jerusalem, 1998). *Israel J. Math.* **120** (2000), part B, 335–359.

[Bes4] A. Besser, On the syntomic regulator for K_1 of a surface, *Israel J. Math.* **190** (2012) 29–66.

[BK] Bannai, Kenichi; Kings, Guido. p-adic elliptic polylogarithm, p-adic Eisenstein series and Katz measure, *Amer. J. Math.* **132** (2010), no. 6, 1609–1654.

[Br] F. Brunault, Valeur en 2 de fonctions L de formes modulaires de poids 2: théorème de Beilinson explicite, *Bull. Soc. Math. France* **135** (2007), no. 2, 215–246.

[Cas1] F. Castella, Heegner cycles and higher weight specializations of big Heegner points, *Math. Annalen* **356** (2013), 1247–1282.

[Cas2] F. Castella, p-adic L-functions and the p-adic variation of Heegner points, submitted.

[Cit] C. Citro, $\mathscr{L}$-invariants of adjoint square Galois representations coming from modular forms, *Int. Math. Res. Not.* (2008), no.14.

[Col] R.F. Coleman, A p-adic Shimura isomorphism and p-adic periods of modular forms, in *p-adic monodromy and the Birch and Swinnerton-Dyer conjecture* (Boston, MA, 1991), 21–51, Contemp. Math. **165**, Amer. Math. Soc., Providence, RI, 1994.

[CW] J. Coates and A. Wiles. On the conjecture of Birch and Swinnerton-Dyer. *Invent. Math.* **39** (3) (1977) 223–251.

[Cz1] P. Colmez. Fonctions L p-adiques, Séminaire Bourbaki, Vol. 1998/99. *Astérisque* **266** (2000), Exp. No. 851, 3, 21–58.

[Cz2] P. Colmez. La conjecture de Birch et Swinnerton-Dyer p-adique, *Astérisque* **294** (2004), ix, 251–319.

[DR1] H. Darmon, V. Rotger, Diagonal cycles and Euler systems I: A p-adic Gross-Zagier formula, *Ann. Sci. Ec. Norm. Supé.*, to appear.

[DR2] H. Darmon, V. Rotger, Diagonal cycles and Euler systems II: the Birch–Swinnerton-Dyer conjecture for Hasse–Weil–Artin L-series, submitted.

[DRS1] H. Darmon, V. Rotger, I. Sols, Iterated integrals, diagonal cycles, and rational points on elliptic curves, *Publications Mathématiques de Besançon* **2** (2012), 19–46.

[Das] S. Dasgupta, Factorization of p-adic Rankin L-series, in preparation.

[Gar] P. Garrett, Decomposition of Eisenstein series: Rankin triple products, *Ann. Math.* **125** (1987), 209–235.

[Gr] B. Gross. On the factorization of p-adic L-series. *Invent. Math.* **57** (1980), no. 1, 83–95.

[GrKu] B. Gross, S. Kudla, Heights and the central critical values of triple product L-functions, *Compositio Math.* **81** (1992), no. 2, 143–209.

[GrSc] B. Gross, C. Schoen, The modified diagonal cycle on the triple product of a pointed curve, *Ann. Inst. Fourier (Grenoble)* **45** (1995), no. 3, 649–679.

[GZ] B.H. Gross and D.B. Zagier. Heegner points and derivatives of L-series. *Invent. Math.* **84** (1986), no. 2, 225–320.

[HaKu] M. Harris, S. Kudla, The central critical value of a triple product L-function, *Ann. Math.* (2) **133** (1991), 605–672.

[HaTi] M. Harris and J. Tilouine, p-adic measures and square roots of special values of triple product L-functions, *Math. Annalen* **320** (2001), 127–147.

[Hi1] H. Hida, A p-adic measure attached to the zeta functions associated with two elliptic modular forms II, *Ann. Inst. Fourier (Grenoble)* **38** (1988) 1–83.

[Hi2] H. Hida, *Elementary theory of L-functions and Eisenstein series*, London Mathematical Society Student Texts **26**, 1993.

[How] B. Howard, Variation of Heegner points in Hida families, *Invent. Math.* **167** (2007), 91–128.

[Kato] K. Kato, p-adic Hodge theory and values of zeta functions of modular forms, Cohomologies p-adiques et applications arithmétiques. III. *Astérisque* No. 295 (2004), ix, 117–290.

[Katz] N.M. Katz, p-adic interpolation of real analytic Eisenstein series, *Ann. Math.* (2) **104** (1976), no. 3, 459–571.

[Ko] V.A. Kolyvagin, Euler systems, *The Grothendieck Festschrift*, Vol. II, 435–483, Progr. Math. **87**, Birkhäuser Boston, Boston, MA, 1990.

[Lau] A. Lauder, Efficient computation of Rankin p-adic L-functions, to appear in the proceedings of the Heidelberg conference "Computations with Modular Forms 2011".

[LLZ] A. Lei, D. Loeffler, S. L. Zerbes, Euler systems for Rankin-Selberg convolutions of modular forms, to appear in *Ann. Math.*

[MR] B. Mazur, K. Rubin, *Kolyvagin systems*, Mem. Amer. Math. Soc. 168, no. 799, 2004.

[MTT] B. Mazur, J. Tate, J. Teitelbaum, On p-adic analogues of the conjectures of Birch and Swinnerton-Dyer, *Invent. Math.* **84** (1986), no. 1, 1–48.

[Nik] M. Niklas, Rigid syntomic regulators and the p-adic L-function of a modular form, Regensburg PhD Thesis, 2010, available at http://epub.uni-regensburg.de/19847/

[PR1] B. Perrin-Riou, Point de Heegner et dérivées de fonctions L p-adiques, *Invent. Math.* **89** (1987), no. 3, 455–510.

[PR2] B. Perrin-Riou, Fonctions L p-adiques d'une courbe elliptique et points rationnels, *Ann. Inst. Fourier (Grenoble)* **43** (1993), no. 4, 945–995.

[PR3] B. Perrin-Riou, Théorie d'Iwasawa des représentations p-adiques sur un corps local, with an appendix by J.-M. Fontaine, *Invent. Math.* **115** (1994), no. 1, 81–161.

[PR4] B. Perrin-Riou, La fonction L p-adique de Kubota-Leopoldt, *Arithmetic geometry* (Tempe, AZ, 1993), 65–93, Contemp. Math. **174**, Amer. Math. Soc., Providence, RI, 1994.

[Ru] K. Rubin, *Euler systems*. Annals of Mathematics Studies **147**, Hermann Weyl Lectures, The Institute for Advanced Study, Princeton University Press, Princeton, NJ, 2000.

[Se] J.-P. Serre, Formes modulaires et fonctions zêta p-adiques, *Modular functions of one variable*, III (Proc. Internat. Summer School, Univ. Antwerp, 1972), pp. 191–268. Lecture Notes in Math. **350**, Springer, Berlin, 1973.

[Sh1] G. Shimura, The special values of the zeta functions associated with cusp forms, *Commun. Pure and Appl. Math.* **29**, (1976) 783–795.

[Sh2] G. Shimura, On a class of nearly holomorphic automorphic forms, *Ann. Math.* **123** (1986), 347–406.

[YZZ] X. Yuan, S. Zhang, W. Zhang, Triple product L-series and Gross–Schoen cycles I: split case, preprint.

[Zh] S. Zhang, Arithmetic of Shimura curves, *Sci. China Math.* **53** (2010), 573–592.

4

Effective local Langlands correspondence

Colin J. Bushnell

This is an overview of an extended programme of joint work with Guy Henniart, aimed at rendering the local Langlands correspondence effective as a tool for local investigations. It is intended to serve as an introduction to recent results in [13].

Throughout, F is a non-Archimedean local field, $\mathfrak{o}_F$ is the discrete valuation ring in F, $\mathfrak{p}_F$ is the maximal ideal of $\mathfrak{o}_F$ and $\Bbbk_F = \mathfrak{o}_F/\mathfrak{p}_F$ is the residue field. The characteristic of $\Bbbk_F$ is denoted by p, but *we make no assumptions concerning the characteristic of* F. We fix a separable algebraic closure $\bar{F}$ of F, and let $\mathcal{W}_F$ be the Weil group of $\bar{F}/F$.

If $n \geqslant 1$ is an integer, then $\mathcal{G}_n(F)$ will denote the set of equivalence classes of irreducible, smooth representations of $\mathcal{W}_F$ of dimension n. (Here, and throughout, we consider only *complex* representations.) On the other side, $\mathcal{A}_n(F)$ is the set of equivalence classes of irreducible *cuspidal* representations of the locally profinite group $\mathrm{GL}_n(F)$. The Langlands correspondence [18], [21], [27] thus gives a canonical bijection $\mathcal{G}_n(F) \to \mathcal{A}_n(F)$, which we denote $\sigma \mapsto {}^L\sigma$.

The theory of *simple characters*, as developed in [15], provides a complete and explicit classification of the elements of $\mathcal{A}_n(F)$. It is therefore natural to ask how the features of this structure theory are reflected by the representations of $\mathcal{W}_F$. The work summarized here reveals a strong and transparent parallelism.

Let $\mathcal{P}_F$ be the wild inertia, or first ramification, subgroup of $\mathcal{W}_F$ and denote by $\widehat{\mathcal{P}}_F$ the set of equivalence classes of irreducible smooth representations of $\mathcal{P}_F$. The group $\mathcal{W}_F$ acts on $\widehat{\mathcal{P}}_F$ by conjugation. The $\mathcal{W}_F$-isotropy group of

Automorphic Forms and Galois Representations, ed. Fred Diamond, Payman L. Kassaei and Minhyong Kim. Published by Cambridge University Press.

$\alpha \in \widehat{\mathcal{P}}_F$ is of the form $\mathcal{W}_E$, where $E = Z_F(\alpha)/F$ is a finite, tamely ramified extension.

Let $\widehat{\mathcal{W}}_F = \bigcup_{n\geqslant 1} \mathcal{G}_n(F)$ be the set of equivalence classes of irreducible smooth representations of $\mathcal{W}_F$. Fixing a representation $\alpha \in \widehat{\mathcal{P}}_F$ and an integer $m \geqslant 1$, let $\mathcal{G}_m(F;\alpha)$ be the set of elements of $\widehat{\mathcal{W}}_F$ which contain α with multiplicity m. Each element σ of $\mathcal{G}_m(F;\alpha)$ then satisfies

$$\dim\sigma = m\,[E{:}F]\,\dim\alpha, \quad E = Z_F(\alpha).$$

Classical Clifford theory yields an explicit description of the elements of $\mathcal{G}_m(F;\alpha)$ in terms of certain (regular) elements of $\mathcal{G}_1(E_m;\alpha)$, where E_m/E is unramified of degree m: see 8.2 below.

On the other hand, the Ramification Theorem of [5] attaches to the pair (α, m) a distinguished conjugacy class $\Theta = \Phi^m_F(\alpha)$ of simple characters in the group $\mathrm{GL}_n(F)$, where $n = m[E{:}F]\dim\alpha$, as above. If $\mathcal{A}_m(F;\Theta)$ denotes the set of $\pi \in \mathcal{A}_n(F)$ which contain Θ, the Langlands correspondence induces a bijection $\mathcal{G}_m(F;\alpha) \to \mathcal{A}_m(F;\Theta)$. The aim is to describe this map. To do this, we use the classification theory from [15] and the theory of tame lifting from [2] and [5]: these give an explicit description of the elements of $\mathcal{A}_m(F;\Theta)$ in terms of (regular) elements of $\mathcal{A}_1(E_m;\Theta_{E_m})$, where $\Theta_{E_m} = \Phi^1_{E_m}(\alpha)$. Combining these two descriptions with the Langlands correspondence $\mathcal{G}_1(E_m;\alpha) \to \mathcal{A}_1(E_m;\Theta_{E_m})$, we get an explicit bijection $\mathcal{G}_m(F;\alpha) \to \mathcal{A}_m(F;\Theta)$. The main result here shows that this bijection differs from the Langlands correspondence by twisting the elements of $\mathcal{G}_1(E_m;\alpha)$ with a fixed, tamely ramified character of $E^\times$. This twisting character is completely computable in many cases; at worst, the Langlands correspondence is determined up to twisting by an unramified character of order dividing the dimension. Such a discrepancy is susceptible to analysis in terms of a finite number of local constant calculations: we summarize that method at the end of §9.

These notes are primarily aimed at giving an overview of recent work, and so are short on detail. The material is based on the theory of simple characters, as laid down in [15] and further developed in [2], [5]. That occupies many pages, so we are constrained to abbreviate it drastically. In the main development in Part I, we have omitted many of the definitions, and concentrated on formal or structural properties illuminated by the simplest useful examples. This serves the main purpose of exhibiting the underlying simplicity of the Langlands correspondence (Main Theorem, 9.4) but obscures the essentially explicit nature of simple characters and the simple strata used to describe them. As a partial remedy, we have added an appendix (§6) to Chapter I, giving a skeletal account of the central definitions and constructions. This may be omitted completely, but may equally be useful as a road-map for the reader wishing to pursue the topic further.

Acknowledgement. The exposition has been improved at several points following suggestions of an anonymous referee, to whom it is a pleasure to give thanks.

I. Cuspidal representations of GL_n

We review the classification of the irreducible cuspidal representations of $\mathrm{GL}_n(F)$. The account is based firmly on [15] and [2], but we have incorporated some more recent insights, mainly from [13].

1. Intertwining and induction

1.1. Intertwining

For a moment, let G be a group, let K_i be a subgroup of G and ρ_i a representation of K_i, for $i = 1, 2$. An element $g \in G$ *intertwines* ρ_1 *with* ρ_2 if

$$\mathrm{Hom}_{K_1^g \cap K_2}(\rho_1^g, \rho_2) \neq 0.$$

Here, K_1^g means $g^{-1}K_1g$ and ρ_1^g is the representation $x \mapsto \rho_1(gxg^{-1})$, $x \in K_1^g$. Surely this property depends only on the double coset K_1gK_2. In the same vein, we write $I_G(\rho_1)$ for the set of $g \in G$ which intertwine ρ_1 with itself. In particular, $I_G(\rho_1)$ invariably contains K_1.

Suppose next that G is *locally profinite,* and that K is an open subgroup of G. Let ρ be a smooth representation of K on a complex vector space W. The space of functions $f : G \to W$, which satisfy

$$f(kg) = \rho(k)\,f(g), \quad k \in K,\ g \in G,$$

and which are compactly supported modulo K, then carries a natural action of G by right translation. The representation of G so obtained is smooth. It is said to be *compactly induced* from ρ, and is denoted $c\text{-}\mathrm{Ind}_K^G\,\rho$.

We specialize to the case where G is the group of F-points of some connected, reductive algebraic group defined over F: we say that G is a connected reductive F-group. Such a group G carries a locally profinite topology, inherited from F.

Proposition. *Let G be a connected reductive F-group, let K be an open subgroup of G, which is compact modulo the centre of G. Let ρ be an* *irreducible smooth representation of K. If $I_G(\rho) = K$, then the induced representation* c-$\mathrm{Ind}_K^G\,\rho$ *is irreducible and cuspidal.*

The proof in [9], while overtly for $G = \mathrm{GL}_2(F)$, remains valid in the stated degree of generality. It applies, in particular, to the case $G = \mathrm{GL}_n(F)$.

The proposition provides the most effective way we have of exhibiting irreducible cuspidal representations of reductive groups. For a given group G, the aim is always to produce a list $\mathcal{D}$ of inducing data (K, ρ) accounting exactly for the irreducible cuspidal representations of G:

(1) *if* $(K, \rho) \in \mathcal{D}$, then $\pi_\rho :=$ c-$\mathrm{Ind}_K^G\,\rho$ *is irreducible and cuspidal;*
(2) *if π is an irreducible cuspidal representation of G, there exists* $(K, \rho) \in \mathcal{D}$ such that $\pi \cong \pi_\rho$;
(3) *if* $(K_i, \rho_i) \in \mathcal{D}$, $i = 1, 2$, *and* $\pi_{\rho_1} \cong \pi_{\rho_2}$, then (K_1, ρ_1) *is G-conjugate to* (K_2, ρ_2).

If $G = \mathrm{GL}_n(F)$, such a list has been obtained [15]. More generally, let D be a central F-division algebra of dimension d^2, $d \geqslant 1$. If $n = md$, for an integer $m \geqslant 1$, the group $G' = \mathrm{GL}_m(D)$ is an *inner form* of G. A similar list for G' is given in [31], implying a classification scheme which is uniform across all of inner forms of G. This uniformity is compatible with the Jacquet–Langlands correspondence, to the extent known [6], [11], [12], [33].

More widely, the inductive approach gives all the cuspidal representations of a classical group G, provided $p \neq 2$ [34], although the structure is more complicated and the classification property (3) is not yet known. J.-K. Yu has used the same sort of method to produce classes of irreducible cuspidal representations of a general connected, reductive F-group G [36]. Yu's method make no pretence at completeness: for example, if $G = \mathrm{GL}_n(F)$, then it yields all irreducible cuspidal representations of G if and only if p does not divide n. In the general case, J.-L. Kim [25] has shown that Yu's construction yields all desired representations provided p is sufficiently large, in a sense depending on G.

1.2. Example

We give the first standard example of the idea of 1.1. It recurs frequently in later pages, despite being rather atypical.

Let V be an F-vector space of dimension n, $A = \mathrm{End}_F(V)$, $G = \mathrm{Aut}_F(V)$. Let L be an $\mathfrak{o}_F$-lattice in V and set

$$\mathfrak{m} = \mathfrak{m}_F(L) = \mathrm{End}_{\mathfrak{o}_F}(L) \cong \mathrm{M}_n(\mathfrak{o}_F).$$

Thus $\mathfrak{m}$ is a *maximal $\mathfrak{o}_F$-order* in A. The ideal $\mathfrak{p}_\mathfrak{m} = \mathfrak{p}_F\mathfrak{m}$ is the Jacobson radical of $\mathfrak{m}$. Set

$$K = F^\times U_{\mathfrak{m}}, \quad U^1_{\mathfrak{m}} = 1+\mathfrak{p}_{\mathfrak{m}}.$$

In particular, K is an open subgroup of G, compact modulo the centre $F^\times$ of G, and $U^1_{\mathfrak{m}}$ is an open normal subgroup of K.

Let λ be an irreducible representation of K such that $\lambda|_{U^1_{\mathfrak{m}}}$ is trivial. The irreducible representation $\lambda|_{U_{\mathfrak{m}}}$ is therefore inflated from an irreducible representation $\tilde\lambda$ of $U_{\mathfrak{m}}/U^1_{\mathfrak{m}} \cong \mathrm{GL}_n(\Bbbk_F)$. We have

$$\boxed{I_G(\lambda) = K \iff \tilde\lambda \text{ is cuspidal.}}$$

So, if $\tilde\lambda$ is cuspidal, then $c\text{-Ind}_K^G\,\lambda$ is irreducible and cuspidal. The boxed statement is a pleasant exercise, *cf.* [9] 14.3. The following assertion, however, lies rather deeper. A proof may be extracted from [15] (6.2 and 8.3.3).

Proposition. *Let $\tilde\lambda$ be an irreducible representation of $U_{\mathfrak{m}}$, trivial on $U^1_{\mathfrak{m}}$. The following are equivalent:*
(1) *$\tilde\lambda$ is not cuspidal;*
(2) *the representation $c\text{-Ind}_K^G\,\lambda$ has no irreducible cuspidal sub-quotient;*
(3) *the representation λ occurs in no cuspidal representation of G.*

Elaborating the first exercise, one may further deduce:

Corollary. *Let (π, V) be an irreducible cuspidal representation of G, having a non-zero fixed point for the group $U^1_{\mathfrak{m}}$. There is a unique irreducible representation λ of K such that $\lambda|_{U_{\mathfrak{m}}}$ is inflated from an irreducible cuspidal representation of $U_{\mathfrak{m}}/U^1_{\mathfrak{m}}$ and $\pi \cong c\text{-Ind}_K^G\,\lambda$.*

2. Simple characters

As before, let V be an F-vector space of finite dimension n, and set $A = \mathrm{End}_F(V)$, $G = \mathrm{Aut}_F(V)$. Fundamental to the classification theory is the concept of "simple character in G". Simple characters are complex and subtle objects defined explicitly but indirectly. The basic theory is rather technical. It occupies the first three chapters of [15], and is further developed in [2]. We want to concentrate on the implications, for the Langlands correspondence, of certain structural features. We have therefore given the briefest possible account of the background material, appending an overview in §6.

2.1. Hereditary orders

We make much use of a family of special sub-rings of A and a system of subgroups of G derived from them.

An F-lattice chain in V is a non-empty set $\mathcal{L}$ of $\mathfrak{o}_F$-lattices in V which is both linearly ordered under inclusion and stable under scalar multiplication: if $x \in F^\times$ and $L_1, L_2 \in \mathcal{L}$, then $xL_i \in \mathcal{L}$ and either $L_1 \subset L_2$ or $L_2 \subset L_1$.

If $\mathcal{L}$ is an F-lattice chain in V, the orbit space $F^\times\backslash\mathcal{L}$ is finite with at most n elements. We set

$$e = e_F(\mathcal{L}) = \left|F^\times\backslash\mathcal{L}\right|.$$

This integer $e_F(\mathcal{L})$ is called the *F-period* of $\mathcal{L}$.

Let $\mathcal{L}$ be an $\mathfrak{o}_F$-lattice chain in V. We set

$$\mathfrak{a} = \mathfrak{a}_F(\mathcal{L}) = \bigcap_{L\in\mathcal{L}} \mathfrak{m}_F(L), \quad \text{where}$$

$$\mathfrak{m}_F(L) = \{x \in A : xL \subset L\} = \mathrm{End}_{\mathfrak{o}_F}(L).$$

The intersection here is finite, with $e = e_F(\mathcal{L})$ distinct factors. The set $\mathfrak{a}$ is an $\mathfrak{o}_F$-order in A. An $\mathfrak{o}_F$-order obtained this way is called *hereditary.* (For a full discussion of hereditary orders, see [30] or the early pages of [15].) Observe that the maximal order $\mathfrak{m}_F(L)$ is the hereditary order defined by the lattice chain $\{xL : x \in F^\times\}$.

For $L \in \mathcal{L}$, let L' be the largest element of $\mathcal{L}$ such that $L' \subset L$ and $L' \neq L$. The set

$$\mathfrak{p}_\mathfrak{a} = \bigcap_{L\in\mathcal{L}} \mathrm{Hom}_{\mathfrak{o}_F}(L, L')$$

is a two-sided ideal of $\mathfrak{a}$. It is the *Jacobson radical* of $\mathfrak{a}$, $\mathfrak{p}_\mathfrak{a} = \mathrm{rad}\,\mathfrak{a}$. It is, moreover, an *invertible* two-sided ideal of $\mathfrak{a}$, its inverse being

$$\mathfrak{p}_\mathfrak{a}^{-1} = \bigcap_{L\in\mathcal{L}} \mathrm{Hom}_{\mathfrak{o}_F}(L', L).$$

One can recover the lattice chain $\mathcal{L}$ from the hereditary order $\mathfrak{a} = \mathfrak{a}_F(\mathcal{L})$, because $\mathcal{L}$ is exactly the set of all $\mathfrak{a}$-lattices in V. When using this viewpoint, the period $e = e(\mathfrak{a}) = e_F(\mathcal{L})$ appears as a sort of ramification index,

$$\mathfrak{p}_F\mathfrak{a} = \mathfrak{p}_\mathfrak{a}^e.$$

Observe that $\mathfrak{a}$ is a maximal order if and only if $e(\mathfrak{a}) = 1$.

Attached to a hereditary $\mathfrak{o}_F$-order $\mathfrak{a}$ in A, we have the unit group $U_\mathfrak{a} = \mathfrak{a}^\times$. This is a compact open subgroup of G. It is a maximal compact subgroup if and only if $\mathfrak{a}$ is a maximal order. Inside $U_\mathfrak{a}$, we have the "standard filtration subgroups"

$$U_\mathfrak{a}^k = 1 + \mathfrak{p}_\mathfrak{a}^k, \quad k \geqslant 1.$$

These are again open in G and normal in $U_\mathfrak{a}$.

Remark. The groups $U_{\mathfrak{a}}$, attached to the hereditary $\mathfrak{o}_F$-order $\mathfrak{a}$, appear in the theory of the *affine building* of G: see, for example, the exposition in [35]. In this context, the groups $U_{\mathfrak{a}}$ are the *parahoric subgroups* of G. From this point of view, a group $U_{\mathfrak{a}}$ also carries many canonical, non-standard filtrations of interest in representation theory [29], [34].

A further concept is useful. Let $\mathfrak{a} = \mathfrak{a}_F(\mathcal{L})$ be a hereditary $\mathfrak{o}_F$-order in A and E a subfield of A containing F. One says that $\mathfrak{a}$ is E-*pure* if $x^{-1}\mathfrak{a}x = \mathfrak{a}$, for all $x \in E^\times$. More expansively, this means that V is an E-vector space and $\mathcal{L}$ is an $\mathfrak{o}_E$-lattice chain in V. If we let $B = \mathrm{End}_E(V)$ be the centralizer of E in A, then $\mathfrak{b} = \mathfrak{a} \cap B$ is the hereditary $\mathfrak{o}_E$-order in B defined by $\mathcal{L}$: in our earlier notation,

$$\mathfrak{b} = \mathfrak{a} \cap B = \mathfrak{a}_E(\mathcal{L}).$$

Observe also that $\mathfrak{p}_{\mathfrak{a}} \cap B = \mathrm{rad}\,\mathfrak{b}$.

2.2. Simple strata

A *simple stratum in* A is a pair $[\mathfrak{a}, \beta]$ consisting of a hereditary $\mathfrak{o}_F$-order $\mathfrak{a}$ in A and an element β of G satisfying the following conditions:

(1) the algebra $E = F[\beta]$ is a field and $\upsilon_E(\beta) < 0$;
(2) $\mathfrak{a}$ is E-pure;
(3) β is *simple over* F.

The effect of (3) is the following. If we have another pair $[\mathfrak{a}, \beta']$ satisfying (1) and (2), such that $\beta' - \beta \in \mathfrak{a}$, then

$$[F[\beta'] : F] \geqslant [F[\beta] : F].$$

The formal definition, which is equivalent to (3), is recalled in 6.3 below. The point needed here is that a simple stratum $[\mathfrak{a}, \beta]$ gives rise, in a canonical and explicit manner, to an open, therefore compact, subgroup $H^1(\beta, \mathfrak{a})$ of $U^1_{\mathfrak{a}}$. At this stage, we note only that the group $H^1(\beta, \mathfrak{a})$ depends on the *equivalence class* of $[\mathfrak{a}, \beta]$: if we have another simple stratum $[\mathfrak{a}, \beta']$ such that $\beta' \equiv \beta \pmod{\mathfrak{a}}$, then $H^1(\beta', \mathfrak{a}) = H^1(\beta, \mathfrak{a})$.

Remark. The property of "simplicity over F" is necessarily expressed via hereditary orders, but it really depends on β alone. Taking $E = F[\beta]$ as in part (1), let V_i be a finite-dimensional E-vector space and $\mathfrak{a}_i$ an E-pure hereditary $\mathfrak{o}_F$-order in $\mathrm{End}_F(V_i)$, $i = 1, 2$. The pair $[\mathfrak{a}_1, \beta]$ is then a simple stratum if and only if $[\mathfrak{a}_2, \beta]$ is a simple stratum.

2.3. Simple characters

To proceed further, we need to choose a smooth character ψ of F *of level one.* That is, ψ is trivial on $\mathfrak{p}_F$ but not on $\mathfrak{o}_F$. If $[\mathfrak{a}, \beta]$ is a simple stratum in A, the choice of ψ gives rise to a finite set $\mathcal{C}(\mathfrak{a}, \beta, \psi)$ of very particular characters of the group $H^1(\beta, \mathfrak{a})$. These are the *simple characters* attached to $[\mathfrak{a}, \beta]$. Again, the set $\mathcal{C}(\mathfrak{a}, \beta, \psi)$ depends only on the equivalence class of β.

The choice of ψ does not affect the definition of simple characters: changing ψ only affects the way simple characters are labelled by simple strata. Explicitly, if ψ' is some other character of F of level one, there is a unit $u \in U_F$ such that $\psi'(x) = \psi(ux)$, $x \in F$. We then have $\mathcal{C}(\mathfrak{a}, \beta, \psi') = \mathcal{C}(\mathfrak{a}, u\beta, \psi)$. For this reason, when treating the relation between simple strata and simple characters, we tend to regard ψ as fixed and use the simpler notation $\mathcal{C}(\mathfrak{a}, \beta) = \mathcal{C}(\mathfrak{a}, \beta, \psi)$.

3. An example

To illuminate the outline of 2.2, 2.3, we give the simplest useful example.

3.1. Minimal elements

Let E/F be a finite field extension, let $\alpha \in E^\times$ and suppose that $E = F[\alpha]$.

Definition. The element α is *minimal* over F if

(1) the integer $\upsilon = \upsilon_E(\alpha)$ is relatively prime to $e = e(E|F)$;
(2) if ϖ is a prime element of F, *the coset* $\alpha^e\varpi^{-\upsilon}+\mathfrak{p}_E \subset U_E$ generates the residue class field extension $\Bbbk_E/\Bbbk_F$.

Condition (2) is, of course, independent of the choice of ϖ. For us, the key property is:

Proposition. *If* α *is minimal over* F *and* $\upsilon_{F[\alpha]}(\alpha) < 0$, *then* α *is simple over* F.

The proof is to be found in [15] 1.4.15.

3.2. Simple characters for minimal elements

Let $[\mathfrak{a}, \alpha]$ be a simple stratum in which α is minimal over F. It is a consequence of parts (1) and (2) in the definition (2.2) that $\alpha^{-1}\mathfrak{a} = \mathfrak{p}_{\mathfrak{a}}^l$, for an integer $l > 0$. Let $E = F[\alpha]$, let B be the A-centralizer of E and set $\mathfrak{b} = \mathfrak{a} \cap B$. In this situation,

$$H^1(\alpha, \mathfrak{a}) = U_{\mathfrak{b}}^1\, U_{\mathfrak{a}}^{[l/2]+1},$$

where $[x]$ denotes the integer part of a real number x.

A character θ of $H^1(\alpha, \mathfrak{a})$ lies in $\mathcal{C}(\mathfrak{a}, \alpha, \psi)$ if and only if $\theta|_{U^1_{\mathfrak{b}}}$ factors through the determinant map $\det_B : B^\times \to E^\times$ and

$$\theta(1+x) = \psi \circ \mathrm{tr}_A(\alpha x), \quad x \in \mathfrak{p}_{\mathfrak{a}}^{[l/2]+1},$$

where $\mathrm{tr}_A : A \to F$ denotes the matrix trace.

3.3. The general case

In general, an element β of G, which is simple over F, is constructed from a finite sequence of pairs (E_i, α_i). Here, the E_i/F are subfields of A of strictly increasing degree, and α_i is minimal over E_i. The definition of $H^1(\beta, \mathfrak{a})$ then follows this sequence step by step, as does the definition of $\mathcal{C}(\mathfrak{a}, \beta)$: see §6 for an overview of the construction.

4. Classification of cuspidal representations

We review the central classification results from [15].

4.1. Intertwining

Let $[\mathfrak{a}, \beta]$ be a simple stratum in $A = \mathrm{End}_F(V)$. Let $E = F[\beta]$, let B be the A-centralizer of E and set $\mathfrak{b} = \mathfrak{a} \cap B$. Let $\theta \in \mathcal{C}(\mathfrak{a}, \beta)$. We attach to θ the following groups:

$$\begin{aligned} \boldsymbol{J}_\theta &= \text{the } G\text{-normalizer of } \theta, \\ J^0_\theta &= \boldsymbol{J}_\theta \cap U_{\mathfrak{a}}, \\ J^1_\theta &= \boldsymbol{J}_\theta \cap U^1_{\mathfrak{a}}. \end{aligned}$$

Lemma. *Let $\mathcal{K}_{\mathfrak{b}}$ denote the group of $y \in B^\times$ such that $y^{-1}\mathfrak{b}y = \mathfrak{b}$.*

(1) *The group $\boldsymbol{J}_\theta$ is open and compact modulo centre in G.*
(2) *The groups $\boldsymbol{J}_\theta$, J^0_θ, J^1_θ depend only on the equivalence class of $[\mathfrak{a}, \beta]$, and satisfy the following relations,*

$$\begin{aligned} \boldsymbol{J}_\theta &= \mathcal{K}_{\mathfrak{b}}\, J^1_\theta, \\ J^0_\theta &= U_{\mathfrak{b}} J^1_\theta, \\ U^1_{\mathfrak{b}} &= U_{\mathfrak{b}} \cap J^1_\theta. \end{aligned}$$

(3) *The set $I_G(\theta)$, of elements of G which intertwine θ, is given by*

$$I_G(\theta) = J^1_\theta\, B^\times\, J^1_\theta.$$

Remarks. We will prefer to label groups by θ rather than the attached simple stratum since, as we shall see in 5.1, the stratum is not a reliable invariant of the situation. So, from now on, we usually write H^1_θ rather than $H^1(\beta, \mathfrak{a})$, for $\theta \in \mathcal{C}(\mathfrak{a}, \beta)$. We observe that J^0_θ is the unique maximal compact subgroup of $\boldsymbol{J}_\theta$ and J^1_θ is the pro-p radical of J^0_θ.

Example. If α is *minimal over* F (as in 3.2), the group J^1_θ is given by

$$J^1_\theta = J^1(\alpha, \mathfrak{a}) = U^1_{\mathfrak{b}}\, U^{[(l+1)/2]}_{\mathfrak{a}}.$$

4.2. Level zero

To get clean statements, we need a variant of the notion of simple character. A *trivial simple character* in G is the trivial character of $U^1_{\mathfrak{a}}$, for a hereditary $\mathfrak{o}_F$-order $\mathfrak{a}$ in A. We use the notation $1^1_{\mathfrak{a}}$ for such a character. If $\theta = 1^1_{\mathfrak{a}}$, the G-normalizer $\boldsymbol{J}_\theta$ of θ is $\mathcal{K}_{\mathfrak{a}}$ (notation as in 4.1 Lemma), while $J^0_\theta = U_{\mathfrak{a}}$ and $H^1_\theta = J^1_\theta = U^1_{\mathfrak{a}}$. The set $I_G(\theta)$ is G itself. All the assertions of 4.1 Lemma thus remain valid in this case, provided we set $E = F$.

4.3. Extended maximal simple types

Let θ be a simple character in G.

Definition. An *extended maximal simple type over* θ is an irreducible representation Λ *of* $\boldsymbol{J}_\theta$ such that $\Lambda|_{H^1_\theta}$ contains θ and $I_G(\Lambda) = \boldsymbol{J}_\theta$.

Let $\mathcal{T}(\theta)$ be the set of equivalence classes of extended maximal simple types over θ.

This concept is only useful for a particular kind of simple character θ. Suppose first that θ is non-trivial, say $\theta \in \mathcal{C}(\mathfrak{a}, \beta)$, for a simple stratum $[\mathfrak{a}, \beta]$ in A. As usual, let B be the A-centralizer of $E = F[\beta]$ and set $\mathfrak{b} = \mathfrak{a} \cap B$. We say that θ is *m-simple* if the hereditary $\mathfrak{o}_E$-order $\mathfrak{b}$ is *maximal,* or, equivalently, if $e(\mathfrak{a}) = e(E|F)$. If, on the other hand, θ is trivial, $\theta = 1^1_{\mathfrak{a}}$ say, then θ is called m-simple if $\mathfrak{a}$ is maximal. The reason for introducing this concept is:

Lemma. *Let θ be a simple character in G. The set $\mathcal{T}(\theta)$ is non-empty if and only if θ is m-simple.*

We have already remarked this property for trivial simple characters, in 1.2 Corollary. In general, one may equally describe the elements of $\mathcal{T}(\theta)$ explicitly: we shall do this in the more suggestive context of 9.3 below.

We may now summarize the main results of [15] concerning the structure of cuspidal representations.

Classification Theorem. *Let π be an irreducible cuspidal representation of G on a complex vector space $\mathcal{V}$.*

(1) *The representation π contains a simple character θ. Any such character is m-simple, and any two are G-conjugate.*
(2) *The natural representation Λ of $\boldsymbol{J}_\theta$ on $\mathcal{V}^\theta$ is irreducible, lies in $\mathcal{T}(\theta)$ and $\pi \cong c\text{-Ind}_{\boldsymbol{J}_\theta}^G \Lambda$.*
(3) *If θ is an m-simple character in G, the map*

$$\Lambda \longmapsto c\text{-Ind}_{\boldsymbol{J}_\theta}^G \Lambda$$

is a bijection between $\mathcal{T}(\theta)$ and the set of equivalence classes of irreducible cuspidal representations of G containing θ.

This theorem yields the desired explicit description of the irreducible cuspidal representations of G.

5. Endo-equivalence classes and lifting

We need some further properties of simple characters, in order to state a fundamental result concerning change of base field. As in 2.3, we work relative to a fixed choice of a smooth character ψ of F, of level one.

5.1. Intertwining and conjugacy

We consider the way in which simple characters $\theta \in \mathcal{C}(\mathfrak{a}, \beta)$ depend on the two associated parameters $\mathfrak{a}$ and β, within appropriate constraints. We start with a rather weak uniqueness property [15] 3.5.1.

Proposition 4.1. *Let $[\mathfrak{a}, \beta]$, $[\mathfrak{a}', \beta']$ be simple strata in A, and suppose that*

$$\mathcal{C}(\mathfrak{a}, \beta) \cap \mathcal{C}(\mathfrak{a}', \beta') \neq \emptyset$$

(whence, in particular, $H^1(\beta, \mathfrak{a}) = H^1(\beta', \mathfrak{a}')$). We then have:

(1) $\mathfrak{a} = \mathfrak{a}'$,
(2) $\mathcal{C}(\mathfrak{a}, \beta) = \mathcal{C}(\mathfrak{a}, \beta')$,
(3) $e(F[\beta]|F) = e(F[\beta']|F)$ *and* $[F[\beta] : F] = [F[\beta'] : F]$.

The hypotheses of Proposition 1 imply no further relation between the fields $F[\beta]$, $F[\beta']$. Indeed, it is easy to find examples where any two fields of given degree and ramification index can give rise to the same sets of simple characters.

The second result [15] 3.5.11 of the sequence deals with intertwining of simple characters attached to the same hereditary order.

Proposition 4.2. *For $i = 1, 2$, let $[\mathfrak{a}, \beta_i]$ be a simple stratum in $A = \mathrm{End}_F(V)$, let $\theta_i \in \mathcal{C}(\mathfrak{a}, \beta_i)$. Suppose that θ_1 intertwines with θ_2 in $G = \mathrm{Aut}_F(V)$. There exists $x \in U_{\mathfrak{a}}$ such that $\theta_2 = \theta_1^x$. Indeed, $\theta \mapsto \theta^x$ is a bijection $\mathcal{C}(\mathfrak{a}, \beta_1) \to \mathcal{C}(\mathfrak{a}, \beta_2)$.*

5.2. Transfer

In another direction, we may fix the element β and vary the order $\mathfrak{a}$. We start from a finite field extension $E = F[\beta]/F$, generated by an element β, of negative valuation and simple over F, as in 2.2.

We suppose given two finite-dimensional E-vector spaces V_1, V_2 and set $A_i = \mathrm{End}_F(V_i)$. Let $\mathfrak{a}_i$ be an E-pure hereditary order in A_i. Thus $[\mathfrak{a}_i, \beta]$ is a simple stratum in A_i. In these circumstances, there is a canonical bijection

$$\tau^{\beta}_{\mathfrak{a}_1,\mathfrak{a}_2} : \mathcal{C}(\mathfrak{a}_1, \beta) \xrightarrow{\approx} \mathcal{C}(\mathfrak{a}_2, \beta).$$

We refer to $\tau^{\beta}_{\mathfrak{a}_1,\mathfrak{a}_2}$ as the *β-transfer* from $\mathfrak{a}_1$ to $\mathfrak{a}_2$. Transfer is natural relative to the orders $\mathfrak{a}_i$: in the obvious notation,

$$\tau^{\beta}_{\mathfrak{a}_1,\mathfrak{a}_3} = \tau^{\beta}_{\mathfrak{a}_2,\mathfrak{a}_3} \circ \tau^{\beta}_{\mathfrak{a}_1,\mathfrak{a}_2}.$$

It may, however, depend on the choice of β, *cf.* 5.1 Proposition 1.

Example. To indicate how $\tau^{\beta}_{\mathfrak{a}_1,\mathfrak{a}_2}$ is constructed, we return to the example of §3, in which the element β is *minimal over F*. Let $\nu = -\upsilon_E(\beta)$, and let $\mathfrak{a}$ be an E-pure hereditary order in some $A = \mathrm{End}_F(V)$. Let B be the A-centralizer of E and $\mathfrak{b} = \mathfrak{a} \cap B$. We have $\beta^{-1}\mathfrak{a} = \mathfrak{p}_{\mathfrak{a}}^l$, where l is the integer $\nu e_F(\mathfrak{a})/e(E|F)$ $= \nu e_E(\mathfrak{b})$. Given $\theta \in \mathcal{C}(\mathfrak{a}, \beta)$, there is a unique character χ_θ of U^1_E such that

$$\theta|_{U^1_{\mathfrak{b}}} = \chi_\theta \circ \det_B.$$

The character χ_θ determines θ uniquely. Given simple strata $[\mathfrak{a}_i, \beta]$ as above, the map $\tau = \tau^{\beta}_{\mathfrak{a}_1,\mathfrak{a}_2}$ is defined by the relation

$$\chi_{\tau\theta} = \chi_\theta, \quad \theta \in \mathcal{C}(\mathfrak{a}_1, \beta).$$

5.3. Endo-equivalence

We start with a pair of finite-dimensional F-vector spaces V_1, V_2. We are given a simple stratum $[\mathfrak{a}_i, \beta_i]$ in $A_i = \mathrm{End}_F(V_i)$, $i = 1, 2$. A *common realization* of

$[\mathfrak{a}_1, \beta_1]$, $[\mathfrak{a}_2, \beta_2]$ consists of a finite-dimensional F-vector space V, a hereditary $\mathfrak{o}_F$-order $\mathfrak{A}$ in $A = \mathrm{End}_F(V)$ and a pair of F-embeddings $f_i : F[\beta_i] \to A$ such that $\mathfrak{A}$ is $f_i(F[\beta_i])$-pure, $i = 1, 2$. Thus each $[\mathfrak{A}, f_i(\beta_i)]$ is a simple stratum in A.

We remark that, for fixed i, any two such embeddings f_i are $U_{\mathfrak{A}}$-conjugate, so the choice of f_i is irrelevant. We therefore speak of the pair $(V, \mathfrak{A})$ as a common realization of the $[\mathfrak{a}_i, \beta_i]$.

Lemma. *Let $[\mathfrak{a}_i, \beta_i]$ be a simple stratum in $A_i = \mathrm{End}_F(V_i)$, and let $\theta_i \in \mathcal{C}(\mathfrak{a}_i, \beta_i)$, $i = 1, 2$. The following are equivalent.*

(1) *There exists a common realization $(V, \mathfrak{A})$ of the strata $[\mathfrak{a}_i, \beta_i]$ such that the simple characters $\tau_{\mathfrak{a}_i,\mathfrak{A}}^{\beta_i}\theta_i$ intertwine in $\mathrm{Aut}_F(V)$.*
(2) *For any common realization $(V, \mathfrak{A})$ of the strata $[\mathfrak{a}_i, \beta_i]$, the simple characters $\tau_{\mathfrak{a}_i,\mathfrak{A}}^{\beta_i}\theta_i$ intertwine in $\mathrm{Aut}_F(V)$.*

As in 5.1 Proposition 2, the characters $\tau_{\mathfrak{a}_i,\mathfrak{A}}^{\beta_i}\theta_i$ intertwine in $\mathrm{Aut}_F(V)$ if and only if they are $U_{\mathfrak{A}}$-conjugate. For a proof of the lemma, see [2] 8.7.

Continuing in the context of the lemma, we say that θ_1 is *endo-equivalent to* θ_2 if the pair (θ_1, θ_2) satisfies the equivalent conditions (1) and (2). This relation of endo-equivalence is an equivalence relation on the class of all non-trivial simple characters in all groups $\mathrm{Aut}_F(V)$, as V ranges over the class of finite-dimensional F-vector spaces.

A trivial simple character never intertwines with a non-trivial one, so we may extend the notion by deeming that all trivial simple characters belong to one endo-equivalence class. Let $\mathcal{E}(F)$ denote the set of endo-equivalence classes of simple characters over F. We denote by $\mathbf{0}_F \in \mathcal{E}(F)$ the class of trivial ones.

We exhibit some useful consequences of these results.

Proposition.

(1) *A simple character is endo-equivalent to any of its transfers.*
(2) *Two simple characters over F, attached to the same hereditary $\mathfrak{o}_F$-order $\mathfrak{a}$, are endo-equivalent if and only if they are $U_{\mathfrak{a}}$-conjugate.*
(3) *Let $\theta_i \in \mathcal{C}(\mathfrak{a}_i, \beta_i)$, $i = 1, 2$. If θ_1 is endo-equivalent to θ_2, then*

$$[F[\beta_1] : F] = [F[\beta_2] : F].$$

Consequently, if $\Theta \in \mathcal{E}(F)$ is the endo-equivalence class of $\theta \in \mathcal{C}(\mathfrak{a}, \beta)$, the integer

$$\deg \Theta = [F[\beta] : F]$$

depends only on Θ. Conventionally, $\deg \mathbf{0}_F = 1$.

Remark. In the context of part (3) of the proposition, the ramification indices $e(F[\beta_i]|F)$ are equal, as are the inertial degrees $f(F[\beta_i]|F)$. The extensions $F[\beta_i]/F$ need not be isomorphic. However, if T_i/F is the maximal tamely ramified sub-extension of $F[\beta_i]/F$, the fields T_i *are* F-isomorphic [13], 2.4. Indeed, there exists $j \in J_\theta^1$ such that $T_2 = T_1^j$. Any two choices of j induce the same isomorphism $x \mapsto j^{-1}xj$ from T_1 to T_2. Thus θ determines the maximal tamely ramified sub-extension uniquely, *up to distinguished isomorphism.*

5.4. Tame lifting

It is apparent from the definitions that a field isomorphism $F \to F'$ induces a bijection $\mathcal{E}(F) \to \mathcal{E}(F')$. In particular, the group Aut F acts on $\mathcal{E}(F)$. The operation $F \mapsto \mathcal{E}(F)$ has a more interesting property: if K/F is a finite, *tamely ramified* field extension, there is a canonical map [2]

$$\mathfrak{i}_{K/F} : \mathcal{E}(K) \longrightarrow \mathcal{E}(F).$$

The definition is outlined in §6. Here, we list only the main properties.

Proposition.

(1) *The map* $\mathfrak{i}_{K/F}$ *is surjective. If* L/K *is finite and tamely ramified, then*

$$\mathfrak{i}_{L/F} = \mathfrak{i}_{K/F} \circ \mathfrak{i}_{L/K}.$$

(2) *The map* $\mathfrak{i}_{K/F}$ *has finite fibres. If* $\Theta \in \mathcal{E}(F)$, *then*

$$\deg \Theta = \sum_{\Phi} \deg \Phi,$$

where Φ *ranges over the elements of* $\mathcal{E}(K)$ *for which* $\mathfrak{i}_{K/F}\Phi = \Theta$. *Moreover,* $\mathfrak{i}_{K/F}\Phi = \mathbf{0}_F$ *if and only if* $\Phi = \mathbf{0}_K$.

(3) *If* K/F *is Galois and* $\Phi \in \mathcal{E}(K)$, *then*

$$\mathfrak{i}_{K/F}^{-1}\big(\mathfrak{i}_{K/F}\Phi\big) = \{\Phi^\gamma : \gamma \in \mathrm{Gal}(K/F)\}.$$

If $\Theta \in \mathcal{E}(F)$, the K/F*-lifts of* Θ are the elements of the fibre $\mathfrak{i}_{K/F}^{-1}\Theta$. If Θ is the endo-equivalence class of $\theta \in \mathcal{C}(\mathfrak{a}, \beta)$, there is a canonical bijection between the set of K/F-lifts of Θ and the simple components of the semisimple K-algebra $K \otimes_F F[\beta]$.

5.5. Relation with automorphic induction

We move briefly to a different situation. Let K/F be a finite, *cyclic* extension of degree d. Let ρ be an irreducible cuspidal representation of $\mathrm{GL}_m(K)$.

The operation of *automorphic induction* attaches to ρ an irreducible smooth representation $\pi = \mathrm{A}_{K/F}\,\rho$ of the group $\mathrm{GL}_{md}(F)$. This is defined in [22] when F has characteristic zero, and in [23] otherwise. For us, the point is that automorphic induction corresponds, via the Langlands correspondence, to the operation of induction to $\mathcal{W}_F$ of smooth representations of $\mathcal{W}_K$.

For the same ρ, the representation $\pi = \mathrm{A}_{K/F}\,\rho$ is parabolically induced from an irreducible cuspidal representation π_L of the Levi factor L of some (not necessarily proper) parabolic subgroup of $\mathrm{GL}_{md}(F)$. The group L is of the form $G_1 \times G_2 \times \cdots \times G_r$, for a divisor r of d and where $G_i \cong \mathrm{GL}_{md/r}(F)$. Thus $\pi_L \cong \pi_1 \otimes \pi_2 \otimes \cdots \otimes \pi_r$, where π_j is an irreducible cuspidal representation of G_j. We use the notation

$$\pi = \pi_1 \boxplus \pi_2 \boxplus \cdots \boxplus \pi_r.$$

By the Classification Theorem 4.3, the representation ρ contains a unique conjugacy class of m-simple characters in $\mathrm{GL}_m(K)$, the endo-class of which we denote $\vartheta(\rho)$. We similarly define $\vartheta(\pi_j) \in \mathcal{E}(F)$, $1 \leqslant j \leqslant r$.

Automorphic Induction Theorem. *Let K/F be cyclic and tamely ramified of degree d. Let ρ be an irreducible cuspidal representation of $\mathrm{GL}_m(K)$ and write*

$$\pi = \mathrm{A}_{K/F}\,\rho = \pi_1 \boxplus \pi_2 \boxplus \cdots \boxplus \pi_r,$$

where $r \geqslant 1$ and π_j is an irreducible cuspidal representation of $\mathrm{GL}_{md/r}(F)$, $1 \leqslant j \leqslant r$. We have

$$\vartheta(\pi_j) = \mathrm{i}_{K/F}\,\vartheta(\rho), \quad 1 \leqslant j \leqslant r.$$

The proof of this theorem is given in [5] but relies heavily on some special cases in [2]. In both of those papers, we assumed that F had characteristic zero since, at the time they were written, automorphic induction was known only in that case. The existence, and relevant properties, of automorphic induction in positive characteristic are established in [23]. Once that theory became available, so did the positive characteristic case of the theorem: the proof requires no modification. We remark also that there is a related result connecting tame lifting with *base change,* in the sense of [1] and [23], but we will not use that here.

6. Appendix: a skeleton of definitions

We extract from [15] and [2] the basic definitions pertaining to simple characters and strata, and state the structure theorems giving them their explicit form. The section may be omitted at first reading: we shall only refer to it once in the pages to follow. However, we are guided by the desire to use the Langlands correspondence as a computational tool, and the material here is essential to any such project. This skeleton should prove adequate for most purposes, and may also serve as a short introduction for a reader unfamiliar with these matters.

Let V be a finite-dimensional F-vector space, and set $A = \mathrm{End}_F(V)$, $G = \mathrm{Aut}_F(V)$.

6.1. Adjoint and co-restriction

This preliminary material is to be found in [15] 1.3, 1.4.

Let E/F be a subfield of A. Thus V is an E-vector space and $B = \mathrm{End}_E(V)$ is the centralizer of E in A. Let $\mathrm{tr}_A : A \to F$ be the reduced trace. Thus $(x, y) \mapsto \mathrm{tr}_A(xy)$ provides a nondegenerate, symmetric bilinear form $A \times A \to F$. Let C be the orthogonal complement of B with respect to this pairing. In particular, C is a (B, B)-bimodule.

Note. If E/F is separable, then A is the orthogonal sum of B and C. Otherwise, $B \subset C$.

Suppose $E = F[\beta]$, for some $\beta \in E^\times$. For $x \in A$, we define $a_\beta(x) = \beta x - x\beta$. Thus a_β is a (B, B)-homomorphism $A \to C$ with kernel B. It follows that $a_\beta(A) = C$.

In the other direction, *a tame co-restriction on A, relative to E/F*, is a (B, B)-homomorphism $s : A \to B$ with the following property. If $\mathfrak{a}$ is an E-pure, hereditary $\mathfrak{o}_F$-order in A, then $s(\mathfrak{a}) = \mathfrak{a} \cap B$. Such a map s exists, and is unique up to multiplication by a unit of E. In particular, we have an infinite exact sequence

$$\dots \to A \xrightarrow{a_\beta} A \xrightarrow{s} A \xrightarrow{a_\beta} A \xrightarrow{s} \dots$$

Now write $\mathfrak{b} = \mathfrak{a} \cap B$, $\mathfrak{p} = \mathrm{rad}\,\mathfrak{a}$ and $\mathfrak{q} = \mathrm{rad}\,\mathfrak{b}$. A tame co-restriction then has the further property

$$s(\mathfrak{p}^k) = \mathfrak{q}^k, \quad k \in \mathbb{Z}.$$

6.2. Relation with duality

The tame co-restriction appears naturally in the context of duality. Let ψ be a smooth character of F, $\psi \neq 1$, and let ψ_A denote the smooth character $x \mapsto \psi(\mathrm{tr}_A\, x)$ of A. For $a \in A$, let $a\psi_A$ denote the character $x \mapsto \psi_A(xa)$. The map $a \mapsto a\psi_A$ then gives a topological isomorphism of A with its group $\widehat{A}$ of smooth characters.

If E/F is a subfield of A with centralizer B, and if $\xi \neq 1$ is a smooth character of E, we may similarly define a character $\xi_B \in \widehat{B}$. This yields an isomorphism $B \to \widehat{B}$, $b \mapsto b\xi_B$. The obvious restriction map $\widehat{A} \to \widehat{B}$ is surjective. So, if we identify A with $\widehat{A}$ and B with $\widehat{B}$ via choices of characters $\psi \in \widehat{F}$, $\xi \in \widehat{E}$, this restriction corresponds to a surjective map $s_{\psi,\xi} : A \to B$. If we take both ψ and ξ to be of level one, then $s_{\psi,\xi}$ is a tame co-restriction on A, relative to E/F.

6.3. Strata

We need a looser definition of stratum, as in Chapter 1 of [15]. We recall (2.1) that the Jacobson radical of a hereditary order $\mathfrak{a}$ is *invertible,* as two-sided ideal of $\mathfrak{a}$.

A *stratum in A* is a quadruple $[\mathfrak{a}, l, m, b]$ as follows. First, $\mathfrak{a}$ is a hereditary $\mathfrak{o}_F$-order in A; we set $\mathfrak{p} = \operatorname{rad}\mathfrak{a}$. The parameters l, m are integers such that $l > m$. Finally, $b \in \mathfrak{p}^{-l}$. Strata $[\mathfrak{a}, l, m, b_i]$, $i = 1, 2$, are deemed *equivalent* if $b_1 \equiv b_2 \pmod{\mathfrak{p}^{-m}}$. We use the notation

$$[\mathfrak{a}, l, m, b_1] \sim [\mathfrak{a}, l, m, b_2].$$

A stratum $[\mathfrak{a}, l, m, \beta]$ is called *pure* if $F[\beta]$ is a field, $\mathfrak{a}$ is $F[\beta]$-pure, and $\beta\mathfrak{a} = \mathfrak{p}^{-l}$.

Let $[\mathfrak{a}, l, m, \beta]$ be a pure stratum in A, and write $E = F[\beta]$. Let B be the A-centralizer of E, and take $\mathfrak{b}$, $\mathfrak{p}$, $\mathfrak{q}$ as in 6.1. Let k be an integer and define

$$\mathfrak{N}_k = \{x \in \mathfrak{a} : a_\beta(x) \in \mathfrak{p}^k\}.$$

For k sufficiently large, we have $\mathfrak{N}_k \subset \mathfrak{b}+\mathfrak{p}$. Assuming $E \neq F$, we define

$$k_0(\beta, \mathfrak{a}) = \max\{k \in \mathbb{Z} : \mathfrak{N}_k \not\subset \mathfrak{b}+\mathfrak{p}\}.$$

In the case $E = F$, it is convenient to set $k_0(\beta, \mathfrak{a}) = -\infty$. Otherwise, we have $k_0(\beta, \mathfrak{a}) \geqslant -l$.

A *simple stratum in A* is a pure stratum $[\mathfrak{a}, l, m, \beta]$ such that

$$m < -k_0(\beta, \mathfrak{a}).$$

To describe the dependence of $k_0(\beta, \mathfrak{a})$ on $\mathfrak{a}$, we note that the matrix algebra $\operatorname{End}_F(E)$ contains a unique E-pure hereditary $\mathfrak{o}_F$-order $\mathfrak{a}(E)$: this is defined by the lattice chain $\{\mathfrak{p}_E^j : j \in \mathbb{Z}\}$ in E. Let $\mathfrak{p}(E) = \operatorname{rad}\mathfrak{a}(E)$, so that $\beta\mathfrak{a}(E) = \mathfrak{p}(E)^{\upsilon}$, where $\upsilon = \upsilon_E(\beta)$. We set

$$k_F(\beta) = k_0(\beta, \mathfrak{a}(E)).$$

We expand a comment made in 2.2.

Proposition. *Let* $[\mathfrak{a}, l, m, \beta]$ *be a pure stratum in* A*, with* $\beta \notin F$.

(1) *The quantity* $k_0(\beta, \mathfrak{a})$ *is given by*

$$k_0(\beta, \mathfrak{a}) = k_F(\beta)\, e_F(\mathfrak{a})/e(F[\beta]|F).$$

(2) *The element* β *is minimal over* F *if and only if* $k_0(\beta, \mathfrak{a}) = -l$*, that is, if and only if* $k_F(\beta) = \upsilon_{F[\beta]}(\beta)$*. In particular, a pure stratum* $[\mathfrak{a}, l, l-1, \beta]$ *is simple if and only if* β *is minimal over* F.

These assertions are proved in [15] 1.4.13, 1.4.15 respectively.

6.4. Synthesis of simple strata

All simple strata are built from minimal elements in a systematic manner.

Theorem 4.3. *Let* $[\mathfrak{a}, l, m, \gamma]$ *be a simple stratum in* A*. Let* B *be the* A*-centralizer of* γ*, let* $\mathfrak{b} = \mathfrak{a} \cap B$*, and let* $s_\gamma : A \to B$ *be a tame co-restriction on* A *relative to* $F[\gamma]/F$*. Let* $[\mathfrak{b}, m, m-1, \alpha]$ *be a simple stratum in* B.

(1) *There is a simple stratum* $[\mathfrak{a}, l, m-1, \beta]$ *in* A *such that*

$$\begin{aligned}[\mathfrak{a}, l, m, \beta] &\sim [\mathfrak{a}, l, m, \gamma] \quad \text{and} \\ [\mathfrak{b}, m, m-1, s_\gamma(\beta-\gamma)] &\sim [\mathfrak{b}, m, m-1, \alpha].\end{aligned}$$

(2) *For any such* β*, we have*

$$\begin{aligned}e(F[\beta]\big|F) &= e(F[\gamma]\big|F)\, e(F[\gamma, \alpha]\big|F[\gamma]),\\ f(F[\beta]\big|F) &= f(F[\gamma]\big|F)\, f(F[\gamma, \alpha]\big|F[\gamma]).\end{aligned}$$

(3) *Moreover,*

$$k_0(\beta, \mathfrak{a}) = \begin{cases} -m & \text{if } \alpha \notin F[\gamma],\\ k_0(\gamma, \mathfrak{a}) & \text{otherwise.}\end{cases}$$

Remark. In the context of Theorem 1, the field $F[\gamma]$ need not be F-isomorphic to a subfield of $F[\beta]$.

All simple strata arise from the construction in Theorem 1. Indeed, let $[\mathfrak{a}, l, m, \beta]$ be a simple stratum in A and set $r = -k_0(\beta, \mathfrak{a})$. We assume $\beta \notin F$, so r is an integer satisfying $m < r \leqslant l$. There is nothing to do if $r = l$, since β is then minimal over F. We therefore assume the contrary.

Theorem 4.4. *There exists a simple stratum* $[\mathfrak{a}, l, r, \gamma]$ *in* A *such that*

$$[\mathfrak{a}, l, r, \gamma] \sim [\mathfrak{a}, l, r, \beta].$$

Moreover, if B is the A-centralizer of γ, if $\mathfrak{b} = \mathfrak{a} \cap B$, and if $s_\gamma : A \to B$ is a tame co-restriction on A relative to $F[\gamma]/F$, then $[\mathfrak{b}, r, r-1, s_\gamma(\beta-\gamma)]$ is equivalent to a simple stratum in B.

This sort of technique also allows one to compare simple strata, step by step.

Proposition. *Let $[\mathfrak{a}, l, m, \beta]$ be a simple stratum in A, let B be the A-centralizer of β and let $\mathfrak{b} = \mathfrak{a} \cap b$. Let $[\mathfrak{a}, l, m, \beta']$ be a simple stratum in A, equivalent to $[\mathfrak{a}, l, m, \beta]$. We then have*

(1) $k_0(\beta', \mathfrak{a}) = k_0(\beta, \mathfrak{a})$;
(2) *if s_β is a tame co-restriction on A relative to $F[\beta]/F$, the stratum $[\mathfrak{b}, m, m-1, s_\beta(\beta'-\beta)]$ is equivalent to either $[\mathfrak{b}, m, m-1, 0]$ or a simple stratum $[\mathfrak{b}, m, m-1, \alpha]$, where $\alpha \in F[\beta]^\times$;*
(3) *the first alternative in (2) holds if and only if $[\mathfrak{a}, l, m-1, \beta']$ is equivalent to a G-conjugate of $[\mathfrak{a}, l, m-1, \beta]$.*

6.5. Groups and characters

We start with a simple stratum $[\mathfrak{a}, l, 0, \beta]$ in A, and attach to it a pair $H^1(\beta, \mathfrak{a}) \subset J^1(\beta, \mathfrak{a})$ of open subgroups of $U^1_\mathfrak{a}$. Set $r = -k_0(\beta, \mathfrak{a})$. Thus r is an integer such that $0 < r \leqslant l$, or else $r = \infty$ (corresponding to the case $\beta \in F^\times$). In the case $r \geqslant l$ (so that β is minimal over F), we use the definition from §3:

$$H^1(\beta, \mathfrak{a}) = U^1_\mathfrak{b}\, U^{[l/2]+1}_\mathfrak{a}, \quad J^1(\beta, \mathfrak{a}) = U^1_\mathfrak{b}\, U^{[(l+1)/2]}_\mathfrak{a},$$

where $\mathfrak{b}$ is the $\mathfrak{a}$-centralizer of β.

We therefore assume $0 < r < l$. We choose a simple stratum $[\mathfrak{a}, l, r, \gamma]$ equivalent to $[\mathfrak{a}, l, r, \beta]$. Let B denote the A-centralizer of β and set $\mathfrak{b} = \mathfrak{a} \cap B$. Inductively, the group $H^1(\gamma, \mathfrak{a})$ has been defined. We put $H^k(\beta, \mathfrak{a}) = H^1(\beta, \mathfrak{a}) \cap U^k_\mathfrak{a}$, $k \geqslant 1$, and similarly for J^k. We set

$$H^1(\beta, \mathfrak{a}) = U^1_\mathfrak{b}\, H^{[r/2]+1}(\gamma, \mathfrak{a}), \quad J^1(\beta, \mathfrak{a}) = U^1_\mathfrak{b}\, J^{[(r+1)/2]}(\gamma, \mathfrak{a}).$$

These groups depend only on the equivalence class of the stratum $[\mathfrak{a}, l, 0, \beta]$.

Next, we choose a smooth character ψ of F of level one. For $a \in A$, we denote by ψ_a the function $x \mapsto \psi(\mathrm{tr}_A(a(x-1)))$ on A. We define a set $\mathcal{C}(\mathfrak{a}, \beta, \psi)$ of characters of $H^1(\beta, \mathfrak{a})$, following the preceding construction. Suppose first that β is minimal over F. As in §3, a character θ lies in $\mathcal{C}(\mathfrak{a}, \beta, \psi)$ if and only if $\theta|_{U^1_\mathfrak{b}}$ factors through $\det_B$ and

$$\theta(y) = \psi_\beta(y), \quad y \in U^{[l/2]+1}_\mathfrak{a}.$$

Otherwise, we take r and γ as before. A character θ of $H^1(\beta, \mathfrak{a})$ lies in $\mathcal{C}(\mathfrak{a}, \beta, \psi)$ if and only if $\theta|_{U^1_{\mathfrak{b}}}$ factors through $\det_B$ and there exists $\phi \in \mathcal{C}(\mathfrak{a}, \gamma, \psi)$ such that

$$\theta(y) = \phi(y)\, \psi_{\beta-\gamma}(y), \quad y \in H^{[r/2]+1}(\gamma, \mathfrak{a}).$$

6.6. Tame lifting

We outline a construction from [2]. For this, we need a simple stratum $[\mathfrak{a}, l, 0, \beta]$ in A, and a subfield K/F of A, commuting with β and such that the algebra $K[\beta]$ is a field.

Proposition 4.5. *Let C denote the A-centralizer of K and $\mathfrak{c} = \mathfrak{a} \cap C$.*

(1) *The quadruple $[\mathfrak{c}, l, 0, \beta]$ is a simple stratum in C.*
(2) *The group $H^1(\beta, \mathfrak{a}) \cap C$ is equal to $H^1(\beta, \mathfrak{c})$.*
(3) *Let $\psi_K = \psi \circ \mathrm{Tr}_{K/F}$. If $\theta \in \mathcal{C}(\mathfrak{a}, \beta, \psi)$, then the restriction*

$$\theta_K := \theta|_{H^1(\beta, \mathfrak{c})}$$

lies in $\mathcal{C}(\mathfrak{c}, \beta, \psi_K)$.

In the situation of the proposition, the field $K[\beta]$ is K-isomorphic to exactly one simple component of the semisimple K-algebra $K \otimes_F F[\beta]$.

Proposition 4.6. *Let K/F be a finite, tamely ramified field extension. Let V be a finite-dimensional K-vector space and set $C = \mathrm{End}_K(V)$, $A = \mathrm{End}_F(V)$. Let $[\mathfrak{c}, l, 0, \beta]$ be a simple stratum in C and let $\mathfrak{a}$ be the unique K-pure hereditary $\mathfrak{o}_F$-order in A such that $\mathfrak{a} \cap C = \mathfrak{c}$. Let $\psi_K = \psi \circ \mathrm{Tr}_{K/F}$.*

There exists a simple stratum $[\mathfrak{c}, l, 0, \beta']$ in C such that

(1) *$\mathcal{C}(\mathfrak{c}, \beta', \psi_K) = \mathcal{C}(\mathfrak{c}, \beta, \psi_K)$, and*
(2) *the quadruple $[\mathfrak{a}, l, 0, \beta']$ is a simple stratum in A.*

Let $\phi \in \mathcal{C}(\mathfrak{c}, \beta, \psi_K)$. For any such β', there exists a unique $\theta \in \mathcal{C}(\mathfrak{a}, \beta', \psi)$ such that $\theta_K = \phi$.

In the context of Proposition 2, the endo-equivalence class $\Theta \in \mathcal{E}(F)$ of θ depends only on the endo-equivalence class $\Phi \in \mathcal{E}(K)$ of ϕ. The process $\Phi \mapsto \Theta$ gives a well-defined map $\mathcal{E}(K) \to \mathcal{E}(F)$ which is independent of the initial choice of ψ. This map is the one denoted $\mathfrak{i}_{K/F}$ in 5.4.

II. Representations of the Weil group

Let $\bar F/F$ be a separable algebraic closure of F, and let $\mathcal{W}_F$ be the Weil group of $\bar F/F$. If E/F is a finite separable field extension with $E \subset \bar F$, we identify the Weil group $\mathcal{W}_E$ of $\bar F/E$ with the subgroup of $\mathcal{W}_F$ which fixes E under the natural action of $\mathcal{W}_F$ on $\bar F$.

Let $\mathcal{P}_F$ denote the wild inertia, or first ramification, subgroup of $\mathcal{W}_F$. Thus $\mathcal{P}_F$ is a closed, normal subgroup of $\mathcal{W}_F$. It is a pro-p group, and may be identified with the Galois group of $\bar F/F^{\mathrm{tr}}$, where F^{tr}/F is the maximal tamely ramified extension of F inside $\bar F$. In particular, if $K \subset \bar F$ and K/F is finite and tamely ramified, then $\mathcal{P}_K = \mathcal{P}_F$.

7. Application of Clifford theory

7.1. Representations

Let $\mathcal{G}_n(F)$ be the set of equivalence classes of irreducible, smooth representations of $\mathcal{W}_F$ of dimension n and set

$$\widehat{\mathcal{W}}_F = \bigcup_{n\geqslant 1} \mathcal{G}_n(F).$$

Analogously, let $\widehat{\mathcal{P}}_F$ be the set of equivalence classes of irreducible smooth representations of $\mathcal{P}_F$. We use elementary Clifford theory to describe $\widehat{\mathcal{W}}_F$ in terms of $\widehat{\mathcal{P}}_F$.

If $\sigma \in \widehat{\mathcal{W}}_F$, the restriction $\sigma|_{\mathcal{P}_F}$, of σ to $\mathcal{P}_F$, is semisimple. It is a direct sum of various $\alpha \in \widehat{\mathcal{P}}_F$, any two of which are $\mathcal{W}_F$-conjugate and occur with the same multiplicity. We enshrine this in the canonical map

$$r_F^1 : \widehat{\mathcal{W}}_F \longrightarrow \mathcal{W}_F\backslash\widehat{\mathcal{P}}_F$$

which sends $\sigma \in \widehat{\mathcal{W}}_F$ to the $\mathcal{W}_F$-orbit of an irreducible component of $\sigma|_{\mathcal{P}_F}$. For an integer $s \geqslant 1$ and $\alpha \in \widehat{\mathcal{P}}_F$, we accordingly define

$$\mathcal{G}_s(F;\alpha) = \{\sigma \in \widehat{\mathcal{W}}_F : \dim \mathrm{Hom}_{\mathcal{P}_F}(\alpha,\sigma) = s\}.$$

For example, let $\mathbf{1}_F$ be the trivial character of $\mathcal{P}_F$. The elements of $\mathcal{G}_s(F;\mathbf{1}_F)$ are then the irreducible, s-dimensional, *tamely ramified* smooth representations of $\mathcal{W}_F$.

Proposition. *Let* $\alpha \in \widehat{\mathcal{P}}_F$.

(1) *The* $\mathcal{W}_F$*-isotropy group of* α *is of the form* $\mathcal{W}_E$, *where* $E = Z_F(\alpha)/F$ *is finite and tamely ramified.*

(2) *There exists* $\rho \in \widehat{\mathcal{W}}_E$ *such that* $\rho|_{\mathcal{P}_F} \cong \alpha$. *If* ρ' *is any other such representation, there is a unique tamely ramified character* ψ *of* $\mathcal{W}_E$ *such that* $\rho' \cong \rho \otimes \psi$.

(3) *Taking* ρ *as in* (2), *let* $\tau \in \mathcal{G}_s(E; \mathbf{1}_F)$. *The representation*

$$\Sigma_\rho(\tau) = \mathrm{Ind}_{E/F}\, \rho \otimes \tau$$

is irreducible, and lies in $\mathcal{G}_s(F; \alpha)$. *The map*

$$\Sigma_\rho : \mathcal{G}_s(E; \mathbf{1}_F) \longrightarrow \mathcal{G}_s(F; \alpha)$$

is a bijection.

All assertions here are straightforward, but a complete proof may be found in §1 of [13].

III. Connections

For an integer $n \geqslant 1$, let $\mathcal{A}_n(F)$ denote the set of equivalence classes of irreducible, smooth, *cuspidal* representations of $\mathrm{GL}_n(F)$. It will also be convenient to have the notation

$$\widehat{\mathrm{GL}}_F = \bigcup_{n \geqslant 1} \mathcal{A}_n(F).$$

Thus the Langlands correspondence $\sigma \mapsto {}^L\sigma$ is a bijection $\widehat{\mathcal{W}}_F \to \widehat{\mathrm{GL}}_F$.

If $\pi \in \mathcal{A}_n(F)$, then π contains a unique G-conjugacy class of simple characters in $G = \mathrm{GL}_n(F)$ (4.3). These simple characters all lie in the same endo-equivalence class, which we have denoted $\vartheta(\pi)$. Thus we have a canonical surjective map

$$\vartheta : \widehat{\mathrm{GL}}_F \longrightarrow \mathcal{E}(F).$$

8. Some basic relations

8.1. Ramification theorem

If K/F is a finite, tamely ramified extension, then $\mathcal{P}_K = \mathcal{P}_F$, and there is a canonical surjection $\mathcal{W}_K \backslash \widehat{\mathcal{P}}_F \to \mathcal{W}_F \backslash \widehat{\mathcal{P}}_F$. The first step in our description of the Langlands correspondence is:

Ramification Theorem.

(1) *There is a unique map* $\Phi_F : \mathcal{W}_F \backslash \widehat{\mathcal{P}}_F \to \mathcal{E}(F)$ *such that*

$$\begin{array}{ccc} \widehat{\mathcal{W}}_F & \xrightarrow{L} & \widehat{\mathrm{GL}}_F \\ {\scriptstyle r_F^1}\downarrow & & \downarrow{\scriptstyle \vartheta} \\ \mathcal{W}_F\backslash\widehat{\mathcal{P}}_F & \xrightarrow[\Phi_F]{} & \mathcal{E}(F) \end{array}$$

commutes. The map Φ_F *is bijective.*

(2) *If* K/F *is a finite, tamely ramified field extension, then*

$$\begin{array}{ccc} \mathcal{W}_K\backslash\widehat{\mathcal{P}}_F & \xrightarrow{\Phi_K} & \mathcal{E}(K) \\ \downarrow & & \downarrow{\scriptstyle \mathrm{i}_{K/F}} \\ \mathcal{W}_F\backslash\widehat{\mathcal{P}}_F & \xrightarrow[\Phi_F]{} & \mathcal{E}(F) \end{array}$$

commutes.

Part (1) is proved in [5] §8, under the restriction that F has characteristic zero. As remarked in 5.5, it holds equally, with the same proof, in positive characteristic. Part (2) is 6.2 of [13] (and follows easily from the Automorphic Induction Theorem of 5.5).

Let $\Theta \in \mathcal{E}(F)$ and let $s \geqslant 1$ be a positive integer. We define $\mathcal{A}_s(F;\Theta)$ to be the set of $\pi \in \mathcal{A}_{s\deg\Theta}(F)$ such that $\vartheta(\pi) = \Theta$. As in [13], we have the following corollary.

Tame Parameter Theorem. *Let* $\alpha \in \widehat{\mathcal{P}}_F$ *and set* $E = Z_F(\alpha)$, $\Theta = \Phi_F(\alpha)$.

(1) *We have*

$$\deg\Theta = [E{:}F]\dim\alpha.$$

(2) *If* Θ *is the endo-equivalence class of* $\theta \in \mathcal{C}(\mathfrak{a},\beta)$, *for a simple stratum* $[\mathfrak{a},\beta]$ *in some matrix algebra, then* E *is* F*-isomorphic to the maximal tamely ramified sub-extension* T/F *of* $F[\beta]/F$.

(3) *The Langlands correspondence induces a bijection*

$$\mathcal{G}_s(F;\alpha) \longrightarrow \mathcal{A}_s(F;\Theta),$$

for all $s \geqslant 1$.

We emphasize that, in part (2), there is no distinguished F-isomorphism of E with T.

Remarks. In the case $\dim\alpha = 1$, part (1) of the Ramification Theorem follows directly from local class field theory: if $\boldsymbol{a}_F : \mathcal{W}_F \to F^\times$ is the Artin Reciprocity map, then $\boldsymbol{a}_F(\mathcal{P}_F) = U_F^1$. In the case $\dim\alpha = p$, one may deduce something of the nature of $\Phi_F(\alpha)$ from Mœglin's treatment [28] of the Langlands correspondence in dimension p, $p \geqslant 5$. For detailed treatment of the case $p = 2$, see [26] or [9], for $p = 3$ see [19]. Otherwise, we have virtually no systematic information concerning the map Φ_F.

8.2. Tamely ramified representations

Let $\mathbf{1}_F$ be the trivial character of $\mathcal{P}_F$, let $n \geqslant 1$, and consider the set $\mathcal{G}_n(F; \mathbf{1}_F)$ of classes of irreducible *tamely ramified* representations of $\mathcal{W}_F$, of dimension n.

Let F_n/F be unramified of degree n, $\Delta = \mathrm{Gal}(F_n/F)$, $X_1(F_n) =$ the group of tamely ramified characters of $F_n^\times$. The group Δ acts on $X_1(F_n)$. We say that $\chi \in X_1(F_n)$ is *Δ-regular* if the characters χ^δ, $\delta \in \Delta$, are distinct. We denote by $X_1(F_n)^{\Delta\text{-reg}}$ the set of Δ-regular elements of $X_1(F_n)$. The map

$$\begin{aligned} \Delta\backslash X_1(F_n)^{\Delta\text{-reg}} &\longrightarrow \mathcal{G}_n(F; \mathbf{1}_F), \\ \chi &\longmapsto \sigma_\chi = \mathrm{Ind}_{F_n/F}\, \chi, \end{aligned}$$

is then a canonical bijection.

Recall that $\mathbf{0}_F \in \mathcal{E}(F)$ is the endo-equivalence class of trivial simple characters over F. As an instance of the Remarks in 8.1, we have $\mathbf{0}_F = \Phi_F(\mathbf{1}_F)$. Let $\mathfrak{m} = \mathrm{M}_n(\mathfrak{o}_F)$. We describe canonical bijections

$$\begin{aligned} \Delta\backslash X_1(F_n)^{\Delta\text{-reg}} &\longrightarrow \mathcal{T}(1_{\mathfrak{m}}^1) \longrightarrow \mathcal{A}_n(F; \mathbf{0}_F), \\ \chi &\longmapsto \Lambda_\chi \longmapsto \pi_\chi. \end{aligned}$$

For the first, let $\boldsymbol{\mu}(F_n)$ denote the group of roots of unity in F_n, of order relatively prime to p. The Galois group Δ acts on $\boldsymbol{\mu}(F_n)$; an element ζ of $\boldsymbol{\mu}(F_n)$ is called *Δ-regular* if the conjugates ζ^δ, $\delta \in \Delta$, are distinct.

We embed F_n in $\mathrm{M}_n(F)$ so that $\mathfrak{m}$ becomes F_n-pure. This embedding identifies $\boldsymbol{\mu}(F_n)$ with a subgroup of $U_{\mathfrak{m}}$. Reduction modulo $\mathfrak{p}_{\mathfrak{m}}$ then identifies $\boldsymbol{\mu}(F_n)$ with a subgroup of $\mathcal{G} = \mathrm{GL}_n(\Bbbk_F)$, the Δ-regular elements of $\boldsymbol{\mu}(F_n)$ becoming elliptic regular in $\mathcal{G}$. Let $\chi \in X_1(F_n)^{\Delta\text{-reg}}$. As in [17] (*cf.* §2 of [10]), there is a unique irreducible cuspidal representation $\tilde{\lambda}_\chi$ of $\mathcal{G}$ such that

$$\mathrm{tr}\, \tilde{\lambda}_\chi(\zeta) = (-1)^{n-1} \sum_{\delta \in \Delta} \chi^\delta(\zeta),$$

for every Δ-regular element ζ of $\boldsymbol{\mu}(F_n)$. We define an irreducible representation Λ_χ of the group $\boldsymbol{J} = F^\times U_{\mathfrak{m}}$ by deeming that $\Lambda_\chi|_{U_{\mathfrak{m}}}$ be the inflation of $\tilde{\lambda}_\chi$ and that $\Lambda_\chi|_{F^\times}$ be a multiple of $\chi|_{F^\times}$. The map $\chi \mapsto \Lambda_\chi$ is then the desired bijection $\Delta\backslash X_1(F_n)^{\Delta\text{-reg}} \to \mathcal{T}(1_{\mathfrak{m}}^1)$. The second bijection above is then $\Lambda_\chi \mapsto c\text{-}\mathrm{Ind}_{\boldsymbol{J}}^G\, \Lambda_\chi$, $G = \mathrm{GL}_n(F)$, as in 1.2 Corollary.

The representation ${}^L\sigma_\chi$, attached to σ_χ by the Langlands correspondence, is *not* π_χ. It is rather

$$ {}^L\sigma_\chi = \pi_{\chi'}, $$

where $\chi' = \omega^{n-1}\chi$ and ω is the unramified character of $F_n^\times$ of order 2.

8.3. A wild lift

We continue in the situation of 8.2. Let Q/F be a finite, totally wildly ramified, field extension. Thus QF_n/Q is unramified of degree n, and we may identify $\mathrm{Gal}(QF_n/Q)$ with Δ. Composition with the field norm N_{QF_n/F_n} gives a Δ-isomorphism $X_1(F_n) \to X_1(QF_n)$, and so leads to a canonical bijection

$$ \mathrm{b}_{Q/F} : \mathcal{A}_n(F; \mathbf{0}_F) \xrightarrow{\approx} \mathcal{A}_n(Q; \mathbf{0}_Q). $$

This map is readily described in terms of types. Let $\pi \in \mathcal{A}_n(F; \mathbf{0}_F)$ be given by a Δ-regular character $\chi \in X_1(F_n)$. Thus π is induced by a representation $\Lambda_\chi \in \mathcal{T}(1^1_{\mathfrak{m}})$ as in 8.2. Likewise, $\mathrm{b}_{Q/F}\,\pi$ is induced by a representation Λ_{χ_Q}, where $\chi_Q = \chi \circ \mathrm{N}_{Q_n/F_n}$. By definition, the representation Λ_{χ_Q} is determined by its restriction to $Q^\times$ (which is a multiple of $\chi_Q|_{Q^\times}$) and the restriction of its character to the set of Δ-regular elements ζ of $\boldsymbol{\mu}(Q_n) = \boldsymbol{\mu}(F_n)$. For such an element ζ,

$$ \mathrm{tr}\,\Lambda_{\chi_Q}(\zeta) = \mathrm{tr}\,\Lambda_\chi(\zeta^{[Q:F]}). $$

Note here that the field degree $[Q{:}F]$ is a power of p.

We remark that, in the case where Q/F is also cyclic, this map $\mathrm{b}_{Q/F}$ is *base change,* in the sense of [1], [23].

9. Main theorem

We return to the context of 4.3, to describe more fully the class of extended maximal types attached to an m-simple character in a group $\mathrm{GL}_n(F)$.

9.1. Notation

We establish notation for the rest of the section. Let θ be a non-trivial m-simple character in $G = \mathrm{GL}_n(F)$. In particular, $\theta \in \mathcal{C}(\mathfrak{a}, \beta, \psi)$, for some simple stratum $[\mathfrak{a}, \beta]$ in $A = \mathrm{M}_n(F)$ and a smooth character ψ of F of level one. We now set $P = F[\beta]$, we let B be the A-centralizer of P and put $\mathfrak{b} = \mathfrak{a} \cap B$.

Attached to θ are the groups $\boldsymbol{J}_\theta$, J^0_θ and J^1_θ of 4.1. Since θ is m-simple, we have $\boldsymbol{J}_\theta = P^\times J^0_\theta = P^\times U_{\mathfrak{b}} J^1_\theta$ and $J^1_\theta \cap P^\times U_{\mathfrak{b}} = U^1_{\mathfrak{b}}$, so the inclusion of $U_{\mathfrak{b}}$ in

J_θ^0 induces an isomorphism $U_{\mathfrak{b}}/U_{\mathfrak{b}}^1 \cong J_\theta^0/J_\theta^1$. Since $\mathfrak{b}$ is a maximal $\mathfrak{o}_P$-order, we have an isomorphism

$$U_{\mathfrak{b}}/U_{\mathfrak{b}}^1 \cong \mathrm{GL}_s(\Bbbk_P),$$

where $s = n/[P{:}F]$, uniquely determined up to conjugation by an element of $U_{\mathfrak{b}}$. Altogether, we have isomorphisms

$$J_\theta^0/J_\theta^1 \cong U_{\mathfrak{b}}/U_{\mathfrak{b}}^1 \cong \mathrm{GL}_s(\Bbbk_P). \qquad (*)$$

Let ξ be an irreducible representation of $P^\times U_{\mathfrak{b}}$, trivial on $U_{\mathfrak{b}}^1$. There is then, by $(*)$, a unique irreducible representation ξ_θ of $\boldsymbol{J}_\theta$ such that $\xi_\theta|_{J_\theta^1}$ is trivial and $\xi_\theta|_{P^\times U_{\mathfrak{b}}} \cong \xi$. The equivalence class of ξ_θ then depends on that of ξ, and not on the choice of isomorphism $(*)$.

9.2. Heisenberg representations

We recall a general result from [15] 5.1.

Lemma. *Let ϕ be a simple character in a group $G' = \mathrm{GL}_r(F)$. There exists a unique irreducible representation $\eta(\phi)$ of J_ϕ^1 such that $\eta(\phi)|_{H_\phi^1}$ contains ϕ.*

Since J_ϕ^1 normalizes ϕ, the restriction of $\eta(\phi)$ to H_ϕ^1 is a multiple of ϕ. One may also show that $I_{G'}(\eta(\phi)) = I_{G'}(\phi)$.

We now revert to the notation of 9.1. The following lies rather deeper, and is proved in [13] 3.2, [15] 5.2.

Proposition 4.7. *Let θ be an m-simple character in $G = \mathrm{GL}_n(F)$. There exists a representation κ of $\boldsymbol{J}_\theta$ such that $\kappa|_{J_\theta^1} \cong \eta(\theta)$ and $I_G(\kappa) = I_G(\theta)$.*

We denote by $\mathcal{H}(\theta)$ the set of equivalence classes of representations κ of $\boldsymbol{J}_\theta$ satisfying the conditions of the proposition.

We elucidate the structure of the space $\mathcal{H}(\theta)$. Let $X_1(\theta)$ be the group of characters ξ of $\boldsymbol{J}_\theta$ with the following properties:

(1) *ξ is trivial on J_θ^1, and*
(2) *ξ is intertwined by every element of $I_G(\theta)$.*

If $\xi \in X_1(\theta)$ and $\kappa \in \mathcal{H}(\theta)$, then surely $\xi \otimes \kappa \in \mathcal{H}(\theta)$. In this manner [15].

Proposition 4.8. *The set $\mathcal{H}(\theta)$ is a principal homogeneous space over the abelian group $X_1(\theta)$.*

The group $X_1(\theta)$ is easy to describe. Let $X_1(P)$ denote the group of tamely ramified characters of $P^\times$ and $X_0(P)_s$ the subgroup of unramified characters ν such that $\nu^s = 1$. Let $\chi \in X_1(P)$ and let $\det_P : B^\times \to P^\times$ be the determinant map. Thus $\chi_B = \chi \circ \det_P|_{P^\times U_\mathfrak{b}}$ provides a character $P^\times U_\mathfrak{b}$, from which we may form the character $(\chi_B)_\theta$ of $\mathbf{J}_\theta$. It is easy to see that $(\chi_B)_\theta$ lies in $X_1(\theta)$ and that the map $\chi \mapsto (\chi_B)_\theta$ gives an isomorphism

$$X_1(P)/X_0(P)_s \xrightarrow{\approx} X_1(\theta).$$

It is sometimes better to view this slightly differently. Let T/F be the maximal tamely ramified sub-extension of P/F. Composition with the field norm $\mathrm{N}_{P/T}$ induces an isomorphism $X_1(T)/X_0(T)_s \to X_1(P)/X_0(P)_s$ and so:

Corollary. *The space $\mathcal{H}(\theta)$ is a principal homogeneous space over the group $X_1(T)/X_0(T)_s$.*

We recall that any two choices of the field T are canonically F-isomorphic, indeed J_θ^1-conjugate. The actions of the groups $X_1(T)$ on $\mathcal{H}(\theta)$ are then related by this conjugation.

9.3. The tensor decomposition

We continue in the same situation. The set $\mathcal{T}(1_\mathfrak{b}^1)$ consists of classes of irreducible representations λ of $P^\times U_\mathfrak{b}$ such that $\lambda|_{U_\mathfrak{b}}$ is the inflation of an irreducible cuspidal representation of $U_\mathfrak{b}/U_\mathfrak{b}^1 \cong \mathrm{GL}_s(\Bbbk_P)$.

Proposition. *Let Θ be the endo-equivalence class of θ. If $\kappa \in \mathcal{H}(\theta)$ and $\lambda \in \mathcal{T}(1_\mathfrak{b}^1)$, then $\kappa \otimes \lambda_\theta \in \mathcal{T}(\theta)$. For any $\kappa \in \mathcal{H}(\theta)$, the map*

$$\begin{aligned} \mathcal{T}(1_\mathfrak{b}^1) &\longrightarrow \mathcal{T}(\theta), \\ \lambda &\longmapsto \kappa \otimes \lambda_\theta, \end{aligned}$$

is a bijection. It induces a bijection

$$\Pi_\kappa^P : \mathcal{A}_s(P; \mathbf{0}_P) \xrightarrow{\approx} \mathcal{A}_s(F; \Theta).$$

The first two assertions come from [13] 3.6 and the final one follows from the Classification Theorem of 4.3.

Let T/F be the maximal tamely ramified sub-extension of P/F. In particular, the extension P/T is totally wildly ramified. Taking account of 8.3, we have a bijection

$$\Pi_\kappa : \mathcal{A}_s(T; \mathbf{0}_T) \xrightarrow{\mathrm{b}_{P/T}} \mathcal{A}_s(P; \mathbf{0}_P) \xrightarrow{\Pi_\kappa^P} \mathcal{A}_s(F; \Theta).$$

As the notation indicates, the map Π_κ does not depend on the choice of parameter field P/T: this follows easily from the remark following the definition of 8.3. Also, the underlying simple character θ determines the tame parameter field T/F uniquely up to a distinguished isomorphism so Π_κ is essentially independent of the choice of T.

9.4. Main theorem

In the notation of Part II, let $\alpha \in \widehat{\mathcal{P}}_F$ and let $s \geqslant 1$ be an integer. Let $\Theta = \Phi_F(\alpha)$. We describe the Langlands correspondence

$$\mathcal{G}_s(F;\alpha) \xrightarrow{\approx} \mathcal{A}_s(F;\Theta).$$

Let $E = Z_F(\alpha)$, so that $\deg\Theta = [E{:}F]\dim\alpha$, by the Tame Parameter Theorem. Set $n = s\deg\Theta$, and let θ be an m-simple character in $G = \mathrm{GL}_n(F)$ of endo-equivalence class Θ: this determines θ uniquely, up to G-conjugation. We choose a simple stratum $[\mathfrak{a},\beta]$ in $\mathrm{M}_n(F)$ such that $\theta \in \mathcal{C}(\mathfrak{a},\beta)$, and use the notation set up in 9.1. By the Tame Parameter Theorem again, the field E is F-isomorphic to the maximal tamely ramified sub-extension T/F of P/F.

One needs to specify an F-isomorphism here. The simple character θ gives rise to a simple character θ_T over T, as in 6.6. Let $\Theta_T \in \mathcal{E}(T)$ be the endo-equivalence class of θ_T. We choose the isomorphism $E \to T$ to carry $\Phi_E(\alpha)$ to Θ_T. This determines it uniquely. We henceforward use this isomorphism to identify E with T.

Let $\rho \in \mathcal{G}_1(E;\alpha)$: thus $\rho \in \widehat{\mathcal{W}}_E$ and $\rho|_{\mathcal{P}_F} \cong \alpha$. Using the proposition of 7.1, any $\sigma \in \mathcal{G}_s(F;\alpha)$ is of the form $\Sigma_\rho(\tau)$, for a uniquely determined representation $\tau \in \mathcal{G}_s(E;\mathbf{1}_E)$. In particular, ${}^L\tau \in \mathcal{A}_s(E;\mathbf{0}_E)$.

Main Theorem. *Let $\rho \in \mathcal{G}_1(E;\alpha)$. There exists a unique $\kappa = \kappa_\rho \in \mathcal{H}(\theta)$ such that*

$${}^L\Sigma_\rho(\tau) = \Pi_\kappa({}^L\tau), \quad \tau \in \mathcal{G}_s(E;\mathbf{1}_E).$$

The map

$$\mathcal{G}_1(E;\alpha) \longrightarrow \mathcal{H}(\theta),$$
$$\rho \longmapsto \kappa_\rho$$

is an isomorphism of $X_1(E)$-spaces.

This summarizes the main results of [13], especially 7.3 and 7.6.

9.5. Comments

9.5.1.

In the special case $E = F, s = 1$, the theorem says essentially nothing. Each of the sets $\mathcal{G}_1(F;\alpha)$, $\mathcal{A}_1(F;\Theta)$ is a principal homogeneous space over the abelian group $X_1(F)$ of tamely ramified characters of $F^\times$. The Langlands correspondence provides an $X_1(F)$-bijection $\mathcal{G}_1(F;\alpha) \to \mathcal{A}_1(F;\Theta)$. Any $X_1(F)$-map $\mathcal{G}_1(F;\alpha) \to \mathcal{A}_1(F;\Theta)$ is therefore bijective, and differs from the Langlands correspondence by a constant $X_1(F)$-translation.

9.5.2.

We return to the general case. Let E_s/E be unramified of degree s and set $\Delta = \mathrm{Gal}(E_s/E)$. Write $\rho_s = \rho|\mathcal{W}_{E_s}$ and $\pi_s = {}^L\rho_s$. In particular, ρ_s is a Δ-fixed point of $\mathcal{G}_1(E_s;\alpha)$ and likewise $\pi_s \in \mathcal{A}_1(E_s;\Theta_s)^\Delta$, where $\Theta_s = \Phi_{E_s}(\alpha)$. The representation π_s contains an m-simple character θ_s of endo-equivalence class Θ_s, lifting an m-simple character θ in $\mathrm{GL}_n(F)$ of endo-equivalence class Θ. The representation π_s contains an extended maximal simple type $\kappa(\rho_s) \in \mathcal{T}(\theta_s) = \mathcal{H}(\theta_s)$ which is fixed by Δ.

The first step of the proof uses an explicit construction, based on the Glauberman correspondence [16] from the representation theory of finite groups, to produce a canonical map $\mathfrak{i}_{E_s/F} : \mathcal{H}(\theta_s)^\Delta \to \mathcal{H}(\theta)$. The representation $\kappa(\rho) = \mathfrak{i}_{E_s/F}\kappa(\rho_s)$ is not the representation κ_ρ required by the theorem. However, for a simple reason as in 9.5.1, there exists $\mu_\rho \in X_1(E_s)^\Delta$ such that

$$\kappa_\rho = \kappa(\mu_\rho \otimes \rho),$$

for all $\rho \in \mathcal{G}_1(E;\alpha)$. The main labour of the proof is in showing that μ_ρ is *independent of* ρ, and so depends only on s, α and the base field F. We therefore denote it $\mu = \mu^F_{s,\alpha}$.

9.5.3.

In the *essentially tame* case, where $\dim\alpha = 1$, the character $\mu^F_{s,\alpha}$ is worked out fully in [7], [8] and [10]. In the general case $\dim\alpha \geqslant 1$, it is constructed as a product following a certain structure tower for the field extension E_s/F:

$$E_s \supset K_0 \supset K_1 \supset \cdots \supset K_r \supset F.$$

Here, K_r/F is unramified and E_s/K_r is totally tamely ramified. Each K_i/K_{i+1}, $0 \leqslant i \leqslant r-1$, is cyclic of prime degree, while E_s has trivial K_0-automorphism group. This yields a decomposition

$$\mu^F_{s,\alpha} = \mu_\alpha^{E_s/K_0} \cdot \mu_\alpha^{K_0/K_1} \cdot \ldots \cdot \mu_\alpha^{K_{r-1}/K_r} \cdot \mu_\alpha^{K_r/F}.$$

The unramified contribution $\mu_\alpha^{K_r/F}$ can be worked out completely, in terms of simple combinatorial invariants. It has order $\leqslant 2$ but may be ramified [13] 10.7. At the other end, $\mu_\alpha^{E_s/K_0}$ is somewhat mysterious: it is unramified of order dividing $2[E_s{:}K_0]\dim\alpha$. The remaining factors are given by various explicit formulae involving transfer factors and certain constants derived from automorphic induction.

9.5.4.

The proof of the Main Theorem exposes a structure of some independent interest. If $\alpha \in \widehat{\mathcal{P}}_F$ and $E = Z_F(\alpha)$, the set $\mathcal{G}_s(F;\alpha)$ carries a natural action of the group $X_1(E)$. Let $\sigma \in \mathcal{G}_s(F;\alpha)$, and write $\sigma = \Sigma_\rho(\tau)$, for some $\rho \in \mathcal{G}_1(E;\alpha)$, $\tau \in \mathcal{G}_s(E;\mathbf{1}_F)$. We set

$$\chi \odot_\alpha \sigma = \Sigma_\rho(\chi \otimes \tau), \quad \chi \in X_1(E).$$

On the other side, let θ be an m-simple character in $G = \mathrm{GL}_n(F)$, say $\theta \in \mathcal{C}(\mathfrak{a},\beta)$. Set $P = F[\beta]$, $s = n/[P{:}F]$ and let Θ be the endo-equivalence class of θ. Let T/F be the maximal tamely ramified sub-extension of P/F. Let $\pi \in \mathcal{A}_s(F;\Theta)$. Thus $\pi = \Pi_\kappa(\xi)$, for $\kappa \in \mathcal{H}(\theta)$ and $\xi \in \mathcal{A}_s(T;\mathbf{0}_T)$. For $\chi \in X_1(T)$, we set

$$\chi \odot_T \pi = \Pi_\kappa(\chi\xi).$$

In the case $\Theta = \Phi_F(\alpha)$, the F-isomorphism $E \cong T$ chosen in 9.4 yields

$${}^L(\chi \odot_\alpha \sigma) = \chi \odot_E {}^L\sigma, \quad \chi \in X_1(E),\ \sigma \in \mathcal{G}_s(F;\alpha).$$

This $\odot_T$-action may be defined more transparently via extended maximal simple types, in the manner of 9.2.

9.5.5.

The version of the Langlands correspondence given by the Main Theorem is well-adapted to describing congruence behaviour, modulo a prime number $l \neq p$. See [14] for a simple treatment of this topic.

9.6. Local constant comparisons

We recall briefly a different method with some claim to effectiveness. It works more generally, but we shall consider only the most interesting case of totally ramified representations.

Let ψ be a non-trivial smooth character of F and s a complex variable. For $\pi_1, \pi_2 \in \widehat{\mathrm{GL}}_F$, let $\varepsilon(\pi_1\times\pi_2, s, \psi)$ be the local constant of [24], [32]. Likewise, for $\sigma \in \widehat{\mathcal{W}}_F$, let $\varepsilon(\sigma, s, \psi)$ be the Langlands–Deligne local constant.

Let $n > 1$. Under the standard characterization [20], if $\sigma \in \mathcal{G}_n(F)$ and $\pi \in \mathcal{A}_n(F)$, then $\pi = {}^L\sigma$ if and only if

$$\varepsilon(\tau \otimes \sigma, s, \psi) = \varepsilon({}^L\tau \times \pi, s, \psi),$$

for all $\tau \in \widehat{\mathcal{W}}_F$ such that $\dim \tau < n$. We use the Tame Parameter Theorem to refine this criterion.

We rely on a result from [3], as follows. Let $c(\psi)$ be the greatest integer k such that $\mathfrak{p}_F^{-k} \subset \operatorname{Ker}\psi$. For $\pi_i \in \mathcal{A}_{n_i}(F)$, $i = 1, 2$, the local constant takes the form

$$\varepsilon(\pi_1 \times \pi_2, s, \psi) = q^{-s(a(\pi_1\times\pi_2)+n_1n_2c(\psi))}\, \varepsilon(\pi_1 \times \pi_2, 0, \psi),$$

where $q = |\Bbbk_F|$ and $a(\pi_1 \times \pi_2)$ is an integer independent of ψ. In particular, if χ is an unramified character of $F^\times$, then

$$\begin{aligned}\varepsilon(\chi\pi_1 \times \pi_2, s, \psi) &= \varepsilon(\pi_1 \times \chi\pi_2, s, \psi)\\ &= \chi(\varpi)^{a(\pi_1\times\pi_2)+n_1n_2c(\psi)}\,\varepsilon(\pi_1 \times \pi_2, s, \psi),\end{aligned}$$

where ϖ is a prime element of F.

For $\pi \in \widehat{\mathrm{GL}}_F$, let $d(\pi)$ be the number of unramified characters χ of $F^\times$ for which $\chi\pi \cong \pi$. We say that π is *totally ramified* if $d(\pi) = 1$. Similarly for representations $\sigma \in \widehat{\mathcal{W}}_F$.

From [3], we obtain:

Lemma. *Let $\pi \in \mathcal{A}_n(F)$ be totally ramified, and let l be a prime divisor of n. There exists a positive divisor n_l of n/l and a totally ramified representation $\pi_l \in \mathcal{A}_{n_l}(F)$ such that $a(\pi_l \times \pi)$ is not divisible by l.*

The defining property of π_l depends only on the endo-equivalence class $\vartheta(\pi_l)$. One can construct an endo-equivalence class, with the desired properties, directly from $\vartheta(\pi)$.

We return to our usual situation with $\alpha \in \widehat{\mathcal{P}}_F$, $E = Z_F(\alpha)$, but we now assume E/F is totally ramified. Set $\Theta = \Phi_F(\alpha)$. If $\pi \in \mathcal{A}_1(F;\Theta)$, then π is totally ramified. We now obtain:

Theorem. *Let S be the set of prime divisors of* $[E{:}F]\dim\alpha$. *There is a subset $\{\sigma_l : l \in S\}$ of $\widehat{\mathcal{W}}_F$ with the following properties.*

(1) *σ_l is totally ramified and* $\dim\sigma_l$ *divides n/l.*
(2) *Let $\sigma \in \mathcal{G}_1(F;\alpha)$, $\pi \in \mathcal{A}_1(F;\Theta)$ and suppose that* $\det\sigma = \omega_\pi$, *the central character of π. The following are equivalent:*
  (a) *$\pi = {}^L\sigma$;*
  (b) *$\varepsilon(\sigma_l \otimes \sigma, s, \psi) = \varepsilon({}^L\sigma_l \times \pi, s, \psi)$, for all $l \in S$.*

The hypothesis in (2) implies that ${}^L\sigma = \chi\pi$, where χ is unramified of order dividing n. The theorem follows on taking π_l as in the lemma and defining σ_l by ${}^L\sigma_l = \pi_l$.

For an application of this result, see [4].

References

[1] J. Arthur and L. Clozel, *Simple algebras, base change, and the advanced theory of the trace formula*, Annals of Math. Studies, vol. 120, Princeton University Press, 1989.

[2] C.J. Bushnell and G. Henniart, Local tame lifting for GL(n) I: simple characters, *Publ. Math. IHES* **83** (1996), 105–233.

[3] ———, Local Rankin-Selberg convolution for $GL(n)$: divisibility of the conductor, *Math. Ann.* **321** (2001), 455–461.

[4] ———, On certain dyadic representations. Appendix to H. Kim and F. Shahidi, Functorial products for $GL(2) \times GL(3)$ and functorial symmetric cube for $GL(2)$, *Annals of Math.* (2) **155** (2002), 883–893.

[5] ———, Local tame lifting for $GL(n)$ IV: simple characters and base change, *Proc. London Math. Soc.* **87** (2003), 337–362.

[6] ———, Local tame lifting for $GL(n)$ III: explicit base change and Jacquet-Langlands correspondence, *J. reine angew. Math.* **508** (2005), 39–100.

[7] ———, The essentially tame local Langlands correspondence, I *J. Amer. Math. Soc.* **18** (2005), 685–710.

[8] ———, The essentially tame local Langlands correspondence, II: totally ramified representations, Compositio Mathematica **141** (2005), 979–1011.

[9] ———, The local Langlands Conjecture for $GL(2)$ Grundlehren der mathematischen Wissenschaften, vol. 335, Springer, 2006.

[10] ———, The essentially tame local Langlands correspondence, III: the general case, *Proc. London Math. Soc.* (3) **101** (2010), 497–553.

[11] ———, The essentially tame local Jacquet-Langlands correspondence, *Pure App. Math. Quarterly* **7** (2011), 469–538.

[12] ———, Explicit functorial correspondences for level zero representations of p-adic linear groups, *J. Number Theory* **131** (2011), 309–331.

[13] ———, To an effective local Langlands correspondence, *Memoirs Amer. Math. Soc.*, to appear. arXiv:1103.5316.

[14] ———, Modular local Langlands correspondence for GL_n, *Int. Math. Res. Not.* (2013), doi: 10.1093/imrn/rnt063.

[15] C.J. Bushnell and P.C. Kutzko, *The admissible dual of $GL(N)$ via compact open subgroups*, Annals of Math. Studies, vol. 129, Princeton University Press, 1993.

[16] G. Glauberman, Correspondences of characters for relatively prime operator groups, *Canadian J. Math.* **20** (1968), 1465–1488.

[17] J.A. Green, The characters of the finite general linear groups, *Trans. Amer. Math. Soc.* **80** (1955), 402–447.

[18] M. Harris and R. Taylor, *On the geometry and cohomology of some simple Shimura varieties*, Annals of Math. Studies, vol. 151, Princeton University Press, 2001.

[19] G. Henniart, La conjecture locale de Langlands pour GL(3), *Mém. Soc. Math. France*, nouvelle série **11/12** (1984).

[20] ———, Caractérisation de la correspondance de Langlands locale par les facteurs ε de paires, *Invent. Math.* **113** (1993), 339–356.

[21] ———, Une preuve simple des conjectures locales de Langlands pour GL_n sur un corps p-adique, *Invent. Math.* **139** (2000), 439–455.

[22] G. Henniart and R. Herb, Automorphic induction for $GL(n)$ (over local non-archimedean fields), *Duke Math. J.* **78** (1995), 131–192.

[23] G. Henniart and B. Lemaire, Changement de base et induction automorphe pour GL_n en caractéristique non nulle, *Mém. Soc. Math. France* **108** (2010).

[24] H. Jacquet, I.I. Piatetskii-Shapiro and J.A. Shalika, Rankin-Selberg convolutions, *Amer. J. Math.* **105** (1983), 367–483.

[25] J.-L. Kim, Supercuspidal representations: an exhaustion theorem, *J. Amer. Math. Soc.* **20** (2007), 273–320.

[26] P.C. Kutzko, The Langlands conjecture for GL_2 of a local field, *Annals of Math.* **112** (1980), 381–412.

[27] G. Laumon, M. Rapoport and U. Stuhler, $\mathcal{D}$-elliptic sheaves and the Langlands correspondence, *Invent. Math.* **113** (1993), 217–338.

[28] C. Mœglin, Sur la correspondance de Langlands-Kazhdan, *J. Math. Pures et Appl. (9)* **69** (1990), 175–226.

[29] A. Moy and G. Prasad, Unrefined minimal K-types for p-adic groups, *Invent. Math.* **116** (1994), 393-408.

[30] I. Reiner, *Maximal orders*, Oxford University Press, 2003.

[31] V. Sécherre and S. Stevens, Représentations lisses de $\mathrm{GL}_m(D)$, IV: représentations supercuspidales. *J. Inst. Math. Jussieu* **7** (2008), 527–574.

[32] F. Shahidi, Fourier transforms of intertwining operators and Plancherel measures for $\mathrm{GL}(n)$, *Amer. J. Math.* **106** (1984), 67–111.

[33] A. Silberger and E.-W. Zink, An explicit matching theorem for level zero discrete series of unit groups of $\mathfrak{p}$-adic simple algebras, *J. reine angew. Math.* **585** (2005), 173–235.

[34] S.A.R. Stevens, The supercuspidal representations of p-adic classical groups, *Invent. Math.* **172** (2008), 289–352.

[35] J. Tits, *Reductive groups over local fields*, Automorphic forms, representations and L-functions (A. Borel and W. Casselman, eds.), *Proc. Symp. Pure Math.*, vol. 33(1), Amer. Math. Soc., 1979, pp. 29–69.

[36] J.-K. Yu, Construction of tame supercuspidal representations, *J. Amer. Math. Soc.* **14** (2001), 579–622.

5

The conjectural connections between automorphic representations and Galois representations

Kevin Buzzard and Toby Gee

Abstract

We state conjectures on the relationships between automorphic representations and Galois representations, and give evidence for them.

Contents

1 Introduction *page* 135
2 L-groups and local definitions 139
3 Global definitions, and the first conjectures 153
4 The case of tori 161
5 Twisting and Gross' η 167
6 Functoriality 176
7 Reality checks 178
8 Relationship with theorems/conjectures in the literature 181
References 185

1. Introduction

1.1.

Given an algebraic Hecke character for a number field F, a classical construction of Weil produces a compatible system of 1-dimensional ℓ-adic representations of $\mathrm{Gal}(\overline{F}/F)$. In the late 1950s, Taniyama's work [Tan57] on L-functions

2000 *Mathematics Subject Classification*. 11F33.

of abelian varieties with complex multiplications led him to consider certain higher-dimensional compatible systems of Galois representations, and by the 1960s it was realised by Serre and others that Weil's construction might well be the tip of a very large iceberg. Serre conjectured the existence of 2-dimensional ℓ-adic representations of $\mathrm{Gal}(\overline{\mathbb{Q}}/\mathbb{Q})$ attached to classical modular eigenforms for the group GL_2 over $\mathbb{Q}$, and their existence was established by Deligne not long afterwards. Moreover, Langlands observed that one way to attack Artin's conjecture on the analytic continuation of Artin L-functions might be via first proving that any non-trivial n-dimensional irreducible complex representation of the absolute Galois group of a number field F came (in some precise sense) from an automorphic representation for GL_n/F, and then analytically continuing the L-function of this automorphic representation instead.

One might ask whether one can associate "Galois representations" to automorphic representations for an arbitrary connected reductive group over a number field. There are several approaches to formalising this problem. Firstly one could insist on working with all automorphic representations and attempt to associate to them complex representations of a "Langlands group", a group whose existence is only conjectural but which, if it exists, should be much bigger than the absolute Galois group of the number field (and even much bigger than the Weil group of the number field) – a nice reference for a rigorous formulation of a conjecture here is [Art02]. Alternatively one could restrict to automorphic representations that are "algebraic" in some reasonable sense, and in this case one might attempt to associate certain complex representations of the fundamental group of some Tannakian category of motives, a group which might either be a pro-algebraic group scheme or a topological group. Finally, following the original examples of Weil and Deligne, one might again restrict to algebraic automorphic representations, and then attempt to associate compatible systems of ℓ-adic Galois representations to such objects (that is, representations of the absolute Galois group of the number field over which the group is defined). The advantage of the latter approach is that it is surely the most concrete.

For the group GL_n over a number field, Clozel gave a definition of what it meant for an automorphic representation to be "algebraic". The definition was, perhaps surprisingly, a non-trivial twist of a notion which presumably had been in the air for many years. Clozel made some conjectures predicting that algebraic automorphic representations should give rise to n-dimensional ℓ-adic Galois representations (so his conjecture encapsulates Weil's result on Hecke characters and Deligne's theorem too). Clozel proved some cases of his conjecture, when he could switch to a unitary group and use algebraic geometry to construct the representations.

The goal of this paper is to generalise (most of) the *statement* of Clozel's conjecture to the case where GL_n is replaced by an arbitrary connected reductive group G. Let us explain the first stumbling block in this programme. The naive conjecture would be of the following form: if an automorphic representation π for G is algebraic (in some reasonable sense) then there should be a Galois representation into the $\overline{\mathbb{Q}}_p$-points of the L-group of G, associated to π. But if one looks, for example, at Proposition 3.4.4 of [CHT08], one sees that they can associate p-adic Galois representations to certain automorphic representations on certain compact unitary groups, but that the Galois representations are taking values in a group $\mathcal{G}_n$ which one can check is *not* the L-group of the unitary group in question (for dimension reasons, for example). In fact there are even easier examples of this phenomenon: if π is the automorphic representation for $\mathrm{GL}_2/\mathbb{Q}$ attached to an elliptic curve over the rationals, then (if one uses the standard normalisation for π) one sees that π has trivial central character and hence descends to an automorphic representation for $\mathrm{PGL}_2/\mathbb{Q}$ which one would surely hope to be algebraic (because it is cohomological). However, the L-group of $\mathrm{PGL}_2/\mathbb{Q}$ is SL_2 and there is no way of twisting the Galois representation afforded by the p-adic Tate module of the curve so that it lands into $\mathrm{SL}_2(\overline{\mathbb{Q}}_\ell)$, because the cyclotomic character has no square root (consider complex conjugation). On the other hand, there do exist automorphic representation for $\mathrm{PGL}_2/\mathbb{Q}$ which have associated Galois representations into $\mathrm{SL}_2(\overline{\mathbb{Q}}_\ell)$; for example one can easily build them from automorphic representations on $\mathrm{GL}_2/\mathbb{Q}$ constructed via the Langlands–Tunnell theorem applied to a continuous even irreducible 2-dimensional representation of $\mathrm{Gal}(\overline{\mathbb{Q}}/\mathbb{Q})$ into $\mathrm{SL}_2(\mathbb{C})$ with solvable image. What is going on?

Our proposed solution is the following. For a general connected reductive group G, we believe that there are *two* reasonable notions of "algebraic". For GL_n these notions differ by a twist (and this explains why this twist appears in Clozel's work). For some groups the notions coincide. But for some others – for example PGL_2 – the notions are disjoint. The two definitions "differ by half the sum of the positive roots". We call the two notions C-algebraic and L-algebraic. It turns out that cohomological automorphic representations are C-algebraic (hence the C), and that given an L-algebraic automorphic representation one might expect an associated Galois representation into the L-group (hence the L). Clozel twists C-algebraic representations into L-algebraic ones in his paper, and hence conjectures that there should be Galois representations attached to C-algebraic representations for GL_n, but this trick is not possible in general. In this chapter we explicitly conjecture the existence of p-adic Galois representations associated to L-algebraic automorphic representations for a general connected reductive group over a number field.

On the other hand, one must not leave C-algebraic representations behind. For example, for certain unitary groups of even rank over the rationals, all automorphic representations are C-algebraic and none are L-algebraic at all! It would be a shame to have no conjecture at all in these cases. We show in Section 5 that given a C-algebraic automorphic representation for a group G, it can be lifted to a C-algebraic representation for a certain covering group $\widetilde{G}$ (a canonical central extension of G by GL_1) where there is enough space to twist C-algebraic representations into L-algebraic ones. After making such a twisting one would then conjecturally expect an associated Galois representation into the L-group not of G but of $\widetilde{G}$. We define the C-group ${}^C G$ of G to be ${}^L\widetilde{G}$. For example, if π is the automorphic representation for the group $\mathrm{PGL}_2/\mathbb{Q}$ attached to an elliptic curve over $\mathbb{Q}$, we can verify that π is C-algebraic, and that $\widetilde{G} = \mathrm{GL}_2/\mathbb{Q}$ in this case, and hence one would expect a Galois representation into $\mathrm{GL}_2(\overline{\mathbb{Q}}_\ell)$ associated to π, which is given by the Tate module of the curve. We also verify the compatibility of the construction with that made by Clozel–Harris–Taylor in the unitary group case.

In this chapter, we explain the phenomena above in more detail. In particular we formulate a conjecture associating p-adic Galois representations to L-algebraic automorphic representations for an arbitrary connected reductive group over a number field, which appears to essentially include all known theorems and conjectures of this form currently in the literature. We initially imagined that such a conjecture was already "known to the experts". However, our experience has been that this is not the case; in fact, it seems that the issues that arise when comparing the definitions of L-algebraic and C-algebraic representations were a known problem, with no clear solution (earlier attempts to deal with this issue have been by means of redefining the local Langlands correspondence and the Satake isomorphism via a twist, as in [Gro99]; however this trick only works for certain groups). In one interesting example in the literature where Galois representations are attached to certain cohomological automorphic representations – the constructions of [CHT08] – the Galois representations take values in a group $\mathcal{G}_n$, whose construction seemed to us to be one whose main motivation was that it was the group that worked, rather than the group that came from a more conceptual argument. We revisit this construction in Section 8.3. Ultimately, we hope that this chapter will clarify once and for all a variety of issues that occur when leaving the relative safety of GL_n, giving a firm framework for further research on Galois representations into groups other than GL_n.

1.2. Acknowledgements

We would like to thank Jeff Adams, James Arthur, Frank Calegari, Brian Conrad, Matthew Emerton, Wee Teck Gan, Dick Gross, Florian Herzig, Robert Langlands, David Loeffler, Ambrus Pál, Richard Taylor and David Vogan for helpful discussions relating to this work. Particular thanks go to Gross, for giving us a copy of Deligne's 2007 letter to Serre and urging us to read it, and to Adams and Vogan for dealing with several questions of ours involving local Langlands at infinity which were apparently "known to the experts" but which we could not extract from the literature ourselves.

The first author was supported by an EPSRC Advanced Research Fellowship, and the second author would like to acknowledge the support of the National Science Foundation (award number DMS-0841491). He would also like to thank the mathematics department of Northwestern University for its hospitality in the early stages of this project.

2. L-groups and local definitions

In this section we give an overview of various standard facts concerning L-groups, the Satake isomorphism, the archimedean local Langlands correspondence, and basic Hodge–Tate theory, often with a specific emphasis on certain arithmetic aspects that are not considered relevant in many of the standard references. In summary: our L-groups will be over $\overline{\mathbb{Q}}$, we will keep track of the two different $\mathbb{Q}$-structures in the Satake isomorphism, and our local Langlands correspondence will concern representations of $G(\mathbb{R})$ or $G(\overline{\mathbb{R}})$, where $\overline{\mathbb{R}}$ is an algebraic closure of the reals which we do not identify with $\mathbb{C}$ (note that on the other hand, all our representations will be on $\mathbb{C}$-vector spaces). This section is relatively elementary but contains all of the crucial local definitions.

2.1. The L-group

We briefly review the notion of an L-group. We want to view the L-group of a connected reductive group as a group over $\overline{\mathbb{Q}}$, rather than the more traditional $\mathbb{C}$, as we shall later on be considering representations into the $\overline{\mathbb{Q}}_p$-points of the L-group. We review the standard definitions from the point of view that we shall be taking.

We take the approach to dual groups explained in section 1 of [Kot84], but work over $\overline{\mathbb{Q}}$. See also section 3.3 of Exposé XXIV of [ABD$^+$66], which is perhaps where the trick of taking limits of based root data is first introduced.

Let k be a field and let G be a connected reductive algebraic group over k. Fix once and for all a separable closure k^{sep} of k, and let Γ_k denote $\mathrm{Gal}(k^{\mathrm{sep}}/k)$. The group G splits over k^{sep}, and if we choose a maximal torus T in $G_{k^{\mathrm{sep}}}$ and a Borel subgroup B of $G_{k^{\mathrm{sep}}}$ containing T, one can associate the based root datum $\Psi_0(G, B, T) := (X^*(T), \Delta^*(B), X_*(T), \Delta_*(B))$ consisting of the character and cocharacter groups of T, and the roots and coroots which are simple and positive with respect to the ordering defined by B. Now let Z_G denote the centre of G; then $(G/Z_G)(k^{\mathrm{sep}})$ acts on $G_{k^{\mathrm{sep}}}$ by conjugation, and if T' and B' are another choice of maximal torus and Borel then there is an element of $(G/Z_G)(k^{\mathrm{sep}})$ sending B' to B and T' to T, and all such elements induce the same isomorphisms of based root data $\Psi_0(G, B, T) \to \Psi_0(G, B', T')$. Following Kottwitz, we define $\Psi_0(G) := (X^*, \Delta^*, X_*, \Delta_*)$ to be the projective limit of the $\Psi_0(G, B, T)$ via these isomorphisms. This means in practice that given a maximal torus T of $G_{k^{\mathrm{sep}}}$, the group X^* is isomorphic to the character group of T but not canonically; however given also a Borel B containing the torus, there is now a canonical map $X^* = X^*(T)$ (and different Borels give different canonical isomorphisms). There is a natural group homomorphism $\mu_G : \Gamma_k \to \mathrm{Aut}(\Psi_0(G))$ (defined for example in §1.3 of [Bor79]) and if $K \subseteq k^{\mathrm{sep}}$ is a Galois extension of k that splits G then μ_G factors through $\mathrm{Gal}(K/k)$.

We let $\widehat{G}$ denote a connected reductive group over $\overline{\mathbb{Q}}$ equipped with a given isomorphism $\Psi_0(\widehat{G}) = \Psi_0(G)^\vee$, the dual root datum to $\Psi_0(G)$. There is a canonical group isomorphism $\mathrm{Aut}(\Psi_0(G)) = \mathrm{Aut}(\Psi_0(G)^\vee)$, sending an automorphism of X^* to its inverse (one needs to insert this inverse to "cancel out" the contravariance coming from the dual construction), and hence a canonical action of Γ_k on $\Psi_0(G)^\vee$. If we choose a Borel, a maximal torus, and a splitting (also called a pinning; see p. 10 of [Spr79] for details and definitions) of $\widehat{G}$ then, as on p.10 of [Spr79], this data induces a lifting $\mathrm{Aut}(\Psi_0(G)^\vee) \to \mathrm{Aut}(\widehat{G})$ and hence (via μ_G) a left action of Γ_k on $\widehat{G}$. We define the L-group ${}^LG$ of G to be the resulting semidirect product, regarded as a group scheme over $\overline{\mathbb{Q}}$ with identity component $\widehat{G}$ and component group Γ_k. For K a field containing $\overline{\mathbb{Q}}$ we have ${}^LG(K) = \widehat{G}(K) \rtimes \Gamma_k$. Often in the literature people use ${}^LG$ to be the group that we call ${}^LG(\mathbb{C})$.

Note that there is a fair amount of "ambiguity" in this definition. The group $\widehat{G}$ is "only defined up to inner automorphisms", as is the lifting of μ_G. So, even if we fix our choice of k^{sep}, points in ${}^LG(K)$ are "only defined up to conjugation by $\widehat{G}(K)$".

If K is an extension of $\overline{\mathbb{Q}}$ and ρ is a group homomorphism $\Gamma_k \to {}^LG(K)$, then we say that ρ is *admissible* if the map $\Gamma_k \to \Gamma_k$ induced by ρ and the surjection ${}^LG(K) \to \Gamma_k$ is the identity. We say two admissible ρs are *equivalent* if they differ by conjugation by an element of $\widehat{G}(K)$.

We remark that readers for whom even the choice of k^{sep} is distasteful can avoid making this choice all together by interpreting $\widehat{G}(K)$ as a scheme of groups over Spec(k) and then interpreting equivalence classes of admissible ρs as above as elements of $H^1(\text{Spec}(k), \widehat{G}(K))$.

We fix once and for all an embedding $\overline{\mathbb{Q}} \to \mathbb{C}$. Later on, when talking about Galois representations, we shall fix a prime number p and an embedding $\overline{\mathbb{Q}} \to \overline{\mathbb{Q}}_p$. This will enable us to talk about the groups ${}^LG(\mathbb{C})$ and ${}^LG(\overline{\mathbb{Q}}_p)$.

2.2. Satake parameters

In this section, k is a non-archimedean local field with integers $\mathcal{O}$, and we again fix a separable closure k^{sep} of k and set $\Gamma_k = \text{Gal}(k^{\text{sep}}/k)$. We normalise the reciprocity map $k^\times \to \Gamma_k^{\text{ab}}$ of local class field theory so that it takes a uniformiser to a geometric Frobenius. We follow Tate's definitions and conventions for Weil groups – in brief, a Weil group $W_k = W_{k^{\text{sep}}/k}$ for k comes equipped with maps $W_k \to \Gamma_k$ and $k^\times \to W_k^{\text{ab}}$ such that the induced map $k^\times \to \Gamma_k^{\text{ab}}$ is the reciprocity homomorphism of class field theory, normalised as above.

Let G/k be a connected reductive group which is furthermore unramified (that is, quasi-split, and split over an unramified extension of k). Then $G(k)$ has hyperspecial maximal compact subgroups (namely $\mathcal{G}(\mathcal{O}) \subseteq G(k)$, where $\mathcal{G}$ is any reductive group over $\mathcal{O}$ with generic fibre G); fix one, and call it K. Nothing we do will depend on this choice, but we will occasionally need to justify this. Let $B = B_k$ be a Borel in G defined over k, let $T = T_k$ be a maximal torus of B, also defined over k, and let T_d be the maximal k-split sub-k-torus of T. Let W_d be the subgroup of the Weyl group of G consisting of elements which map T_d to itself. Let ${}^oT$ denote the maximal compact subgroup of $T(k)$. It follows from an easy cohomological calculation (done for example in §9.5 of [Bor79]) that the inclusion $T_d \to T$ induces an isomorphism of groups $T_d(k)/T_d(\mathcal{O}) \to T(k)/{}^oT$. We normalise Haar measure on $G(k)$ so that K has measure 1 (and remark that by 3.8.2 of [Tit79] this normalisation is independent of the choice of hyperspecial maximal compact K). If R is a field of characteristic zero then let $H_R(G(k), K)$ denote the Hecke algebra of bi-K-invariant R-valued functions on $G(k)$ with compact support, and with multiplication given by convolution. Similarly let $H_R(T(k), {}^oT)$ denote the analogous Hecke algebra for $T(k)$ (with Haar measure normalised so that ${}^oT$ has measure 1).

The Satake isomorphism (see for example §4.2 of [Car79]) is a canonical isomorphism $H_{\mathbb{C}}(G(k),K) = \mathbb{C}[X_*(T_d)]^{W_d} = H_{\mathbb{C}}(T_d(k), T_d(\mathcal{O}))^{W_d} = H_{\mathbb{C}}(T(k), {}^oT)^{W_d}$, where $X_*(T_d)$ is the cocharacter group of T_d (the Satake isomorphism is the first of these equalities; the others are easy). We normalise the Satake isomorphism in the usual way, so that it does not depend on the choice of the Borel subgroup containing T; this is the only canonical way to do things. This standard normalisation is however not in general "defined over $\mathbb{Q}$" – for example if $k = \mathbb{Q}_p$ and $G = \mathrm{GL}_2$ and $K = \mathrm{GL}_2(\mathbb{Z}_p)$ then the Satake isomorphism sends the characteristic function of $K\left(\begin{smallmatrix} p & 0 \\ 0 & 1 \end{smallmatrix}\right)K$ to a function on $T(\mathbb{Q}_p)$ taking the value $\sqrt{p}$ on the matrix $\left(\begin{smallmatrix} p & 0 \\ 0 & 1 \end{smallmatrix}\right)$. This square root of p appears because the definition of the Satake isomorphism involves a twist by half the sum of the positive roots of G (see formula (19) of section 4.2 of [Car79]) and because of this twist, the isomorphism does *not* in general induce a canonical isomorphism $H_{\mathbb{Q}}(G(k), K) = \mathbb{Q}[X_*(T_d)]^{W_d}$.

In [Gro99] and [Clo90] this issue of square roots is avoided by renormalising the Satake isomorphism. Let us stress that we shall *not* do this here, and we shall think of $H_{\mathbb{Q}}(G(k), K)$ and $\mathbb{Q}[X_*(T_d)]^{W_d}$ as giving two possibly distinct $\mathbb{Q}$-structures on the complex algebraic variety $\mathrm{Spec}(H_{\mathbb{C}}(G(k), K))$ which shall perform two different functions – they will give us two (typically distinct) notions of being defined over a subfield of $\mathbb{C}$. We note however that if half the sum of the positive roots of G is in the weight lattice $X^*(T)$ (this occurs for example if G is semi-simple and simply connected, or a torus) then the map $\delta^{1/2} : T(k) \to \mathbb{R}_{>0}$ mentioned in formula (19) of [Car79] is $\mathbb{Q}$-valued (see formula (4) of [Car79] for the definition of δ) and the proof of Theorem 4.1 of [Car79] makes it clear that in this case the Satake isomorphism does induce an isomorphism $H_{\mathbb{Q}}(G(k), K) = \mathbb{Q}[X_*(T_d)]^{W_d}$.

Next we recall how the Satake isomorphism above leads us to an unramified local Langlands correspondence. Say π is a smooth admissible irreducible complex representation of $G(k)$, and assume that $\pi^K \neq 0$ for our given choice of K (so in particular, π is unramified). Then π^K is a 1-dimensional representation of $H_{\mathbb{C}}(G, K)$ and hence gives rise to a $\mathbb{C}$-valued character of $H_{\mathbb{C}}(G, K)$. Now results of Gantmacher and Langlands in section 6 of [Bor79] enable one to canonically associate to π an unramified continuous admissible representation $r_\pi : W_k \to {}^LG(\mathbb{C})$, where "admissible" in this context means a group homomorphism such that if one composes it with the canonical map ${}^LG(\mathbb{C}) \to \Gamma_k$ then one obtains the canonical map $W_k \to \Gamma_k$ (part of the definition of the Weil group) and "unramified" means that the resulting 1-cocycle $W_k \to \widehat{G}(\mathbb{C})$ is trivial on the inertia subgroup of W_k. In fact there are two ways of normalising the construction: if we follow section 6 of [Bor79] then in the crucial Proposition 6.7 Borel has chosen the σ that appears there to be an *arbitrary* generator of the Galois group of a finite unramified extension of k which splits G, and we are free to choose σ to be either an arithmetic or a geometric Frobenius. We shall let σ denote a geometric Frobenius: now an easy check (unravelling the definitions in [Bor79] and [Car79]) shows that if $G = \mathrm{GL}_1$ and π is an unramified representation of $\mathrm{GL}_1(k)$ then the corresponding Galois representation $W_k \to \mathrm{GL}_1(\mathbb{C})$ is the one induced by our given isomorphism $k^\times = W_k^{\mathrm{ab}}$. See Remark 5.20 for some comments about what would have happened had we chosen an arithmetic Frobenius here, and normalised our class field theory isomorphisms so that they sent arithmetic Frobenii to uniformisers.

Remark 5.1. The local Langlands correspondence behaves in a natural way under certain unramified twists. Let $\xi : G \to \mathbb{G}_m$ be a character of G defined over k. Then $\xi(K)$ is a compact subset of $k^\times$ and is hence contained in $\mathcal{O}^\times$. If $s \in \mathbb{C}$ then we can define $\chi : G(k) \to \mathbb{C}^\times$ by $\chi(g) = |\xi(g)|^s$; then $\chi(K) = 1$. Hence for an irreducible π with a K-fixed vector, $\pi \otimes \chi$ is also irreducible with a K-fixed vector. The map χ induces an automorphism of $H_{\mathbb{C}}(G(k), K)$ sending a function f to the function $g \mapsto \chi(g)f(g)$, and similarly it induces automorphisms of $H_{\mathbb{C}}(T(k)/{}^oT)$ and $H_{\mathbb{C}}(T_d(k)/T_d(\mathcal{O}))$. These latter two automorphisms are equivariant for the action of W_d, because χ is constant on $G(k)$-conjugacy classes. Furthermore, the Satake isomorphism commutes with these automorphisms – this follows without too much trouble from the formula defining the Satake isomorphism in §4.2 of [Car79] and the observation that the kernel of ξ contains the unipotent radical of any Borel in G. Now if $Y = \mathrm{Spec}(H_{\mathbb{C}}(T_d(k), T_d(\mathcal{O})))$ is the complex torus dual to T_d, then given an irreducible π with a K-fixed vector the Satake isomorphism produces a W_d-orbit of elements of Y. Twisting π by χ amounts to changing this orbit by adding to it the W_d-stable element y_χ of Y corresponding to the character of $T_d(k)/T_d(\mathcal{O})$ induced by χ. Let us define an element t_χ of $\widehat{T} = X^*(T) \otimes \mathbb{C}^\times$, the complex torus dual to T, by $t_\chi = \xi \otimes q^{-s} = \xi \otimes |\varpi|^s$ (with q the size of the residue field of k and ϖ a uniformiser). The inclusion $T_d \to T$ induces a surjection $\widehat{T} \to Y$ and some elementary unravelling shows that it sends t_χ to y_χ. It follows that if $r_\pi : W_k \to {}^LG$ is the representation of W_k corresponding to π, then $r_{\pi\otimes\chi}(w) = \hat{\xi}(|w|^s)r_\pi(w)$, with $\hat{\xi} : \mathbb{C}^\times \to \widehat{T} \subseteq \widehat{G}$ the cocharacter of $\widehat{T}$ corresponding to the character ξ of T.

We now come back to the **Q**-structures on $H_{\mathbb{C}}(G(k), K)$. As we have already mentioned, we have a natural $\mathbb{Q}$-structure on $H_{\mathbb{C}}(G(k), K)$ coming from the $\mathbb{Q}$-valued functions $H_{\mathbb{Q}}(G(k), K)$, and we have another one coming from $\mathbb{Q}[X_*(T_d)]^{W_d}$ via the Satake isomorphism. This means that, for a smooth irreducible admissible representation π of $G(k)$ with a K-fixed vector, there are two (typically distinct) notions of what it means to be "defined over E", for E a subfield of $\mathbb{C}$. Indeed, if π is a smooth admissible irreducible representation of $G(k)$ with a K-fixed vector, then π^K is a 1-dimensional complex vector space on which $H_{\mathbb{C}}(G(k), K)$ acts, and this action induces maps

$$H_{\mathbb{Q}}(G(k), K) \to \mathbb{C}$$

and

$$\mathbb{Q}[X_*(T_d)]^{W_d} \to \mathbb{C}.$$

Definition 5.2. Let π be smooth, irreducible and admissible, with a K-fixed vector. Let E be a subfield of $\mathbb{C}$.

(i) We say that π *is defined over* E if the induced map $H_{\mathbb{Q}}(G(k), K) \to \mathbb{C}$ has image lying in E.
(ii) We say that *the Satake parameter of* π *is defined over* E if the induced map $\mathbb{Q}[X_*(T_d)]^{W_d} \to \mathbb{C}$ has image lying in E.

If half the sum of the positive roots of G is in the lattice $X^*(T)$, then these notions coincide. However there is no reason for them to coincide in general and we shall shortly see examples for $\mathrm{GL}_2(\mathbb{Q}_p)$ where they do not.

Note also that it is not immediately clear that these notions are independent of the choice of K: perhaps there is some π with a K-fixed vector and a K'-fixed vector for two non-conjugate hyperspecial maximal compacts (for example, the trivial 1-dimensional representation of $\mathrm{SL}_2(\mathbb{Q}_p)$ has this property), and which is defined over E (or has Satake parameter defined over E) for one choice but not for the other. The reader should bear in mind that for the time being these notions depend on the choice of K, although we will soon see (in Corollary 5.4 and Lemma 5.5) that they are in fact independent of this choice.

We now discuss some other natural notions of being "defined over E" for E a subfield of $\mathbb{C}$, and relate them to the notions above. So let E be a subfield of $\mathbb{C}$ and let π be a smooth irreducible admissible complex representation of $G(k)$ with a K-fixed vector, for our fixed choice of K. Let V be the underlying vector space for π.

Lemma 5.3. *The following are equivalent:*

(i) *The representation π is defined over E.*
(ii) *There is an E-subspace V_0 of V which is $G(k)$-stable and such that $V_0 \otimes_E \mathbb{C} = V$.*
(iii) *For any (possibly discontinuous) field automorphism σ of $\mathbb{C}$ which fixes E pointwise, we have $\pi \cong \pi^\sigma = \pi \otimes_{\mathbb{C},\sigma} \mathbb{C}$ as $\mathbb{C}$-representations.*

Proof. That (ii) implies (iii) is clear – it's an abstract representation-theoretic fact. Conversely if (iii) holds, then (ii) follows from Lemma I.1 of [Wal85] (note: his E is not our E), because V^K is 1-dimensional. This latter lemma of Waldspurger also shows that if (ii) holds then V_0^K is 1-dimensional over E, and hence (ii) implies (i). To show that (i) implies (ii) we look at the explicit construction giving π from the algebra homomorphism $H_{\mathbb{Q}}(G(k), K) \to \mathbb{C}$ given in [Car79]. Given a homomorphism $H_{\mathbb{Q}}(G(k), K) \to \mathbb{C}$ with image landing in E, the resulting spherical function $\Gamma : G(k) \to \mathbb{C}$ defined in equation (30) of [Car79] is also E-valued. Now if we define V_0 to be the E-valued functions on $G(k)$ of the form $f(g) = \sum_{i=1}^{n} c_i \Gamma(gg_i)$ for $c_i \in E$ and $g_i \in G(k)$, then $G(k)$ acts on V_0 by right translations, $V_0 \otimes_E \mathbb{C}$ is the V_Γ of §4.4 of [Car79], and the arguments in §4.4 of [Car79] show that $\pi \cong V_0 \otimes_E \mathbb{C}$. □

Corollary 5.4. *If π is a smooth irreducible admissible unramified representation of $G(k)$, then the notion of being "defined over E" is independent of the choice of hyperspecial maximal compact K for which π^K is non-zero.*

Proof. This is because condition (iii) of Lemma 5.3 is independent of this choice. □

We now prove the analogous result for Satake parameters.

Lemma 5.5. *If π is a smooth irreducible admissible unramified representation of $G(k)$, then the notion of π having Satake parameter being defined over E is independent of the choice of hyperspecial maximal compact K for which $\pi^K \neq 0$.*

Proof. Say π is an unramified smooth irreducible admissible representation of $G(k)$ with a K-fixed vector. The Satake isomorphism associated to K gives us a character of the algebra $H_{\mathbb{C}}(T(k), {}^oT)^{W_d}$ and hence a W_d-orbit of complex characters of $T(k)$. Now by p. 45 of [Bor79] and sections 3 and 4 of [Car79], π is a subquotient of the principal series representation attached to any one of these characters, and Theorem 2.9 of [BZ77] then implies that this W_d-orbit of complex characters are the only characters for which π occurs as a subquotient of the corresponding induced representations. Hence the W_d-orbit of characters, and hence the map $\mathbb{Q}[X_*(T_d)]^{W_d} \to \mathbb{C}$ attached to π, does not depend on the choice of K in the case when π has fixed vectors for more than one conjugacy class of hyperspecial maximal compact. In particular the image of $\mathbb{Q}[X_*(T_d)]^{W_d}$ in $\mathbb{C}$ is well-defined independent of the choice of K, and hence the notion of having Satake parameter defined over E is also independent of the choice of K. □

To clarify the meaning of having a Satake parameter defined over E, we now explain that in the case of $G = \mathrm{GL}_n$ the notion becomes a more familiar one. If π is an unramified representation of $\mathrm{GL}_n(k)$ and we choose our Borel to be the upper triangular matrices and our torus to be the diagonal matrices, then the formalism above associates to π an algebra homomorphism $\mathbb{C}[X_*(T_d)]^{W_d} \to \mathbb{C}$. But here $T = T_d$ is the diagonal matrices, and W_d is the usual Weyl group W of G. The ring $\mathbb{C}[X_*(T_d)] = \mathbb{C}[X_*(T)]$ is then just the ring of functions on the dual torus $\widehat{T}$, and hence an unramified π gives rise to a W_d-orbit on $\widehat{T}$, which can be interpreted as a semisimple conjugacy class S_π in $\mathrm{GL}_n(\mathbb{C})$.

Lemma 5.6. *Let G be the group GL_n/k and let π be an unramified representation of $G(k)$. Let E be a subfield of $\mathbb{C}$. Then the following are equivalent:*

(i) *The Satake parameter of π is defined over E.*
(ii) *The conjugacy class S_π is defined over E.*
(iii) *The conjugacy class S_π contains an element of $\mathrm{GL}_n(E)$.*

Proof. The statement that the Satake parameter is defined over E is precisely the statement that the induced map $\mathbb{Q}[X_*(T)]^W \to \mathbb{C}$ takes values in E, which is the statement that the characteristic polynomial of an element of S_π has coefficients in E. Hence (i) and (ii) are equivalent. Furthermore, (ii) is equivalent to (iii) because given a monic polynomial with coefficients in E it is easy to construct a semisimple matrix with this polynomial as its characteristic polynomial. □

We leave to the reader the following elementary checks. Let G_1 and G_2 be unramified connected reductive groups over k, and let π_1, π_2 be unramified representations of $G_1(k)$, $G_2(k)$. Then $\pi := \pi_1 \otimes \pi_2$ is an unramified representation of $(G_1 \times G_2)(k)$. One can check that π is defined over E iff π_1 and π_2 are defined over E, and that π has Satake parameter defined over E iff π_1 and π_2 do. Now say k_1/k is a finite unramified extension of non-archimedean local fields, and G/k_1 is unramified connected reductive, and set $H = \mathrm{Res}_{k_1/k}(G)$. Then H is unramified over k_1, and if π is a representation of $G(k_1) = H(k)$ then π is unramified as a representation of $G(k_1)$ if and only if it is unramified as a representation of $H(k)$. Furthermore, the two notions of being defined over E (one for G and one for H) coincide. Moreover, the two notions of having Satake parameter defined over E – one for G and one for H – also coincide; we give the argument for this as it is a little trickier. Let T_d denote a maximal split torus in G and let T denote its centraliser. The Satake homomorphism for G is an injective ring homomorphism from an unramified Hecke algebra for G into $\mathbb{C}[T(k_1)/U]$, with U a maximal compact subgroup of $T(k_1)$. The Satake homomorphism for H is a map between the same two rings, and it can be easily checked from the construction in Theorem 4.1 of [Car79] that it is in fact the same map. The homomorphism for G is an isomorphism onto the subring $\mathbb{C}[T(k_1)/U]^{W(G)}$ of $\mathbb{C}[T(k_1)/U]$, with $W(G)$ the relative Weyl group for the pair (G, T_d). The homomorphism for H is an isomorphism onto $\mathbb{C}[T(k_1)/U]^{W(H)}$, and hence $\mathbb{C}[T(k_1)/U]^{W(G)} = \mathbb{C}[T(k_1)/U]^{W(H)}$. Now intersecting with $\mathbb{Q}[T(k_1)/U]$ we deduce that $\mathbb{Q}[T(k_1)/U]^{W(G)} = \mathbb{Q}[T(k_1)/U]^{W(H)}$ and hence the two $\mathbb{Q}$-structures – one coming from G and one from H – coincide.

We finish this section by noting that the notion of being defined over E does not coincide with the notion of having Satake parameter defined over E, if $k = \mathbb{Q}_p$ and $G = \mathrm{GL}_2$. For example, if π is the trivial 1-dimensional representation of $\mathrm{GL}_2(\mathbb{Q}_p)$ then π is defined over $\mathbb{Q}$ but the Satake parameter attached to π has eigenvalues $\sqrt{p}$ and $1/\sqrt{p}$, so the Satake parameter is not defined over $\mathbb{Q}$ (consider traces) but only over $\mathbb{Q}(\sqrt{p})$. Similarly if π is the character $|\det|^{1/2}$ of $\mathrm{GL}_2(\mathbb{Q}_p)$ then π_p is not defined over $\mathbb{Q}$ but the Satake parameter of π has characteristic polynomial $(X-1)(X-p)$ and hence is defined over $\mathbb{Q}$. This issue of the canonical normalisation of the Satake isomorphism "introducing a square root of p" is essentially the reason that one sees two normalisations of local Langlands for GL_n in the literature – one used for local questions and one used for local–global compatibility. We are not attempting to unify these two notions – indeed, one of the motivations of this paper is to draw the distinction between the two notions and explain what each is good for.

2.3. Local Langlands at infinity

We recall the statements and basic properties of the local Langlands correspondence for connected reductive groups over the real or complex field. In fact we work in slightly more generality, for the following reason: the groups that we will apply the definitions and results of this section to are groups defined over completions of number fields at infinite places, so the following subtlety arises: the completion of a number field at a real infinite place is canonically isomorphic to the reals, but the completion of a number field at a complex place is isomorphic to the complex numbers, but not canonically. Hence we actually work with groups defined over either $\mathbb{R}$ or a degree two extension of $\mathbb{R}$ which will be isomorphic to $\mathbb{C}$ but may or may not be canonically isomorphic to $\mathbb{C}$. Note however that all our representations will be on $\mathbb{C}$-vector spaces – there is no ambiguity in our coefficient fields.

Let k be either the real numbers or an algebraic closure of the real numbers. Let G be a connected reductive group over k. Fix an algebraic closure $\overline{k}$ of k and let $T \subseteq B$ be a maximal torus and a Borel subgroup of $G_{\overline{k}}$. If π is an irreducible admissible complex representation of $G(k)$ then Langlands associates to π, in a completely canonical way, a $\widehat{G}(\mathbb{C})$-conjugacy class of admissible homomorphisms $r = r_\pi$ from the Weil group $W_k = W_{\overline{k}/k}$ of k to ${}^LG(\mathbb{C})$. For simplicity let us choose a maximal torus $\widehat{T}$ in $\widehat{G}_{\mathbb{C}}$; this is just for notational convenience. The group W_k contains a finite index subgroup canonically isomorphic to $\overline{k}^\times$; let us assume that $r(\overline{k}^\times) \subseteq \widehat{T}(\mathbb{C})$ (which can always be arranged, possibly after conjugating r by an element of $\widehat{G}(\mathbb{C})$). If σ and τ denote the two $\mathbb{R}$-isomorphisms $\overline{k} \to \mathbb{C}$ then one checks easily that for $z \in \overline{k}^\times$ we have $r(z) = \sigma(z)^{\lambda_\sigma}\tau(z)^{\lambda_\tau}$ for $\lambda_\sigma, \lambda_\tau \in X_*(\widehat{T}) \otimes \mathbb{C}$ such that $\lambda_\sigma - \lambda_\tau \in X_*(\widehat{T})$. Note that because we may not want to fix a preferred choice of isomorphism $\overline{k} = \mathbb{C}$, we might sometimes "have no preference between λ_σ and λ_τ"; this makes our presentation diverge slightly from other standard references, where typically one isomorphism is preferred.

Because $\widehat{T}(\mathbb{C})$ is usually not its own normaliser in $\widehat{G}(\mathbb{C})$, there is usually more than one way of conjugating $r(\overline{k}^\times)$ into $\widehat{T}(\mathbb{C})$, with the consequence that the pair $(\lambda_\sigma, \lambda_\tau) \in (X_*(\widehat{T}) \otimes \mathbb{C})^2$ is not a well-defined invariant of r_π; it is only well-defined up to the (diagonal) action of the Weyl group $W = W(G, T)$ on $(X_*(\widehat{T}) \otimes \mathbb{C})^2$. For notational convenience however we will continue to refer to the elements λ_σ and λ_τ of $X_*(\widehat{T}) \otimes \mathbb{C}$ and will check that none of our important later definitions depend on the choice we have made. If $k = \mathbb{R}$ then recall from the construction of the L-group that there is an action of Γ_k on $X^*(T) \otimes \mathbb{C}$, and the non-trivial element of this group sends the W-orbit of λ_σ to the W-orbit of λ_τ. If k is isomorphic to $\mathbb{C}$ then λ_σ and λ_τ are in general unrelated, subject to their difference being in $X^*(T)$.

The Weyl group orbit of $(\lambda_\sigma, \lambda_\tau)$ in $(X^*(T) \otimes \mathbb{C})^2$ is naturally an invariant attached to the Weil group representation r_π rather than to π itself, but we can access a large part of it (however, not quite all of it) more intrinsically from π using the Harish-Chandra isomorphism. We explain the story when $k = \mathbb{R}$; the analogous questions in the case $k \cong \mathbb{C}$ can then be resolved by restriction of scalars.

So, for this paragraph only, we assume $k = \mathbb{R}$. If we regard $G(k)$ as a real Lie group with Lie algebra $\mathfrak{g}$, then our maximal torus T of $G_{\overline{k}}$ gives rise to a Cartan subalgebra $\mathfrak{h}$ of $\mathfrak{g} \otimes_{\mathbb{R}} \overline{k}$. If we now break the symmetry and use σ to identify $\overline{k}$ with $\mathbb{C}$, we can interpret the Lie algebra of $T \times_{\overline{k},\sigma} \mathbb{C}$ as a complex Cartan subalgebra $\mathfrak{h}_\sigma^{\mathbb{C}}$ of the complex Lie algebra $\mathfrak{g}^{\mathbb{C}} := \mathfrak{g} \otimes_{\mathbb{R}} \mathbb{C}$. We have a canonical isomorphism $\mathfrak{h}_\sigma^{\mathbb{C}} = X_*(T) \otimes_{\mathbb{Z}} \mathbb{C}$ (this isomorphism implicitly also uses σ, because $X_*(T) = \mathrm{Hom}(\mathrm{GL}_1/\overline{k}, T)$ was computed over $\overline{k}$). Now via the Harish-Chandra isomorphism (normalised in the usual way, so it is independent of the choice of Borel) one can interpret the infinitesimal character of π as a W-orbit in $\mathrm{Hom}_{\mathbb{C}}(\mathfrak{h}_\sigma^{\mathbb{C}}, \mathbb{C}) = X^*(T) \otimes_{\mathbb{Z}} \mathbb{C} = X_*(\widehat{T}) \otimes_{\mathbb{Z}} \mathbb{C}$. Furthermore, this W-orbit contains λ_σ (this seems to be well-known; see Proposition 7.4 of [Vog93] for a sketch proof). On the other hand, we note that applying this to both σ and τ gives us a pair of W-orbits in $X^*(T) \otimes \mathbb{C}$, whereas our original construction of $(\lambda_\sigma, \lambda_\tau)$ gives us the W-orbit of a pair, which is a slightly finer piece of information (which should not be surprising: there are reducible principal series representations of $\mathrm{GL}_2(\mathbb{R})$ whose irreducible subquotients (one discrete series, one finite-dimensional) have the same infinitesimal character but rather different associated Weil representations).

We go back now to the general case $k = \mathbb{R}$ or $k \cong \mathbb{C}$. We have a W-orbit $(\lambda_\sigma, \lambda_\tau)$ in $(X^*(T) \otimes_{\mathbb{Z}} \mathbb{C})^2$ attached to π. One obvious "algebraicity" criterion that one could impose on π is that $\lambda_\sigma \in X_*(\widehat{T}) = X^*(T)$. Note that λ_σ is only well-defined up to an element of the Weyl group, but the Weyl group of course preserves $X_*(\widehat{T}) = X^*(T)$, so the notion is well-defined. Also λ_σ depends on the isomorphism $\sigma : \overline{k} \to \mathbb{C}$, but if we use τ instead then the notion remains unchanged, because $\lambda_\sigma - \lambda_\tau \in X_*(\widehat{T})$ and hence $\lambda_\sigma \in X_*(\widehat{T})$ if and only if $\lambda_\tau \in X_*(\widehat{T})$. This notion of algebraicity is frequently used in the literature – one can give the connected component of the Weil group of k the structure of the real points of an algebraic group $\mathcal{S}$ over $\mathbb{R}$ and one is asking here that the Weil representation associated to π restricts to a map $\mathcal{S}(\mathbb{R}) \to {}^LG(\mathbb{C})$ induced by a morphism of algebraic groups $\mathcal{S}_{\mathbb{C}} \to {}^LG_{\mathbb{C}}$ via the inclusion $\mathcal{S}(\mathbb{R}) \subset \mathcal{S}(\mathbb{C})$.

Definition 5.7. We say that an admissible Weil group representation $r : W_k \to {}^LG(\mathbb{C})$ is *L-algebraic* if $\lambda_\sigma \in X^*(T)$. We say that an irreducible representation π of $G(k)$ is *L-algebraic* if the Weil group representation associated to it by Langlands is L-algebraic.

Note that the notion of L-algebraicity for a Weil group representation r depends only on the restriction of r to $\overline{k}^\times$, and the notion of L-algebraicity for a representation of $G(k)$ depends only on the infinitesimal character of this representation when $k = \mathbb{R}$ (and we shall shortly see that the same is true when $k \cong \mathbb{C}$).

Later on we will need the following easy lemma. Say $k = \mathbb{R}$ and $(\lambda_\sigma, \lambda_\tau)$ is a representative of the W-orbit on $X_*(\widehat{T})^2$ associated to an L-algebraic π_∞. Regard λ_σ and λ_τ as maps $\mathbb{C}^\times \to \widehat{T}$.

Lemma 5.8. *If i is a square root of -1 in $\overline{k}$ and j is the usual element of order 4 in W_k then the element $\alpha_\infty := \lambda_\sigma(i)\lambda_\tau(i)r_{\pi_\infty}(j) \in {}^LG(\mathbb{C})$ has order dividing 2, and its $\widehat{G}(\mathbb{C})$-conjugacy class is well-defined independent of (a) the choice of order of σ and τ, (b) the choice of representative $(\lambda_\sigma, \lambda_\tau)$ of the W-orbit and (c) the choice of square root of -1 in $\overline{k}$.*

Proof. Set $r := r_\pi$. We have $\lambda_\sigma(z)r(j) = r(j)\lambda_\tau(z)$, and $\lambda_\sigma(z)$ commutes with $\lambda_\tau(z')$, and from this it is easy to check that $(\alpha_\infty)^2 = 1$ and that α_∞ is unchanged if we switch σ and τ. Changing representative of the W-orbit just amounts to conjugating r by an element of $\widehat{G}(\mathbb{C})$ and hence conjugating α_∞ by this same element. Finally one checks easily that conjugating α_∞ by $\lambda_\sigma(-1)$ gives us the analogous element with i replaced by $-i$. □

The notion of L-algebraicity will be very important to us later; however it is not hard to find automorphic representations that "appear algebraic in nature" but whose infinite components are not L-algebraic. For example one can check that if E is an elliptic curve over $\mathbb{Q}$ and π is the associated automorphic representation of $\mathrm{PGL}_2/\mathbb{Q}$, then the local component π_∞, when considered as a representation of $\mathrm{PGL}_2(\mathbb{R})$, is not L-algebraic: the element λ_σ above is in $X_*(\widehat{T}) \otimes_{\mathbb{Z}} \frac{1}{2}\mathbb{Z}$ but not in $X_*(\widehat{T})$. What has happened is that the canonical normalisation of the Harish-Chandra homomorphism involves (at some point in the definition) a twist by half the sum of the positive roots, and it is this twist that has taken us out of the lattice in the elliptic curve example.

This observation motivates a *second* notion of algebraicity – which it turns out is the one used in Clozel's paper for the group GL_n. Let us go back to the case of a general connected reductive G over k, either the reals or a field isomorphic to the complexes. Recall that we have fixed $T \subseteq B \subseteq G_{\overline{k}}$ and hence we have the notion of a positive root in $X^*(T)$. Let $\delta \in X^*(T) \otimes \mathbb{C}$ denote half the sum of the positive roots. We observed above that the assertion "$\lambda_\sigma \in X^*(T)$" was independent of the choice of B and of the isomorphism $\overline{k} \cong \mathbb{C}$. But the assertion "$\lambda_\sigma - \delta \in X^*(T)$" is also independent of such choices, for if $\lambda_\sigma - \delta \in X^*(T)$ and w is in the Weyl group, then $w.\lambda_\sigma - \delta = w(\lambda_\sigma - \delta) - (\delta - w.\delta) \in X^*(T)$, and also $\lambda_\tau - \delta = (\lambda_\sigma - \delta) + (\lambda_\tau - \lambda_\sigma) \in X^*(T)$.

Definition 5.9. We say that the admissible Weil group representation $r : W_k \to {}^LG(\mathbb{C})$ is *C-algebraic* if $\lambda_\sigma - \delta \in X^*(T)$. We say that the irreducible admissible representation π of $G(k)$ is *C-algebraic* if the Weil group representation associated to π via Langlands' construction is C-algebraic.

Again, C-algebraicity for r only depends on the restriction of r to $\overline{k}^\times$, and C-algebraicity for π only depends on its infinitesimal character when $k = \mathbb{R}$ (and as we are about to see, the same is true for $k \cong \mathbb{C}$).

Here are some elementary remarks about these definitions. If $\delta \in X^*(T)$ then the notions of L-algebraic and C-algebraic coincide. If G_1 and G_2 are connected reductive over k, if r_i $(i = 1, 2)$ are admissible representations $r_i : W_k \to {}^LG_i(\mathbb{C})$, then there is an obvious notion of a product $r_1 \times r_2 : W_k \to {}^L(G_1 \times G_2)(\mathbb{C})$ and $r_1 \times r_2$ is L-algebraic (resp. C-algebraic) iff r_1 and r_2 are. One can furthermore check that if k denotes an algebraic closure of the reals and G/k is connected reductive, and if $H = \mathrm{Res}_{k/\mathbb{R}}(G)$, and if π is an irreducible admissible representation of $G(k) = H(\mathbb{R})$, then π is L-algebraic (resp. C-algebraic) when considered as a representation of $G(k)$ if and only if it is L-algebraic (resp. C-algebraic) when considered as a representation of $H(\mathbb{R})$. This assertion comes from a careful reading of sections 4 and 5 of [Bor79]. Indeed, if T is a maximal torus of G/k then $\mathrm{Res}_{k/\mathbb{R}}(T)$ is a maximal torus of $H/\mathbb{R}$, and if $\lambda_\sigma, \lambda_\tau \in X^*(T)\otimes\mathbb{C}$ are the parameters attached to a representation of $G(k)$, then $\lambda_\sigma \oplus \lambda_\tau$ and $\lambda_\tau \oplus \lambda_\sigma \in (X^*(T) \oplus X^*(T))\otimes\mathbb{C}$ are the parameters attached to the corresponding representation of $H(\mathbb{R})$ (identifying $\widehat{H}(\mathbb{C})$ with $\widehat{G}(\mathbb{C})^2$), and if δ is half the sum of the positive roots for G then $\delta \oplus \delta$ is half the sum of the positive roots for H. As a consequence, we see that both L-algebraicity and C-algebraicity of a representation π of $G(k)$ are conditions that only depend on the infinitesimal character of the representation of the underlying real reductive group.

Let us again attempt to illustrate the difference between the two notions of algebraicity by considering the trivial 1-dimensional representation of $\mathrm{GL}_2(\mathbb{R})$. The Local Langlands correspondence associates to this the 2-dimensional representation $|.|^{1/2} \oplus |.|^{-1/2}$ of the Weil group of the reals. If we choose the diagonal torus in GL_2 and identify its character group with $\mathbb{Z}^2$ in the obvious way, then we see that $\lambda_\sigma = \lambda_\tau = \delta = (\frac{1}{2}, -\frac{1}{2})$. In particular, λ_σ is not in $X^*(T)$, but $\lambda_\sigma - \delta$ is, meaning that this representation is C-algebraic but not L-algebraic. Another example would be the character $|\det|^{1/2}$ of $\mathrm{GL}_2(\mathbb{R})$; this is associated to the representation $|.| \oplus 1$ of the Weil group, and so $\lambda_\sigma = \lambda_\tau = (1, 0)$ (or $(0, 1)$, allowing for the Weyl group action) and on this occasion λ_σ is in $X^*(T)$ but $\lambda_\sigma - \delta$ is not, hence the representation is L-algebraic but not C-algebraic. Finally let us consider the discrete series representation of $\mathrm{GL}_2(\mathbb{R})$ with trivial central character associated to a weight 2 modular form. The associated representation of the Weil group sends an element z of $\overline{\mathbb{R}}^\times$ to a matrix with eigenvalues $\sqrt{z.\bar{z}}/z$ and $\sqrt{z.\bar{z}}/\bar{z}$, the square root being the positive square root. We see that the set $\{\lambda_\sigma, \lambda_\tau\}$ equals the set $\{(\frac{1}{2}, -\frac{1}{2}), (-\frac{1}{2}, \frac{1}{2})\}$ (with ambiguities due to both the Weyl group action and the two choices of identification of $\overline{\mathbb{R}}$ with $\mathbb{C}$) and neither λ_σ nor λ_τ are in $X^*(T)$, but both of $\lambda_\sigma - \delta$ and $\lambda_\tau - \delta$ are, so again the representation is C-algebraic but not L-algebraic.

2.4. The Hodge–Tate cocharacter

In this subsection, let k be a finite extension of the p-adic numbers $\mathbb{Q}_p$. Let H be a (not necessarily connected) reductive algebraic group over a fixed algebraic closure $\overline{\mathbb{Q}}_p$ of $\mathbb{Q}_p$. Note that we do not fix an embedding $k \to \overline{\mathbb{Q}}_p$. Let $\overline{k}$ denote an algebraic closure of k and let $\rho : \mathrm{Gal}(\overline{k}/k) \to H(\overline{\mathbb{Q}}_p)$ denote a continuous group homomorphism. We say that ρ is crystalline/de Rham/Hodge–Tate if for some (and hence any) faithful representation $H \to \mathrm{GL}_N$ over $\overline{\mathbb{Q}}_p$, the resulting N-dimensional Galois representation is crystalline/de Rham/Hodge–Tate. Let C denote the completion of $\overline{k}$. Then for any injection of fields $i : \overline{\mathbb{Q}}_p \to C$ there is an associated Hodge–Tate cocharacter $\mu_i : (\mathrm{GL}_1)_C \to H_C$ (where the base extension from H to H_C is via i). We know of no precise reference for the construction of μ_i in this generality; if H were defined over $\mathbb{Q}_p$ and ρ took values in $H(\mathbb{Q}_p)$ then μ_i is constructed in [Ser79]. The general case can be reduced to this case in the following way: H descends to group H_0 defined over a finite extension E of $\mathbb{Q}_p$, and a standard Baire category theorem argument shows that ρ takes values in $H_0(E')$ for some finite extension E' of E. Now let $H_1 = \mathrm{Res}_{E'/\mathbb{Q}_p} H_0$, so ρ takes values in $H_1(\mathbb{Q}_p)$, and Serre's construction of μ then yields μ_i as above which can be checked to be well-defined independent of the choice of H_0 and so on via an elementary calculation (do the case $H = \mathrm{GL}_n$ first).

Note that there is a choice of sign that one has to make when defining μ_i; we follow Serre so, for example, the cyclotomic character gives rise to the identity map $\mathrm{GL}_1 \to \mathrm{GL}_1$.

Now the conjugacy class of μ_i arises as the base extension (via i) of a cocharacter $\nu_i : (\mathrm{GL}_1)_{\overline{\mathbb{Q}}_p} \to H$ over $\overline{\mathbb{Q}}_p$. Now any $i : \overline{\mathbb{Q}}_p \to C$ is an injection whose image contains k and hence induces an injection $j =$"i^{-1}": $k \to \overline{\mathbb{Q}}_p$. Another careful calculation, which again we omit, shows that the conjugacy class of ν_i depends only on j, so we may set $\nu_j := \nu_i$, a conjugacy class of maps $\mathrm{GL}_1 \to H$ over $\overline{\mathbb{Q}}_p$.

In applications, H will be related to an L-group as follows. If G is connected and reductive over k, and $\rho : \mathrm{Gal}(\overline{k}/k) \to {}^L G(\overline{\mathbb{Q}}_p)$ is an admissible representation, then, because G splits over a finite Galois extension k' of k, ρ will descend to a representation $\rho : \mathrm{Gal}(\overline{k}/k) \to \widehat{G}(\overline{\mathbb{Q}}_p) \rtimes \mathrm{Gal}(k'/k)$. The target group can be made into the $\overline{\mathbb{Q}}_p$-points of an algebraic group H over $\overline{\mathbb{Q}}_p$, and if the associated representation is Hodge–Tate then the preceding arguments associate a conjugacy class of maps $\nu_j : \mathrm{GL}_1 \to H$ to each $j : k \to \overline{\mathbb{Q}}_p$. If $\widehat{T}$ is a torus in $\widehat{G}$ as usual, then ν_j gives rise to an element of $X_*(\widehat{T})/W$, with W the Weyl group of $G_{\overline{k}}$.

3. Global definitions, and the first conjectures

3.1. Algebraicity and arithmeticity

Let G be a connected reductive group defined over a number field F. Fix an algebraic closure $\overline{F}$ of F and form the L-group ${}^LG = \widehat{G} \rtimes \mathrm{Gal}(\overline{F}/F)$ as in the previous section. For each place v of F, fix an algebraic closure $\overline{F_v}$ of F_v, and an embedding $\overline{F} \hookrightarrow \overline{F_v}$. Nothing we do depends in any degree of seriousness on these choices – changing them will just change things "by an inner automorphism".

Let π be an automorphic representation of G. Then we may write $\pi = \otimes'_v \pi_v$, a restricted tensor product, where v runs over all places (finite and infinite) of F. Recall that in the previous section we defined notions of L-algebraic and C-algebraic for certain representations of real and complex groups. We now globalise these definitions.

Definition 5.10. We say that π is *L-algebraic* if π_v is L-algebraic for all infinite places v of F.

Definition 5.11. We say that π is *C-algebraic* if π_v is C-algebraic for all infinite places v of F.

Note that, for $G = \mathrm{GL}_n$, the notion of C-algebraic coincides (in the isobaric case) with Clozel's notion of algebraic used in [Clo90], although for GL_2 this choice of normalisation goes back to Hecke. Note also that restriction of scalars preserves both notions: if K/F is a finite extension of number fields and π is an automorphic representation of G/K then π is L-algebraic (resp. C-algebraic) when considered as a representation of $G(\mathbb{A}_K)$ if and only if π is L-algebraic (resp. C-algebraic) when considered as a representation of $\mathrm{Res}_{K/F}(G)(\mathbb{A}_F)$. Indeed, this is a local statement and we indicated the proof in Section 2.3.

As examples of these notions, we observe that for Hecke characters of number fields, our notions of L-algebraic and C-algebraic both coincide with the classical notion of being algebraic or of type A_0. For GL_2 the notions diverge: the trivial 1-dimensional representation of $\mathrm{GL}_2(\mathbb{A}_\mathbb{Q})$ is C-algebraic but not L-algebraic, whereas the representation $|\det|^{1/2}$ of $\mathrm{GL}_2(\mathbb{A}_\mathbb{Q})$ is L-algebraic but not C-algebraic. For $\mathrm{GL}_3/\mathbb{Q}$ the notions of L-algebraic and C-algebraic coincide again (because half the sum of the positive roots is in the weight lattice); indeed they coincide for GL_n over a number field if n is odd, and differ by a non-trivial twist if n is even.

The above definitions depend only on the behaviour of π at infinite places. The ones below depend only on π at the finite places; we remind the reader that the crucial local definitions are given in Definition 5.2.

Definition 5.12. We say that π is *L-arithmetic* if there is a finite subset S of the places of F, containing all infinite places and all places where π is ramified, and a number field $E \subset \mathbb{C}$, such that for each $v \notin S$, the Satake parameter of π_v is defined over E.

Definition 5.13. We say that π is *C-arithmetic* if there is a finite subset S of the places of F, containing all infinite places and all places where π is ramified, and a number field $E \subset \mathbb{C}$, such that π_v is defined over E for all $v \notin S$.

Again we note that for K/F a finite extension and π an automorphic representation of G/K, π is L-arithmetic (resp. C-arithmetic) if and only if π considered as an automorphic representation of $\mathrm{Res}_{K/F}(G)$ is.

Let us consider some examples. An automorphic representation π of GL_n/F will be L-arithmetic if there is a number field such that all but finitely many of the Satake parameters attached to π have characteristic polynomials with coefficients in that number field. So, for example, the trivial 1-dimensional representation of $\mathrm{GL}_2(\mathbb{A}_\mathbb{Q})$ would not be L-arithmetic, because the trace of the Satake parameter at a prime p is $p^{1/2} + p^{-1/2} = \frac{p+1}{\sqrt{p}}$, and any subfield of $\mathbb{C}$ containing $(p+1)/\sqrt{p}$ for infinitely many primes p would also contain $\sqrt{p}$ for infinitely many primes p and hence cannot be a number field. However it would be C-arithmetic, because for all primes p, π_p is the base extension to $\mathbb{C}$ of a representation of $\mathrm{GL}_2(\mathbb{A}_\mathbb{Q})$ on a vector space over $\mathbb{Q}$. Similarly, the representation $|\det|^{1/2}$ of $\mathrm{GL}_2(\mathbb{A}_\mathbb{Q})$ is L-arithmetic, because all Satake parameters are defined over $\mathbb{Q}$. However this representation is not C-arithmetic: each individual π_p is defined over a number field but there is no number field over which infinitely many of the π_p are defined, again because such a number field would have to contain the square root of infinitely many primes.

Now let π be an arbitrary automorphic representation for an arbitrary connected reductive group G over a number field.

Conjecture 5.14. *π is L-arithmetic if and only if it is L-algebraic.*

Conjecture 5.15. *π is C-arithmetic if and only if it is C-algebraic.*

These conjectures are seemingly completely out of reach. The general ideas behind them (although perhaps not the precise definitions we have given) seem to be part of the folklore nowadays, although it is worth pointing out that as far as we know the first person to raise such conjectures explicitly was Clozel in [Clo90] in the case $G = \mathrm{GL}_n$.

For the group GL_1 over a number field both of the conjectures are true; indeed in this case both conjectures say the same thing, the "algebraic implies arithmetic" direction being relatively standard, and the "arithmetic implies algebraic" direction being a non-trivial result in transcendence theory proved by Waldschmidt in [Wal81]. We prove both conjectures for a general torus in Section 4, for the most part by reducing to the case of GL_1. On the other hand, neither direction of either conjecture is known for the group $GL_2/\mathbb{Q}$, although in this case the conjectures turn out to be equivalent and there are some partial results in both directions. In particular, Sarnak has shown ([Sar02]) that for a Maass form with coefficients in $\mathbb{Z}$, the associated L-arithmetic automorphic representation is necessarily L-algebraic, and this result was generalised to the case of coefficients in certain quadratic fields in [Bru03]. Furthermore, if π is a cuspidal automorphic representation for $GL_2/\mathbb{Q}$ which is discrete series at infinity, then we show in §3.3 that both conjectures hold for π. However if π is principal series at infinity then both directions of both conjectures are in general open.

If one makes Conjectures 5.14 and 5.15 for all groups G simultaneously, then they are in fact equivalent, by the results of Section 5 below; for groups with a twisting element (see Section 5 for this terminology) this follows from Propositions 5.35 and 5.36, and the general case reduces to this one by passage to the covering groups of Section 5 – see Proposition 5.42.

3.2. Galois representations attached to automorphic representations

We now fix a prime number p and turn to the notion of associating p-adic Galois representations to automorphic representations. Because automorphic representations are objects defined over $\mathbb{C}$ and p-adic Galois representations are defined over p-adic fields, we need a method of passing from one field to the other. We have already fixed an injection $\overline{\mathbb{Q}} \to \mathbb{C}$; now we fix once and for all a choice of algebraic closure $\overline{\mathbb{Q}}_p$ of $\mathbb{Q}_p$ and, reluctantly, an isomorphism $\iota : \mathbb{C} \to \overline{\mathbb{Q}}_p$ of "coefficient fields". Recall that our L-groups are defined over our fixed algebraic closure $\overline{\mathbb{Q}}$ of $\mathbb{Q}$; our fixed inclusion $\overline{\mathbb{Q}} \to \mathbb{C}$ then induces, via ι, an embedding $\overline{\mathbb{Q}} \to \overline{\mathbb{Q}}_p$. Ideally we should only be fixing an embedding $\overline{\mathbb{Q}} \to \overline{\mathbb{Q}}_p$, and all our constructions should only depend on the restriction of ι to $\overline{\mathbb{Q}}$, but of course we cannot prove this. Our choice of ι does affect matters, in the usual way: for example, if $f = \sum a_n q^n$ is one of the holomorphic cuspidal newforms for $GL_2/\mathbb{Q}$ of level 1 and weight 24 then 13 splits into two prime ideals in the coefficient field of f, and a_{13} is in one of these prime ideals but not the other; hence f will be ordinary with respect to some choices of ι but not for others. For notational simplicity we drop ι from our notation but our conjectural association of p-adic Galois representations attached to automorphic representations will depend very much on this choice.

We now state two conjectures on the existence of Galois representations attached to L-algebraic automorphic representations, the second stronger than the first (in that it specifies a precise set of places at which the Galois representation is unramified/crystalline – this is the only difference between the two conjectures). The first version is the more useful one when formulating conjectures about functoriality. Note that both conjectures depend implicitly on our choice of isomorphism $\iota : \mathbb{C} \to \overline{\mathbb{Q}}_p$ which we use to translate complex parameters to p-adic ones.

Conjecture 5.16. *If π is L-algebraic, then there is a finite subset S of the places of F, containing all infinite places, all places dividing p, and all places where π is ramified, and a continuous Galois representation $\rho_\pi = \rho_{\pi,\iota} : \mathrm{Gal}(\overline{F}/F) \to {}^L G(\overline{\mathbb{Q}}_p)$, which satisfies*

- *The composite of ρ_π and the natural projection ${}^L G(\overline{\mathbb{Q}}_p) \to \mathrm{Gal}(\overline{F}/F)$ is the identity map.*
- *If $v \notin S$, then $\rho_\pi|W_{F_v}$ is $\widehat{G}(\overline{\mathbb{Q}}_p)$-conjugate to $\iota(r_{\pi_v})$.*
- *If v is a finite place dividing p then $\rho_\pi|_{\mathrm{Gal}(\overline{F_v}/F_v)}$ is de Rham, and the Hodge–Tate cocharacter of this representation can be explicitly read off from π via the recipe in Remark 5.18.*
- *If v is a real place, let $c_v \in G_F$ denote a complex conjugation at v. Then $\rho_{\pi,\iota}(c_v)$ is $\widehat{G}(\overline{\mathbb{Q}}_p)$-conjugate to the element $\iota(\alpha_v) = \iota(\lambda_{\sigma_v}(i)\lambda_{\tau_v}(i)r_{\pi_v}(j))$ of Lemma 5.8.*

Conjecture 5.17. *Assume that π is L-algebraic. Let S be the set of the places of F consisting of all infinite places, all places dividing p, and all places where π is ramified. Then there is a continuous Galois representation $\rho_{\pi,\iota} : \mathrm{Gal}(\overline{F}/F) \to {}^L G(\overline{\mathbb{Q}}_p)$, which satisfies*

- *The composite of $\rho_{\pi,\iota}$ and the natural projection ${}^L G(\overline{\mathbb{Q}}_p) \to \mathrm{Gal}(\overline{F}/F)$ is the identity map.*
- *If $v \notin S$, then $\rho_{\pi,\iota}|W_{F_v}$ is $\widehat{G}(\overline{\mathbb{Q}}_p)$-conjugate to $\iota(r_{\pi_v})$.*
- *If v is a finite place dividing p then $\rho_{\pi,\iota}|_{\mathrm{Gal}(\overline{F_v}/F_v)}$ is de Rham, and the Hodge–Tate cocharacter associated to this representation is given by the recipe in Remark 5.18. Furthermore, if π_v is unramified then $\rho_{\pi,\iota}|_{\mathrm{Gal}(\overline{F_v}/F_v)}$ is crystalline.*
- *If v is a real place, let $c_v \in G_F$ denote a complex conjugation at v. Then $\rho_{\pi,\iota}(c_v)$ is $\widehat{G}(\overline{\mathbb{Q}}_p)$-conjugate to the element $\iota(\alpha_v) = \iota(\lambda_{\sigma_v}(i)\lambda_{\tau_v}(i)r_{\pi_v}(j))$ of Lemma 5.8.*

Remark 5.18. The recipe for the Hodge–Tate cocharacter in the conjectures above is as follows. Say $j : F \to \overline{\mathbb{Q}}$ is an embedding of fields. We have fixed $\overline{\mathbb{Q}} \to \mathbb{C}$ and (via ι) $\overline{\mathbb{Q}} \to \overline{\mathbb{Q}}_p$, so j induces $F_v \to \overline{\mathbb{Q}}_p$ for some place $v|p$ and $F_w \to \mathbb{C}$ for some place $w|\infty$. The inclusion $F_w \to \mathbb{C}$ enables us to identify $\mathbb{C}$ as an algebraic closure of F_w and now attached to π_w and the identity $\sigma : \mathbb{C} \to \mathbb{C}$ we have constructed an element $\lambda_\sigma \in X^*(T)/W$. Our conjecture is that this element λ_σ is the Hodge–Tate cocharacter associated to the embedding $F_v \to \overline{\mathbb{Q}}_p$.

Remark 5.19. The representation $\rho_{\pi,\iota}$ is not necessarily unique up to $\widehat{G}(\overline{\mathbb{Q}}_p)$-conjugation. One rather artificial reason for this is that if π is a non-isobaric L-algebraic automorphic representation of $\mathrm{GL}_2/\mathbb{Q}$ such that π_v is 1-dimensional for almost all v, then there are often many non-semisimple 2-dimensional Galois representations that one can associate to π (as well as a semisimple one). But there are other more subtle reasons too. For example if G is a torus over F then the admissible Galois representations into the L-group of G are parametrised by $H^1(F, \widehat{G})$ (with the Galois group acting on $\widehat{G}$ via the action used to form the L-group), and there may be non-zero elements of this group which restrict to zero in $H^1(F_v, \widehat{G})$ for all places v of F. If this happens then there is more than one Galois representation that can be associated to the trivial 1-dimensional automorphic representation of G. We are grateful to Hendrik Lenstra and Bart de Smit for showing us the following explicit example of a rank 3 torus over $\mathbb{Q}$ where this phenomenon occurs. If Γ is the group $(\mathbb{Z}/2\mathbb{Z})^2$ and Q is the quaternion group of order 8 then Q gives a non-zero element of $H^2(\Gamma, \pm 1)$ whose image in $H^2(\Gamma, \mathbb{C}^\times)$ is non-zero but whose restriction to $H^2(D, \mathbb{C}^\times)$ is zero for any cyclic subgroup D of Γ (consider the corresponding extension of Γ by $\mathbb{C}^\times$ to see these facts). We now "dimension shift". The rank 4 torus $\mathbb{Z}[\Gamma] \otimes_{\mathbb{Z}} \mathbb{C}^\times$ has no cohomology in degree greater than zero and has $\mathbb{C}^\times$ as a subgroup, so the quotient group T is a complex torus with an action of Γ and with the property that there's an element of $H^1(\Gamma, T)$ whose restriction to any cyclic subgroup is zero. Finally, Γ is isomorphic to $\mathrm{Gal}(\mathbb{Q}(\sqrt{13}, \sqrt{17})/\mathbb{Q})$ (a non-cyclic group all of whose decomposition groups are cyclic) and T with its Galois action can be realised as the complex points of the dual group of a torus over $\mathbb{Q}$, giving us our example: there is more than one Galois representation associated to the trivial 1-dimensional representation of this torus.

Remark 5.20. We have normalised local class field theory so that geometric Frobenius elements correspond to uniformisers, and defined our Weil groups accordingly. Had we normalised things the other way (associating arithmetic Frobenius to uniformisers) then the natural thing to do when formulating our unramified local Langlands dictionary would have been to use an arithmetic Frobenius as a generator of our Galois group. In particular our unramified local

Langlands dictionary at good finite places would be changed by a non-trivial involution. Had we made this choice initially, Conjectures 5.16 and 5.17 need to be modified: one needs to change the Hodge–Tate cocharacter μ to $-\mu$. However these new conjectures are equivalent to the conjectures as stated, because the required Galois representation predicted by the new conjecture may be obtained directly from ρ_π by applying the Chevalley involution of ${}^LG$ (the Chevalley involution of $\widehat{G}$ extends to ${}^LG$ and induces the identity map on the Galois group), or indirectly as $\rho_{\tilde{\pi}}$ where $\tilde{\pi}$ is the contragredient of π. We omit the formal proof that these constructions do the job – in fact, although the arguments at the finite places are not hard, we confess that we were not able to find a precise published reference for the statements at infinity that we need. The point is that we need to know how the local Langlands correspondence for real and complex reductive groups behaves under taking contragredients. The involution on the π side induced by contragredient corresponds on the Galois side to the involution on the local Weil representations induced by an involution of the Weil group of the reals/complexes sending $z \in \overline{k}^\times \cong \mathbb{C}^\times$ to z^{-1}. It also corresponds to the involution on the Weil representations induced by the Chevalley involution. Both these facts seem to be well-known to the experts but the proof seems not to be in the literature. Jeff Adams and David Vogan inform us that they are working on a manuscript called "The Contragredient" in which these issues will be addressed.

3.3. Example: the groups $\mathrm{GL}_2/\mathbb{Q}$ and $\mathrm{PGL}_2/\mathbb{Q}$

The following example illustrates the differences between the C- and L-notions in two situations, one where things can be "fixed by twisting" and one where they cannot. The proofs of the assertions made here only involve standard unravelling of definitions and we shall omit them.

Let $\mathbb{A}$ denote the adeles of $\mathbb{Q}$. For N a positive integer, let $K_0(N)$ denote the subgroup of $\mathrm{GL}_2(\widehat{\mathbb{Z}})$ consisting of matrices which are upper triangular modulo N. If $\mathrm{GL}_2^+(\mathbb{R})$ denotes the matrices in $\mathrm{GL}_2(\mathbb{R})$ with positive determinant then $\mathrm{GL}_2(\mathbb{A}) = \mathrm{GL}_2(\mathbb{Q})K_0(N)\mathrm{GL}_2^+(\mathbb{R})$. Now let f be a modular form of weight $k \geq 2$ which is a normalised cuspidal eigenform for the subgroup $\Gamma_0(N)$ of $\mathrm{SL}_2(\mathbb{Z})$, and let s denote a complex number. We think of f as a function on the upper half plane. Associated to f and s we define a function ϕ_s on $\mathrm{GL}_2(\mathbb{A})$ by writing an element of $\mathrm{GL}_2(\mathbb{A})$ as $\gamma\kappa u$ with $\gamma \in \mathrm{GL}_2(\mathbb{Q})$, $\kappa \in K_0(N)$ and $u = \left(\begin{smallmatrix} a & b \\ c & d \end{smallmatrix}\right) \in \mathrm{GL}_2^+(\mathbb{R})$, and defining $\phi_s(\gamma\kappa u) = (\det u)^{k-1+s}(ci+d)^{-k}f((ai+b)/(ci+d))$. This function is well-defined and is a cuspidal automorphic form, which generates an automorphic representation π_s of $\mathrm{GL}_2(\mathbb{A})$. The element s is just a twisting factor; if

s is a generic complex number then π_s will not be algebraic or arithmetic for either of the "C" or "L" possibilities above.

First we consider the arithmetic side of the story. For p a prime not dividing N, let a_p be the coefficient of q^p in the q-expansion of f. It is well-known that the subfield of $\mathbb{C}$ generated by the a_p is a number field E. An elementary but long explicit calculation shows the following. If $\pi_{s,p}$ denotes the local component of π_s at p, then $\pi_{s,p}$ has a non-zero invariant vector under the group $\mathrm{GL}_2(\mathbb{Z}_p)$ and the action of the Hecke operators T_p and S_p on this 1-dimensional space are via the complex numbers $p^{2-k-s}a_p$ and p^{2-k-2s}. The Satake parameter associated to $\pi_{s,p}$ is the semisimple conjugacy class of $\mathrm{GL}_2(\mathbb{C})$ consisting of the semisimple elements with characteristic polynomial $X^2 - a_p p^{3/2-k-s}X + p^{2-k-2s}$. Hence π_s is L-arithmetic if $s \in \frac{1}{2} + \mathbb{Z}$. In fact one can go further. By the six exponentials theorem of transcendental number theory one sees easily that a complex number c with the property that p^c is algebraic for at least three prime numbers p must be rational. Hence if π_s is L-arithmetic then s is rational and (because a number field is only ramified at finitely many primes and hence cannot contain the tth root of infinitely many prime numbers for any $t > 1$) one can furthermore deduce that $2s \in \mathbb{Z}$. Next one observes that a_p must be non-zero for infinitely many primes $p \nmid N$ (because one can apply the Cebotarev density theorem to the mod $\ell > 2$ Galois representation associated to f and to the identity matrix) and deduce (again because a number field cannot contain the square root of infinitely many primes) that π_s is L-arithmetic iff $s \in \frac{1}{2} + \mathbb{Z}$. Now by Proposition 5.36 (whose proof uses nothing that we haven't already established) we see that π_s is C-arithmetic iff $s \in \mathbb{Z}$.

We now consider the algebraic side of things. If $\pi_{s,\infty}$ denotes the local component of π_s at infinity and we choose the Cartan subalgebra $\mathfrak{h}^{\mathbb{C}}$ of $\mathfrak{gl}_2(\mathbb{C})$ spanned by $H := \left(\begin{smallmatrix} 1 & 0 \\ 0 & -1 \end{smallmatrix}\right)$ and $Z := \left(\begin{smallmatrix} 1 & 0 \\ 0 & 1 \end{smallmatrix}\right)$ then the infinitesimal character of $\pi_{s,\infty}$ (thought of as a Weil group orbit in $\mathrm{Hom}_{\mathbb{C}}(\mathfrak{h}^{\mathbb{C}}, \mathbb{C})$) sends H to $\pm(k-1)$ and Z to $2s + k - 2$. The characters of the torus in $\mathrm{GL}_2(\mathbb{R})$ give rise to the lattice $X^*(T)$ in $\mathrm{Hom}(\mathfrak{h}^{\mathbb{C}}, \mathbb{C})$ consisting of linear maps that send H and Z to integers of the same parity. Hence π_s is C-algebraic iff $s \in \mathbb{Z}$ and L-algebraic iff $s \in \frac{1}{2} + \mathbb{Z}$. In particular π_s is L-algebraic iff it is L-arithmetic, and π_s is C-algebraic iff it is C-arithmetic.

We now play the same game for Maass forms. If f (a real analytic function on the upper half plane) is a cuspidal Maass form of level $\Gamma_0(N)$ which is an eigenform for the Hecke operators, and $s \in \mathbb{C}$ then one can define a function ϕ_s on $\mathrm{GL}_2(\mathbb{A})$ by writing an element of $\mathrm{GL}_2(\mathbb{A})$ as $\gamma\kappa u$ as above, writing $u = \left(\begin{smallmatrix} a & b \\ c & d \end{smallmatrix}\right)$, and defining $\phi_s(\gamma\kappa u) = \det(u)^s f((ai+b)/(ci+d))$. If we now assume that f is the Maass form associated by Langlands and Tunnell to a

Galois representation $\rho : \mathrm{Gal}(\overline{\mathbb{Q}}/\mathbb{Q}) \to \mathrm{SL}_2(\mathbb{C})$ with solvable image, and if $p \nmid N$ is prime and $a_p = \mathrm{tr}(\rho(\mathrm{Frob}_p))$, then the a_p generate a number field E (an abelian extension of $\mathbb{Q}$ in this case) and a similar explicit calculation, which again we omit, shows that π_s is L-arithmetic iff π_s is L-algebraic iff $s \in \mathbb{Z}$, and that π_s is C-arithmetic iff π_s is C-algebraic iff $s \in \frac{1}{2} + \mathbb{Z}$.

Note in particular that the answer in the Maass form case is different to the holomorphic case in the sense that $s \in \mathbb{Z}$ corresponded to the C-side in the holomorphic case and the L-side in the Maass form case.

An automorphic representation for $\mathrm{PGL}_2/\mathbb{Q}$ is just an automorphic representation for $\mathrm{GL}_2/\mathbb{Q}$ with trivial central character. One checks that the π_s corresponding to the holomorphic modular form has trivial central character iff $s = 1 - \frac{k}{2}$ (this is because the form was assumed to have trivial Dirichlet character) and, again because the form has trivial character, k must be even so in particular the π_s which descends to $\mathrm{PGL}_2/\mathbb{Q}$ is C-algebraic and C-arithmetic. However, the π_s corresponding to the Maass form with trivial character has trivial central character iff $s = 0$, which is L-algebraic and L-arithmetic. Hence, when applied to the group $\mathrm{PGL}_2/\mathbb{Q}$, our conjecture above says that there should be a Galois representation to $\mathrm{SL}_2(\overline{\mathbb{Q}}_\ell)$ associated to the Maass form but it says nothing about the holomorphic form. However, the holomorphic form is clearly algebraic in some sense and indeed there is a Galois representation associated to the holomorphic form – namely the Tate module of the elliptic curve. Note however that the determinant of the Tate module of an elliptic curve is the cyclotomic character, which is not the square of any 1-dimensional Galois representation (complex conjugation would have to map to an element of order 4) and hence no twist of the Tate module of an elliptic curve can take values in $\mathrm{SL}_2(\overline{\mathbb{Q}}_\ell)$. This explains why we have thus far restricted to L-algebraic representations for our conjecture attaching Galois representations to automorphic representations.

Finally, we note that it is easy to check that the automorphic representations corresponding to Hilbert modular forms with weights that are *not* all congruent modulo 2, are neither C-algebraic nor L-algebraic (cf. pp. 91–92 of [Clo90]).

3.4. Why C-algebraic?

Our conjecture above only attempts to associate Galois representations to L-algebraic automorphic representations. So why consider C-algebraic representations at all? For GL_n the issue is only one of twisting: π is L-algebraic iff $\pi.|\det(.)|^{(n-1)/2}$ is C-algebraic. Furthermore, for groups such as SL_2 in which half the sum of the positive roots is in $X^*(T)$, the notions of L-algebraic and C-algebraic coincide. On the other hand, as the previous example of

$\mathrm{PGL}_2/\mathbb{Q}$ attempted to illustrate, one does not always have this luxury of being able to pass easily between L-algebraic and C-algebraic representations for a given group G. Furthermore a lot of naturally occurring representations are C-algebraic: for example any cohomological automorphic representation will always be C-algebraic (see Lemma 5.52 below) and, as the case of $\mathrm{PGL}_2/\mathbb{Q}$ illustrated, there may be natural candidates for Galois representations associated to these automorphic representations, but they may not take values in the L-group of G! In fact, essentially all known examples of Galois representations attached to automorphic representations ultimately come from the cohomology of Shimura varieties (although in some cases the constructions also use congruence arguments), and this cohomology is naturally decomposed in terms of cohomological automorphic representations. Much of the rest of this chapter is devoted to examining the relationship between C-algebraic and L-algebraic in greater detail, and defining a "C-group", which should conjecturally receive the Galois representations attached to C-algebraic automorphic representations.

4. The case of tori

4.1.

In this section we prove conjectures 5.14, 5.15, 5.16 and 5.17 when G is a torus over a number field F. That we can do this should not be considered surprising. Indeed, if $G = \mathrm{GL}_1$ then the results have been known for almost 30 years, and the general case can be reduced to the GL_1 case via base change and a p-adic version of results of Langlands on the local and global Langlands correspondence for tori. Unfortunately we have not been able to find a reference which does what we want so we include some of the details here.

First note that if G is a torus then half the sum of the positive roots is zero, so the Satake isomorphism preserves $\mathbb{Q}$-structures and hence the notions of C-arithmetic and L-arithmetic coincide and we can use the phrase "arithmetic" to denote either of these notions. Furthermore, again because half the sum of the positive roots is zero, the notions of C-algebraic and L-algebraic also coincide, and we can use the phrase "algebraic" to mean either of these two notions (and in the case $G = \mathrm{GL}_1/F$ this coincides with the classical definition, and with Weil's notion of being of type (A_0)).

Recall that to give a torus G/F is to give its character group, which (after choosing an $\overline{F}$) is a finite free $\mathbb{Z}$-module equipped with a continuous action of $\mathrm{Gal}(\overline{F}/F)$. Let $K \subset \overline{F}$ denote a finite Galois extension of F which splits G; then this action of $\mathrm{Gal}(\overline{F}/F)$ factors through $\mathrm{Gal}(K/F)$. An automorphic representation of G/F is just a continuous group homomorphism $G(F)\backslash G(\mathbb{A}_F) \to \mathbb{C}^\times$.

Let BC denote the usual base change map from automorphic representations of G/F to automorphic representations of G/K, induced by the norm map $N : G(\mathbb{A}_K) \to G(\mathbb{A}_F)$.

Lemma 5.21. *If π is an automorphic representation of G/F then π is algebraic if and only if $BC(\pi)$ is.*

Proof. This is a local statement, and if we translate it over to a statement about representations of Weil groups then it says that if k is an algebraic closure of $\mathbb{R}$ then $r : W_{\mathbb{R}} \to {}^LG(\mathbb{C})$ is algebraic iff its restriction to W_k is, which is clear because our definition of algebraicity of r only depended on the restriction of r to W_k. □

Now let T denote a torus over a local field k, and assume T splits over an unramified extension of k. The topological group $T(k)$ has a unique maximal compact subgroup U. Let χ be a continuous group homomorphism $T(k) \to \mathbb{C}^\times$ with U in its kernel. Note that U is hyperspecial and hence χ is unramified.

Lemma 5.22. *For E a subfield of $\mathbb{C}$, the following are equivalent:*

(i) *χ is defined over E.*
(ii) *The Satake parameter of χ is defined over E.*
(iii) *The image of χ is contained in $E^\times$.*

Proof. The Satake isomorphism in this situation is simply the identity isomorphism $\mathbb{C}[T(k)/U] = \mathbb{C}[T(k)/U]$, which induces the identity isomorphism $\mathbb{Q}[T(k)/U] = \mathbb{Q}[T(k)/U]$, so (i) and (ii) are equivalent. The equivalence of (i) and (iii) follows from the statement that $\chi : T(k)/U \to \mathbb{C}^\times$ is $E^\times$-valued if and only if the induced ring homomorphism $\mathbb{Q}[T(k)/U] \to \mathbb{C}$ is E-valued. □

If k_1/k is a finite extension of local fields and if T/k is a torus then we also use the notation BC to denote the map $\mathrm{Hom}(T(k), \mathbb{C}^\times) \to \mathrm{Hom}(T(k_1), \mathbb{C}^\times)$ induced by the norm map $N : T(k_1) \to T(k)$. Now suppose again that T is an unramified torus over k and $\chi : T(k) \to \mathbb{C}^\times$ is an unramified character.

Corollary 5.23. *If χ is defined over E and if k_1/k is a finite extension, then* $\mathrm{BC}(\chi)$ *is defined over E.*

Proof. The image of $\mathrm{BC}(\chi)$ is contained in the image of χ. □

We now go back to the global situation. Let π denote an automorphic representation of G/F, with G a torus, and let $\mathrm{BC}(\pi)$ denote its base change to G/K, where K is a finite Galois extension of F which splits G.

Corollary 5.24. *If π is arithmetic then* $\mathrm{BC}(\pi)$ *is arithmetic.*

Proof. Immediate from the previous corollary. □

Theorem 5.25. *If G is a split torus over a number field, then the notions of arithmetic and algebraic coincide.*

Proof. The fact that algebraic implies arithmetic is standard; the other implication is Théorème 5.1 of [Wal82] (which uses a non-trivial result in transcendence theory). □

Corollary 5.26. *If G is a torus over a number field F and π is an automorphic representation of G, and π is arithmetic, then π is algebraic.*

Proof. If π is arithmetic then its base change to K (a splitting field for G) is arithmetic (by Corollary 5.24), and hence algebraic by the previous theorem. Hence π is algebraic by Lemma 5.21. □

To show that algebraic automorphic representations for G are arithmetic, we give a re-interpretation of what it means for an automorphic representation of a torus to be algebraic; we are grateful to Ambrus Pál for pointing out to us that such a re-interpretation should exist. First some notation. Let $F_\infty := F \otimes_{\mathbb{Q}} \mathbb{R}$. Let Σ denote the set of embeddings $\sigma : F \hookrightarrow \overline{\mathbb{Q}}$. Recall that because we have fixed an embedding $\overline{\mathbb{Q}} \to \mathbb{C}$, each $\sigma \in \Sigma$ can be regarded as an embedding $F \to \mathbb{C}$, and hence induces maps $F_\infty \to \mathbb{C}$ and $\sigma_\infty : G(F_\infty) \to G_\sigma(\mathbb{C})$, where G_σ is the group over $\overline{\mathbb{Q}}$ induced from G via base extension via σ.

Proposition 5.27. *The representation π is algebraic if and only if for each $\sigma \in \Sigma$ there is an algebraic character $\lambda_\sigma : G_\sigma \to \mathrm{GL}_1/\overline{\mathbb{Q}}$ such that π agrees with $\prod_{\sigma\in\Sigma} \lambda_\sigma \circ \sigma_\infty$ on $G(F_\infty)^0$ (the identity component of the Lie group $G(F_\infty)$).*

Proof. This statement is local at infinity, and can be checked by "brute force", explicitly working out what the local Langlands correspondence for tori over the reals and complexes is and noting that it is true in every case. □

Corollary 5.28. *If π is an algebraic automorphic representation of G/F, and if we write $\pi = \pi_f \times \pi_\infty$, with $\pi_\infty : G(F_\infty) \to \mathbb{C}^\times$, then $\pi_\infty(G(F))$ is contained within a number field.*

Proof. By the preceding proposition we know that $\pi_\infty|_{G(F)}$ is the product of a character of order at most 2 by a continuous group homomorphism

$G(F) \to \mathbb{C}^\times$ which is the product of maps $\phi_\sigma : G(F) \xrightarrow{\sigma} G_\sigma(\mathbb{C}) \to \mathbb{C}^*$ given by composing an algebraic character with an embedding $\sigma : F \hookrightarrow \mathbb{C}$. Hence it suffices to prove that $\phi(G(F))$ is contained within a number field for such a ϕ. However both T and $\mathbb{G}_m$ are defined over F, so the character descends to some number field L, which we may assume splits T and contains the images of all embeddings $F \hookrightarrow \mathbb{C}$. But then $\phi(F) \subset \phi(L) \subset L^\times$, as required. □

Theorem 5.29. *The notions of arithmetic and algebraic coincide for automorphic representations of tori over number fields.*

Proof. Let G/F be a torus over a number field, and let π be an automorphic representation of G. By Corollary 5.26 we know that if π is arithmetic then π is algebraic, so we only have to prove the converse. We will make repeated use of the trivial observation (already used above) that if X is a finite index subgroup of an abelian group Y, then the image of a character of Y is contained in a number field if and only if the image of its restriction to X is contained in a (possibly smaller) number field.

Let $\pi = \otimes_v \pi_v$ be algebraic. If K is a finite Galois extension of F splitting G then $BC_{K/F}(\pi)$ is algebraic by Lemma 5.21 and hence arithmetic by Theorem 5.25. Hence there is a number field E and some finite set S_K of places of K, containing all the infinite places, such that for $w \notin S_K$, $\mathrm{BC}(\pi)_w$ is defined over E, and hence $\mathrm{BC}(\pi)_w$ has image in $E^\times$. By increasing S_K if necessary, we can assume that S_K is precisely the set of places of K lying above a finite set S of places of F.

Let $N : G(\mathbb{A}_K) \to G(\mathbb{A}_F)$ denote the norm map. Standard results from global class field theory (see for example p. 244 of [Lan97] for the crucial argument) imply that $G(F)N(G(\mathbb{A}_K))$ is a closed and open subgroup of finite index in $G(\mathbb{A}_F)$. Hence if $\mathbb{A}_F^S$ denotes the restricted product of the completions of F at places not in S, and $\mathbb{A}_K^{S_K}$ denotes the analogous product for K, then $G(F)N(G(\mathbb{A}_K^{S_K}))$ has finite index in $G(\mathbb{A}_F^S)$. Let $\pi^S : G(\mathbb{A}_F^S) \to \mathbb{C}^\times$ denote the restriction of π to $G(\mathbb{A}_F^S)$. Then $\pi = \pi^S.\prod_{v\in S, v\nmid\infty} \pi_v.\pi_\infty$. We know that π is trivial on $G(F)$ (by definition) and that π_∞ sends $G(F)$ to a number field (by Corollary 5.28). We also know that $\pi_v(G(F_v))$ (and thus $\pi_v(G(F))$ is contained within a number field for each finite place $v \in S$ (because $BC_{K/F}(\pi)$ is arithmetic, we know that $\pi_v(N(K_w))$ is contained in a number field, where $w|v$ is a place of K, and $N(K_w)$ has finite index in F_v). Hence $\pi^S(G(F))$ is contained within a number field. Then since $G(F)N(G(\mathbb{A}_K^{S_K}))$ has finite index in $G(\mathbb{A}_F^S)$, we deduce that $\pi(G(\mathbb{A}_F^S))$ is contained within a number field, and hence π is arithmetic, as required. □

What remains now is to prove Conjecture 5.17 (which of course implies Conjecture 5.16). This follows straightforwardly from Langlands' proof of the Langlands correspondence for tori, and the usual method for associating Galois representations to algebraic representations of $\mathbb{G}_m$. Take π, G, F and K as above, and again let Σ denote the field embeddings $F \to \overline{\mathbb{Q}}$, noting now that because of our fixed embedding $\overline{\mathbb{Q}} \subset \overline{\mathbb{Q}}_p$ we can also interpret each element of Σ as a field embedding $F \to \overline{\mathbb{Q}}_p$.

The first step is to associate a "p-adic automorphic representation" – a continuous group homomorphism $\pi_p : G(F)\backslash G(\mathbb{A}_F) \to \overline{\mathbb{Q}}_p$ – to π, which we do by mimicking the standard construction in the split case. For $\sigma \in \Sigma$ recall that σ_∞ is the induced map $G(F_\infty) \to G_\sigma(\mathbb{C})$. By Proposition 5.27 there are characters $\lambda_\sigma \in X^*(G_\sigma)$ (regarded here as maps $G_\sigma(\mathbb{C}) \to \mathbb{C}^\times$) for each $\sigma \in \Sigma$ with the property that

$$\pi|_{G(F_\infty)^0} = \prod_{\sigma \in \Sigma} \lambda_\sigma \circ \sigma_\infty.$$

The right-hand side of the above equation can be regarded as a character of $G(F_\infty)$ and hence as a character λ_∞ of $G(\mathbb{A}_F)$, trivial at the finite places. Define $\pi^{\mathrm{alg}} = \pi/\lambda_\infty$, a continuous group homomorphism $G(\mathbb{A}_F) \to \mathbb{C}^\times$ trivial on $G(F_\infty)^0$ but typically non-trivial on $G(F)$. However $\pi^{\mathrm{alg}}(G(F))$ is contained within a number field by Corollary 5.28, and it is now easy to check that the image of π^{alg} is contained within $\overline{\mathbb{Q}}$. We now regard π^{alg} as taking values in $\overline{\mathbb{Q}}_p$ via our fixed embedding $\overline{\mathbb{Q}} \subset \overline{\mathbb{Q}}_p$.

Now, let $F_p = F \otimes_{\mathbb{Q}} \mathbb{Q}_p$, and note that every $\sigma : F \to \overline{\mathbb{Q}}_p$ in Σ induces a map $F_p \to \overline{\mathbb{Q}}_p$ and hence a map $\sigma_p : G(F_p) \to G_\sigma(\overline{\mathbb{Q}}_p)$. Each λ_σ can be regarded as a map $G_\sigma(\overline{\mathbb{Q}}_p) \to \overline{\mathbb{Q}}_p^\times$, and hence the product

$$\lambda_p := \prod_\sigma \lambda_\sigma \circ \sigma_p$$

is a continuous group homomorphism $G(F_p) \to \overline{\mathbb{Q}}_p^\times$ and can also be regarded as a continuous group homomorphism $G(\mathbb{A}_F) \to \overline{\mathbb{Q}}_p^\times$, trivial at all places other than those above p. The crucial point, which is easy to check, is that the product $\pi_p := \pi^{\mathrm{alg}}\lambda_p$ is a continuous group homomorphism $G(\mathbb{A}_F) \to \overline{\mathbb{Q}}_p^\times$ which is trivial on $G(F)$.

Now, in Theorem 2(b) of [Lan97], Langlands proves that there is a natural surjection with finite kernel from the set of $\widehat{G}(\mathbb{C})$-conjugacy classes of continuous homomorphisms from the Weil group W_F to ${}^LG(\mathbb{C})$ to the set of continuous homomorphisms from $G(F)\backslash G(\mathbb{A}_F)$ to $\mathbb{C}^\times$ (that is, the set of automorphic representations of G/F), compatible with the local Langlands correspondence at every place. His proof starts by establishing a natural surjection from the analogous sets with the continuity conditions removed, and then checking that continuity on one side is equivalent to continuity on the other. However, $\overline{\mathbb{Q}}_p \cong \mathbb{C}$ as abstract fields, and one can check that the calculations on pages 243ff make no use of any particular features of the topology of $\mathbb{C}^\times$, and hence apply equally well to continuous homomorphisms $W_F \to {}^LG(\overline{\mathbb{Q}}_p)$ and continuous characters $G(F)\backslash G(\mathbb{A}_F) \to \overline{\mathbb{Q}}_p^\times$. Thus π_p gives a continuous homomorphism (or perhaps several, in which case we simply choose one)

$$r_\pi : W_F \to {}^LG(\overline{\mathbb{Q}}_p),$$

and by construction we see that for each finite place $v \nmid p$ at which $\pi_v = (\pi_p)_v$ is unramified, $r_\pi|_{W_{F_v}}$ is $\widehat{G}(\overline{\mathbb{Q}}_p)$-conjugate to r_{π_v}. Again, by construction the composite of r_π and the natural projection ${}^LG(\overline{\mathbb{Q}}_p) \to \mathrm{Gal}(\overline{F}/F)$ is just the natural surjection $W_F \to \mathrm{Gal}(\overline{F}/F)$.

Lemma 5.30. *The representation r_π of W_F factors through the natural surjection $W_F \to \mathrm{Gal}(\overline{F}/F)$.*

Proof. The kernel of the natural surjection $W_F \to \mathrm{Gal}(\overline{F}/F)$ is the connected component of the identity in W_F; but r_π must vanish on this, because ${}^LG(\overline{\mathbb{Q}}_p)$ is totally disconnected. □

We let ρ_π denote the representation of $\mathrm{Gal}(\overline{F}/F)$ determined by r_π.

Lemma 5.31. *The representation ρ_π satisfies all the properties required in the statement of Conjecture 5.17.*

Proof. We need to check the claimed properties at places dividing p and at real places. For the former, we must firstly check that ρ_π is de Rham with the correct Hodge–Tate weights. However, it is sufficient to check this after restriction to any finite extension of F, and in particular we may choose an extension which splits G. The evident compatibility of the construction of ρ_π with base change then easily reduces us to the split case, which is standard. Similarly, the property of being crystalline may be checked over any unramified extension, and if π_v is unramified then by definition G splits over an unramified extension of F_v, and we may again reduce to the split case.

Suppose now that v is a real place of F. Recall that the natural surjection $W_{F_v} \to \mathrm{Gal}(\overline{F_v}/F_v)$ sends j to complex conjugation, so we need to determine $r_\pi|_{W_{F_v}}(j)$. Let $\sigma_v : F \hookrightarrow \mathbb{C}$ denote the embedding corresponding to v (it is unique because v is a real place) and let λ_v denote the character λ_{σ_v} of G_{σ_v}. Let χ_v denote the map $G(F_v) \to \mathbb{C}^\times$ induced by σ_v and λ_v. Then $(\pi_p)_v = \pi_v/\chi_v$. Applying local Langlands we see that the cohomology class in $H^1(W_{F_v}, \widehat{G})$ associated to $(\pi_p)_v$ is the difference of those associated to π_v and χ_v (because local Langlands is an isomorphism of groups in this abelian setting). Furthermore, one can check (either by the construction of the local Langlands correspondence for real tori in section 9.4 of [Bor79], or an explicit case-by-case check) that the cohomology class attached to χ_v is represented by a cocycle which sends $j \in W_{F_v}$ to the element $\lambda_v(-1)$ of $\widehat{G}$ (where we now view λ_v as a cocharacter $\mathbb{C}^\times \to \widehat{G}$). Our assertion about $r_\pi|_{W_{F_v}}(j)$ now follows immediately from an explicit calculation on cocycles. □

5. Twisting and Gross' η

5.1. Algebraicity and arithmeticity under central extensions

Definition 5.32. We will call a central extension

$$1 \to \mathbb{G}_m \to G' \to G \to 1$$

of algebraic groups over F a $\mathbb{G}_m$*-extension* of G.

Note that one can (after making compatible choices of maximal compact subgroups at infinity, and using the fact that $H^1(F, \mathbb{G}_m) = 0$ by Hilbert 90) identify automorphic representations on G with automorphic representations on G' which are trivial on $\mathbb{G}_m$. We abuse notation slightly and speak about the $\mathbb{G}_m$-extension $G' \to G$ interchangeably with the $\mathbb{G}_m$-extension $1 \to \mathbb{G}_m \to G' \to G \to 1$.

We now consider how our various notions of arithmeticity and algebraicity behave under pulling back via a $\mathbb{G}_m$-extension. So say $G' \to G$ is a $\mathbb{G}_m$-extension. The induced map $G'(\mathbb{A}_F) \to G(\mathbb{A}_F)$ is a surjection, and if π is an automorphic representation of $G(\mathbb{A}_F)$ then the induced representation π' of $G'(\mathbb{A}_F)$ is also an automorphic representation. Furthermore, we have the following compatibilities between π and π'.

Lemma 5.33. *π is L-algebraic (resp. C-algebraic, resp. L-arithmetic, resp. C-arithmetic) if and only if π' is.*

Proof. Let us start with the L- and C-algebraicity assertions. These assertions follow from purely local assertions at infinity: one needs to check that if k is an archimedean local field (a completion of F in the application), and if π is a representation of $G(k)$, with π' the induced representation of $G'(k)$, then π is L-algebraic (resp. C-algebraic) if and only if π' is. These statements can easily be checked using infinitesimal characters. Indeed a straightforward calculation (using an explicit description of the Harish-Chandra isomorphism) shows that if T' is a maximal torus in G' over the complexes (where we base change G' to the complexes via the map $k \to \mathbb{C}$ induced from $\sigma : \overline{k} \to \mathbb{C}$), and if the image of T' in G is T (a maximal torus of G), and if λ_σ and λ'_σ are the elements of $X^*(T) \otimes_{\mathbb{Z}} \mathbb{C}$ and $X^*(T') \otimes_{\mathbb{Z}} \mathbb{C}$ corresponding to π and π' as in Section 2.3, then the natural map $X^*(T) \otimes_{\mathbb{Z}} \mathbb{C} \to X^*(T') \otimes_{\mathbb{Z}} \mathbb{C}$ sends λ_σ to λ'_σ and both results follow easily.

It remains to prove the arithmeticity statements. Again these statements follow from purely local assertions. Let k denote a non-archimedean local field (a non-archimedean completion of F at which everything is unramified in the application) and let π be an irreducible smooth representation of $G(k)$, with π' the corresponding representation of $G'(k)$. Assume π' and π (and hence G' and G) are unramified. The C-arithmeticity assertion of the Proposition follows from the assertion that π is defined over a subfield E of $\mathbb{C}$ iff π' is; this is however immediate from Lemma 5.3. The L-arithmeticity statement follows from the assertion that the Satake parameter of π is defined over E iff the Satake parameter of π' is defined over E, which is then what remains to be proved. So let T' be the centraliser of a maximal split torus in G' over k, and let T be its image in G. Then $T'(k) \to T(k)$ is a surjection by Hilbert 90. As noted in the proof of Lemma 5.5, Theorem 2.9 of [BZ77] shows that the W_d-orbit of complex characters of $T(k)$ determined by the Satake isomorphism applied to π are precisely the characters for which π occurs as a subquotient of the corresponding induced representations, and the analogous assertion also holds for π'. It follows that the orbit of characters of $T'(k)$ corresponding to π' is precisely the orbit induced from the characters of $T(k)$ via the surjection $T'(k) \to T(k)$. This implies that the Satake parameter of π' (thought of as a character of $\mathbb{Q}[X_*(T'_d)]^{W_d}$) is induced from the Satake parameter of π via a map between the corresponding unramified Hecke algebras which is in fact the obvious map, and our assertion now follows easily. □

5.2. Twisting elements

We now explain the relationship between L-algebraic and C-algebraic automorphic representations for a connected reductive group G over a number field F. In particular, we examine the general question of when L-algebraic representations can be twisted to C-algebraic representations, following an idea of Gross (see [Gro99]). We show that in general it is always possible to replace G by a $\mathbb{G}_m$-extension for which this twisting is possible, and in this way one can formulate general conjectures about the association of Galois representations to C-algebraic (and in particular cohomological by Lemma 5.52 below) automorphic representations. As usual, let X^* denote the character group in the based root datum for G, with its Galois action. Let us stress that we always equip X^* with the Galois action coming from the construction used to define the L-group (which might well not be the same as the "usual" Galois action on $X^*(T)$ induced by the Galois action on $T(\overline{F})$, if T is a maximal torus in G which happens to be defined over F).

Definition 5.34. We say that an element $\theta \in X^*$ is a *twisting element* if θ is $\mathrm{Gal}(\overline{F}/F)$-stable and $\langle \theta, \alpha^\vee \rangle = 1 \in \mathbb{Z}$ for all simple coroots $\alpha^\vee$.

For some groups G there are no twisting elements; for example, if $G = \mathrm{PGL}_2$. On the other hand, if G is semi-simple and simply-connected then half the sum of the positive roots is a twisting element. Another case where twisting elements exist are groups G that are split and have simply-connected derived subgroup, for example $G = \mathrm{GL}_n$, although in this case half the sum of the positive roots might not be in X^*.

If Q is the quotient of G by its derived subgroup, then $X^*(Q) \subseteq X^*$ and the arguments in section II.1.18 of [Jan03] show that $X^*(Q) = (X^*)^W$, where W is the Weyl group of $G_{\overline{F}}$. Furthermore, $X^*(Q)$ is $\mathrm{Gal}(\overline{F}/F)$-stable, and the induced action of $\mathrm{Gal}(\overline{F}/F)$ on $X^*(Q)$ is precisely the usual action, induced by the Galois action on $Q(\overline{F})$.

Now let δ denote half the sum of the positive roots of G. If $\delta \in X^*$ then δ is a twisting element; but in general we only have $\delta \in \frac{1}{2}X^*$. Let S' denote the maximal split torus quotient of G, so that

$$X^*(S') = (X^*)^{W,\mathrm{Gal}(\overline{F}/F)}.$$

Then if θ is a twisting element, we see that

$$\theta - \delta \in \frac{1}{2}X^*(S').$$

Thus we have a character $|\cdot|^{\theta-\delta}$ of $G(F)\backslash G(\mathbb{A}_F)$, defined as the composite

$$G(\mathbb{A}_F) \longrightarrow S'(\mathbb{A}_F) \xrightarrow{2(\theta-\delta)} \mathbb{A}_F^\times \xrightarrow{|\cdot|} \mathbb{R}_{>0} \xrightarrow{x \mapsto \sqrt{x}} \mathbb{R}_{>0}$$

The main motivation behind the notion of twisting elements is the following two propositions.

Proposition 5.35. *If θ is a twisting element, then an automorphic representation π is C-algebraic if and only if $\pi \otimes |\cdot|^{\theta-\delta}$ is L-algebraic.*

Proof. This is a consequence of condition 10.3(2) of [Bor79], although formally one has to "reverse-engineer" the construction of (using the notation of §10.2 of [Bor79]) $\alpha \mapsto \pi_\alpha$. We sketch the argument using the notation there. The question is local at each infinite place, so let k denote a completion of F

at an infinite place. Let $\widetilde{G}$ be the central extension of G described in §10.2 of [Bor79]. The character $|\cdot|^{\theta-\delta}$ induces a character of $G(k)$ and of $\widetilde{G}(k)$. If Q denotes the maximal torus quotient of $\widetilde{G}$ then this character can be extended to a character of $Q(k)$. The associated element of $H^1(W_k, \widehat{Q})$ is the image of an element $\alpha \in H^1(W_k, Z_L)$, with Z_L the centre of $\widehat{G}$, and one checks easily that the character π_α of $G(k)$ in §10.2 of [Bor79] coincides with $|\cdot|^{\theta-\delta}$. If T_0 is a maximal torus in $\widetilde{G}$, and T is its image in G, then the restriction of α to $\overline{k}^\times$ is a Z_L-valued character which, when considered as a $\widehat{T_0}$-valued character of $\overline{k}^\times$, has image in $\widehat{T}$ and which (via an easy diagram chase) coincides with the restriction to $\overline{k}^\times$ of the cohomology class associated via local Langlands to the restriction of $|\cdot|^{\theta-\delta}$ to $T(k)$. Hence a on $\overline{k}^\times$ is the composite of the norm map down to $\mathbb{R}_{>0}$, the square root map, and the cocharacter of $\widehat{T}$ associated to $2(\theta-\delta)$. Twisting π by $|\cdot|^{\theta-\delta}$ corresponds to twisting r_π by a by 10.3(2) of [Bor79] and the result follows easily. □

Proposition 5.36. *If θ is a twisting element, then π is C-arithmetic if and only if $\pi \otimes |\cdot|^{\theta-\delta}$ is L-arithmetic.*

Proof. Again this is a local issue: by Definitions 5.12 and 5.13 it suffices to check that if k (a completion of F) is a non-archimedean local field, if χ denotes the restriction of $|\cdot|^{\theta-\delta}$ to $G(k)$ and if π is an unramified representation of $G(k)$, then π is defined over a subfield E of $\mathbb{C}$ iff $\pi \otimes \chi$ has Satake parameter defined over the same subfield E. This is relatively easy to check: we sketch the details (using the notation of section 2.2). Let T be a maximal torus of G/k, with maximal compact subgroup ${}^oT$. Then χ induces an automorphism i of $H_{\mathbb{C}}(T(k), {}^oT)$ sending $[{}^oTt{}^oT]$ to $\chi(t)[{}^oTt{}^oT]$, and i commutes with the action of the Weyl group W_d and hence induces an automorphism i of the W_d-invariants of this complex Hecke algebra. If m_π is the complex character of $H(T(k), {}^oT)^{W_d}$ associated to π and $m_{\pi\otimes\chi}$ is the character associated to $\pi \otimes \chi$ then one checks easily that $m_{\pi\otimes\chi} = m_\pi \circ i$. The other observation we need is that if K is a hyperspecial maximal compact subgroup of $G(k)$ and if S denotes the Satake isomorphism $S : H_{\mathbb{C}}(G(k), K) \to H_{\mathbb{C}}(T(k), {}^oT)^{W_d}$ then $i \circ S$ maps $H_{\mathbb{Q}}(G(k), K)$ into $H_{\mathbb{Q}}(T(k), {}^oT)$ (this follows immediately from formula (19) of section 4.2 of [Car79]), and hence into $H_{\mathbb{Q}}(T(k), {}^oT)^{W_d}$, and an injection between $\mathbb{Q}$-vector spaces which becomes an isomorphism after tensoring with $\mathbb{C}$ must itself be an isomorphism. Hence $i \circ S : H_{\mathbb{Q}}(G(k), K) \cong H_{\mathbb{Q}}(T(k), {}^oT)^{W_d}$ and now composing with m_π the result follows easily. □

Thus for groups with a twisting element, our L-notions and C-notions can be twisted into each other.

5.3. Adjoining a twisting element

If G has a twisting element then we have just seen that one can, by twisting, pass between our L- and C- notions. What can one do when G has no twisting element (for example if $G = \mathrm{PGL}_2$)? In this case we will show that G has a $\mathbb{G}_m$-extension $\widetilde{G}$ which *does* have a twisting element. Here we use some ideas that we learnt from reading a 2007 letter from Deligne to Serre; we thank Dick Gross for drawing our attention to this letter. We remark that previous versions of this manuscript contained a slightly messier construction involving a two-step process, reducing first via a z-extension to the the case where G had simply-connected derived subgroup and then making another extension from there (this procedure of reducing to the case of a simply-connected derived subgroup seems to be often used in the literature but it turned out not to be necessary in this case).

Proposition 5.37. *(a) There is a $\mathbb{G}_m$-extension*

$$1 \to \mathbb{G}_m \to \widetilde{G} \to G \to 1$$

such that $\widetilde{G}$ has a canonical twisting element θ.

(b) If G has a twisting element then $\widetilde{G} \cong G \times \mathbb{G}_m$. More generally, there is a natural bijection between the set of splittings of $1 \to \mathbb{G}_m \to \widetilde{G} \to G \to 1$ and the set of twisting elements of G (and in particular, the sequence splits over F if and only if G has a twisting element).

Proof. (a) Let G^{ad} denote the quotient of G by its centre and let G^{sc} denote the simply-connected cover of G^{ad}. Over $\overline{F}$ we can choose compatible (unnamed) Borels and tori T^{sc} and T^{ad} in G^{sc} and G^{ad}, and hence define the notion of a positive root in $X^*(T^{\mathrm{sc}})$ and $X^*(T^{\mathrm{ad}})$.

Now G^{sc} is simply-connected, and hence half the sum of the positive roots for G^{sc} is a character η of T^{sc} and hence induces a character γ of $Z := \ker(G^{\mathrm{sc}} \to G^{\mathrm{ad}})$. This character γ is independent of the notion of positivity because all Borels in $G^{\mathrm{sc}}_{\overline{F}}$ are conjugate. It takes values in μ_2 because η^2, the sum of the positive roots, is a character of T^{ad} and is hence trivial on Z. Furthermore γ is independent of the choice of T^{sc}, and defined over F.

Pushing the diagram

$$0 \to Z \to G^{\mathrm{sc}} \to G^{\mathrm{ad}} \to 0$$

out along $\gamma : Z \to \mathbb{G}_m$ gives us an extension

$$0 \to \mathbb{G}_m \to G^1 \to G^{\mathrm{ad}} \to 0$$

of groups over F and now pulling back along $G \to G^{\mathrm{ad}}$ gives us an extension

$$0 \to \mathbb{G}_m \to \widetilde{G} \to G \to 0.$$

The group $\widetilde{G}$ is the extension we seek. One can define it "all in one go" as a subquotient of $G \times G^{\mathrm{sc}} \times \mathbb{G}_m$: it is the elements (g, h, k) such that the images of g and h in G^{ad} are equal, modulo the image of the finite group Z under the map sending z to $(1, z, \gamma(z))$.

If T^1 and $\widetilde{T}$ are maximal tori in G^1 and $\widetilde{G}$ then the character groups of these tori fit into the following Galois-equivariant commutative diagram

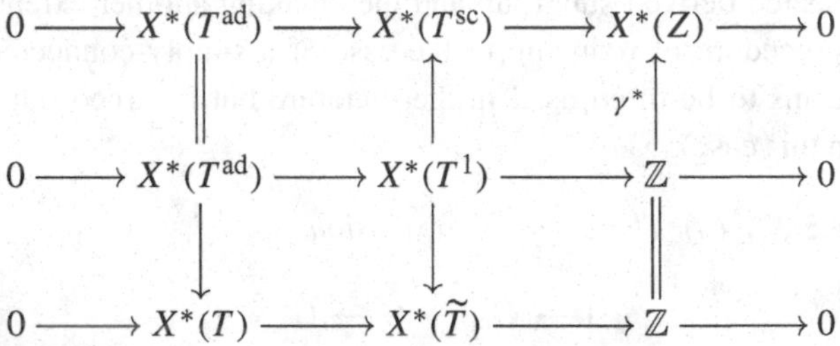

$$\begin{array}{ccccccccc}
0 & \longrightarrow & X^*(T^{\mathrm{ad}}) & \longrightarrow & X^*(T^{\mathrm{sc}}) & \longrightarrow & X^*(Z) & \longrightarrow & 0 \\
 & & \| & & \uparrow & & \uparrow{\scriptstyle \gamma^*} & & \\
0 & \longrightarrow & X^*(T^{\mathrm{ad}}) & \longrightarrow & X^*(T^1) & \longrightarrow & \mathbb{Z} & \longrightarrow & 0 \\
 & & \downarrow & & \downarrow & & \| & & \\
0 & \longrightarrow & X^*(T) & \longrightarrow & X^*(\widetilde{T}) & \longrightarrow & \mathbb{Z} & \longrightarrow & 0
\end{array}$$

where $X^*(T^1)$ can be thought of as a pullback – the subgroup of $X^*(T^{\mathrm{sc}}) \oplus \mathbb{Z}$ consisting of elements whose images in $X^*(Z)$ coincide, and $X^*(\widetilde{T})$ can be thought of as a pushforward – the quotient of $X^*(T^1) \oplus X^*(T)$ by $X^*(T^{\mathrm{ad}})$ embedded anti-diagonally.

We define θ thus: there is an element θ^1 of $X^*(T^1)$ whose image in $X^*(T^{\mathrm{sc}})$ is η, and whose image in $\mathbb{Z}$ is 1. We let θ be the image of θ^1 in $X^*(\widetilde{T})$.

We claim that θ is a twisting element for $\widetilde{G}$, and this will suffice to prove part (a). Note that the snake lemma implies that $X^*(T^1)$ injects into $X^*(\widetilde{T})$ and hence it suffices to show that θ^1 is a twisting element for G^1. It is a standard fact that η pairs to one with each simple coroot, and it follows immediately that θ^1 pairs to one with each simple coroot. Furthermore, η is a Galois-stable element of $X^*(T^{\mathrm{sc}})$, and Galois acts trivially on $\mathbb{Z}$, from which one can deduce that Galois acts trivially on θ, and so θ is indeed a twisting element.

(b) Recall that we have a short exact sequence

$$0 \to X^*(T) \to X^*(\widetilde{T}) \to \mathbb{Z} \to 0.$$

If t is a twisting element for G, then $t \in X^*(T)$ can be regarded as an element of $X^*(\widetilde{T})$ whose image in $\mathbb{Z}$ is zero. Recall that $\theta \in X^*(\widetilde{T})$ from (a) is a twisting element for $\widetilde{G}$, whose image in $\mathbb{Z}$ is 1. Now consider the difference $\chi := \theta - t$. This is an element in $X^*(\widetilde{T})$ which is Galois stable and pairs to zero with each simple coroot. Hence if $\widetilde{Q}$ denotes the maximal split torus quotient of $\widetilde{G}$, then χ is an element of the subgroup $X^*(\widetilde{Q})$ of $\widetilde{T}$. In particular, χ induces a map $\widetilde{G} \to \mathbb{G}_m$ defined over F, and the composite $\mathbb{G}_m \to \widetilde{G} \to \mathbb{G}_m$ is the identity map. It follows easily that the induced map $\widetilde{G} \to G \times \mathbb{G}_m$ induced by χ and the canonical map $\widetilde{G} \to G$ is an isomorphism.

This same argument shows that there is a bijection between the splitting elements for G and the splittings of the short exact sequence. To give a splitting of the exact sequence is to give a map $\widetilde{G} \to \mathbb{G}_m$ such that the induced map $\mathbb{G}_m \to \widetilde{G} \to \mathbb{G}_m$ is the identity. If $\chi : \widetilde{G} \to \mathbb{G}_m$ is such a map then χ gives an element of $X^*(\widetilde{T})$ which is Galois stable, pairs to zero with each simple coroot, and whose image in $\mathbb{Z}$ is 1. Conversely any such element gives a splitting. Now one checks easily that $\chi \in X^*(\widetilde{T})$ has these properties then $\theta - \chi$ (θ our fixed twisting element for $\widetilde{G}$ coming from part (a)) has image zero in $\mathbb{Z}$, so can be regarded as an element of $X^*(T)$ that can easily be checked to be a twisting element for G, and conversely for any twisting element $t \in X^*(T)$, we have already seen that $\theta - t$ gives a splitting. □

The importance of $\widetilde{G}$ is that if π is a C-algebraic representation for G, we can pull it back to $\widetilde{G}$ and then twist as in Proposition 5.35 to get an L-algebraic representation, for which we predict the existence of a Galois representation – but into the L-group of $\widetilde{G}$ rather than the L-group of G. The construction of $\widetilde{G}$ is completely canonical, and this motivates the following definition:

Definition 5.38. The *C-group of G* is defined to be the L-group of $\widetilde{G}$. We will denote this group by ${}^CG$.

The C-group is "as functorial as the L-group is", and naturally receives the Galois representations conjecturally (and in some cases provably) attached to C-algebraic automorphic representations for G. One can think of the C-group as doing, in a canonical way, the job of unravelling all the square roots that appear in the Satake isomorphism normalised a la Langlands when applied to a cohomological representation (for we shall see later on in Lemma 5.52 that cohomological representations are C-algebraic). Note that the dimension of the C-group is one more than the dimension of the L-group, but this extra degree of freedom is cancelled out by the fact that the inclusion $\mathbb{G}_m \to \widetilde{G}$ gives rise to a map $d : {}^CG \to \mathbb{G}_m$ and we will see later on that our conjectures imply that if π is C-algebraic for G then the associated Galois representation, when composed with the map $d : {}^CG(\overline{\mathbb{Q}}_p) \to \mathbb{G}_m(\overline{\mathbb{Q}}_p)$, is always the cyclotomic character.

It would be nice to see in a concrete manner how the C-group is related to the L-group and we achieve this in the next proposition. Let χ denote the sum of the positive roots for G (after choosing some Borel and torus); then χ can be thought of as a cocharacter of a maximal torus $\widehat{T}$ of $\widehat{G}$, that is, a map $\mathbb{G}_m \to \widehat{T}$. Set $e = \chi(-1)$.

Proposition 5.39. *The element e is a central element of $\widehat{G}$ and is independent of all choices. There is a canonical surjection*

$$\widehat{G} \times \mathbb{G}_m \to \widehat{\widetilde{G}}$$

with kernel central and of order 2, generated by $(e, -1)$. This surjection is equivariant for the action of Galois used to define the L-groups of G and $\widetilde{G}$.

Proof. We have a canonical map $c : \widetilde{G} \to G$; let us beef this up to an isogeny $\widetilde{G} \to G \times \mathbb{G}_m$ with kernel of order 2. To do this we need to construct an appropriate map $\xi : \widetilde{G} \to \mathbb{G}_m$, which we do thus: recall that the intermediate group G^1 used in the construction of $\widetilde{G}$ was a push-out of $G^{\mathrm{sc}} \times \mathbb{G}_m$ along Z; we define a map $G^{\mathrm{sc}} \times \mathbb{G}_m \to \mathbb{G}_m$ by sending (g, λ) to λ^2; the image of Z in $\mathbb{G}_m$ has order at most 2, and thus this induces a map $G^1 \to \mathbb{G}_m$ and hence a map $\xi : \widetilde{G} \to \mathbb{G}_m$. The restriction of ξ to the subgroup $\mathbb{G}_m$ of $\widetilde{G}$ is the map $\lambda \mapsto \lambda^2$, and hence the kernel of the induced map $(c, \xi) : \widetilde{G} \to G \times \mathbb{G}_m$ has order 2 and hence this map is an isogeny for dimension reasons. This isogeny induces a dual isogeny

$$\widehat{G} \times \mathbb{G}_m \to \widehat{\widetilde{G}}$$

with kernel of order 2, and so to check that the kernel is generated by $(e, -1)$ it suffices to prove that $(e, -1)$ is in the kernel. But one easily checks that $(\chi, 1)$ maps to 2θ under the natural map

$$X_*(\widehat{T}) \oplus \mathbb{Z} \to X_*(\widehat{\widetilde{T}})$$

induced by the isogeny, and the result follows immediately by evaluating this cocharacter at -1. Furthermore (c, ξ) is defined over F so the dual isogeny commutes with the Galois action used to define the L-groups. □

The interested reader can use the preceding Proposition to compute examples of C-groups. For example the C-group of $(\mathrm{GL}_n)_F$ is isomorphic to the group $(\mathrm{GL}_n \times \mathrm{GL}_1)_{\overline{\mathbb{Q}}} \times \mathrm{Gal}(\overline{F}/F)$, as $e = (-1)^{n-1}$ and the surjection $\mathrm{GL}_n \times \mathrm{GL}_1 \to \mathrm{GL}_n \times \mathrm{GL}_1$ sending (g, μ) to $(g\mu^{n-1}, \mu^2)$ has kernel $(e, -1)$. As another example, the C-group of $(\mathrm{PGL}_2)_F$ is isomorphic to $(GL_2)_{\overline{\mathbb{Q}}} \times \mathrm{Gal}(\overline{F}/F)$ because the kernel of the natural map $\mathrm{SL}_2 \times \mathrm{GL}_1 \to \mathrm{GL}_2$ is $(e, -1)$. Note however that the C-group of $(\mathrm{PGL}_n)_F$ is not in general $(\mathrm{GL}_n)_{\overline{\mathbb{Q}}} \times \mathrm{Gal}(\overline{F}/F)$; its connected component is $(\mathrm{SL}_n \times \mathrm{GL}_1)/\langle(-1)^{n-1}, -1\rangle$, which is isomorphic to $\mathrm{SL}_n \times \mathrm{GL}_1$ if n is odd, and to a central extension of GL_n by a cyclic group of order $n/2$ if n is even.

One can use the above constructions and Conjecture 5.16 to formulate a conjecture associating ${}^C G$-valued Galois representations to C-algebraic automorphic representations (and hence, by Lemma 5.52 below, to cohomological automorphic representations) for an arbitrary connected reductive group G over a number field. One uses Lemma 5.33 to pull the C-algebraic representation back to a C-algebraic representation on $\widetilde{G}$, twists using θ and the recipe in the statement of Proposition 5.35 to get an L-algebraic representation on $\widetilde{G}$, and then uses Conjecture 5.16 on this bigger group. The map denoted $|.|^{\theta-\delta}$ in the aforementioned proposition can be checked to be the map $\widetilde{G}(\mathbb{A}_F) \to \mathbb{R}_{>0}$ sending g to $|\xi(g)|^{1/2}$, with $\xi : \widetilde{G} \to \mathbb{G}_m$ the map defined in the proof of Proposition 5.39. Note finally that the map $\mathbb{G}_m \to \widetilde{G}$ induces a map $d : {}^C G \to \mathbb{G}_m$.

Unravelling, and in particular using Remark 5.1 with the G there being our $\widetilde{G}$, we see that Conjecture 5.16 implies the following conjecture.

Conjecture 5.40. *If π is C-algebraic, then there is a finite subset S of the places of F, containing all infinite places, all places dividing p, and all places where π is ramified, and a continuous Galois representation $\rho_\pi = \rho_{\pi,\iota}$: $\mathrm{Gal}(\overline{F}/F) \to {}^C G(\overline{\mathbb{Q}}_p)$, which satisfies*

- *The composite of ρ_π and the natural projection ${}^C G(\overline{\mathbb{Q}}_p) \to \mathrm{Gal}(\overline{F}/F)$ is the identity map.*
- *The composite of ρ_π and d is the cyclotomic character.*
- *If $v \notin S$, then $\rho_\pi|W_{F_v}$ is $\widehat{\widetilde{G}}(\overline{\mathbb{Q}}_p)$-conjugate to the representation sending $w \in W_{F_v}$ to $\iota(r_{\pi_v}(w)\hat{\xi}(|w|^{1/2}))$, where $\hat{\xi}$ is the map $\mathbb{C}^\times \to \widehat{\widetilde{G}}(\mathbb{C})$ dual to ξ, and where the norm on the Weil group sends a geometric Frobenius to the reciprocal of the size of the residue field.*
- *If v is a finite place dividing p then $\rho_\pi|_{\mathrm{Gal}(\overline{F_v}/F_v)}$ is de Rham, and the Hodge–Tate cocharacter of this representation can be explicitly read off from π via the recipe of Remark 5.41 below.*
- *If v is a real place, let $c_v \in G_F$ denote a complex conjugation at v, let λ_{σ_v} and $\lambda_{\tau_v} \in X^*(T) \otimes \mathbb{C}$ be the elements associated to π_v in Section 2.3, and let $\tilde{\lambda}_{\sigma_v}$ and $\tilde{\lambda}_{\tau_v}$ denote their images in $X^*(\widetilde{T}) \otimes \mathbb{C}$. Then $\tilde{\lambda}_{\sigma_v} + \frac{1}{2}\xi$, $\tilde{\lambda}_{\tau_v} + \frac{1}{2}\xi \in X^*(\widetilde{T}) = X_*(\widehat{\widetilde{T}})$ and $\rho_{\pi,\iota}(c_v)$ is $\widehat{G}(\overline{\mathbb{Q}}_p)$-conjugate to the element*
$$\iota(\alpha_v) = \iota\left(\left(\tilde{\lambda}_{\sigma_v} + \frac{1}{2}\xi\right)(i)\left(\tilde{\lambda}_{\tau_v} + \frac{1}{2}\xi\right)(i) r_{\pi_v}(j)\right)$$
associated to the twist of the lift of π, as in Lemma 5.8.

Remark 5.41. The recipe for the Hodge–Tate cocharacter in the conjectures above is as follows. As in Remark 5.18, any $j : F \to \overline{\mathbb{Q}}$ gives rise to an infinite place w of F equipped with a fixed map $F_w \to \mathbb{C}$, which we use to identify $\mathbb{C}$ with an algebraic closure of F_w. The identity $\sigma : \mathbb{C} \to \mathbb{C}$ and the representation π_w give rise to λ_σ as in Remark 5.18, except that this time $\lambda_\sigma \in (X^*(T) \otimes_{\mathbb{Z}} \frac{1}{2}\mathbb{Z})/W$. We thus obtain an element $\tilde{\lambda}_\sigma \in (X^*(\widetilde{T}) \otimes_{\mathbb{Z}} \frac{1}{2}\mathbb{Z})/W$. Then $\tilde{\lambda}_\sigma + \frac{1}{2}\xi \in X^*(\widetilde{T})/W$, and our conjecture is that this element $\tilde{\lambda}_\sigma + \frac{1}{2}\xi$ is the Hodge–Tate cocharacter associated to the embedding $F_v \to \overline{\mathbb{Q}}_p$.

We end this section by showing that the results in it imply the equivalence of Conjectures 5.14 and 5.15 (made for all groups simultaneously).

Proposition 5.42. *Let G be a connected reductive group over a number field. If Conjecture 5.14 is true for $\widetilde{G}$ then Conjecture 5.15 is true for G. Similarly if Conjecture 5.15 is true for $\widetilde{G}$ then Conjecture 5.14 is true for G.*

Proof. We prove the first assertion; the second one is similar. Say G is connected and reductive, and π is C-arithmetic (resp. C-algebraic). Let π' be the pullback of π to $\widetilde{G}$. Then π' is C-arithmetic (resp. C-algebraic) by Lemma 5.33. By Proposition 5.36 (resp. Proposition 5.35) $\pi' \otimes |.|^{\theta-\delta}$ is L-arithmetic (resp. L-algebraic). Applying Conjecture 5.14 to $\widetilde{G}$ we deduce that $\pi' \otimes |.|^{\theta-\delta}$ is L-algebraic (resp. L-arithmetic). Running the argument backwards now shows us that π is C-algebraic (resp. C-arithmetic). □

6. Functoriality

6.1.

Suppose that G, G' are two connected reductive groups over F, and that we have an *algebraic* L-group homomorphism

$$r : {}^L G \to {}^L G',$$

i.e. a homomorphism of algebraic groups over $\overline{\mathbb{Q}}$ which respects the projections to $\mathrm{Gal}(\overline{F}/F)$ (recall that we are using the Galois group rather than the Weil group when forming the L-group). Assume that G' is quasi-split over F. Then we have the following weak version of Langlands' functoriality conjecture (note that we are only demanding compatibility with the local correspondence at a subset of the unramified places, and at infinity).

Conjecture 5.43. *If π is an automorphic representation of G, then there is an automorphic representation π' of G', called a* functorial transfer *of π, such that*

- *For all infinite places v, and for all finite places v at which π and G' are unramified, $r_{\pi'_v}$ is $\widehat{G}'(\mathbb{C})$-conjugate to $r \circ r_{\pi_v}$.*

A trivial consequence of the definitions is

Lemma 5.44. *If π is L-algebraic, then any functorial transfer of π is also L-algebraic.*

We also have the only slightly less trivial

Lemma 5.45. *If π is L-arithmetic, then any functorial transfer of π is also L-arithmetic.*

Proof. This result follows from a purely local assertion. If v is a finite place where G and G' are unramified and if k is the completion of F at v then the morphism r of L-groups induces a morphism $\widehat{G} \to \widehat{G}'$ which commutes with the action of the Frobenius at v. If T_d (resp. T'_d) denotes a maximal k-split torus in G/k (resp. G'/k) with centraliser T (resp. T') then the map $\widehat{G} \to \widehat{G}'$ induces a map $\widehat{T} \to \widehat{T}'$ (well-defined up to restricted Weyl group actions) which commutes with the action of Frobenius, and hence maps $\widehat{T}_d \to \widehat{T}'_d$ and $X^*(\widehat{T}'_d) = X_*(T'_d) \to X^*(\widehat{T}_d) = X_*(T_d)$. Now looking at the explicit definition of the Satake isomorphism in Proposition 6.7 of [Bor79] we see, after unravelling, that the map $\mathbb{Q}[X_*(T'_d)]^{W'_d} \to \mathbb{Q}[X_*(T_d)]^{W_d}$ induced from $X_*(T'_d) \to X_*(T_d)$ above has the property that, after tensoring up to $\mathbb{C}$ and taking spectra, it sends the point in $\mathrm{Spec}(\mathbb{C}[X_*(T_d)]^{W_d})$ corresponding to r_{π_v} to the point in $\mathrm{Spec}(\mathbb{C}[X_*(T'_d)]^{W'_d})$ corresponding to $r \circ r_{\pi_v}$. We now deduce that if the Satake parameter of π_v is defined over a subfield E of $\mathbb{C}$ then so is the Satake parameter of π'_v (because the homomorphism $\mathbb{Q}[X_*(T'_d)]^{W'_d} \to \mathbb{C}$ corresponding to π'_v factors through $\mathbb{Q}[X_*(T_d)]^{W_d}$ and hence through E) and the result follows. □

In addition,

Proposition 5.46. *If Conjecture 5.16 holds for π, then it holds for any functorial transfer of π.*

Proof. With notation as above, one easily checks that $\rho_{\pi',\iota} := r \circ \rho_{\pi,\iota}$ satisfies all the conditions of Conjecture 5.16. □

Note that functoriality relies on things normalised in Langlands' canonical way; the natural analogues of the results above in the C-algebraic and C-arithmetic cases are not true in general, because a morphism of algebraic groups does not send half the sum of the positive roots to half the sum of the positive roots in general.

7. Reality checks

7.1.

By Proposition 2 of [Lan79] any automorphic representation π on G is a subquotient of an induction $\mathrm{Ind}_{P(\mathbb{A}_F)}^{G(\mathbb{A}_F)}\sigma$, where P is a parabolic subgroup of G with Levi quotient M, and σ is a cuspidal representation of M. If π' is another automorphic subquotient of $\mathrm{Ind}_{P(\mathbb{A}_F)}^{G(\mathbb{A}_F)}\sigma$, then π_v and π'_v are equal for all but finitely many places, so π is C-arithmetic (respectively L-arithmetic) if and only if π' is C-arithmetic (respectively L-arithmetic). The following lemma shows that π is C-algebraic (respectively L-algebraic) if and only if π' is C-algebraic (respectively L-algebraic).

Lemma 5.47. *Suppose that π and π' are subquotients of a common induction $\mathrm{Ind}_{P(\mathbb{A}_F)}^{G(\mathbb{A}_F)}\sigma$. Then π and π' have the same infinitesimal character.*

Proof. This is immediate from the calculation of the infinitesimal character of an induction – see for example Proposition 8.22 of [Kna01]. □

Furthermore, we can check the compatibility of Conjecture 5.16 for π and π' (note that we cannot check the compatibility for Conjecture 5.17 because π and π' may be ramified at different places).

Proposition 5.48. *Suppose that π and π' are subquotients of a common induction $\mathrm{Ind}_{P(\mathbb{A}_F)}^{G(\mathbb{A}_F)}\sigma$. Suppose that π is L-algebraic. If Conjecture 5.16 is valid for π then it is valid for π'.*

Proof. Suppose that Conjecture 5.16 is valid for π. We wish to show that $\rho_{\pi',\iota} := \rho_{\pi,\iota}$ satisfies all the conditions in Conjecture 5.16. Since for all but finitely many places π_v and π'_v are unramified and isomorphic, the first two conditions are certainly satisfied. The third condition is satisfied by Lemma 5.47. It remains to check that if v is a real place, then (with obvious notation) $\lambda_\sigma(i)\lambda_\tau(i)r_{\pi_v}(j)$ and $\lambda'_\sigma(i)\lambda'_\tau(i)r_{\pi'_v}(j)$ are $\widehat{G}(\mathbb{C})$-conjugate. As explained to us by David Vogan, it follows from the results of [ABV92] (specifically from Theorem 1.24 and Proposition 6.16) that $\lambda_\sigma(-1)r_{\pi_v}(j)$ and $\lambda'_\sigma(-1)r_{\pi'_v}(j)$ are $\widehat{G}(\mathbb{C})$-conjugate. It is easy to check that these are $\widehat{G}(\mathbb{C})$-conjugate to $\lambda_\sigma(i)\lambda_\tau(i)r_{\pi_v}(j)$ and $\lambda'_\sigma(i)\lambda'_\tau(i)r_{\pi'_v}(j)$ respectively (one conjugates by $r_{\pi_v}(e^{-i\pi/4})$, $r_{\pi'_v}(e^{-i\pi/4})$), as required. □

More generally, if π, π' are nearly equivalent (that is, π_v and π'_v are isomorphic for all but finitely many v), then π is C-arithmetic (respectively L-arithmetic) if and only if π' is C-arithmetic (respectively L-arithmetic). We would like to be able to prove as above that π and π' have the same infinitesimal character, and we would like to obtain the analogue of Proposition 5.48.

Unfortunately, these seem in general to be beyond the reach of current techniques. However, we can prove these results for GL_n, and we can then deduce them for general groups under the assumption of functoriality.

Proposition 5.49. *If $G = \mathrm{GL}_n$ and π, π' are nearly equivalent, then π, π' have the same infinitesimal character. Suppose further that π is L-algebraic. If Conjecture 5.16 is valid for π then it is valid for π'.*

Proof. By the strong multiplicity one theorem for isobaric representations (Theorem 4.4 of [JS81]), π, π' are both subquotients of a common induction $\mathrm{Ind}_{P(\mathbb{A}_F)}^{G(\mathbb{A}_F)} \sigma$. The result follows from Lemma 5.47 and Proposition 5.48. □

Proposition 5.50. *Let G be arbitrary. Assume Conjecture 5.43. If π and π' are nearly equivalent automorphic representations of G, then for any infinite place v, π_v and π'_v have the same infinitesimal characters. Suppose further that π is L-algebraic. If Conjecture 5.16 is valid for π then it is valid for π'.*

Proof. To begin with, note that for each infinite place v we have a natural injection ${}^LG_v \to {}^LG$, where G_v is the base change of G to F_v. Since $r_{\pi_v}|_{\mathbb{C}^\times}$ is valued in ${}^LG_v$, as is $c_v := \lambda_\sigma(i)\lambda_\tau(i)r_{\pi_v}(j)$ if v is real, we see that their $\widehat{G}(\mathbb{C})$-conjugacy classes in ${}^LG$ are determined by their $\widehat{G}(\mathbb{C})$-conjugacy classes in ${}^LG_v$.

Now, the $\widehat{G}(\mathbb{C})$-conjugacy classes of semisimple elements of ${}^LG_v(\mathbb{C})$ are determined by the knowledge of the conjugacy classes of their images under all representations of ${}^LG_v(\mathbb{C})$. To see this, note that since the formation of the L-group is independent of the choice of inner form, it suffices to check this in the case where G_v is quasi-split; but the result then follows immediately from Proposition 6.7 of [Bor79].

Let $r : {}^LG \to \mathrm{GL}_n \times \mathrm{Gal}(\overline{F}/F)$ be a homomorphism of L-groups. Then by Conjecture 5.43, there are automorphic representations Π, Π' on GL_n which are functorial transfers of π, π' respectively. By Proposition 5.49, Π and Π' have the same infinitesimal characters. Thus for each infinite place v, $r_{\Pi_v}|_{\mathbb{C}^\times}$ and $r_{\Pi'_v}|_{\mathbb{C}^\times}$ are conjugate, i.e. $r \circ r_{\pi,\iota}|_{\mathbb{C}^\times}$ and $r \circ r_{\pi',\iota}|_{\mathbb{C}^\times}$ are conjugate. Since this is true for all r, we see that $r_{\pi,\iota}|_{\mathbb{C}^\times}$ and $r_{\pi',\iota}|_{\mathbb{C}^\times}$ are conjugate, whence π and π' have the same infinitesimal character.

As in the proof of Proposition 5.48, it remains to check that if v is a real place of F, then $c_v := \lambda_\sigma(i)\lambda_\tau(i)r_{\pi_v}(j)$ and $c'_v := \lambda'_\sigma(i)\lambda'_\tau(i)r_{\pi'_v}(j)$ are $\widehat{G}(\mathbb{C})$-conjugate. By a similar argument to that used in the first half of this proof, we see that if $r : {}^LG \to \mathrm{GL}_n \times \mathrm{Gal}(\overline{F}/F)$ is a homomorphism of L-groups, then $r(c_v)$ and $r(c'_v)$ are conjugate in $\mathrm{GL}_n(\mathbb{C})$. Furthermore, c_v and c'_v are both semisimple. Thus c_v and c'_v are $\widehat{G}(\mathbb{C})$-conjugate. □

7.2. Cohomological representations

Cohomological automorphic representations provide a good testing ground for our conjectures. It follows easily (see below) that any cohomological representation is C-algebraic, and one can often show that they are C-arithmetic, too (it would not surprise us if Shimura variety experts could prove they were always C-arithmetic with relative ease). In the case $G = \mathrm{GL}_n$ these arguments are due to Clozel, who also shows that for GL_n any regular C-algebraic representation is cohomological after possibly twisting by a quadratic character (see Lemme 3.14 of [Clo90]).

Let v be an infinite place of F, and let K_v be the fixed choice of a maximal compact subgroup of $G(F_v)$ used in the definition of automorphic forms on G. Let $\mathfrak{g}_v$ be the complexification of the Lie algebra of $G(F_v)$. Recall that π_v may be thought of as a $(\mathfrak{g}_v, K_v)$-module, with underlying $\mathbb{C}$-vector space V_v, say.

Definition 5.51. We say that π_v is *cohomological* if there is an algebraic complex representation U of $G(F_v)$ and a non-negative integer i such that

$$H^i(\mathfrak{g}_v, K_v; U \otimes V_v) \neq 0.$$

We say that π is cohomological if π_v is cohomological for all archimedean places v.

Lemma 5.52. *If π is cohomological, then it is C-algebraic.*

Proof. By Corollary 4.2 of [BW00], if π is cohomological then for each archimedean place v there is a continuous finite-dimensional representation U_v of $G(F_v)$ such that π_v and U_v have the same infinitesimal characters. The result then follows from Lemma 5.53 below. □

Lemma 5.53. *If v is an archimedean place of F and U is a continuous finite dimensional representation of $G(F_v)$ with infinitesimal character χ_v, identified with an element of $X^*(T) \otimes_{\mathbb{Z}} \mathbb{C}$ as in Section 2.3, then $\chi_v - \delta \in X^*(T)$.*

Proof. This follows almost at once from the definition of the Harish-Chandra isomorphism; see for example (5.43) in [Kna02]. □

We note that a cuspidal cohomological unitary automorphic representation π is also C-arithmetic, at least when π_∞ is cohomological for the trivial representation; for we can restrict scalars down to $\mathbb{Q}$ and then follow the argument in §2.3 of [BR94]. This argument presumably works in some greater generality.

8. Relationship with theorems/conjectures in the literature

8.1.

In [Clo90], Clozel makes a number of conjectures about certain C-algebraic automorphic representations for GL_n. We now examine the compatibility of these conjectures with those of this chapter. Clozel calls an automorphic representation of GL_n *algebraic* if it is C-algebraic and isobaric; his principal reason for restricting to isobaric representations is that he wishes to use the language of Tannakian categories.

Let $\pi = \otimes' \pi_v$ be an algebraic (in Clozel's sense) representation of GL_n over F. Then Clozel conjectures (see conjectures 3.7 and 4.5 of [Clo90]) that

Conjecture 5.54. *Let $\pi_f = \otimes'_{v \nmid \infty} \pi_v$. Then there is a number field $E \subset \mathbb{C}$ such that π_f is defined over E (that is, such that $\pi_f \otimes_{\mathbb{C},\sigma} \mathbb{C} \cong \pi_f$ for all automorphisms σ of $\mathbb{C}$ which fix E pointwise). In addition, Conjecture 5.17 holds for $\pi \otimes |\cdot|^{(n-1)/2}$.*

(In fact, Clozel conjectures much more than this – he conjectures that there is a motive whose local L-factors agree with those of $\pi \otimes |\cdot|^{(n-1)/2}$ at all finite places; the required Galois representation is then obtained as the p-adic realisation of this motive.)

By Proposition 5.49 we see, since any automorphic representation of GL_n is nearly equivalent to an isobaric one, that Conjecture 5.54 implies Conjecture 5.16 for GL_n, and in fact an examination of the proof shows that it implies Conjecture 5.17. We claim that it also implies that π is C-arithmetic; in fact, this follows at once from Proposition 3.1(iii) of [Clo90]. Thus for GL_n our conjectures follow from those of Clozel.

The reason that our conjectures are weaker than Clozel's conjectures is that for groups other than GL_n we do not have as good an understanding of the local Langlands correspondence – for a general group G we cannot even *formulate* a version of Conjecture 5.17 which includes behaviour at the bad places without also formulating a precise local Langlands conjecture in full generality. Even for GL_n Clozel had to be careful, restricting to isobaric representations in order not to make a conjecture which was trivially false, and such phenomena would also show up in the general case, and are typically even less well-understood here.

8.2.

In [Gro99] Gross presents a conjecture which assigns a Galois representation to an automorphic representation on a group G with the property that any arithmetic subgroup is finite (in fact Gross gives six conditions equivalent to this in Proposition 1.4 of [Gro99]). We now discuss the relationship of this conjecture to our conjectures. Gross' assumptions imply that G splits over a CM field L, and he also assumes that G has a twisting element η in the sense of Definition 5.34. In fact, Gross has informed us that one should in addition assume that the group G is semisimple and simply connected, so we make this assumption from now on. This assumption in fact implies that one can take $\eta = \delta$ in the below, but for those who want to be more optimistic than Gross we have kept the two notations distinct in the below.

Let V be an absolutely irreducible representation of G over $\mathbb{Q}$ with trivial central character. Let S be a finite set of primes of size at least 2, containing all primes at which G is ramified. For each $\ell \notin S$ we let K_ℓ be a hyperspecial maximal compact subgroup of $G(\mathbb{Q}_\ell)$, and for each $\ell \in S$ we let K_ℓ be an Iwahori subgroup of $G(\mathbb{Q}_\ell)$. Let K be the product of the K_ℓ. Then $M(V, K)$ is the space of algebraic modular forms given by

$$M(V,K) := \{f : G(\mathbb{A}_\mathbb{Q})/(G(\mathbb{R})_+ \times K) \to V :\\ f(\gamma g) = \gamma f(g) \text{ for all } \gamma \in G(\mathbb{Q}),\, g \in G(\mathbb{A}_\mathbb{Q})\}.$$

Let H_S be the unramified Hecke algebra – the restricted tensor product of the unramified Hecke algebras H_ℓ for each $\ell \notin S$. Let H_K be the full Hecke algebra, the tensor product of H_S and the Iwahori Hecke algebras H_ℓ at places ℓ in S. Let A be $H_K \otimes \mathbb{Q}[\pi_0(G(\mathbb{R}))]$. This acts on $M(V, K)$ (see section 6 of [Gro99]), and we let N be a simple A-submodule of $M(V, K)$. We assume that N gives the Steinberg character on H_ℓ for all $\ell \in S$ (see section 12 of [Gro99]), and if V is trivial and $\prod_{l \in S} G(\mathbb{Q}_l)$ is compact, we exclude the case that N is trivial.

By Proposition 12.3 of [Gro99], $\mathrm{End}_A(N)$ is a CM field, and by (7.4) of [Gro99], $\pi_0(G(\mathbb{R}))$ acts on N through a character

$$\phi_\infty : \pi_0(G(\mathbb{R})) \to \{\pm 1\} \subset E^\times.$$

By Proposition 8.5 of [Gro99], the simple submodules of $N \otimes \mathbb{C}$ may be identified (compatibly with the actions of H_K) with irreducible automorphic representations $\pi = \pi_f \otimes \pi_\infty$ with $\pi_\infty \xrightarrow{\sim} V \otimes \mathbb{C}$, and π_l Steinberg for all $l \in S$. For all $l \notin S$, the unramified local Langlands correspondence (i.e. the Satake isomorphism) identifies the character of T_l on N with a homomorphism $r_{N,l} : W_{\mathbb{Q}_l} \to {}^L G(\mathbb{C})$, and π_l corresponds to this parameter under the local Langlands correspondence. Fix such a representation π.

Gross then makes the following conjecture (see Conjecture 17.2 as well as (15.3) and (16.8) of [Gro99]) (note that while Gross normalises his Weil groups so that an arithmetic Frobenius element corresponds to a uniformiser, he also normalises his Satake isomorphisms so that they are constructed with arithmetic Frobenii, so the comments of Remark 5.20 apply):

Conjecture 5.55. *If p is a prime, and $\iota : \mathbb{C} \xrightarrow{\sim} \overline{\mathbb{Q}}_p$, then there is a continuous Galois representation*

$$\rho_{N,\iota} : \mathrm{Gal}(\overline{\mathbb{Q}}/\mathbb{Q}) \to {}^L G(\overline{\mathbb{Q}}_p)$$

satisfying

- *If $l \notin S$, then $\rho_{N,\iota}|W_{\mathbb{Q}_l}$ is $\widehat{G}(\overline{\mathbb{Q}}_p)$-conjugate to $\iota(r_{N,l}) \otimes |\cdot|^{\eta-\delta}$.*
- *If s_∞ is a complex conjugation in $\mathrm{Gal}(\overline{\mathbb{Q}}/\mathbb{Q})$, then $\rho_{N,\iota}(s_\infty)$ is $\widehat{G}(\overline{\mathbb{Q}}_p)$-conjugate to $(\iota(\eta(-1)\phi_\infty(-1)), s_\infty)$.*

This conjecture follows from Conjecture 5.17. Indeed, the representation π is C-algebraic, so by Proposition 5.35 $\pi \otimes |\cdot|^{\eta-\delta}$ is L-algebraic. Applying Conjecture 5.17 gives everything in Conjecture 5.55 (for the description of complex conjugation, see [Gro07]).

Note that Gross in fact conjectures something slightly stronger; he shows that π is C-algebraic, and in fact that π is defined over E, and conjectures that for any place $\lambda|p$ of E there is a natural Galois representation $\rho_{N,\lambda} : \mathrm{Gal}(\overline{\mathbb{Q}}/\mathbb{Q}) \to {}^L G(E_\lambda)$. As Gross has explained to us, this rationality conjecture should follow from the hypothesis that π is Steinberg at two places, together with local-global compatibility for the Galois representations at these places.

8.3.

We now discuss an example drawn from [CHT08]. Let F be a totally real field, and let E be a quadratic totally imaginary extension of F. Let G be an n-dimensional unitary group over F which splits over E, and which is compact (that is, isomorphic to $U(n)$) at all infinite places. Then the dual group of G is GL_n, and if we let $\mathrm{Gal}(E/F) = \{1, c\}$, then the L-group of G is given by

$${}^L G = \mathrm{GL}_n \rtimes \mathrm{Gal}(\overline{F}/F)$$

where $\mathrm{Gal}(\overline{F}/F)$ acts on GL_n via its projection to $\mathrm{Gal}(E/F) = \{1, c\}$, with

$$x^c := \Phi_n x^{-t} \Phi_n^{-1}$$

where Φ_n is an anti-diagonal matrix with alternating entries $1, -1$. Note that $\Phi_n^{-1} = \Phi_n^t = (-1)^{n-1}\Phi_n$.

By Proposition 5.39, we have

$$^C G = \Big((\mathrm{GL}_n \times \mathbb{G}_m)/((-1)^{n-1}, -1) \Big) \rtimes \mathrm{Gal}(\overline{F}/F)$$

with $\mathrm{Gal}(\overline{F}/F)$ acting by

$$(g, \mu)^c = (\Phi_n g^{-t} \Phi_n^{-1}, \mu).$$

In Section 1 of [CHT08] there is a definition of a group $\mathcal{G}_n$. This group is a semidirect product $(\mathrm{GL}_n \times \mathbb{G}_m) \rtimes \mathrm{Gal}(\overline{F}/F)$, but with $\mathrm{Gal}(\overline{F}/F)$ acting by

$$(g, \mu)^c = (\mu g^{-t}, \mu).$$

There is a morphism $j : {}^C G \to \mathcal{G}_n$ defined thus. For $(g, \mu) \in \mathrm{GL}_n \times \mathbb{G}_m$ and $\gamma \in \mathrm{Gal}(\overline{F}/E)$ we set

$$j((g, \mu) \times \gamma) = (g\mu^{1-n}, \mu^{2(1-n)}) \times \gamma$$

(and note that $j(((-1)^{n-1}, -1) \times \gamma) = (1 \times \gamma)$. If $\tilde{c} \in \mathrm{Gal}(\overline{F}/F)$ has image $c \in \mathrm{Gal}(E/F)$ then we set

$$j(1 \times \tilde{c}) = (\Phi_n, (-1)^{n-1}) \times \tilde{c}.$$

It is easily checked these determine a unique homomorphism $j : {}^C G \to \mathcal{G}_n$. For $n > 1$, j is an isogeny with kernel of order $n - 1$. Since G is compact at infinity, any automorphic representation π of $G(\mathbb{A}_F)$ is cohomological and thus by Lemma 5.52 is C-algebraic. Conjecture 5.40 predicts the existence of a Galois representation

$$\rho_\pi : \mathrm{Gal}(\overline{F}/F) \to {}^C G(\overline{\mathbb{Q}}_p)$$

with the property that $d \circ \rho_\pi$ is the cyclotomic character ε (with d as in the discussion preceding Conjecture 5.40). One checks easily that the composite

$$j \circ \rho_\pi : \mathrm{Gal}(\overline{F}/F) \to \mathcal{G}_n(\overline{\mathbb{Q}}_p)$$

has multiplier ε^{1-n} (the multiplier of a representation into $\mathcal{G}_n$ is its projection onto the $\mathbb{G}_m$ factor).

Now, under certain mild hypotheses on G and π, we note that a Galois representation r'_π satisfying the properties imposed on $j \circ r_\pi$ by Conjecture 5.40 is proved to exist in [CHT08] and [Tay08]. Specifically, everything except the form of complex conjugation follows from Proposition 3.4.4 of [CHT08] (although see also Theorems 4.4.2 and Theorems 4.4.3 of [CHT08] for related results on GL_n whose notation may be easier to compare to the notation used in this paper), and the form of complex conjugation follows from Theorem 4.1 of [Tay08]. (Note when comparing the unramified places that by definition the local Langlands correspondence $\mathrm{rec}(\pi_v)$ used in [CHT08] is our r_{π_v}.)

Conversely, the constructions of [CHT08], when they apply, actually imply our Conjecture 5.40 for π (up to Frobenius semisimplification at unramified finite places). To see this, note there is a morphism $j' : \mathcal{G}_n \times \mathrm{GL}_1 \to {}^C G$ such that $j' \circ (j \times d)$ is the identity; concretely, j' is defined on the identity component by

$$((h, \mu), \lambda) \mapsto (h\lambda^{(n-1)/2}, \lambda^{1/2})$$

(which is well-defined independent of the choice of square root by the definition of ${}^C G$). Then we may set $r_\pi = j' \circ (r'_\pi \times \varepsilon)$, which is easily checked to have the required properties.

Note: Florian Herzig points out that in Section 2.4, the element ν_j only gives rise to an element of $X_*(\widehat{T})$ modulo the normaliser of $\widehat{T}$ in H, which might be bigger than the Weyl group. The fix is to attach an element of $X_*(\widehat{T})/W$ not to a map $k \to \overline{\mathbb{Q}}_p$ but instead to a map $\overline{k} \to \overline{\mathbb{Q}}_p$, which is indeed what we have in the setting of Conjectures 5.16 and 5.17. More details can be found in `arXiv:1009.0785`, the ArXiv version of this paper.

References

[ABD+66] M. Artin, J. E. Bertin, M. Demazure, P. Gabriel, A. Grothendieck, M. Raynaud, and J.-P. Serre, *Schémas en groupes. Fasc. 7: Exposés 23 à 26*, Séminaire de Géométrie Algébrique de l'Institut des Hautes Études Scientifiques, vol. 1963/64, Institut des Hautes Études Scientifiques, Paris, 1965/1966. MR 0207710 (34 #7525)

[ABV92] Jeffrey Adams, Dan Barbasch, and David A. Vogan, Jr., *The Langlands classification and irreducible characters for real reductive groups*, Progress in Mathematics, vol. 104, Birkhäuser Boston Inc., Boston, MA, 1992. MR MR1162533 (93j:22001)

[Art02] James Arthur, A note on the automorphic Langlands group, *Canad. Math. Bull.* **45** (2002), no. 4, 466–482, Dedicated to Robert V. Moody. MR MR1941222 (2004a:11120)

[Bor79] A. Borel, Automorphic L-functions, *Automorphic forms, representations and L-functions* (Proc. Sympos. Pure Math., Oregon State Univ., Corvallis, Ore., 1977), Part 2, Proc. Sympos. Pure Math., XXXIII, Amer. Math. Soc., Providence, R.I., 1979, pp. 27–61. MR MR546608 (81m:10056)

[BR94] Don Blasius and Jonathan D. Rogawski, Zeta functions of Shimura varieties, *Motives* (Seattle, WA, 1991), Proc. Sympos. Pure Math., vol. 55, Amer. Math. Soc., Providence, RI, 1994, pp. 525–571. MR MR1265563 (95e:11051)

[Bru03] Farrell Brumley, Maass cusp forms with quadratic integer coefficients, *Int. Math. Res. Not.* (2003), no. 18, 983–997. MR 1962012 (2004d:11025)

[BW00] A. Borel and N. Wallach, *Continuous cohomology, discrete subgroups, and representations of reductive groups*, second edn., Mathematical Surveys and Monographs, vol. 67, American Mathematical Society, Providence, RI, 2000. MR MR1721403 (2000j:22015)

[BZ77] I. N. Bernstein and A. V. Zelevinsky, Induced representations of reductive p-adic groups. *I, Ann. Sci. École Norm. Sup. (4)* **10** (1977), no. 4, 441–472. MR MR0579172 (58 #28310)

[Car79] P. Cartier, Representations of p-adic groups: a survey, *Automorphic forms, representations and L-functions* (Proc. Sympos. Pure Math., Oregon State Univ., Corvallis, Ore., 1977), Part 1, Proc. Sympos. Pure Math., XXXIII, Amer. Math. Soc., Providence, R.I., 1979, pp. 111–155. MR MR546593 (81e:22029)

[CHT08] Laurent Clozel, Michael Harris, and Richard Taylor, Automorphy for some l-adic lifts of automorphic mod l Galois representations, *Pub. Math. IHES* **108** (2008), 1–181.

[Clo90] Laurent Clozel, Motifs et formes automorphes: applications du principe de fonctorialité, *Automorphic forms, Shimura varieties, and L-functions*, Vol. I (Ann Arbor, MI, 1988), Perspect. Math., vol. 10, Academic Press, Boston, MA, 1990, pp. 77–159. MR MR1044819 (91k:11042)

[Gro99] Benedict H. Gross, Algebraic modular forms, *Israel J. Math.* **113** (1999), 61–93. MR MR1729443 (2001b:11037)

[Gro07] ———, Odd Galois representations, preprint, 2007.

[Jan03] Jens Carsten Jantzen, *Representations of algebraic groups*, second edn., Mathematical Surveys and Monographs, vol. 107, American Mathematical Society, Providence, RI, 2003. MR MR2015057 (2004h:20061)

[JS81] H. Jacquet and J. A. Shalika, On Euler products and the classification of automorphic forms. II, *Amer. J. Math.* **103** (1981), no. 4, 777–815. MR MR623137 (82m:10050b)

[Kna01] Anthony W. Knapp, *Representation theory of semisimple groups*, Princeton Landmarks in Mathematics, Princeton University Press, Princeton, NJ, 2001, An overview based on examples, Reprint of the 1986 original. MR MR1880691 (2002k:22011)

[Kna02] ———, *Lie groups beyond an introduction*, second edn., Progress in Mathematics, vol. 140, Birkhäuser Boston Inc., Boston, MA, 2002. MR MR1920389 (2003c:22001)

[Kot84] Robert E. Kottwitz, Stable trace formula: cuspidal tempered terms, *Duke Math. J.* **51** (1984), no. 3, 611–650. MR MR757954 (85m:11080)

[Lan79] R. P. Langlands, On the notion of an automorphic representation, *Automorphic forms, representations and L-functions* (Proc. Sympos. Pure Math., Oregon State Univ., Corvallis, Ore., 1977), Part 1, Proc. Sympos. Pure Math., XXXIII, Amer. Math. Soc., Providence, R.I., 1979, pp. 203–207.

[Lan97] ———, Representations of abelian algebraic groups, *Pacific J. Math.* (1997), Special Issue, 231–250, Olga Taussky-Todd: in memoriam. MR MR1610871 (99b:11125)

[Sar02] Peter Sarnak, Maass cusp forms with integer coefficients, *A panorama of number theory or the view from Baker's garden* (Zürich, 1999),

Cambridge University Press, Cambridge, 2002, pp. 121–127. MR 1975448 (2004c:11053)

[Ser79] Jean-Pierre Serre, Groupes algébriques associés aux modules de Hodge-Tate, *Journées de Géométrie Algébrique de Rennes.* (Rennes, 1978), Vol. III, Astérisque, vol. 65, Soc. Math. France, Paris, 1979, pp. 155–188. MR MR563476 (81j:14027)

[Spr79] T. A. Springer, Reductive groups, *Automorphic forms, representations and L-functions* (Proc. Sympos. Pure Math., Oregon State Univ., Corvallis, Ore., 1977), Part 1, Proc. Sympos. Pure Math., XXXIII, Amer. Math. Soc., Providence, R.I., 1979, pp. 3–27. MR MR546587 (80h:20062)

[Tan57] Yutaka Taniyama, L-functions of number fields and zeta functions of abelian varieties, *J. Math. Soc. Japan* **9** (1957), 330–366. MR 0095161 (20 #1667)

[Tay08] Richard Taylor, Automorphy for some l-adic lifts of automorphic mod l Galois representations. II, *Pub. Math. IHES* **108** (2008), 183–239.

[Tit79] J. Tits, Reductive groups over local fields, *Automorphic forms, representations and L-functions* (Proc. Sympos. Pure Math., Oregon State Univ., Corvallis, Ore., 1977), Part 1, Proc. Sympos. Pure Math., XXXIII, Amer. Math. Soc., Providence, R.I., 1979, pp. 29–69. MR MR546588 (80h:20064)

[Vog93] David A. Vogan, Jr., The local Langlands conjecture, *Representation theory of groups and algebras*, Contemp. Math., vol. 145, Amer. Math. Soc., Providence, RI, 1993, pp. 305–379. MR MR1216197 (94e:22031)

[Wal81] Michel Waldschmidt, Transcendance et exponentielles en plusieurs variables, *Invent. Math.* **63** (1981), no. 1, 97–127. MR MR608530 (82k:10042)

[Wal82] ———, Sur certains caractères du groupe des classes d'idèles d'un corps de nombres, *Seminar on Number Theory, Paris 1980-81* (Paris, 1980/1981), Progr. Math., vol. 22, Birkhäuser Boston, Boston, MA, 1982, pp. 323–335. MR MR693328 (85c:11097)

[Wal85] J.-L. Waldspurger, Quelques propriétés arithmétiques de certaines formes automorphes sur GL(2), *Compositio Math.* **54** (1985), no. 2, 121–171. MR MR783510 (87g:11061a)

6

Geometry of the fundamental lemma

Pierre-Henri Chaudouard

Contents

1	Some words about the fundamental lemma	*page* 188
2	Local orbital integrals and their geometric interpretation	190
3	Hitchin fibration	196
4	Geometric description of Hitchin fibers	201
5	Examples of Hitchin fibers in rank 2	205
6	A truncated Hitchin fibration	210
7	The main cohomological result	213
References		219

1. Some words about the fundamental lemma

1.1. Origin

The global Langlands correspondence predicts deep relations between automorphic forms on a reductive group over a number field and representations of the absolute Galois group of this number field. It implies the Langlands functoriality, a family of transfers between automorphic representations on different groups. A general strategy to attack Langlands functoriality is still missing. However, in some specific but basic cases, the *endoscopic* ones, Langlands suggested to use the Arthur–Selberg trace formula. Roughly speaking, the trace formula relates the trace of a test function on the automorphic spectrum to more geometric distributions, namely the (global) orbital integrals attached to

Automorphic Forms and Galois Representations, ed. Fred Diamond, Payman L. Kassaei and Minhyong Kim. Published by Cambridge University Press.

rational conjugacy class. The point is that there exists an explicit transfer of (regular semisimple) conjugacy classes from an endoscopic group to the group itself and this transfer is expected to be dual to the transfer of automorphic forms. In particular, there should be deep relations between orbital integrals on a group and on its endoscopic groups. Conversely, if one knows these relations, one should get, by the trace formula, character identities between automorphic forms which should characterize the automorphic transfer. The global orbital integrals are products of local ones and we expect also relations between local orbital integrals. The simplest relation and the most important one is the fundamental lemma, a combinatorial identity between orbital integrals for the units of the Hecke algebra. It appears in the works [17] of Labesse–Langlands and [18] of Langlands, and it is stated in general in the work [19] of Langlands–Shelstad.

1.2. Geometry and cohomology

Ngô proved the fundamental lemma in [23]. More precisely, Ngô proved a variant of this statement for Lie algebra over local fields of positive characteristic. It is known by the work of Waldspurger that it suffices to prove this variant (cf. [25],[26], and cf. [9] for different methods). The advantage of this latter situation is that orbital integrals (both local and global) then have a geometric meaning: they count the number of rational points of some varieties over finite fields. By Grothendieck–Lefschetz trace formula, the fundamental lemma admits a cohomological interpretation. Ngô indeed proves the fundamental lemma through a deep cohomological study of the elliptic part of the Hitchin fibration.

1.3. The fundamental lemma for $GL(n)$

In the rest of the chapter, we will focus only on the group $GL(n)$. In this case, the fundamental lemma is a tautological statement. Nonetheless, Ngô's main cohomological theorem is still very deep in this situation. Moreover, his geometric and cohomological arguments become easier to understand. We will also say something about the extension of Ngô's work by Laumon and myself (cf. [7] and [8]) which is the key of our proof of the *weighted* fundamental lemma (a generalized form of the fundamental lemma stated by Arthur which is also needed in the endoscopic program).

Let's describe quickly how the chapter is organized. We begin in Section 2 by the geometric interpretation of local orbital integrals through affine Springer fibers. Then we introduce Hitchin's fibration in Section 3. We give a geometric description of Hitchin fibers in Section 4 using the Hitchin–Beauville–Narasimhan–Ramanan correspondence. Then in Section 5 we study an example of a non-separated Hitchin fiber. In Section 6, we explain the truncation of the Hitchin fibration we used in [7]. In the more technical Section 7, we state and explain Ngô's main cohomological theorem and its extension. We discuss at length some good open subset of the base of the Hitchin fibration. We hope that this makes the constructions of Section 9 of [8] more accessible.

1.4. Acknowledgement

This expository chapter is largely based on the one hand on the papers [22] and [23] of Ngô Bao Châu and on the other hand on my work [7] and [8] with Gérard Laumon on the weighted fundamental lemma. This text also benefits greatly from many talks Laumon or me gave on this subject. I thank Gérard Laumon for having shared with me some of his notes and for his help for the figures. Finally I would like to thank the referee for carefully reading the chapter.

2. Local orbital integrals and their geometric interpretation

2.1. Notation

Let $\mathbb{F}_q$ be a finite field with q elements. Let n be an integer. We assume that the characteristic of $\mathbb{F}_q$ satisfies the following inequality

$$\mathrm{char}(\mathbb{F}_q) > n.$$

Let

$$G = GL(n)$$

over $\mathbb{F}_q$ and let $\mathfrak{g} = \mathfrak{gl}(n)$ be its Lie algebra. The adjoint action of G on $\mathfrak{g}$ is simply the action by conjugation of $GL(n)$ on $\mathfrak{gl}(n)$.

Let

$$\mathcal{O} = \mathbb{F}_q[[\varepsilon]]$$

be the ring of power series with coefficients in $\mathbb{F}_q$ and let F be its fraction field. So

$$F = \mathbb{F}_q((\varepsilon))$$

is the field of Laurent power series with coefficients in $\mathbb{F}_q$.

We will frequently use left quotients denoted by \.

2.2. Orbital integrals

Let $\gamma \in \mathfrak{g}(F)$. We assume that γ is *regular semisimple* which means that its characteristic polynomial has n distinct roots in a suitable extension of F. The centralizer of γ in G (by abuse of notation, we will not distinguish G and $G \times_{\mathbb{F}_q} F$) is then a F-maximal subtorus of G. It is denoted by T_γ or simply T.

Let

$$\mathbf{1}_{\mathfrak{g}(\mathcal{O})}$$

be the characteristic function of $\mathfrak{g}(\mathcal{O})$. We can attach to the element γ and the function $\mathbf{1}_{\mathfrak{g}(\mathcal{O})}$ the following *orbital integral*

$$\mathcal{O}_\gamma = \int_{T(F)\backslash G(F)} \mathbf{1}_{\mathfrak{g}(\mathcal{O})}(g^{-1}\gamma g)\, dg$$

where the measure dg is the quotient of Haar measures on $G(F)$ and $T(F)$ respectively normalized by

$$\mathrm{vol}(G(\mathcal{O})) = 1$$

and

$$\mathrm{vol}(X_*(T)\backslash T(F)) = 1$$

where $X_*(T)$ is the group of F-rational cocharacters of T. Here the choice of the uniformizer ε gives an injective morphism

$$\begin{array}{ccc} X_*(T) & \to & T(F) \\ \lambda & \mapsto & \lambda(\varepsilon) \end{array}$$

so that we can view the group $X_*(T)$ as a subgroup of $T(F)$.

Remark 6.1. The orbital integral is finite: since the function $\mathbf{1}_{\mathfrak{g}(\mathcal{O})}$ is compactly supported and the orbit of γ is closed, the integral can be taken over a compact subset of $T(F)\backslash G(F)$.

The integral $\mathcal{O}_\gamma$ is the Lie algebra analog of an orbital integral attached to the unit function of the spherical algebra of $G(F)$.

2.3. Orbital integral as a counting function

A subset $\mathcal{L} \subset F^n$ is called a lattice if it is a sub-$\mathcal{O}$-module free of rank n. For example, $\mathcal{L}_0 = \mathcal{O}^n$ is a such a lattice and it will be called "standard" in the sequel. Let us denote by $\mathfrak{X}$ the set of lattices.

The group $G(F)$ acts on F^n and thus acts on the set $\mathfrak{X}$ of lattices. It is easy to see that the action is transitive. Moreover the stabilizer of the standard lattice is $G(\mathcal{O})$.

For any endomorphism $\gamma \in \mathfrak{g}(F)$ of F^n, we say that a lattice is γ-stable if $\gamma(\mathcal{L}) \subset \mathcal{L}$. Let

$$\mathfrak{X}_\gamma := \{\mathcal{L} \in \mathfrak{X} \mid \gamma(\mathcal{L}) \subset \mathcal{L}\}$$

be the set of γ-stable lattices. For example, the standard lattice $\mathcal{L}_0$ is γ-stable if and only if $\gamma \in \mathfrak{g}(\mathcal{O})$. If $g \in G(F)$ and $\mathcal{L} = g \cdot \mathcal{L}_0$ then $\mathcal{L}$ is γ-stable if and only if $\mathcal{L}_0$ is $g^{-1}\gamma g$-stable that is if and only if $g^{-1}\gamma g \in \mathfrak{g}(\mathcal{O})$.

Proposition 6.2. *The map $g \mapsto g\mathcal{O}^n$ gives bijections*

- $G(F)/G(\mathcal{O}) \simeq \mathfrak{X}$
- $\{g \in G(F)/G(\mathcal{O}) \mid g^{-1}\gamma g \in \mathfrak{g}(\mathcal{O})\} \simeq \mathfrak{X}_\gamma$.

From now on, let us assume that $\gamma \in \mathfrak{g}(F)$ is *regular semisimple*. Remember we have defined the orbital integral $\mathcal{O}_\gamma$.

Proposition 6.3. *The group $X_*(T)$ acts on $\mathfrak{X}_\gamma$ without fixed points and*

$$\mathcal{O}_\gamma = |X_*(T)\backslash\mathfrak{X}_\gamma|.$$

Proof. The subgroup $T(F)$ of $G(F)$ acts also on $\mathfrak{X}$. Since any element of $T(F)$ commutes with γ, the action preserves $\mathfrak{X}_\gamma$. By restriction, we get an action of the subgroup $X_*(T)$. The stabilizer of a lattice in $X_*(T)$ is a compact discrete subgroup thus a finite group: it must be trivial since the group $X_*(T)$ is a free $\mathbb{Z}$-module. Hence the action of $X_*(T)$ does not have fixed points. For the remaining equality, using our normalization of measures, we can write

$$\begin{aligned}\mathcal{O}_\gamma &= \int_{X_*(T)\backslash G(F)} \mathbf{1}_{\mathfrak{g}(\mathcal{O})}(g^{-1}\gamma g)\, dg \\ &= \sum_{g \in X_*(T)\backslash G(F)/G(\mathcal{O})} \int_{H_g\backslash G(\mathcal{O})} \mathbf{1}_{\mathfrak{g}(\mathcal{O})}((gk)^{-1}\gamma gk)\, dk\end{aligned}$$

where $H_g = G(\mathcal{O}) \cap g^{-1}X_*(T)g$ is a torsion-free, discrete and compact group and hence must be trivial. Moreover, using $\mathrm{vol}(G(\mathcal{O})) = 1$ and the fact that the map $g \mapsto \mathbf{1}_{\mathfrak{g}(\mathcal{O})}(g^{-1}\gamma g)$ is clearly invariant on the left by $G(\mathcal{O})$, we get

$$\begin{aligned}\mathcal{O}_\gamma &= \sum_{g \in X_*(T)\backslash G(F)/G(\mathcal{O})} \int_{G(\mathcal{O})} \mathbf{1}_{\mathfrak{g}(\mathcal{O})}(((gk)^{-1}\gamma gk) \\ &= \sum_{g \in X_*(T)\backslash G(F)/G(\mathcal{O})} \mathbf{1}_{\mathfrak{g}(\mathcal{O})}(g^{-1}\gamma g) \\ &= |X_*(T)\backslash\mathfrak{X}_\gamma|.\end{aligned}$$

□

2.4. Affine Grassmannian

We would like to view the set $\mathfrak{X}$ of lattices as the set of $\mathbb{F}_q$-points of an algebraic variety. This is possible for each set $\mathfrak{X}^{i,j}$ for integers $i \geqslant 0$ and $j \in \mathbb{Z}$ where

$$\mathfrak{X}^{i,j} = \{\mathcal{L} \in \mathfrak{X} \mid \varepsilon^i \mathcal{O}^n \subset \mathcal{L} \subset \varepsilon^{-i} \mathcal{O}^n \text{ and } \wedge^n \mathcal{L} = \varepsilon^j \mathcal{O}\}.$$

Here $\wedge^n \mathcal{L}$ is the maximal exterior power of $\mathcal{L}$: this is a fractional ideal of $\mathcal{O}$. Let V_i be the left quotient $\varepsilon^i \mathcal{O}^n \backslash \varepsilon^{-i} \mathcal{O}^n$: it is a $\mathbb{F}_q$-vector space. Note that the uniformizer induces on V_i a nilpotent endomorphism (still denoted by ε) which satisfies $\varepsilon^{2i} = 0$. The map

$$\mathcal{L} \to V_{\mathcal{L}} = \varepsilon^i \mathcal{O}^n \backslash \mathcal{L}$$

induces a bijection of $\mathfrak{X}^{i,j}$ onto the set

$$\{W \subset V_i \mid \dim(W) = ni - j \text{ and } \varepsilon(W) \subset W\}$$

of subspaces of V_i of dimension $ni - j$ which are stable under the nilpotent endomorphism ε. This latter set is the set of $\mathbb{F}_q$-points of a projective variety. Indeed, we can consider the grassmanniann of linear subspaces W of dimension $ni - j$ in V (this is a projective variety) and inside the grassmannian, the condition $\varepsilon(W) \subset W$ defines a closed subvariety which is classically known as a Springer fiber. But we have more: the inclusion $\mathfrak{X}^{i,j} \subset \mathfrak{X}^{i+1,j}$ gives a closed immersion of the corresponding Springer fibers. Thus the set

$$\mathfrak{X}^j = \{\mathcal{L} \in \mathfrak{X} \mid \wedge^n \mathcal{L} = \varepsilon^j \mathcal{O}\}$$

is the set of $\mathbb{F}_q$-points of an ind-variety. Finally, $\mathfrak{X}$ is the set of $\mathbb{F}_q$-points of the disjoint union (over $j \in \mathbb{Z}$) of the ind-varieties corresponding to $\mathfrak{X}^j$. This ind-variety is called the affine Grassmannian.

In the sequel, we change slightly our notation: $\mathfrak{X}$ denotes the affine Grassmannian. The set of $\mathbb{F}_q$-points is denoted by $\mathfrak{X}(\mathbb{F}_q)$.

2.5. Affine Springer fiber

Let $\gamma \in \mathfrak{g}(F)$ be a regular semisimple element. The condition of being γ-stable defines a closed ind-subvariety still denoted $\mathfrak{X}_\gamma \subset \mathfrak{X}$. The variety $\mathfrak{X}_\gamma$ was introduced by Kazhdan and Lusztig in [16] and it is now called an affine Springer fiber.

Theorem 6.4. *(Kazhdan–Lusztig, cf. [16])*

– *The reduced ind-scheme of $\mathfrak{X}_\gamma$ is represented by a variety, locally of finite type and of finite dimension.*

– *The quotient $X_*(T)\backslash\mathfrak{X}_\gamma$ is a projective variety (where T is the centralizer of γ in G).*

Remark 6.5. A formula for the dimension of $\mathfrak{X}_\gamma$ was stated by Kazhdan–Lusztig and proved by Bezrukavnikov in [5].

2.6. An example of affine Springer fiber

Take $G = GL(2)$ and $\gamma = \begin{pmatrix} \varepsilon & 0 \\ 0 & -\varepsilon \end{pmatrix}$. The set of connected components of the affine Springer fiber $\mathfrak{X}_\gamma$ is indexed by $\mathbb{Z}$. For $j \in \mathbb{Z}$, the connected component $\mathfrak{X}_\gamma^j$ is such that $\mathfrak{X}_\gamma^j(\mathbb{F}_q)$ is the set of lattices $\mathcal{L}$ in F^2 which are γ-stable and of index $\wedge^2\mathcal{L} = \varepsilon^j\mathcal{O}$. Let $p_1 : F^2 \to F$ (resp. p_2) be the first (resp. second) projection. For any lattice $\mathcal{L} \subset F^2$, let us consider the free $\mathcal{O}$-sub-modules of F defined by $\mathcal{L}_1 = p_1(\mathcal{L})$, $\mathcal{L}_2 = p_2(\mathcal{L})$, $\mathcal{L}^1 = \mathcal{L} \cap \mathrm{Ker}(p_2)$ and $\mathcal{L}^2 = \mathcal{L} \cap \mathrm{Ker}(p_1)$. We have $\wedge^2\mathcal{L} \simeq \mathcal{L}_1 \otimes_\mathcal{O} \mathcal{L}^2 \simeq \mathcal{L}_2 \otimes_\mathcal{O} \mathcal{L}^1$ and a commutative diagram

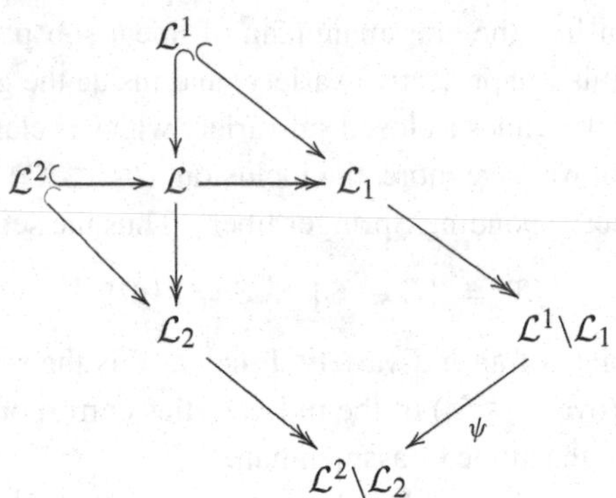

where the arrow ψ is an isomorphism of $\mathcal{O}$-module defined by $\psi(x_1) = x_2$ for any $x_1 \in \mathcal{L}_1$ and $x_2 \in \mathcal{L}_2$ such that $x_1 + x_2 \in \mathcal{L}$. If moreover the lattice $\mathcal{L}$ is γ-stable, the morphism ψ is also γ-equivariant for the actions induced by γ, namely multiplication by ε on $\mathcal{L}^1\backslash\mathcal{L}_1$, resp. $-\varepsilon$ on $\mathcal{L}^2\backslash\mathcal{L}_2$. It follows that $\mathcal{L}^1\backslash\mathcal{L}_1$ is killed by ε. Let $i \in \mathbb{Z}$ be such that $\mathcal{L}^1 = \varepsilon^i\mathcal{O}$. Since we have

$$\mathcal{L}^1 \oplus \mathcal{L}^2 \subset \mathcal{L} \subset \mathcal{L}_1 \oplus \mathcal{L}_2,$$

we also have

$$\varepsilon^i\mathcal{O} \oplus \varepsilon^{j-i+1}\mathcal{O} \subset \mathcal{L} \subset \varepsilon^{i-1}\mathcal{O} \oplus \varepsilon^{j-i}\mathcal{O}. \tag{6.1}$$

Let $\mathbb{P}_i$ be the set of lattices $\mathcal{L}$ which satisfies (6.1) and $\wedge^2\mathcal{L} = \varepsilon^j\mathcal{O}$. Clearly, $\mathbb{P}_i$ is the set of $\mathbb{F}_q$-points of a projective line. Any lattice $\mathcal{L} \in \mathbb{P}_i$ is γ-stable. The

Figure 1 Infinite chain of projective lines

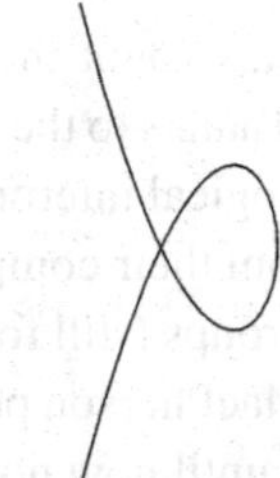

Figure 2 A projective line with a node

intersection $\mathbb{P}_i \cap \mathbb{P}_{i'}$ is empty unless $i' \in \{i+1, i, i-1\}$. Moreover $\mathbb{P}_i \cap \mathbb{P}_{i+1}$ is reduced to a single split lattice $\varepsilon^i \mathcal{O} \oplus \varepsilon^{j-i}\mathcal{O}$ (which is the point ∞ in $\mathbb{P}_i$ and the point 0 in $\mathbb{P}_{i+1}$).

In this way, we see that $\mathfrak{X}_\gamma^j$ is an infinite chain of projective lines as in Figure 1. The black nodes are the split lattices $\varepsilon^i \mathcal{O} \oplus \varepsilon^{j-i}\mathcal{O}$.

The centralizer of γ in G is a split torus of rank 2. We have $X_*(T) \simeq \mathbb{Z}^2$ and the action of $\mathbb{Z}^2$ on $\mathfrak{X}_\gamma$ permutes the connected components. The stabilizer of a connected component is isomorphic to $\mathbb{Z}$ and acts on it by translation on the chain. The quotient $X_*(T)\backslash\mathfrak{X}_\gamma$ looks like a projective line with a node (cf. Figure 2 below).

2.7. The work of Goresky–Kottwitz–MacPherson

In some cases, Goresky, Kottwitz and MacPherson were able to compute the cohomology of the affine Springer fiber $\mathfrak{X}_\gamma$. In their work [14], they assume that the regular semisimple element $\gamma \in \mathfrak{g}(F)$ is "equivalued" and unramified. Let us explain their first hypothesis "equivalued". Technically this essentially means the following: over an algebraic closure, the element γ is conjugated to a diagonal matrix $diag(\lambda_1, \ldots, \lambda_n)$ and there exists $r \in \mathbb{Q}$ such that $\mathrm{val}(\lambda_i - \lambda_j) = r$ for any $i \neq j$. The point is that they can prove that the cohomology of affine Springer fibers $\mathfrak{X}_\gamma$ associated to equivalued elements γ is always pure (cf. [15]). A first consequence of the purity is that it is possible to deduce the cohomology from the equivariant cohomology for some torus action. Here enters their second hypothesis "unramified". Technically γ is unramified means that γ can be conjugated to a diagonal matrix on the maximal unramified extension of F. In particular, its centralizer is a split torus of rank n

over such an extension. From a geometric point of view, this implies that, after extension of scalars to an algebraic closure k of $\mathbb{F}_q$, a torus of rank n over k acts on $\mathfrak{X}_\gamma$. Moreover, there's a combinatorial way to get the equivariant cohomology of $\mathfrak{X}_\gamma$ for this action from the knowledge of orbits of dimension less or equal to one.

We have seen that orbital integrals count the number of rational points of quotients of affine Springer fibers. Thanks to the Grothendieck–Lefschetz fixed point formula, this gives a cohomological interpretation to orbital integrals and also to the fundamental lemma. From their computation of the cohomology of affine Springer fibers of reductive groups (still for "equivalued" and unramified elements), Goresky, Kottwitz and MacPherson proved the fundamental in these cases. In fact, it is conjectured (but until now not known) that the cohomology of affine Springer fibers is always pure. So their hypothesis of equivaluation could be removed. Nonetheless their use of equivariant cohomology needs a "big" torus action which appears only in the unramified case.

3. Hitchin fibration

3.1. Notation

Let $k = \overline{\mathbb{F}_q}$ be an algebraic closure of the finite field $\mathbb{F}_q$. Let C be a connected, smooth, projective curve over $k = \overline{\mathbb{F}_q}$ of genus g_C. Let $D = 2D'$ be an even and effective divisor on C. We assume that

$$\deg(D) > 2g_C.$$

Let n be an integer such that $\mathrm{char}(\mathbb{F}_q) > n$.

3.2. Hitchin bundles

A *Hitchin bundle* is a pair $(\mathcal{E}, \theta)$ where

- $\mathcal{E}$ is a vector bundle on C of rank n and degree 0
- $\theta : \mathcal{E} \to \mathcal{E}(D) = \mathcal{E} \otimes_{\mathcal{O}_C} \mathcal{O}_C(D)$ is a homomorphism of $\mathcal{O}_C$-modules.

Remark 6.6. When D is a canonical divisor (which we do not assume), a Hitchin pair is called a Higgs bundle in the literature. The setting we consider appears in Ngô's paper [22].

3.3. Characteristic polynomial

Let $(\mathcal{E}, \theta)$ be a Hitchin bundle. The trace of θ is the section of $\mathcal{O}(D)$ defined by

$$\mathrm{trace}(\theta) : \mathcal{O}_C \xrightarrow{id} \mathcal{E}nd(\mathcal{E}) \xrightarrow{\theta} \mathcal{O}_C(D) \in H^0(C, \mathcal{O}_C(D)),$$

where the first arrow is the identity section and the second one is θ viewed as an element of $\mathrm{Hom}(\mathcal{E}nd(\mathcal{E}), \mathcal{O}_C(D))$. For any $1 \leqslant i \leqslant n$, we also get sections

$$\mathrm{trace}(\wedge^i\theta) \in H^0(C, \mathcal{O}_C(iD))$$

and the *characteristic polynomial* of θ is the polynomial

$$\chi_\theta(X) = X^n - \mathrm{trace}(\theta)X^{n-1} + \ldots + (-1)^n \,\mathrm{trace}(\wedge^n\theta).$$

3.4. Hitchin fibration

Let $\mathbb{M}$ be the *algebraic k-stack* which classifies Hitchin bundles $(\mathcal{E}, \theta)$.

Let $\mathbb{A}$ be the *affine space* of characteristic polynomials

$$X^n - a_1X^{n-1} + \ldots + (-1)^n a_n$$

with $a_i \in H^0(C, \mathcal{O}_C(iD))$. By Riemann–Roch theorem, it is easy to compute its dimension

$$\begin{aligned}\dim_k(\mathbb{A}) &= \sum_{i=1}^{n} \dim_k(H^0(C, \mathcal{O}_C(iD))) \qquad (6.2)\\ &= \sum_{i=1}^{n}(1 - g_C + i\deg(D))\\ &= \frac{n(n+1)}{2}\deg(D) + n(1 - g_C).\end{aligned}$$

The *Hitchin fibration* is the morphism

$$f : \mathbb{M} \to \mathbb{A}$$

defined by

$$f(\mathcal{E}, \theta) = \chi_\theta.$$

3.5. Adelic description of Hitchin fibers

Let F be the function field of C. Let $|C|$ be the set of closed points of C. For any $c \in |C|$, let $\mathcal{O}_c$ be the completion of the local ring at c and let F_c be the fraction field of $\mathcal{O}_c$. Let

$$\mathbb{A}_F = \varinjlim_{S} \prod_{c\in S} F_c \prod_{c\notin S} \mathcal{O}_c$$

be the ring of adèles of C where the direct limit is taken over finite subsets $S \subset |C|$. The field F is diagonally embedded in $\mathbb{A}_F$. Let

$$\mathcal{O} = \prod_{c \in |C|} \mathcal{O}_c \subset \mathbb{A}_F.$$

To the divisor $D = \sum_{c \in |C|} n_c[c]$, we attach an idèle $\varpi_D = (\varpi_c^{n_c})_c \in \mathbb{A}_F^\times$. We have a degree morphism

$$\deg : \mathbb{A}_F^\times \to \mathbb{Z}$$

which is trivial on $\mathcal{O}^\times$ and $F^\times$. We have

$$\deg(\varpi_D) = \deg(D).$$

Let

$$\chi = X^n - a_1 X^{n-1} + \ldots + (-1)^n a_n$$

be a characteristic polynomial in the Hitchin base $\mathbb{A}(k)$. We can view it as a polynomial with coefficients in F. Let $G = GL(n)$ and $\mathfrak{g} = \mathfrak{gl}(n)$ its Lie algebra. Let $\mathcal{H}_\chi$ be the set of pairs

$$(g, \gamma) \in G(\mathbb{A}_F)/G(\mathcal{O}) \times \mathfrak{g}(F)$$

such that

1. $\deg(\det(g)) = 0$;
2. the characteristic polynomial of γ is χ;
3. we have the following integral condition

$$g^{-1}\gamma g \in \varpi_D^{-1}\mathfrak{g}(\mathcal{O}).$$

The map

$$\begin{array}{ccc} G(F) \times G(\mathbb{A}_F)/G(\mathcal{O}) \times \mathfrak{g}(F) & \to & G(\mathbb{A}_F)/G(\mathcal{O}) \times \mathfrak{g}(F) \\ (\delta, g, \gamma) & \mapsto & \delta \cdot (g, \gamma) := (\delta g, \delta\gamma\delta^{-1}) \end{array}$$

defines an action on the left of the group $G(F)$ on the set $\mathcal{H}_\chi$. We can form the quotient groupoid $[G(F)\backslash\mathcal{H}_\chi]$. It is a small category where the objects are elements of $\mathcal{H}_\chi$ and for (g, γ) and (g', γ') the set $\mathrm{Hom}((g, \gamma), (g', \gamma'))$ is the set of $\delta \in G(F)$ such that $(g', \gamma') = \delta \cdot (g, \gamma)$. The next proposition gives a description “à la Weil” of a Hitchin fiber.

Proposition 6.7. *The Hitchin fiber $f^{-1}(\chi)(k)$ is equivalent to the quotient groupoid $[G(F)\backslash\mathcal{H}_\chi]$.*

Let us denote

$$\mathbb{A}^{\mathrm{rss}} \tag{6.3}$$

the open subset of $\chi \in \mathbb{A}$ which are *square-free* in the ring $F[X]$ of polynomials with coefficients in F. The exponent rss means regular semi-simple since the characteristic polynomial of $\gamma \in \mathfrak{g}(F)$ is square-free if and only γ is regular semi-simple. Let $\chi \in \mathbb{A}^{\mathrm{rss}}$ and $\gamma \in \mathfrak{g}(F)$ with characteristic polynomial $\chi_\gamma = \chi$. Then the centralizer of γ in G is a maximal torus T and the set of $\gamma' \in \mathfrak{g}(F)$ such that $\chi_{\gamma'} = \chi$ is simply the conjugacy class of γ under $G(F)$. That's why the Hitchin fiber $f^{-1}(\chi)(k)$ is also equivalent to the quotient groupoid

$$[T(F)\backslash\mathcal{H}_\gamma]$$

where $\mathcal{H}_\gamma$ is the set of $g \in G(\mathbb{A}_F)/G(\mathcal{O})$ such that $\deg(\det(g)) = 0$ and $g^{-1}\gamma g \in \varpi_D^{-1}\mathfrak{g}(\mathcal{O})$. There is a more suggestive way to write this quotient. Let

$$(\gamma_c)_{c\in|C|} = \varpi_D\gamma \in \mathfrak{g}(\mathbb{A}_F). \tag{6.4}$$

The integrality condition 3 for $g = (g_c)_{c\in|C|} \in G(\mathbb{A})/G(\mathcal{O})$ is equivalent to

$$g_c^{-1}\gamma_c g_c \in \mathfrak{g}(\mathcal{O}_c).$$

Thanks to Proposition 6.2, such a coset g_c is nothing else but a k-point in the affine Springer fiber $\mathfrak{X}_{\gamma_c}$. The Hitchin fiber $f^{-1}(\chi)(k)$ is thus equivalent to the quotient groupoid

$$[T(F)\backslash \prod_{c\in|C|}{}' \mathfrak{X}_{\gamma_c}(k)]$$

where the restricted product $\prod'$ means that at almost every point c we take the standard lattice. Note that since $T(F)$ centralizes γ it also centralizes γ_c and it does act on $\mathfrak{X}_{\gamma_c}$.

3.6. Counting points of Hitchin fibers

Suppose in this section that the curve C and the divisor D come from a curve C_0 and a divisor D_0 defined over a finite field $\mathbb{F}_q$. In the same way as in Section 3.5, we get an adelic description of the category of $\mathbb{F}_q$-points of a Hitchin fiber: one has to replace the objects relative to C, D and k by the analogous objects relative to C_0, D_0 and $\mathbb{F}_q$ which are denoted by a subscript 0.

We can count the number of points: this is precisely the sum over isomorphism classes weighted by the inverse of the order of the group of automorphisms. Like in Proposition 6.8, the result is expressed as an orbital integral. Let $\chi \in \mathbb{A}^{\mathrm{rss}}(\mathbb{F}_q)$ and $\gamma \in \mathfrak{g}(F_0)$ such that $\chi_\gamma = \chi$. Let T be the centralizer of γ in G. For any $c \in |C_0|$, the groups $G(F_c)$ and $T(F_c)$ are provided with Haar measures respectively normalized by $\mathrm{vol}(G(\mathcal{O}_c)) = 1$ and the volume of the maximal compact subgroup of $T(F_c)$ is 1. Let

$$T(\mathbb{A}_0)^0 = \{t \in T(\mathbb{A}_0) \mid \deg(t) = 0\}.$$

The following proposition is the starting observation of Ngô (cf. [22]).

Proposition 6.8. *The number of $\mathbb{F}_q$-points of the Hitchin fiber $f^{-1}(\chi)$ is equal to the following product of orbital integrals*

$$\mathrm{vol}(T(F_0)\backslash T(\mathbb{A}_0)^0) \prod_{c\in|C_0|} \int_{T(F_c)\backslash G(F_c)} \mathbf{1}_{\mathfrak{g}(\mathcal{O}_c)}(g^{-1}\gamma_c g)\, dg_c, \tag{6.5}$$

where the element γ_c is defined as in (6.4), *the measure is the quotient of Haar measures on $G(F_c)$ and $T(F_c)$, the function $\mathbf{1}_{\mathfrak{g}(\mathcal{O}_c)}$ is the characteristic function of $\mathfrak{g}(\mathcal{O}_c)$. The expression* (6.5) *is finite if and only if the Hitchin fiber $f^{-1}(\chi)$ has finitely many $\mathbb{F}_q$-points.*

Remark 6.9. For almost every $c \in |C_0|$, the element γ_c belongs to $\mathfrak{g}(\mathcal{O}_c)$ and moreover its image in the residue field is still regular semisimple. This implies that the orbital integral

$$\int_{T(F_c)\backslash G(F_c)} \mathbf{1}_{\mathfrak{g}(\mathcal{O}_c)}(g^{-1}\gamma_c g)\, dg_c$$

is 1. At the other places, the orbital integral is finite (since the conjugacy class of a semisimple element is closed, the integral is in fact taken over a compact subset of $T(F_c)\backslash G(F_c)$). Thus the product is finite. This is not always the case for the volume $\mathrm{vol}(T(F)\backslash T(\mathbb{A}_0)^0)$. In general, one can write

$$F_0[X]/(\chi) = \prod_{i=1}^{r} E_i$$

as a product of finite extensions E_i of F_0. We have

$$T(F)\backslash T(\mathbb{A}_0) = \prod_{i=1}^{r} E_i^{\times}\backslash \mathbb{A}_{E_i}^{\times}.$$

The degree morphism gives a surjection of the right-hand side onto $\mathbb{Z}^r$. It restricts to a surjection of $T(F)\backslash T(\mathbb{A}_0)^0$ onto the sublattice of $(n_1, \ldots, n_r) \in \mathbb{Z}^r$ such that $n_1 + \ldots + n_r = 0$. Thus $T(F)\backslash T(\mathbb{A}_0)^0$ cannot be of finite volume unless $r = 1$. In this case it is compact. The condition $r = 1$ means that the polynomial χ is irreducible over F_0.

4. Geometric description of Hitchin fibers

4.1. A slight variant of the Hitchin fibration

In the following, we will focus on the (open) regular semisimple locus of the Hitchin fibration

$$\mathbb{M}^{\mathrm{rss}} = \mathbb{M} \times_{\mathbb{A}} \mathbb{A}^{\mathrm{rss}}$$

where the open set $\mathbb{A}^{\mathrm{rss}}$ of the base has been defined in (6.3). In fact, we are going to introduce some etale open subset of $\mathbb{M}^{\mathrm{rss}}$. For this, we introduce another datum, namely a closed point of C, denoted by ∞. We assume that the point ∞ does not belong to the support of the divisor D. Let

$$\mathbb{A}^{\infty} \subset \mathbb{A}^{\mathrm{rss}}$$

be the open subset of characteristic polynomials $\chi = X^n - a_1X^{n-1} + \ldots + (-1)^n a_n \in \mathbb{A}$ such that the polynomial

$$\chi_\infty = X^n - a_1(\infty)X^{n-1} + \ldots + (-1)^n a_n(\infty) \in k[X]$$

has only simple roots. Let

$$\mathcal{A} \to \mathbb{A}^{\infty}$$

be the étale Galois cover of $\mathbb{A}^{\infty}$ of group the symmetric group $\mathfrak{S}_n$ given by

$$\mathcal{A} = \{(\chi, \tau) \in \mathbb{A}^{\infty} \times k^n \mid \chi_\infty = \prod_{i=1}^{n}(X - \tau_i)\}.$$

The fiber product $\mathbb{M} \times_{\mathbb{A}} \mathcal{A}$ classifies quadruples $(\mathcal{E}, \theta, \chi, \tau)$ where $(\mathcal{E}, \theta)$ is a Hitchin bundle and $(\chi, \tau) \in \mathcal{A}$ is such that $\chi = f(\mathcal{E}, \theta)$. This implies that the endomorphism θ_∞ of the k-vector-space $\mathcal{E}_\infty$ must have n distinct eigenvalues. In particular, θ_∞ is regular semisimple and θ is generically regular semisimple.

Let

$$\mathcal{M} \to \mathbb{M} \times_{\mathbb{A}} \mathcal{A}$$

be the $\mathbb{G}_m$-torsor obtained by choosing an eigenvector e_1 in the line Ker $(\theta_\infty - \tau_1 \mathrm{Id}_{\mathcal{E}_\infty})$. We get a Hitchin morphism (still denoted by f) by base change

$$f : \mathcal{M} \longrightarrow \mathbb{M} \times_{\mathbb{A}} \mathcal{A} \longrightarrow \mathcal{A}$$

By an argument from deformation theory, we can show the following theorem.

Theorem 6.10. *(Biswas–Ramanan [6], cf. also [23] §4.14). The algebraic stack $\mathcal{M}$ is smooth over k.*

Remark 6.11. The main point in the above constructions is to avoid the singularities of the global nilpotent cone. The additional datum $\tau \in k^n$ will later play more or less the role of a parabolic structure. The datum e_1 is not at all essential but it is convenient here in order to rigidify the situation.

4.2. The spectral curve of Hitchin–Beauville–Narasimhan–Ramanan

Let

$$\pi_\Sigma : \Sigma_D = \mathbf{Spec}\Big(\bigoplus_{i=0}^{\infty} \mathcal{O}_C(-iD)X^i\Big) \to C$$

be the total space of the line bundle $\mathcal{O}_C(D)$.

Let $a = (\chi, \tau) \in \mathcal{A}$. We write

$$\chi(X) = X^n - a_1 X^{n-1} + \ldots + (-1)^n a_n$$

with $a_i \in H^0(C, \mathcal{O}_C(iD))$.

The *spectral curve* Y_a (cf. [3]) is the closed curve in Σ_D defined by

$$Y_a = \mathbf{Spec}\Big(\bigoplus_{i=0}^{\infty} \mathcal{O}_C(-iD)X^i\Big)/\mathcal{I}_a)$$

where the sheaf of ideals $\mathcal{I}_a$ of $\mathcal{O}_\Sigma$ is generated by $\mathcal{O}_C(-nD)\chi$. The canonical projection

$$\pi_a : Y_a \to C$$

is a finite cover of degree n, which is *étale over* the point ∞. But we have the datum $\tau = (\tau_1, \ldots, \tau_n)$ which is an ordering on the (simple) roots of χ_∞. Thus we have also an ordering on the fiber of π_a over ∞:

$$\pi_a^{-1}(\infty) = \{\infty_1, \ldots, \infty_n\}.$$

Proposition 6.12. *The spectral curve Y_a is*

1. *reduced;*
2. *connected;*
3. *not always irreducible: there is a $1-1$ correspondence between the set of irreducible components of Y_a and the set of irreducible factors of the characterictic polynomial $\chi \in F[X]$.*

The assertion 1 is due to the fact that χ belongs to the regular semisimple locus $\mathbb{A}^{\text{rss}}$ by construction of $\mathcal{A}$. Let

$$\mathcal{A}^{\text{ell}}$$

be the open subset of (χ, τ) such that $\chi \in F[X]$ is irreducible. By the assertion 3 above, it is also the open subset of $a \in \mathcal{A}$ such that Y_a is integral.

The arithmetic genus of Y_a is defined by

$$q_{Y_a} = \dim_k(H^1(Y_a, \mathcal{O}_{Y_a})) = \dim_k(H^1(C, \pi_{a,*}\mathcal{O}_{Y_a})).$$

One can compute,

$$\pi_{a,*}\mathcal{O}_{Y_a} = \mathcal{O}_C \oplus \mathcal{O}_C(-D) \oplus \ldots \oplus \mathcal{O}((-n+1)D)$$

and by the theorem of Riemann–Roch, one gets

Proposition 6.13. *The arithmetic genus of Y_a does not depend on a and it is equal to*

$$q_{Y_a} = \frac{n(n-1)}{2}\deg(D) + n(g_C - 1) + 1.$$

4.3. Hitchin–Beauville–Narasimhan–Ramanan correspondence

This the following theorem (cf. [3]).

Theorem 6.14. *Let $a \in \mathcal{A}$. The Hitchin fiber $\mathcal{M}_a = f^{-1}(a)$ is isomorphic to the stack of torsion-free coherent $\mathcal{O}_{Y_a}$-modules $\mathcal{F}$ of degree 0 and rank 1 at generic points of Y_a, equipped with a trivialization of their stalk at ∞_1.*

Let us recall briefly how one can construct a Hitchin bundle from a torsion-free $\mathcal{O}_{Y_a}$-module of rank 1. The multiplication by X on $\mathcal{O}_\Sigma$ gives the universal section

$$\mathcal{O}_\Sigma \to \pi_\Sigma^*\mathcal{O}_C(D)$$

and for any $a \in \mathcal{A}$ a section

$$\mathcal{O}_{Y_a} \to \pi_a^*\mathcal{O}_C(D).$$

By tensoring by a coherent $\mathcal{O}_{Y_a}$-modules $\mathcal{F}$, we get a morphism $\mathcal{F} \to \mathcal{F} \otimes_{\mathcal{O}_{Y_a}} \pi_a^*\mathcal{O}_C(D)$ and by the projection formula a twisted endomorphism of $\pi_{a,*}\mathcal{F}$

$$\theta : \pi_{a,*}\mathcal{F} \to \pi_{a,*}(\mathcal{F} \otimes_{\mathcal{O}_{Y_a}} \pi_a^*\mathcal{O}_C(D)) = \pi_{a,*}(\mathcal{F})(D).$$

If $\mathcal{F}$ is moreover torsion-free of rank 1, then $\pi_{a,*}\mathcal{F}$ is torsion-free $\mathcal{O}_C$-module of rank n. But since C is smooth, it is a locally free $\mathcal{O}_C$-module of rank n thus a vector bundle of rank n on C. So we have a pair $(\pi_{a,*}\mathcal{F}, \theta)$. We can compute its degree :

$$\begin{aligned}\deg(\pi_{a,*}\mathcal{F}) &= \chi(C, \pi_{a,*}\mathcal{F}) + n(g_C - 1)\\ &= \chi(Y_a, \mathcal{F}) + n(g_C - 1)\\ &= \deg(\mathcal{F}) + \chi(Y_a, \mathcal{O}_{Y_a}) + n(g_C - 1)\\ &= \deg(\mathcal{F}) + 1 - q_{Y_a} + n(g_C - 1)\end{aligned}$$

If we assume $\deg(\mathcal{F}) = 0$, using Proposition 6.13, we get

$$\deg(\pi_{a,*}\mathcal{F}) = -\frac{n(n-1)}{2}\deg(D).$$

But remember that the divisor $D = 2D'$ is even. So the pair $(\pi_{a,*}\mathcal{F} \otimes_{\mathcal{O}_C} \mathcal{O}_C((n-1)D'), \theta)$ is a Hitchin bundle (the underlying vector bundle is now of degree 0). Finally the trivialization at ∞_1 gives our fourth datum e_1.

4.4. First consequences of Theorem 4.3.1

Let $\mathcal{A}^{\mathrm{sm}}$ the open set of a such that Y_a is smooth. In fact, one can show that $\mathcal{A}^{\mathrm{sm}}$ is not empty. When Y_a is smooth, a torsion free $\mathcal{O}_{Y_a}$-module of rank 1 is a line bundle. Thus we get :

Corollary 6.15. *For $a \in \mathcal{A}^{sm}$, the Hitchin fiber $\mathcal{M}_a$ is the Jacobian of Y_a. In particular, it is an abelian variety (and as such a scheme).*

For any $a \in \mathcal{A}$, we can consider the smooth commutative group scheme

$$\mathrm{Pic}^0(Y_a)$$

of line bundles on Y_a of degree 0, equipped with a trivialization of their stalk at ∞_1.

By tensor product, this groups acts on (and is an open substack of) the stack of torsion-free $\mathcal{O}_{Y_a}$-modules of degree 0 and rank 1. By Theorem 6.14, we get an action of $\mathrm{Pic}^0(Y_a)$ on the Hitchin fiber $\mathcal{M}_a$.

Let $\mathcal{M}_a^{\mathrm{reg}} \subset \mathcal{M}_a$ be the open substack $(\mathcal{E}, \theta, \tau, e_1) \in \mathcal{M}_a$ such that θ_c is regular for any $c \in C$. Recall that an endomorphism of k^n is regular if its centralizer is of minimal dimension, namely n. In the correspondence of Theorem 6.14, this open substack corresponds to $\mathrm{Pic}^0(Y_a)$. We can state :

Corollary 6.16. *$\mathcal{M}_a^{\mathrm{reg}}$ is a $\mathrm{Pic}^0(Y_a)$-torsor.*

4.5. Dimension of Hitchin fibers

As a consequence of the work [1] of Altman–Iarrobino–Kleiman on the compactified Jacobian, we have the following theorem

Theorem 6.17. *Let $a \in \mathcal{A}$.*

1. *The open substack $\mathcal{M}_a^{\mathrm{reg}}$ is dense in $\mathcal{M}_a$.*
2. *We have*

$$\dim(\mathcal{M}_a) = \dim(\mathcal{M}_a^{\mathrm{reg}}) = \dim(\mathrm{Pic}^0(Y_a))$$

and this dimension is q_{Y_a}, the arithmetic genus of Y_a. In particular, $\dim(\mathcal{M}_a)$ does not depend on a.

3. *The set of irreducible components of $\mathcal{M}_a$ is a torsor under the abelian group of connected component of $\mathrm{Pic}^0(Y_a)$ and we have*

$$\pi_0(\mathrm{Pic}^0(Y_a)) \simeq \{(n_i) \in \mathbb{Z}^{Irr(Y_a)} \mid \sum_i n_i = 0\}$$

where $Irr(Y_a)$ is the set of irreducible components of Y_a.

Recall that Y_a is irreducible if and only if a belongs to the elliptic set $\mathcal{A}^{\mathrm{ell}}$. From the assertion 3 of Theorem 6.17, we get the corollary :

Corollary 6.18. *The Hitchin fiber $\mathcal{M}_a$ is irreducible if and only if a belongs to the elliptic set $\mathcal{A}^{\mathrm{ell}}$.*

One can also compute the dimension of the stack $\mathcal{M}$. We have

$$\dim(\mathcal{M}) = \dim(\mathcal{A}) + \dim(f).$$

The dimensions of $\mathcal{A}$ and f are respectively given by (6.2) and the combination of Proposition 6.13 and the assertion 2 of Theorem 6.17. We get :

Corollary 6.19. *We have*

$$\dim(\mathcal{M}) = n^2 \deg(D) + 1.$$

5. Examples of Hitchin fibers in rank 2

5.1. The situation

In this section, we take

$$C = \mathbb{P}_k^1$$

We write $\mathbb{P}_k^1 = \mathrm{Spec}(k[y]) \cup \{0\}$ and in the affine chart $\mathrm{Spec}(k[y])$, the point ∞ is defined by the equation $y = 0$. We take

$$D = 2[0].$$

5.2. The base $\mathcal{A}$

The scheme $\mathcal{A}$ classifies pairs

$$(X^2 - a_1(y)X + a_2(y), (\tau_1, \tau_2))$$

where, for $i = 1, 2$

$$a_i(y) \in H^0(\mathbb{P}^1_k, \mathcal{O}(2i[0])) = \{a_i(y) \in k[y] \mid \deg(a_i) \leqslant 2i\}$$

are such that the discriminant $(a_1)^2 - 4a_2$ does not vanish at $y = 0$ and $(\tau_1, \tau_2) \in k^2$ is the ordered pair of distinct root of

$$X^2 - a_1(0)X + a_2(0).$$

In the following, we will restrict ourselves to the case

$$a_1 = 0,$$

that is to the case of Hitchin pairs with traceless twisted endomorphism.

Remember the open subsets

$$\mathcal{A}^{\mathrm{ell}} \subset \mathcal{A}^{\mathrm{sm}} \subset \mathcal{A}.$$

Let $a \in \mathcal{A}$ with $a_1 = 0$. Then

1. $a \in \mathcal{A}^{\mathrm{sm}}$ if and only if a_2 has only simple roots.
2. $a \in \mathcal{A}^{\mathrm{ell}}$ if and only if a_2 is not a square in $k[y]$.

5.3. Spectral curves

In our situation, all spectral curves are of arithmetic genus

$$q_{Y_a} = 1.$$

In Figure 3 below, one finds some examples of spectral curve. From the left to the right, one has the following cases

- a_2 has only simple roots ; the curve Y_a is a smooth projective curve of genus 1: it is an elliptic curve.
- a_2 has a double root and two simple roots. Then Y_a is integral and has only one singularity which is a node.
- a_2 has a triple root. Then Y_a is integral and has only one singularity which is a cusp.
- a_2 is the square of a polynomial with two distinct roots. Then Y_a has two irreducible components which intersect transversally.

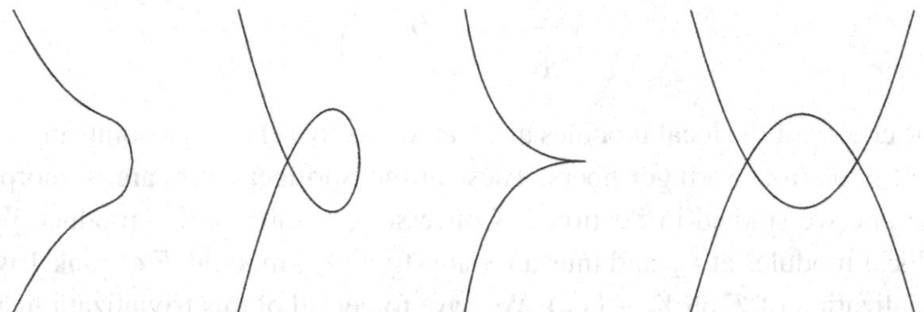

Figure 3 Examples of spectral curve

5.4. Examples of elliptic Hitchin fibers

When $a \in \mathcal{A}^{\mathrm{ell}}$, the spectral curve is an integral curve of arithmeric genus 1 thus an elliptic curve or a plane cubic with a node or a cusp. Theorem 6.14 identifies the Hitchin fiber $\mathcal{M}_a$ with the compactified Jacobian of Y_a (which is simply the Jacobian when Y_a is smooth). But it is known (cf. [2]) that the compactified Jacobian for an curve integral curve of genus 1 is (non-canonically) isomorphic to the curve. So the first three pictures in Figure 3 are also typical examples of elliptic Hitchin fibers.

5.5. A non-elliptic Hitchin fiber

In this paragraph, we take $a_1(y) = 0$ and

$$a_2(y) = (y^2 - 1)^2.$$

In this case, the spectral curve Y_a is the union of two projective lines which intersect transversally. Let

$$U = \mathbb{P}^1_k - \{-1, 1\} = \mathrm{Spec}(k[(\frac{y+1}{y-1})^{\pm 1}]).$$

Let $y_\pm \in Y_a$ be the points over ± 1. To understand the Hitchin fiber $\mathcal{M}_a$, we will use Theorem 6.14. So let $\mathcal{F}$ be a torsion-free $\mathcal{O}_{Y_a}$-module of generic ranks 1 and degree 0. Outside the points $y_\pm$, the $\mathcal{O}_{Y_a}$-module $\mathcal{F}$ is invertible and may be trivialized. We fix such a trivialization. Then the $\mathcal{O}_{Y_a}$-module $\mathcal{F}$ is determined by its restriction to the formal neighbourhoods of $y_\pm$, which is a torsion-free

$$k[[y \pm 1]][X]/(X^2 - a_2(y))$$

-module of generic rank 1. By the obvious local analog of Theorem 6.14, this is nothing else but a free $k[[y \pm 1]]$-submodule of $k((y \pm 1))$ of rank 2 stable under the endomorphism given by the matrix

$$\begin{pmatrix} y^2-1 & 0 \\ 0 & 1-y^2 \end{pmatrix}.$$

So the choice of the local modules at $y_\pm$ amounts to a choice of points in a product of two affine Springer fibers. These affine Springer fibers are isomorphic to the one we studied in Section 2. Conversely, a point in this product gives two local modules at $y_\pm$ and thus a torsion-free $\mathcal{O}_{Y_a}$-module $\mathcal{F}$ of rank 1 with a trivialization of $\mathcal{F}$ on $Y_a - \{y_\pm\}$. We have to get rid of this trivialization. The group

$$H^0(U, \mathbb{G}_{m,k}^2) = (k^\times)^2((\frac{y+1}{y-1})^{\mathbb{Z}})^2 \simeq (k^\times)^2 \times (\mathbb{Z}^2),$$

which is the group of automorphism of the trivial line bundle on $Y_a - \{y_\pm\}$, acts on the product of affine Springer fibers. But we also want $\mathcal{F}$ to be of degree 0 and be trivialized at the marked point $\infty_1 \in Y_a$. For the first condition, we can assume that the point belongs to a fixed connected component of the product of affine Springer fibers, namely the product of connected components of "degree 0" of the two affine Springer fibers: this is a product of two infinite chains of projective lines (cf. Figure 4 below).

The stabilizer in $H^0(U, \mathbb{G}_{m,k}^2)$ of both the point ∞_1 and the fixed connected component is identified with

$$\mathbb{G}_{m,k} \times \mathbb{Z}.$$

The group $\mathbb{Z}$ acts on each chain of projective line by translation (the generator $1 \in \mathbb{Z}$ sends a line to the next one) and the group $\mathbb{Z}$ acts anti-diagonally on the product (cf. the bold arrow in Figure 4). The action of $\mathbb{G}_{m,k}$ on a chain of projective lines fixes the split lattices and the connected components of the complement are precisely the orbits of dimension 1. The group $\mathbb{G}_{m,k}$

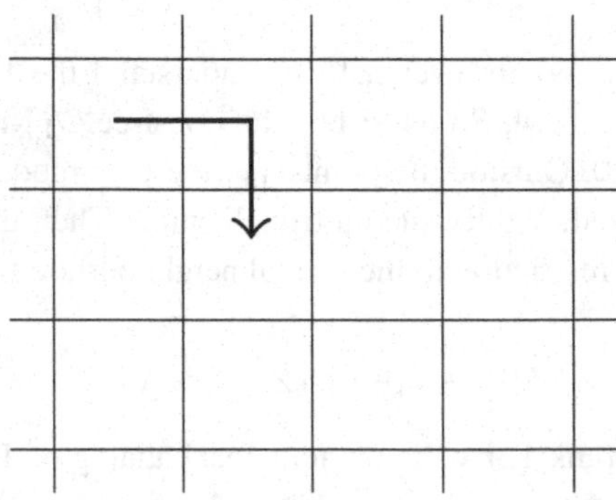

Figure 4 A product of two infinite chains of $\mathbb{P}^1$In bold the action of $\mathbb{Z}$

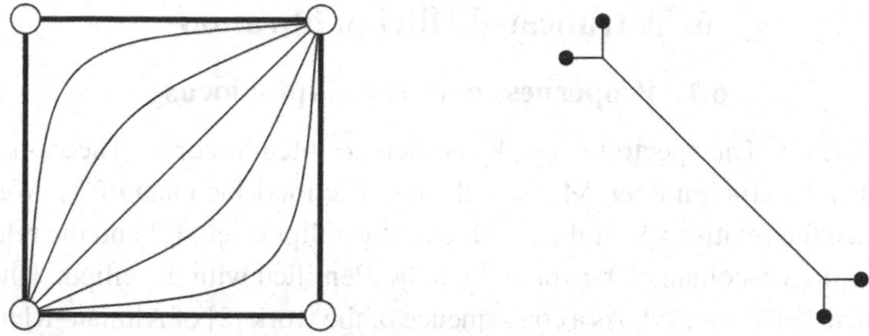

Figure 5 On the left: action of $\mathbb{G}_{m,k}$ on a square. On the right: the quotient (up to some $B\mathbb{G}_{m,k}$).

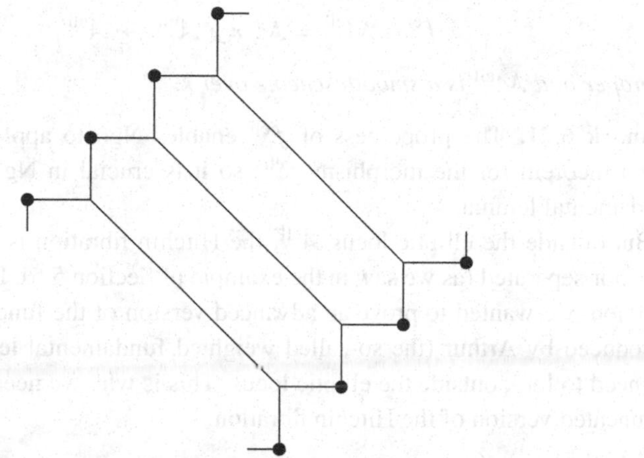

Figure 6 The Hitchin fiber $\mathcal{M}_a$ (up to $B\mathbb{G}_{m,k}$)
An infinite chain of non-separated projective lines

acts diagonally on the product of the two chains. It preserves each square in Figure 4. On the left in Figure 5 below, we extracted a square from Figure 4 and we drew with a circle the four fixed points for the $\mathbb{G}_{m,k}$-action. Moreover we drew some 1-dimensional orbits for this action. The main point to observe here is that upper orbits tend to the *two* bold orbits in the upper left corner. Similarly, lower orbits tend to the *two* bold orbits in the lower right corner. So when we take the quotient of the square by the action of $\mathbb{G}_{m,k}$, each fixed point gives a $B\mathbb{G}_{m,k}$ and the quotient of the complement is a projective line with two non-separated 0 and two non-separated ∞ (pictured in Figure 5 on the right).

Finally, the Hitchin fiber $\mathcal{M}_a$ can be identified with the quotient of the product of the two chains by the action of $\mathbb{G}_{m,k} \times \mathbb{Z}$. So, up to some $B\mathbb{G}_{m,k}$, the Hitchin fiber $\mathcal{M}_a$ is an infinite chain of non-separated projective lines (as pictured in Figure 6). Note that it is neither of finite type nor separated.

6. A truncated Hitchin fibration

6.1. Properness over the elliptic locus

Let $a \in \mathcal{A}^{\mathrm{ell}}$. The spectral curve Y_a is then an integral curve. Theorem 6.14 identifies the Hitchin fiber $\mathcal{M}_a$ with the compactified Jacobian of Y_a. We can introduce the relative spectral curve Y over the elliptic set $\mathcal{A}^{\mathrm{ell}}$ and the relative compactified Jacobian of Y over $\mathcal{A}^{\mathrm{ell}}$ can be identified with the elliptic Hitchin morphism $\mathcal{M}^{\mathrm{ell}} \to \mathcal{A}^{\mathrm{ell}}$. As a consequence of the work [2] of Altman–Kleiman on the compactified Jacobian, we have the following theorem.

Theorem 6.20. *The elliptic Hitchin morphism*

$$f^{\mathrm{ell}} : \mathcal{M}^{\mathrm{ell}} = \mathcal{M} \times_{\mathcal{A}} \mathcal{A}^{\mathrm{ell}} \to \mathcal{A}^{\mathrm{ell}}$$

is proper and $\mathcal{M}^{\mathrm{ell}}$ is a smooth scheme over k.

Remark 6.21. The properness of f^{ell} enables Ngô to apply the decomposition theorem for the morphism f^{ell}: so it is crucial in Ngô's proof of the fundamental lemma.

But outside the elliptic locus $\mathcal{A}^{\mathrm{ell}}$, the Hitchin fibration is neither of finite type nor separated (as we saw in the example of Section 5.5). In our work with Laumon, we wanted to prove an advanced version of the fundamental lemma introduced by Arthur (the so-called weighted fundamental lemma). For this, we need to look outside the elliptic locus. This is why we needed to introduce a truncated version of the Hitchin fibration.

6.2. The notion of ξ-stability

We first introduce a parameter of stability: let $\xi = (\xi_1, \dots, \xi_n) \in \mathbb{R}^n$ such that

$$\sum_{i=1}^{n} \xi_i = 0.$$

Let us introduce the following definition.

Definition 6.22. Let $m = (\mathcal{E}, \theta, \tau, e_1) \in \mathcal{M}$. One says that m is ξ-stable if for any subbundle

$$0 \subsetneq \mathcal{F} \subsetneq \mathcal{E}$$

such that $\theta(\mathcal{F}) \subset \mathcal{F}(D)$ one has

$$\deg(\mathcal{F}) + \sum_i \xi_i < 0 \tag{6.6}$$

where the sum is over $1 \leqslant i \leqslant n$ such that τ_i is an eigenvalue of $\theta_{|\mathcal{F}_\infty}$.

Remark 6.23. When $\xi = 0$, one gets the usual stability for the underlying Hitchin pairs $(\mathcal{E}, \theta)$. But here for generic ξ, we take advantage of our third datum τ.

The notion of ξ-stability is reminiscent (through Theorem 6.14) of the work [13] of Estèves on the compactified Jacobian.

Remark 6.24. The subbundles $0 \subsetneq \mathcal{F} \subsetneq \mathcal{E}$ such that $\theta(\mathcal{F}) \subset \mathcal{F}(D)$ are determined by their generic fiber which must be a θ-stable linear subspace of the generic fiber of $\mathcal{E}$. However, by construction of $\mathcal{M}$, the endomorphism θ is generically regular semi-simple and there is only a finite number of θ-stable linear subspaces. Hence there is only a finite number of subbundles $\mathcal{F}$ satisfying $\theta(\mathcal{F}) \subset \mathcal{F}(D)$. When $m \in \mathcal{M}$ is elliptic, the characteristic polynomial of θ is generically irreducible and there is no such subspace and no such subbundle $\mathcal{F}$. So, any $m \in \mathcal{M}^{\text{ell}}$ is ξ-stable for any ξ.

6.3. Properness of $\mathcal{M}^\xi$

Let $\mathcal{M}^\xi$ be the ξ-stable substack of $\mathcal{M}$. We say that ξ is generic if

$$\sum_{i \in I} \xi_i \notin \mathbb{Z}$$

for any $\emptyset \neq I \subsetneq \{1, \ldots, n\}$. Concretely, for generic ξ, there is no difference between the notions of ξ-stability and ξ-semistability (defined by large inequality in (6.6)).

Through Theorem 6.14, we can deduce the following theorem from the work [13] of Estèves (the properness can also be proved directly from methods of the paper [20] by Langton).

Theorem 6.25. *1. The stack $\mathcal{M}^\xi$ is a smooth open substack of $\mathcal{M}$ which contains $\mathcal{M}^{\text{ell}}$. It is even an algebraic space.*

2. If ξ is generic, the ξ-stable Hitchin fibration

$$f^\xi : \mathcal{M}^\xi \to \mathcal{A}$$

is proper.

Remark 6.26. As we shall see in the next example, if we take $\xi = 0$, the stack $\mathcal{M}^\xi$ is of finite type but one cannot check the existence part of the valuative criterium of properness for f^ξ. If we use a substack defined by the usual semi-stability (for Hitchin pairs) then we get a stack of finite type which is not separeted.

Remark 6.27. Over a finite field we can compute the number of rational points of a truncated Hitchin fiber $\mathcal{M}_a^{\xi}$. Perhaps a more surprising fact is that, for ξ generic, we get essentially a global weighted orbital integral constructed by Arthur (cf. section 11 of [7]). This generalizes proposition 6.8.

6.4. An example of a truncated Hitchin fiber

Let's go back to the example of Section 5.5. In Figures 7, 8 and 9, we draw on the left the effect of stability, semi-stability and ξ-stability (for a generic ξ). In Figure 7, we find that the stability condition isolates in the figure 4 a stair of width one (open) square. The stable Hitchin fiber $\mathcal{M}_a^{\xi=0}$ is a $\mathbb{G}_{m,k}$, thus not proper. In Figure 8, the semi-stability condition defines a stair of width three squares ; the bold lines (which are $\mathbb{G}_{m,k}$) and also the vertices common to four bold lines belong to the stair. Even up to a $B\mathbb{G}_{m,k}$ corresponding to

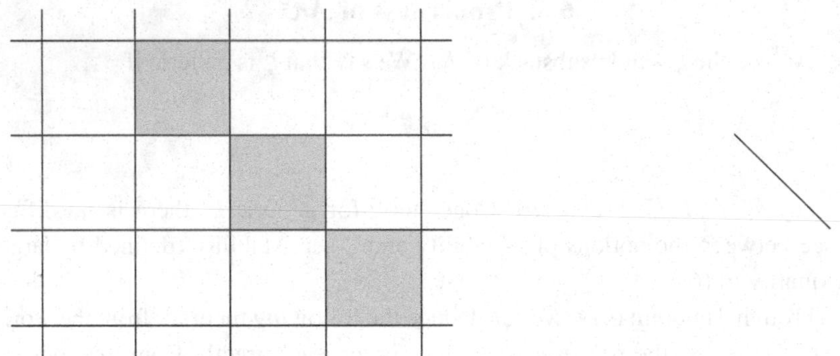

Figure 7 On the left, in grey the stable region and on the right the quotient ($\simeq \mathbb{G}_{m,k}$)

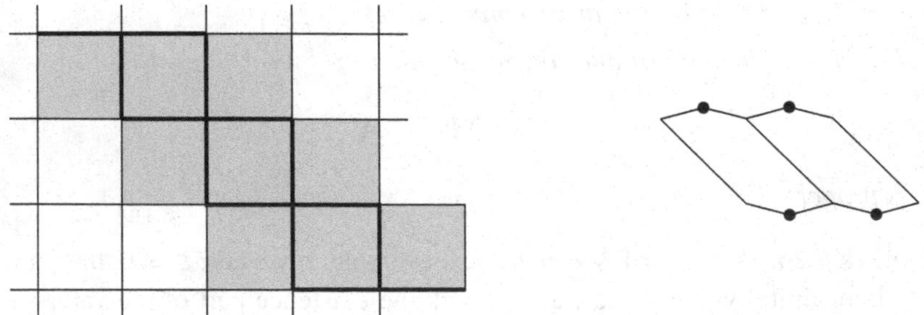

Figure 8 On the left, in grey the stable region and on the right the (non-separated) quotient
There are two pairs of non-separated bold points

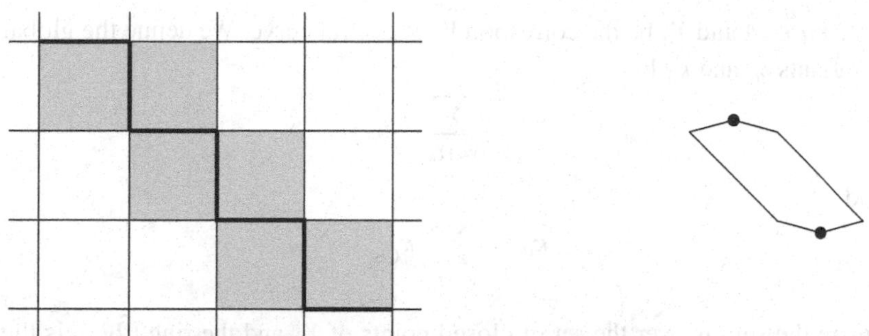

Figure 9 On the left, in grey the ξ-stable region (for generic ξ) and on the right the quotient

these vertices, the quotient is not separated. Finally, in Figure 9, the effect of ξ-stability (for ξ generic) is to take a stair of width two squares; the stair contains the bold lines but no vertex. Then the quotient has no more non-separated points and the truncated Hitchin fiber $\mathcal{M}_a^\xi$ is isomorphic to two projective lines intersecting transversally. Not surprisingly, the Hitchin fiber $\mathcal{M}_a^\xi$ is thus isomorphic to the spectral curve.

7. The main cohomological result

7.1. Invariant δ

Let $\mathrm{Spf}(A)$ be a formal germ of a reduced curve. Let $\tilde{A}$ be its normalization in the total ring of fractions of A. We can attach to A the following invariants:

1. Serre's invariant
$$\delta_A = \mathrm{length}(\tilde{A}/A);$$
2. The number r_A of branches (which is also the number of connected components of $\mathrm{Spf}(\tilde{A})$);
3. The multiplicity m_A;
4. The κ-invariant defined by
$$\kappa_A = \mathrm{length}(\Omega^1_{\tilde{A}/A}).$$

Note that both δ_A and κ_A vanish for a non-singular germ. Note also that we have $\kappa_A = m_A - r_A$, at least when the characteristic of A is greater than the multiplicity m_A.

Let $a \in \mathcal{A}$ and Y_a be the corresponding spectral curve. We define the global invariants δ_a and κ_a by

$$\delta_a = \sum_{y \in |Y_a|} \delta_{\hat{\mathcal{O}}_{Y_a,y}}$$

and

$$\kappa_a = \sum_{y \in |Y_a|} \kappa_{\hat{\mathcal{O}}_{Y_a,y}}$$

where the sum is over the set of closed points of Y_a and the ring $\hat{\mathcal{O}}_{Y_a,y}$ is the completion of the local ring of Y_a at y. Thus we get two constructible functions $a \mapsto \delta_a$ and $a \mapsto \kappa_a$ on $\mathcal{A}$. The first one is moreover upper semi-continuous. Let $\delta \in \mathbb{N}$ and $\mathcal{A}^\delta$ be the locally closed subset of $a \in \mathcal{A}$ such that $\delta_a = \delta$. For $\delta = 0$, we have $\mathcal{A}^0 = \mathcal{A}^{\mathrm{sm}}$ the open dense subset of $a \in \mathcal{A}$ such that Y_a is smooth.

7.2. The codimension of $\mathcal{A}^\delta$

For fields k of characteristic 0, in [12] Diaz and Harris have shown the equality for the codimension of $\mathcal{A}^\delta$:

$$\mathrm{codim}_{\mathcal{A}}(\mathcal{A}^\delta) = \delta, \qquad (6.7)$$

but their argument which uses the tangent cone does not work in positive characteristic (see also [21] sect. 3.3). Nonetheless, to get his main cohomological result, Ngô needs at least the inequality $\geqslant$ in (6.7) (which is unknown in positive characteristic). In his paper, Ngô is able to prove the desired inequality when δ is smaller than $\deg(D)$. But, outside the elliptic set, this last condition is never satisfied. In fact, in our work with Laumon, we use an open subset of $\mathcal{A}$ where the codimension of $\mathcal{A}^\delta$ can be computed by deformation theory. Let us sketch briefly our approach. In [8] section 9, we treated the analogous problem for the so-called cameral curve. We hope that the following discussion will make the arguments in [8] more transparent.

It is possible to consider a slight variant $\mathcal{B}$ of the moduli space of pairs

$$(a, \tilde{Y}_a \to_\varphi \Sigma_D)$$

such that $a \in \mathcal{A}$, the curve $\tilde{Y}_a$ is smooth and projective and the morphism φ is the normalization of the spectral curve $Y_a \subset \Sigma_D$. On any connected component of $\mathcal{B}$, the invariant δ_a of Y_a must be constant. Conversely, by the work of Teissier (cf. [11]), $\mathcal{A}^\delta$ is in one-one bijection with some connected components of $\mathcal{B}$. In general, one does not know how to compute the dimension of $\mathcal{B}$. However, when one has

$$\kappa_a < \deg(D) - 2g_C + 2, \tag{6.8}$$

the space $\mathcal{B}$ is smooth at $(a, \tilde{Y}_a \to_\varphi \Sigma_D)$. Indeed it suffices to check that

$$\mathrm{Ext}^2(L, \mathcal{O}_{\tilde{Y}_a}) = 0 \tag{6.9}$$

where L is the cotangent complex (in degrees -1 and 0)

$$L = [\varphi^*\Omega^1_{\Sigma_D/C} \to \Omega^1_{\tilde{Y}_a/C}].$$

One can write $\mathrm{Ext}^2(L, \mathcal{O}_{\tilde{Y}_a}) = \mathrm{Ext}^1(\mathcal{H}^{-1}(L), \mathcal{O}_{\tilde{Y}_a})$ where $\mathcal{H}^{-1}(L)$ is the sheaf of cohomology of L in degree -1. By Serre duality, it suffices to check that

$$H^0(\tilde{Y}_a, \mathcal{H}^{-1}(L) \otimes \Omega^1_{\tilde{Y}_a/k}) = 0. \tag{6.10}$$

Take $\pi = \pi_\Sigma \circ \varphi$. One can compute the degree

$$\begin{aligned}\deg(\mathcal{H}^{-1}(L) \otimes \Omega^1_{\tilde{Y}_a/k}) &= \deg(\varphi^*\Omega^1_{\Sigma_D/C}) - \deg(\Omega^1_{\tilde{Y}_a/C}) \\ &\quad + \deg(\Omega^1_{\tilde{Y}_a/Y_a}) + \deg(\Omega^1_{\tilde{Y}_a/k}) \\ &= \deg(\pi^*\mathcal{O}(-D)) + \kappa_a + \deg(\pi^*\Omega^1_{C/k}) \\ &= -n\deg(D) + \kappa_a + n(2g_C - 2)\end{aligned}$$

which is negative if

$$\kappa_a < n(\deg(D) - 2g_C + 2).$$

This condition is satisfied by (6.8) and gives the vanishing assertion (6.9) when $\tilde{Y}_a$ is connected. In general, a similar computation for each connected component of $\tilde{Y}_a$ gives the same result under the condition (6.8).

Once we know the smoothness at $(a, \tilde{Y}_a \to_\varphi \Sigma_D)$, the dimension is given by

$$\dim(\mathrm{Ext}^1(L, \mathcal{O}_{\tilde{Y}_a})) = \dim(\mathrm{Ext}^0(\mathcal{H}^{-1}(L), \mathcal{O}_{\tilde{Y}_a}) + \dim(\mathrm{Ext}^1(\mathcal{H}^0(L), \mathcal{O}_{\tilde{Y}_a})). \tag{6.11}$$

The sheaf $\mathcal{H}^0(L) = \Omega^1_{\tilde{Y}_a/Y_a}$ is a torsion sheaf of length κ_a. Denoting by r_a the number of connected components of $\tilde{Y}_a$, we can compute (6.11) by Serre duality, the Riemann–Roch formula, the vanishing property (6.10) and the dimension (6.2) of $\mathcal{A}$: this gives

$$\begin{aligned}\dim(\mathrm{Ext}^1(L, \mathcal{O}_{\tilde{Y}_a})) &= n(\deg(D) - 2g_C + 2) - \kappa_a + g_{\tilde{Y}_a} - r_a + \kappa_a \\ &= n(\deg(D) - 2g_C + 2) + g_{\tilde{Y}_a} - r_a \\ &= \dim(\mathcal{A}) - (\frac{n(n-1)}{2}\deg(D) + n(g_C - 1) - g_{\tilde{Y}_a} + r_a) \\ &= \dim(\mathcal{A}) - q_a + 1 + g_{\tilde{Y}_a} - r_a \\ &= \dim(\mathcal{A}) - \delta_a.\end{aligned}$$

The last two equalities come on the one hand from the formula for the arithmetic genus q_a of Y_a (cf. proposition 6.13) and on the other hand from the long exact sequence in cohomology associated to

$$0 \to \mathcal{O}_{Y_a} \to \mathcal{O}_{\tilde{Y}_a} \to \mathcal{O}_{\tilde{Y}_a}/\mathcal{O}_{Y_a} \to 0. \tag{6.12}$$

This is precisely expected by (6.7).

7.3. The open subset $\mathcal{A}^{\text{good}}$

It is defined as the biggest open subset of $\mathcal{A}$ such that for any closed point $a \in \mathcal{A}^{\text{good}}$ the inequality (6.8) is true. It contains for example all a such that the irreducible components of the spectral curve Y_a are smooth.

7.4. The decomposition theorem

Let ℓ be a prime number *invertible* in k. Let ξ be a generic stability parameter (cf. Sections 6.2 and 6.3). By Theorem 6.25, the algebraic space $\mathcal{M}^\xi$ is smooth over k and the Hitchin morphism f^ξ is proper. By Deligne theorem (cf. [10]), this implies that the complexe of ℓ-adic sheaves

$$Rf_*^\xi \overline{\mathbb{Q}_l}$$

is pure. By the decomposition theorem of Beilinson–Bernstein–Deligne–Gabber [4], the direct sum of perverse cohomology sheaves

$$^p\mathcal{H}^\bullet(Rf_*^\xi \overline{\mathbb{Q}_l}) = \bigoplus_i {}^p\mathcal{H}^i(Rf_*^\xi \overline{\mathbb{Q}_l})$$

is *semi-simple*. So we can write

$$^p\mathcal{H}^\bullet(Rf_*^\xi \overline{\mathbb{Q}_l}[\dim(\mathcal{M})]) = \bigoplus_{a \in \mathcal{A}} i_{a,*} j_{a,!*} \mathcal{F}_a^\bullet[\dim(a)], \tag{6.13}$$

where

- the sum is over Zariski points in $\mathcal{A}$;
- $i_a : \overline{\{a\}} \hookrightarrow \mathcal{A}$ is the canonical inclusion of the closure of a in $\mathcal{A}$;
- $\mathcal{F}_a^\bullet$ is a graded local system on a smooth open subset of $\overline{\{a\}}$ and $j_{a,!*}\mathcal{F}_a^\bullet$ is its middle extension to $\overline{\{a\}}$.

7.5. Properties of the socle

The socle of $^p\mathcal{H}^\bullet(Rf_*^\xi \overline{\mathbb{Q}_l}[\dim(\mathcal{M})])$ is the finite set of $a \in \mathcal{A}$ such that $\mathcal{F}_a^\bullet \neq 0$. A fundamental problem is to determine the socle. We shall see in the next paragraph that it is possible to determine it at least if the Hitchin morphism f^ξ is restricted to the open subset $\mathcal{A}^{\text{good}}$. Besides Ngô's article [23], we refer also the reader to [24] another article of Ngô.

Meanwhile, we would like to give some properties of the socle. Let a be an element of the socle. The amplitude of a is defined by

$$\text{Ampl}(a) = m_a - m'_a$$

where m_a, resp. m'_a, is the maximum, resp. the minimum, of integers m such that $\mathcal{F}_a^m \neq 0$. By Poincaré duality, we have

$$\mathcal{F}_a^{-m} = (\mathcal{F}_a^m)^\vee[\dim(a)]$$

so $m'_a = -m_a$ and

$$\text{Ampl}(a) = 2m_a.$$

Moreover since $\mathcal{F}_a^{m_a}$ appears in $R^{\dim(\mathcal{M}^\xi)+m_a-\dim(a)} f_*^\xi \overline{\mathbb{Q}_l}$, we must have

$$\dim(\mathcal{M}^\xi) + m_a - \dim(a) \leqslant 2\dim(f^\xi) \tag{6.14}$$

and we have equality in (6.14) if and only if $\mathcal{F}_a^{m_a}$ appears in $R^{2\dim(f^\xi)} f_*^\xi \overline{\mathbb{Q}_l}$. We get

$$m_a \leqslant \dim(f^\xi) - \dim(\mathcal{A}) + \dim(a). \tag{6.15}$$

So the amplitude satisfies the first inequality (with the same case of equality as that for (6.14))

$$\text{Ampl}(a) \leqslant 2(\dim(f^\xi) - \dim(\mathcal{A}) + \dim(a)). \tag{6.16}$$

In what follows we will be a little sketchy (for example, we do not specify when we have to take geometric points instead of Zariski ones). Recall (cf. Section 4.4) that the Picard group scheme $\text{Pic}^0(Y_a)$ acts on the Hitchin fiber $\mathcal{M}_a$. In fact, this action does not preserve the open substack $\mathcal{M}_a^\xi$. But the neutral component of $\text{Pic}^0(Y_a)$, denoted by P_a, does act on $\mathcal{M}_a^\xi$. It is then possible to deduce from it an action of the homology of P_a on the stalk of $\mathcal{F}_a^\bullet$ at a. But we have the usual Chevalley dévissage

$$0 \to P_a^{\text{aff}} \to P_a \to P_a^{\text{ab}} \to 0$$

where P_a^{aff} is affine and the quotient is an abelian scheme. By a weight argument, (cf. [23] §7.4.8), the action of the homology of P_a factors through an action of the homology of P_a^{ab}; moreover the stalk of $\mathcal{F}_a^\bullet$ at a is a free graded module over this homology (proposition 7.4.10 of [23], cf. also [8] proposition 10.3.1 and proof of theorem 10.5.1). So the amplitude satisfies the second inequality

$$\text{Ampl}(a) \geqslant 2\dim(P_a^{\text{ab}}). \tag{6.17}$$

In fact, the group scheme $P_a^{\rm ab}$ can be identified with the the neutral component of the Picard group scheme of the normalization $\tilde{Y}_a$ of Y_a. We thus have

$$\dim(P_a^{\rm ab}) = \dim(H^1(\tilde{Y}_a, \mathcal{O}_{\tilde{Y}_a}))$$

and by the long exact sequence in cohomology associated to the short exact sequence (6.12) and Theorem 6.17, we get

$$\begin{aligned}\dim(P_a^{\rm ab}) &= -1 + r_a - \delta_a + \dim(P_a) \\ &= -1 + r_a - \delta_a + \dim(f^\xi)\end{aligned} \tag{6.18}$$

where r_a is the number of connected components of $\tilde{Y}_a$. Combining (6.17) and (6.16), we get

$$\mathrm{codim}_{\mathcal{A}}(a) \leqslant \delta_a - r_a + 1. \tag{6.19}$$

7.6. The support theorem on $\mathcal{A}^{\rm good}$

From now on, we assume that we are working on the open subset $\mathcal{A}^{\rm good}$. So we should introduce new notation $\mathcal{M}^{\xi,\rm good} = \mathcal{M}^\xi \times_{\mathcal{A}} \mathcal{A}^{\rm good}$ and so on. By abuse, we will keep the former ones. We can state the main cohomological theorem which is the key of the fundamental lemma.

Theorem 6.28. *The socle of* ${}^p\mathcal{H}^\bullet(Rf_*^\xi\overline{\mathbb{Q}_l})$ *contains a single element which is the generic point of* $\mathcal{A}$.

In other words, the only support of a simple constituent of ${}^p\mathcal{H}^\bullet(Rf_*^\xi\overline{\mathbb{Q}_l})$ is $\mathcal{A}$ itself.

Remark 6.29. This theorem is due to Ngô on the elliptic set and to Laumon and myself for the extension outside the elliptic set. By the theorem, the perverse cohomology of the Hitchin fiber is determined by its restriction to any open dense subset of $\mathcal{A}$. Thanks to the Grothendieck–Lefschetz trace formula and the countings of Proposition 6.8 and Remark 6.27, the stalks of the perverse cohomology are related to global (weighted) orbital integrals. So the theorem gives a technical sense to the vague assertion that global (weighted) orbital integrals are "limits" of the simplest orbital integrals (those associated to smooth spectral curves which can be computed "by hands").

Remark 6.30. For general groups, the support theorem is not true as stated: in general, there are other supports besides $\mathcal{A}$. But all new supports fit perfectly in the theory of endoscopy: they are the bases of Hitchin fibration associated to endoscopic groups. The determination of the supports is the key to the solution of the fundamental lemma. Indeed, Ngô first checks by hand a global variant of the fundamental lemma on a smaller set. Then, by his support theorem, the identity extends to a larger set. Finally he gets from it the local statement by local–global methods.

Let us briefly explain how to get Theorem 6.28. Let a be an element in the socle. If a does not belong to the elliptic set $\mathcal{A}^{\mathrm{ell}}$, the spectral curve has at least two irreducible components so we have $r_a > 1$. Thus (6.19) gives

$$\operatorname{codim}_{\mathcal{A}}(a) < \delta_a$$

in contradiction with (6.7) (here we use that $a \in \mathcal{A}^{\mathrm{good}}$). If $a \in \mathcal{A}^{\mathrm{ell}}$ then $r_a = 1$. By (6.7), the inequalities (6.19) and (6.14) must be equalities. Hence $\mathcal{F}_a^{m_a}$ must appear in $R^{2\dim(f^\xi)} f_*^\xi \overline{\mathbb{Q}_l}$. But on the elliptic set, the Hitchin fibers are irreducible (cf. Theorem 6.17) and this sheaf is simply the constant sheaf $\overline{\mathbb{Q}_l}$ on $\mathcal{A}^{\mathrm{ell}}$. So a must be the generic point of $\mathcal{A}$.

Remark 6.31. For Hitchin fibrations for general reductive groups, the elliptic fibers are in general not irreducible : this explains in part why there are new supports besides $\mathcal{A}$ in the decomposition theorem.

References

[1] A. Altman, A. Iarrobino, and S. Kleiman. Irreducibility of the compactified Jacobian. In *Real and complex singularities (Proc. Ninth Nordic Summer School/NAVF Sympos. Math., Oslo, 1976)*, pages 1–12. Sijthoff and Noordhoff, Alphen aan den Rijn, 1977.

[2] A. Altman and S. Kleiman. Compactifying the Picard scheme. *Adv. in Math.*, 35(1) :50–112, 1980.

[3] A. Beauville, M. Narasimhan, and S. Ramanan. Spectral curves and the generalised theta divisor. *J. Reine Angew. Math.*, 398 :169–179, 1989.

[4] A. Beĭlinson, J. Bernstein, and P. Deligne. Faisceaux pervers. In *Analysis and topology on singular spaces, I (Luminy, 1981)*, volume 100 of *Astérisque*, pages 5–171. Soc. Math. France, Paris, 1982.

[5] R. Bezrukavnikov. The dimension of the fixed point set on affine flag manifolds. *Math. Res. Lett.*, 3(2) :185–189, 1996.

[6] I. Biswas and S. Ramanan. An infinitesimal study of the moduli of Hitchin pairs. *J. London Math. Soc. (2)*, 49(2) :219–231, 1994.

[7] P.-H. Chaudouard and G. Laumon. Le lemme fondamental pondéré. I. Constructions géométriques. *Compos. Math.*, 146(6) :1416–1506, 2010.

[8] P.-H. Chaudouard and G. Laumon. Le lemme fondamental pondéré. II. Enoncés cohomologiques. *Ann. of Math.*, 176(3) :1647–1781, 2012.

[9] R. Cluckers, T. Hales, and F. Loeser. Transfer principle for the fundamental lemma. In *On the stabilization of the trace formula*, volume 1 of *Stab. Trace*

Formula Shimura Var. Arith. Appl., pages 309–347. Int. Press, Somerville, MA, 2011.

[10] P. Deligne. La conjecture de Weil. II. *Inst. Hautes Études Sci. Publ. Math.*, (52) :137–252, 1980.

[11] M. Demazure, H. Pinkham, and B. Teissier, editors. *Séminaire sur les Singularités des Surfaces*, volume 777 of *Lecture Notes in Mathematics*. Springer, Berlin, 1980. Held at the Centre de Mathématiques de l'École Polytechnique, Palaiseau, 1976–1977.

[12] S. Diaz and J. Harris. Ideals associated to deformations of singular plane curves. *Trans. Amer. Math. Soc.*, 309(2) :433–468, 1988.

[13] E. Esteves. Compactifying the relative Jacobian over families of reduced curves. *Trans. Amer. Math. Soc.*, 353(8) :3045–3095 (electronic), 2001.

[14] M. Goresky, R. Kottwitz, and R. MacPherson. Homology of affine Springer fibers in the unramified case. *Duke Math. J.*, 121(3) :509–561, 2004.

[15] M. Goresky, R. Kottwitz, and R. MacPherson. Purity of equivalued affine Springer fibers. *Represent. Theory*, 10 :130–146 (electronic), 2006.

[16] D. Kazhdan and G. Lusztig. Fixed point varieties on affine flag manifolds. *Israel J. Math.*, 62(2) :129–168, 1988.

[17] J.-P. Labesse and R. P. Langlands. L-indistinguishability for SL(2). *Canad. J. Math.*, 31(4) :726–785, 1979.

[18] R. Langlands. *Les débuts d'une formule des traces stable*, volume 13 of *Publications Mathématiques de l'Université Paris VII*. Université de Paris VII U.E.R. de Mathématiques, Paris, 1983.

[19] R. Langlands and D. Shelstad. On the definition of transfer factors. *Math. Ann.*, 278 :219–271, 1987.

[20] S. Langton. Valuative criteria for families of vector bundles on algebraic varieties. *Ann. of Math. (2)*, 101 :88–110, 1975.

[21] G. Laumon. Fibres de Springer et jacobiennes compactifiées. In *Algebraic geometry and number theory*, volume 253 of *Progr. Math.*, pages 515–563. Birkhäuser Boston, Boston, MA, 2006.

[22] B. C. Ngô. Fibration de Hitchin et endoscopie. *Invent. Math.*, 164(2) :399–453, 2006.

[23] B. C. Ngô. Le lemme fondamental pour les algèbres de Lie. *Publ. Math. Inst. Hautes Études Sci.*, (111) :1–169, 2010.

[24] B. C. Ngô. Decomposition theorem and abelian fibration. In *On the stabilization of the trace formula*, volume 1 of *Stab. Trace Formula Shimura Var. Arith. Appl.*, pages 253–264. Int. Press, Somerville, MA, 2011.

[25] J.-L. Waldspurger. Endoscopie et changement de caractéristique. *J. Inst. Math. Jussieu*, 5(3) :423–525, 2006.

[26] J.-L. Waldspurger. L'endoscopie tordue n'est pas si tordue. *Mem. Amer. Math. Soc.*, 194(908) :x+261, 2008.

7

The p-adic analytic space of pseudocharacters of a profinite group and pseudorepresentations over arbitrary rings

Gaëtan Chenevier

Abstract

Let[1] G be a profinite group which is topologically finitely generated,[2] p a prime number and $d \geq 1$ an integer. We show that the functor from rigid analytic spaces over $\mathbb{Q}_p$ to sets, which associates to a rigid space Y the set of continuous d-dimensional pseudocharacters $G \longrightarrow \mathcal{O}(Y)$, is representable by a quasi-Stein rigid analytic space X, and we study its general properties.

Our main tool is a theory of *determinants* extending the one of pseudocharacters but which works over an arbitrary base ring; an independent aim of this chapter is to expose the main facts of this theory. The moduli space X is constructed as the generic fiber of the moduli formal scheme of continuous formal determinants on G of dimension d.

As an application to number theory, this provides a framework to study rigid analytic families of Galois representations (e.g. eigenvarieties) and generic fibers of pseudodeformation spaces (especially in the "residually reducible" case, including when $p \leq d$).

Introduction

Let G be a group, A a commutative ring with unit and let

$$T : G \longrightarrow A$$

[1] The author is supported by the CNRS, as well as by the ANR project ANR-10-BLAN 0114.

[2] Actually, we only assume that for any normal open subgroup $H \subset G$, there are only finitely many continuous group homomorphisms $H \longrightarrow \mathbb{Z}/p\mathbb{Z}$.

Automorphic Forms and Galois Representations, ed. Fred Diamond, Payman L. Kassaei and Minhyong Kim. Published by Cambridge University Press.

be a map such that $T(gh) = T(hg)$ for all $g, h \in G$. For $n \geq 1$ an integer and $\sigma \in \mathfrak{S}_n$, set[3] $T^\sigma(g_1, g_2, \ldots, g_n) = T(g_{i_1} g_{i_2} \cdots g_{i_r})$ if σ is the cycle $(i_1 i_2 \ldots i_r)$, and in general

$$T^\sigma = \prod T^{c_i}$$

if $\sigma = c_1 \ldots c_s$ is the cycle decomposition of σ. The *n-dimensional pseudocharacter identity* is the relation

$$\forall g_1, g_2, \ldots, g_n, g_{n+1} \in G, \quad \sum_{\sigma \in \mathfrak{S}_{n+1}} \varepsilon(\sigma) T^\sigma(g_1, g_2, \ldots, g_{n+1}) = 0, \quad (7.1)$$

where $\varepsilon(\sigma)$ denotes the signature of the permutation σ. We say that T is a *d-dimensional pseudocharacter* of G with values in A if T satisfies the d-dimensional pseudocharacter identity, if $T(1) = d$ and if $d!$ is invertible in A.

The main interest of pseudocharacters lies in the close relations they share with traces of representations: by an old result of Frobenius [Fr, p. 50], the trace of a representation $G \longrightarrow \mathrm{GL}_d(A)$ is a d-dimensional pseudocharacter,[4] and it is known that the converse holds when A is an algebraically closed field with $d! \in A^*$ (Procesi [P3], Taylor [T] for $\mathbb{Q}$-algebras, [Rou] in general[5]) as well as in various other situations (see below). In particular, we obtain this way an interesting parameterization of the isomorphism classes of semisimple representations of G over such algebraically closed fields. As the covariant functor from the category of $\mathbb{Z}[1/d!]$–commutative algebras with unit to the category Ens of sets, which associates to A the set of d-dimensional pseudocharacters $G \rightarrow A$, is obviously representable,[6] it turned out to be an interesting substitute for the quotient functor $\mathrm{Hom}(G, \mathrm{GL}_d(-))/\mathrm{PGL}_d(-)$ of

3 This expression has the following important interpretation, due to Kostant. Assume that $T : \mathrm{GL}_m(A) \rightarrow A$ is the trace map, if $g_1, \ldots, g_n \in \mathrm{GL}_m(A)$ and if $\sigma \in \mathfrak{S}_n$, then $T^\sigma(g_1, \ldots, g_n)$ coincides with the trace of the element $(g_1, \ldots, g_n)\sigma$ acting on $V^{\otimes_A n}$, where $V := A^m$.

4 In view of the previous footnote, this simply expresses the fact that $\Lambda^{n+1} A^n = 0$.

5 As already observed in [BC, §1], let us warn the reader that although Rouquier does not require $d!$ to be invertible in A in [Rou], there is a gap in the proof of his Lemma 4.1, hence of his Theorem 4.2, without this assumption. Indeed, it is not clear that each element of his ring R is algebraic over k, as asserted on p. 580 line 2, because the polynomial P_x given by his Lemma 2.13 might be identically zero if $d!$ is not invertible in A.

6 Consider the ring B_0 which is the quotient of the polynomial $\mathbb{Z}[1/d!]$-algebra over the indeterminates X_g for all $g \in G$, by the ideal generated by the elements $X_{gh} - X_{hg}$ for all $g, h \in G$. For each $\sigma \in \mathfrak{S}_n$ and $g_1, \cdots, g_n \in G^n$, we have a well defined element $X^\sigma(g_1, \ldots, g_n) \in B_0$ defined as the class of $X_{g_{i_1} g_{i_2} \cdots g_{i_r}}$ if σ is the cycle $(i_1 i_2 \ldots i_r)$, and of $\prod_i X^{c_i}(g_1, \cdots, g_n)$ in general if $\sigma = \prod_i c_i$ is the cycle decomposition of σ. Define B_1 as the quotient of B_0 by the ideal generated by $X_1 - d$ and the elements $\sum_{\sigma \in \mathfrak{S}_{d+1}} \varepsilon(\sigma) X^\sigma(g_1, g_2, \ldots, g_{d+1})$ for all $g_1, \cdots, g_{d+1} \in G^{d+1}$. The map $G \rightarrow B_1, g \mapsto \overline{X_g}$, is the universal d-dimensional pseudocharacter of G.

isomorphism classes of d-dimensional representations of G. Indeed, since they have been introduced in number theory by Wiles [W] (when $d = 2$), and by Taylor [T] under the form above (sometimes under the name of *pseudorepresentations*), they have proved to be a successful tool, first to actually construct some (Galois) representations, and then to study Galois representations and Hecke-algebras.

Over $\mathbb{Q}$-algebras, most of the basic properties of pseudocharacters follow actually from earlier work of Procesi on invariants of n-tuples of $d \times d$-matrices [P2] and on the very close subject of Cayley–Hamilton algebras [P3]. In relation to deformation theory, pseudocharacters over local rings have also been studied by Nyssen [N] and Rouquier [Rou] in the residually irreducible case, and by Bellaïche–Chenevier [BC, Ch. 1] in the residually multiplicity free case.

The first part of this chapter addresses the problem of setting a definition for an A-valued pseudocharacter of dimension d which works for an *arbitrary ring*, i.e. without the assumption that $d! \in A^*$, and to extend to this setting most of the aforementioned results. When $d!$ is invertible, the pseudocharacter identity of degree d is very close to the Cayley–Hamilton identity of degree d defined by the pseudocharacter T and this is actually the key to most of the interesting properties of pseudocharacters;[7] it is certainly not surprising that the definition above of pseudocharacters does not work well in general. The key notion turned out to be the one of *multiplicative homogeneous polynomial laws* on algebras, which have been studied by Roby ([Ro1],[Ro2]), Ziplies [Z1], Ferrand [Fe] and more recently by Vaccarino ([V1],[V2],[V3]), and which immediately leads to a definition for a "generalized" pseudocharacter. To avoid confusion, we rather call them *(law-) determinants*. Up to the language of polynomial laws of Roby [Ro1] that we recall in a preliminary § 1.1, our definition is surprisingly simple:

Definition. An A-valued determinant on G of dimension d is an A-polynomial law $D : A[G] \longrightarrow A$ which is homogeneous of degree d and multiplicative.

Of course, usual determinants of true A-algebra representations $A[G] \longrightarrow M_d(A)$ are determinants in this setting, and we shall prove various converse results. By definition, it is equivalent to give a determinant as above and a finite collection of maps $G^d \longrightarrow A$ satisfying various identities, which are in general much more complicated than the pseudocharacter identity. We make explicit this point of view in the special case $d = 2$ (§ 7.6). Thanks to Roby's work, there is also an equivalent general definition for a determinant in terms

[7] Precisely, relation (7.1) is exactly $T(\mathrm{CH}_T(g_1, g_2, \ldots, g_d)g_{d+1}) = 0$ (see [P2]) where CH_T is the multi-linearization of the "characteristic polynomial of degree d associated to T", which is the homogeneous polynomial $x^d - T(x)x^{d-1} + \frac{T(x)^2 - T(x^2)}{2}x^{d-2} + \cdots + \det(x)$.

of the divided power ring $\Gamma^d_{\mathbb{Z}}(\mathbb{Z}[G])$ of degree d ([Ro2]), which is naturally isomorphic to the, maybe more standard, ring of invariants $(\mathbb{Z}[G]^{\otimes d})^{\mathfrak{S}_d}$: an A-valued determinant on G of dimension d is simply a ring homomorphism

$$\Gamma^d_{\mathbb{Z}}(\mathbb{Z}[G]) \longrightarrow A.$$

In particular, *the natural functor associating to A the set of d-dimensional A-valued determinants is representable* by the ring $\Gamma^d_{\mathbb{Z}}(\mathbb{Z}[G])^{\mathrm{ab}} \overset{\sim}{\rightarrow} ((\mathbb{Z}[G]^{\otimes d})^{\mathfrak{S}_d})^{\mathrm{ab}}$.

Thanks to works of many people (Amitsur, Procesi, Donkin, Zubkov, Vaccarino and certainly others), most of the deepest properties of determinants are actually known, although it is hard to extract from the literature a unified picture.[8] In the first half of this chapter, which may be viewed mostly as an introduction to the subject, we make an attempt to expose the theory from the narrow point of view of determinants, trying to remain as self-contained (and coherent) as possible.

In the first section, we develop the most basic properties of determinants (§ 1.3,§ 1.4): polynomial identities, *Kernel of a determinant*, *faithful and Cayley–Hamilton quotients*, and properties with respect to base change. An important role is played there and in the whole theory by a polynomial identity which is formally analogue to *Amitsur's formula* (7.5), which expresses the determinant of a sum of elements $x_1 + \cdots + x_r$ in terms of the coefficients of the characteristic polynomials of some explicit monomials in x_i; we give an elementary proof for this formula for any determinant by mimicking an elegant proof in the matrix case due to Reutenauer–Schützenberger[9] [RS]. Another important step is to show that the faithful quotient $A[G]/\ker(D)$ satisfies the Cayley–Hamilton identity, and for that we have to rely for the moment on an important result of Vaccarino [V1] describing $\Gamma^d_{\mathbb{Z}}(\mathbb{Z}\{X\})^{\mathrm{ab}}$ when $\mathbb{Z}\{X\}$ is the free ring over the finite set X (actually, we only use that this ring is torsion free, but Vaccarino's result is much stronger, see § 1.3). Using results of Procesi, we show also that *over $\mathbb{Q}$-algebras, determinants and pseudocharacters coincide*, but we do not know if this holds under the weaker assumption that $d! \in A^*$ (we do however prove it when $d = 2$ and in several other cases, see Remark 7.22). These last two points are actually the only places in the chapter where we are not self-contained (see Remarks 7.13 and 7.22).

In the second section, we prove the analogue for determinants of the standard aforementioned results of the theory of pseudocharacters. The approach

8 We thank a referee for pointing out the recent paper [DCPRR] for a collection of results on Cayley–Hamilton algebras.

9 As F. Vaccarino pointed out to us, very similar results had also been obtained by Ziplies in [Z2].

we follow is inspired from the one in [BC, Ch. 1], but we have to face several extra difficulties inherent to the use of polynomial laws and also from the presence of some inseparable extensions which occur in characteristic $p \leq d$.

Theorem A. *Let k be an algebraically closed field and $D : k[G] \longrightarrow k$ be a determinant of dimension d. There exists a unique semisimple representation $\rho : G \longrightarrow \mathrm{GL}_d(k)$ such that for any $g \in G$, $\det(1 + t\rho(g)) = D(1 + tg)$.*

(See Theorem 7.34) Of course, "unique" here means "unique up to k-isomorphism". In fact, *if k is any perfect field, or any field of characteristic $p > 0$ such that either $p > d$ or $[k : k^p] < \infty$*, we show the stronger fact that *$k[G]/\ker(D)$ is a semisimple finite dimensional k-algebra* (Theorem 7.38).

Theorem B. *Let A be a henselian local ring with algebraically closed residue field k, $D : A[G] \longrightarrow A$ a determinant of dimension d, and let ρ be a semisimple representation attached to $D \otimes_A k$ by Theorem A. If ρ is irreducible, then there exists a unique representation $\tilde{\rho} : G \longrightarrow \mathrm{GL}_d(A)$ such that for any $g \in G$, $\det(1 + t\tilde{\rho}(g)) = D(1 + tg)$.*

(See Theorem 7.43) Actually, we show the stronger fact that the biggest Cayley–Hamilton quotient of $A[G]$ is the faithful one, and is isomorphic to $(M_d(A), \det)$. We consider also the more general case where, under the assumption of Theorem B, ρ is only assumed to be multiplicity free, and we show then that any Cayley–Hamilton quotient of $A[G]$ is a generalized matrix algebra in the sense of [BC, Ch. 1], extending a result there.

Let us stress here that Theorems A and B should not be considered as original, as they could probably be deduced from earlier works of Procesi ([P1],[P4]) via the relations between determinants and generic matrices established by Vaccarino, Donkin and Zubkov.

The last part of the second section deals with the problem deforming a given determinant D_0 to $A[\epsilon]$ with $\epsilon^2 = 0$ (§ 2.4). The set of such deformations of D_0 appears naturally as a relative tangent space and has a natural structure of A-module. When G is a topological group and A a topological ring, we say that an A-valued determinant on G is continuous if the coefficients of the characteristic polynomial $D(t - g)$ are continuous functions of $g \in G$. The main result here is the following (Prop 7.63):

Proposition C. *Let k be a discrete algebraically closed field, G a profinite group, $\rho : G \longrightarrow \mathrm{GL}_d(k)$ a semisimple continuous representation, and $D_0 = \det \circ \rho$. Let $p \geq 0$ be the characteristic of the field k. The space of continuous deformations of D_0 to $k[\epsilon]$ is finite dimensional in the following two cases:*

(a) $p = 0$ or $p > d$, and the continuous cohomology group $H^1_c(G, \mathrm{ad}(\rho))$ is finite dimensional over k,

(b) $0 < p \leq d$ *and for each open subgroup* $H \subset G$*, there are only finitely many continuous homomorphisms* $H \to \mathbb{Z}/p\mathbb{Z}$*.*

All of this being done, we are perfectly well equipped to study rigid analytic families of pseudocharacters. Let us assume from now on that G is a profinite topological group, fix $d \geq 1$ an integer, and let p be a prime number. Assume moreover that G satisfies the following finiteness condition: *For any normal open subgroup* $H \subset G$*, there are only finitely many continuous group homomorphisms* $H \longrightarrow \mathbb{Z}/p\mathbb{Z}$. This holds for instance when G is topologically finitely generated, when G is the absolute Galois group of a local field of characteristic $\neq p$ (e.g. $\mathbb{Q}_p$), or when G is the absolute Galois group of a number field with finite restricted ramification.

Let An be the category of rigid analytic spaces over $\mathbb{Q}_p$ in the sense of Tate (see [BGR]). If X is such a space, we shall denote by $\mathcal{O}_X$ its structural sheaf and by $\mathcal{O}(X)$ the $\mathbb{Q}_p$-algebra of global sections of $\mathcal{O}_X$. We equip $\mathcal{O}(X)$ with the topology of uniform convergence on the open affinoids of X. The main aim of this chapter is to study the contravariant functor E^{an} : An $\longrightarrow$ Ens, which associates to a rigid space X the set $E^{\mathrm{an}}(X)$ of continuous d-dimensional pseudocharacters $G \longrightarrow \mathcal{O}(X)$.

Theorem D. E^{an} *is representable by a quasi-Stein rigid analytic space.*

(See Theorem 7.72) This rigid analytic space might be called the *p-adic character variety of G in dimension d*. To show this theorem we actually start with studying other natural functors. First, we fix a continuous semisimple representation

$$\bar{\rho} : G \longrightarrow \mathrm{GL}_d(\overline{\mathbb{F}}_p)$$

and whose determinant D takes values in some finite field $k \subset \overline{\mathbb{F}}_p$. We consider the continuous deformation functor F of D to discrete artinian local $W(k)$-algebras with residue field k. Here $W(k)$ denotes the Witt ring of k. We prove first the following (Prop. 7.59):

Proposition E. *F is prorepresentable by a complete local noetherian* $W(k)$*-algebra* $A(\bar{\rho})$ *with residue field k.*

Of course, for the noetherian property we rely on Proposition C. The ring $A(\bar{\rho})$ is constructed as a certain profinite completion of

$$\Gamma^d_{\mathbb{Z}}(\mathbb{Z}[G])^{\mathrm{ab}} \otimes_{\mathbb{Z}} W(k).$$

We consider then the functor E from the category of formal schemes over $\mathrm{Spf}(\mathbb{Z}_p)$ to sets, which associates to $\mathcal{X}$ the set of continuous d-dimensional $\mathcal{O}(\mathcal{X})$-valued determinants on G. We can attach to each such formal determinant a subset of "residual determinants". The set $|G(d)|$ of all residual determinants is in natural bijection with the set of (determinants of the) continuous semisimple representations $\bar{\rho}$ as above, taken up to isomorphism and Frobenius actions on coefficients. It turns out that that *the subfunctor* $E_{\bar{\rho}} \subset E$ *parameterizing determinants which are residually constant and "equal to"* $\det \circ \bar{\rho}$ *is representable* and isomorphic to the affine formal scheme $\mathrm{Spf}(A(\bar{\rho}))$ over $\mathrm{Spf}(\mathbb{Z}_p)$ (where $A(\bar{\rho})$ is equipped with the m-adic topology given by its maximal ideal m). Our main second result is then the following (§ 3.2.2), which implies Thm. D:

Theorem F. *The functor E is representable by the disjoint union of the* $\mathrm{Spf}(A(\bar{\rho}))$, *for* $\bar{\rho} \in |G(d)|$. *The functor E^{an} is canonically isomorphic to the generic fiber of E in the sense of Berthelot.*

If we fix an isomorphism $W(k)[[t_1, \ldots, t_h]]/I \overset{\sim}{\rightarrow} A(\bar{\rho})$, then we get a closed immersion

$$\mathrm{Spf}(A(\bar{\rho}))^{\mathrm{rig}} \hookrightarrow \mathbb{B}^h_{[0,1[}$$

as the closed subspace of the open h-dimensional unit ball defined by $I = 0$, and X is then a disjoint union of such spaces.

In Section 4, we give some general complements about the rigid analytic space X representing E^{an}. For instance, consider the functor

$$E^{\mathrm{irr}} : \mathrm{An} \rightarrow \mathrm{Ens}$$

which associates to any rigid space X the set of isomorphism classes of pairs (R, ρ) where R is an Azumaya $\mathcal{O}_X$-algebra of rank d^2 and $\rho : G \rightarrow R^*$ is a continuous group homomorphism such that for all closed points $x \in X$, the evaluation $\rho_x : G \rightarrow R_x^*$ is absolutely irreducible (see § 4.2).

Proposition G. *E^{irr} is representable by a Zariski-open subspace of E^{an} equipped with its universal Cayley–Hamilton representation.*

In the last section 5, we give an application of some of the previous results to Galois representations. Let G be the Galois group of a maximal algebraic extension of $\mathbb{Q}$ unramified outside $\{2, \infty\}$, and let X be the 2-adic analytic space parameterizing the 2-dimensional rigid analytic pseudocharacters of G (so $p = d = 2$). This space X is an admissible disjoint union of three open subspaces X^{odd}, X^+ and X^- over which the trace of a complex conjugation of G is respectively 0, 2 and -2.

Theorem H. *X^{odd} (resp. $X^{\pm}$) is the open unit ball of dimension* 3 *(resp. dimension* 2*) over* $\mathbb{Q}_2$.

The author would like to thank Emmanuel Breuillard, Claudio Procesi and Francesco Vaccarino for some useful discussions, a referee for his careful reading, as well as Jean-Pierre Serre and Joël Bellaïche for their remarks.

1. Determinants of algebras

1.1. Homogeneous multiplicative A-polynomial laws

We need some preliminaries about polynomial laws between two modules. We refer to [Ro1] and [Ro2] for the proofs of all the results stated below.

Let A be a commutative unital ring, and let M and N be two A-modules. Let $\mathcal{C}_A$ be the category of commutative A-algebras. Each A-module M gives rise to a functor $\underline{M} : \mathcal{C}_A \longrightarrow \text{Ens}$ via the formula $B \mapsto M \otimes_A B$. An *A-polynomial law* $P : M \longrightarrow N$ is a natural transformation $\underline{M} \longrightarrow \underline{N}$. In other words, it is a collection of maps

$$P_B : M \otimes_A B \longrightarrow N \otimes_A B,$$

where B is any commutative A-algebra, which commute with any scalar extension $B \to B'$ over A. By a slight abuse of notation, if B is a commutative A-algebra and $m \in M \otimes_A B$ we shall often write $P(m)$ for $P_B(m)$. When $B = A[T_1, \ldots, T_s]$, we shall write $M[T_1, \ldots, T_s]$ for $M \otimes_A A[T_1, \ldots, T_s]$.

We refer to [Ro1] for the basic operations that we can do with polynomial laws. If B is a commutative A-algebra and $P : M \longrightarrow N$ is an A-polynomial law, we will denote by $P \otimes_A B : M \otimes_A B \longrightarrow N \otimes_A B$ the natural induced B-polynomial law.[10]

We say that P is homogeneous of degree n (an integer ≥ 0) if $P(bx) = b^n P(x)$ for all object B in $\mathcal{C}_A$, $b \in B$ and $x \in M \otimes_A B$.

Example 7.1. Let $P : M \longrightarrow N$ be a homogeneous A-polynomial law of degree n.

(i) When $n = 1$ (resp. $n = 0$), P_A is an A-linear[11] map and $P_B = P_A \otimes_A B$ (resp. $P_B = P_A(0) \otimes 1$ is a constant), and $P \mapsto P_A$ induces a bijection between A-polynomial laws of degree 1 (resp. 0) and $\text{Hom}_A(M, N)$ (resp. N).

(ii) When $n = 2$, P_B is again uniquely determined by P_A, which is any map $q : M \longrightarrow N$ such that $q(am) = a^2 q(m)$ for all $a \in A$, $m \in M$, and such that $(m, m') \mapsto q(m + m') - q(m) - q(m')$ is A-bilinear.

(iii) When $n \geq 3$, P_A does not determine P_B in general. For instance, let A be the finite field $\mathbb{F}_q$ with q elements, $M = \mathbb{F}_q^2$ and let X, Y be an A-basis of $\text{Hom}_A(M, A)$. The A-polynomial law $P : M \to A$ defined by

[10] By definition, if C is a commutative B-algebra, $(P \otimes_A B)_C = P_C$ via the isomorphism $(- \otimes_A B) \otimes_B C = - \otimes_A C$.

[11] Let X, Y, T be indeterminates. If $u, v \in M$, then $P(uX + vY) \in N[X, Y]$. As P has degree 1, sending (X, Y) to (XT, YT) shows that $P(uX + vY)$ is of the form $a(u, v)X + b(u, v)Y$ for some well-defined functions $a, b : M^2 \to N$. By evaluating (X, Y) at $(1, 0)$, $(0, 1)$ and $(1, 1)$, we obtain respectively $a(u, v) = P(u)$, $b(u, v) = P(v)$ and $P(u + v) = P(u) + P(v)$.

$P = XY^q - X^qY$ is homogeneous of degree $q+1$, we have $P_A = 0$ but $UV^q - U^qV \in P(M[U,V]) \neq 0$.

In any cases, a homogeneous P of degree n is uniquely determined by $P_{A[T_1,\dots,T_n]} : M[T_1,\dots,T_n] \longrightarrow N[T_1,\dots,T_n]$. Precisely, if $X \subset M$ generates M as A-module, then such a P is uniquely determined by the (finite) set of functions

$$P^{[\alpha]} : X^n \longrightarrow N,$$

with $\alpha \in I_n = \{(\alpha_1,\alpha_2,\dots,\alpha_n) \in \mathbb{N}^n,\ \alpha_1+\cdots+\alpha_n = n\}$, defined by the relation

$$P(\sum_{i=1}^n T_i x_i) = \sum_{\alpha \in I_n} P^{[\alpha]}(x_1,\dots,x_n)T^\alpha,$$

where $T^\alpha = \prod_{i=1}^n T_i^{\alpha_i}$.

We denote by $\mathcal{P}_A^n(M,N)$ the A-module of homogeneous A-polynomial laws of degree n from M to N. The functor $\mathcal{P}_A^n(M,-) : \mathrm{Mod}(A) \to \mathrm{Mod}(A)$ is representable by the usual A-module $\Gamma_A^n(M)$ of divided powers of order n on M relative to A ([Ro1, Thm. 4.1]). Let us recall that $\Gamma_A^n(M)$ is naturally isomorphic to the nth-graded piece of the commutative A-algebra $\Gamma_A(M)$ which is generated by the symbols $m^{[i]}$ for $m \in M$ and $i \geq 0$, with the usual homogeneous relations:

- $m^{[0]} = 1$ for all $m \in M$,
- $(am)^{[i]} = a^i m^{[i]}$ for all $a \in A$ and $m \in M$,
- $m^{[i]}m^{[j]} = \frac{(i+j)!}{i!j!} m^{[i+j]}$ for all $i, j \geq 0$ and $m \in M$,
- $(m+m')^{[i]} = \sum_{p+q=i} m^{[p]}m'^{[q]}$ for all $i \geq 0$ and $m, m' \in M$.

The natural map $P^{\mathrm{univ}} : m \mapsto m^{[n]}$, $M \longrightarrow \Gamma_A^n(M)$, induces *the universal homogeneous A-polynomial law of degree n*. For $\alpha \in I_n$ as above, $(P^{\mathrm{univ}})^{[\alpha]}(m_1,\dots,m_d) = \prod_{j=1}^d m_j^{[\alpha_j]}$.

Let R and S be two A-algebras[12], and $P : R \longrightarrow S$ be a homogeneous A-polynomial law of degree n. We say that P is *multiplicative* if $P(1) = 1$ and if $P(xy) = P(x)P(y)$ for all B and $x, y \in R \otimes_A B$. For example, the homogeneous multiplicative A-polynomial laws of degree 1 are the A-algebra homomorphisms. By [Ro2], the structure of A-algebra on R

[12] By an A-algebra we shall always mean an associative and unital A-algebra (but not necessarily commutative).

induces an A-algebra structure on[13] $\Gamma^n_A(R)$, and it turns out that the functor $\mathcal{M}^n_A(R,-)$, from A-algebras to sets, that associates to any A-algebra S the set of $\mathcal{M}^n_A(R,S)$ of n-homogeneous multiplicative A-polynomial laws from R to S, is representable by the A-algebra $\Gamma^n_A(R)$ ([Ro2, Théorème]). In particular, the universal homogeneous A-polynomial law

$$P^{\mathrm{univ}} : R \longrightarrow \Gamma^n_A(R), \quad r \mapsto r^{[n]},$$

is multiplicative.

Remark 7.2. Let M be an A-module and let $\mathrm{TS}^n_A(M)$ be the A-submodule of $M^{\otimes^n_A}$ invariant by the symmetric group $\mathfrak{S}_n$. The natural map $M \longrightarrow \mathrm{TS}^n_A(M)$, $m \mapsto m^{\otimes n}$, induces a homogeneous A-polynomial law of degree n, hence there is a natural A-linear map

$$\Gamma^n_A(M) \longrightarrow \mathrm{TS}^n_A(M), \tag{7.2}$$

which is actually an isomorphism if M is free as A-module ([Ro1, Prop. IV.5]). When $M = R$ is an A-algebra, $\mathrm{TS}^n_A(R)$ has an obvious A-algebra structure and $r \mapsto r^{\otimes n}$ is clearly multiplicative, so (7.2) is actually an A-algebra homomorphism. In particular, if R is free as A-module, then

$$\Gamma^n_A(R) \overset{\sim}{\to} \mathrm{TS}^n_A(R)$$

is an A-algebra isomorphism.

Remark 7.3. If B is a commutative A-algebra and M an A-module, the homogeneous A-polynomial law of degree n

$$M \longrightarrow \Gamma^n_B(M \otimes_A B), \; m \mapsto (m \otimes 1)^{[n]},$$

induces an isomorphism ([Ro1, Thm. III.3]) $\Gamma^n_A(M) \otimes_A B \overset{\sim}{\to} \Gamma^n_B(M \otimes_A B)$. When $M = R$ is an A-algebra, this latter isomorphism is a B-algebra homomorphism as the polynomial law above is multiplicative.

1.2. Definition of a determinant

Let R be any A-algebra and $d \geq 1$ an integer.

Definition. A d-dimensional A-valued determinant on R is an element of $\mathcal{M}^d_A(R,A)$, i.e. a multiplicative A-polynomial law $\mathrm{D} : R \longrightarrow A$ which is homogeneous of degree d. When $R = A[G]$ for some group G (or unital monoid), we say also that D is a determinant on G.

[13] This structure is not to be confused with the A-algebra structure on $\Gamma_A(R)$, which is always graded and commutative. The A-algebra $\Gamma^n_A(R)$ is commutative if R is, its neutral element is $1_R^{[n]}$, and $\Gamma^1_A(R) = R$.

Of course, if $R = M_d(A)$ (resp. any Azumaya algebra of rank d^2 over its center A), the usual determinant $\det : M_d(A) \longrightarrow A$ (resp. the reduced norm) induces in the obvious way[14] a determinant of dimension d. In particular, for any A-algebra homomorphism $\rho : R \longrightarrow M_d(A)$,

$$\mathrm{D} := \det \circ \rho$$

is a d-dimensional A-valued determinant on R. In Section 7.43, we will prove some converse to this construction. For example, we will show that when A is an algebraically closed field, any determinant of R is of the form above, and we will also study the case when A is a local henselian ring. When $d = 1$, a determinant $\mathrm{D} : R \longrightarrow A$ of dimension 1 is by definition the same as an A-algebra homomorphism (see Example 7.1 (i)).

Let $\det_A(R, d) : \mathcal{C}_A \longrightarrow \mathrm{Ens}$ be the covariant functor associating to any commutative A-algebra B, the set of B-valued determinants $R \otimes_A B \longrightarrow B$ of dimension d, which is the same as the set of multiplicative homogeneous A-polynomial laws $R \longrightarrow B$ of dimension d (recall that $\mathcal{M}^d_A(R, B) \overset{\sim}{\rightarrow} \mathcal{M}^d_B(R \otimes_A B, B)$ by Remark 7.3). It is equivalent to give such a law or an A-algebra homomorphism $\Gamma^d_A(R) \longrightarrow B$, which necessarily factors through its abelianization[15] $\Gamma^d_A(R)^{\mathrm{ab}}$, hence we get the:

Proposition 7.4. *$\det_A(R, d)$ is representable by the A-algebra $\Gamma^d_A(R)^{\mathrm{ab}}$.*

In particular, when $R = \mathbb{Z}[G]$, then $\det_{\mathbb{Z}}(\mathbb{Z}[G], d)$ is representable by

$$\Gamma^d_{\mathbb{Z}}(\mathbb{Z}[G])^{\mathrm{ab}} \overset{\sim}{\rightarrow} \mathrm{TS}^d_{\mathbb{Z}}(\mathbb{Z}[G])^{\mathrm{ab}},$$

that we shall simply denote by $\mathbb{Z}(G, d)$. This ring is nonzero thanks to the trivial representation of dimension d of G. We will set also

$$X(G, d) = \mathrm{Spec}(\mathbb{Z}(G, d)).$$

Of course, if S is any scheme, we may define a determinant of dimension d on G over S as an $\mathcal{O}(S)$-valued determinant of dimension d on G, and $X(G, d)$ obviously still represents this extended determinant functor.

Example 7.5. (i) When $R = A[X]$ is a polynomial ring in one variable, then $\mathrm{TS}^d_A(R) = A[X_1, \ldots, X_d]^{\mathfrak{S}_d} = A[\Sigma_1, \ldots, \Sigma_d]$ by the classical theorem on symmetric polynomials (with the obvious notations for X_i and Σ_j). In particular, $\Gamma^d_A(R) = \Gamma^d_A(R)^{\mathrm{ab}} \simeq A[\Sigma_1, \ldots, \Sigma_d]$ by Remark 7.2. As

14 For any commutative A-algebra B, define $\det_B$ as the determinant $M_d(B) \rightarrow B$.

15 By definition, the abelianization of a ring R is the quotient of R by the two-sided ideal generated by the $xy - yx$ with $x, y \in R$.

we will see in § 1.3, the universal determinant is the determinant of the regular representation of $A[X]$ on

$$\Gamma^d_A(R)[X]/(X^d - \Sigma_1 X^{d-1} + \Sigma_2 X^{d-2} - \cdots + (-1)^d \Sigma_d).$$

(ii) When R is an Azumaya algebra of rank d^2 over its center A, a result of Ziplies [Z1] (see also Ex. 7.28) shows that the reduced norm is the unique A-valued determinant of dimension d of R, and even that the reduced norm induces an A-algebra isomorphism $\Gamma^d_A(R)^{\mathrm{ab}} \xrightarrow{\sim} A$.

(iii) When G is a finite group $\mathbb{Z}(G,d)$ is a finite $\mathbb{Z}$-algebra (as $\Gamma^d_{\mathbb{Z}}(\mathbb{Z}[G])$ is free of finite type as $\mathbb{Z}$-module).

(iv) Using Remark 7.3, we get that if B is any commutative A-algebra, the natural A-algebra homomorphism $B \otimes_A \Gamma^d_A(R)^{\mathrm{ab}} \longrightarrow \Gamma^d_B(B \otimes_A R)^{\mathrm{ab}}$ is an isomorphism.

In the case $R= A[G]$, it is equivalent to give a determinant $A[G] \longrightarrow A$ of dimension d and a d-homogeneous multiplicative polynomial law $\mathbb{Z}[G] \longrightarrow A$. Such a law is uniquely determined by the set of functions

$$\mathrm{D}^{[\alpha]} : G^d \longrightarrow A, \ \alpha \in I_d,$$

which satisfy a finite number of identities coming from the requirement that the map

$$\prod_{j=1}^{d} g_j^{[\alpha_j]} \mapsto \mathrm{D}^{[\alpha]}(g_1, \ldots, g_d) \in A$$

extends to a ring homomorphism $\Gamma^d_{\mathbb{Z}}(\mathbb{Z}[G]) \longrightarrow A$.

Example 7.6. *(Determinants of dimension* 2 *on a group* G *(or a unital monoid))* As an example, let us specify a bit those relations when $d = 2$. In this case, we may write

$$\mathrm{D}(gU + hV) = D(g)U^2 + f(g,h)UV + D(h)V^2$$

for some functions $D = \mathrm{D}_{|G} : G \longrightarrow A$ and $f : G \times G \longrightarrow A$. As we are in degree 2, any pair of such functions determines a unique homogeneous $\mathbb{Z}$-polynomial law of degree 2 from $\mathbb{Z}[G]$ to A, under the (obviously necessary) assumptions:

$$\forall g \in G, \ \ f(g,g) = 2\,D(g), \quad \forall g,h \in G, \ \ f(g,h) = f(h,g).$$

We claim that given D and f satisfying this condition, the axiom of multiplicativity of D is equivalent to the following set of conditions:

(i) D is a group homomorphism $G \longrightarrow A^*$ (in particular $D(1) = 1$),
(ii) for all $g, h, h' \in G$, $f(hg, h'g) = f(h, h')D(g)$,
(iii) for all $g, g', h, h' \in G$, $f(hg, h'g') + f(hg', h'g) = f(h, h')f(g, g')$.

Indeed, assuming that D is a group homomorphism, condition (ii) means that $\mathrm{D}(xg) = \mathrm{D}(x)\mathrm{D}(g)$ for all $x \in \mathbb{Z}[G]$ and $g \in G$. Assuming this relation, condition (iii) means that $\mathrm{D}(xy) = \mathrm{D}(x)\mathrm{D}(y)$ for all $x, y \in R$. Obviously, this multiplicativity property extends automatically to D_B for all commutative A-algebras B.

We can write these conditions in a slightly different way. Define $T : G \longrightarrow A$ by the formula

$$T(g) = f(g, 1).$$

Applying (iii) to $g' = h' = 1$, we see that $T(1) = 2$ and for all $g, h \in G$

$$f(g, h) = T(h)T(g) - T(hg),$$

and in particular $T(gh) = T(hg)$. Morever, $f(g, h) = D(h)T(gh^{-1})$ by (ii).

Lemma 7.7. *The above map* $\mathrm{D} \mapsto (T, D)$ *induces a bijection between the set of* A*-valued determinants of* G *of dimension* 2 *and the set of pairs of functions* $(T, D) : G \to A$ *such that* $D : G \to A^*$ *is a group homomorphism,* $T : G \to A$ *is a function with* $T(1) = 2$, *and which satisfy for all* $g, h \in G$:

(a) $T(gh) = T(hg)$,
(b) $D(g)T(g^{-1}h) - T(g)T(h) + T(gh) = 0$.

The lemma follows easily once we observe that assuming (ii), it is enough to check (iii) for $g' = h' = 1$. Note that applying (iii) to $(h, h', g, g') = (g_1, 1, g_2, g_3)$ we obtain $\forall g_1, g_2, g_3 \in G$

$$T(g_1)T(g_2)T(g_3) - T(g_1)T(g_2g_3) - T(g_2)T(g_1g_3) - T(g_3)T(g_1g_2) + T(g_1g_2g_3) + T(g_1g_2g_3) = 0,$$

which is the pseudocharacter relation of dimension 2 for T. We will see in Proposition 7.23 the following converse result: *Assume that* 2 *is invertible in* A. *Let* $T : G \longrightarrow A$ *be a map such that* $T(1) = 2$, $T(gh) = T(hg)$ *for all* $g, h \in G$, *and that satisfies the* 2*-dimensional pseudocharacter identity. If we set* $D(g) = \frac{T(g)^2 - T(g^2)}{2}$, *then* (D, T) *defines a determinant of* G *of dimension* 2. The non-trivial part is to show that D is a group homomorphism.

1.3. Some polynomial identities

Let R be an A-algebra, B a commutative A-algebra, and $\mathrm{D} \in \mathcal{M}_A^d(R, B)$. For each $r \in R$, we define the *characteristic polynomial* $\chi(r, t) \in B[t]$ of r by the formula

$$\chi(r,t) := \mathrm{D}(t-r) =: \sum_{i=0}^{d} (-1)^i \Lambda_i(r) t^{d-i}.$$

This formula defines A-polynomial laws $\Lambda_i : R \longrightarrow B$ which are homogeneous of degree i, for $i \geq 0$. We have $\Lambda_0 = 1$, $\Lambda_d = \mathrm{D}$, $\Lambda_i = 0$ for $i \geq d+1$, and Λ_1 is an A-linear map, that we shall also denote by Tr and call the *trace* associated to D.

When $B = A$, in which case D is a determinant, this defines as well a homogeneous A-polynomial law of degree d

$$\chi(r) : R \longrightarrow R, r \mapsto r^d - \Lambda_1(r) r^{d-1} + \Lambda_2(r) r^{d-2} + \cdots + (-1)^d \Lambda_d(r).$$

If $n \geq 0$ is an integer, we shall denote by $I_{n,d}$ the set of $\alpha = (\alpha_1, \alpha_2, \ldots, \alpha_n) \in \mathbb{N}^n$ such that $\sum_{i=1}^n \alpha_i = d$. We will need to consider for each $\alpha \in I_{n,d}$ the A-polynomial law $\chi_\alpha : R^n \longrightarrow R$ defined by the following identity in $R[t_1, \ldots, t_n]$:

$$\chi(t_1 r_1 + \cdots + t_n r_n) = \sum_\alpha \chi_\alpha(r_1, \ldots, r_n) t^\alpha,$$

where $t^\alpha = \prod_{i=1}^n t_i^{\alpha_i}$.

Example 7.8. (i) Let us go back to the case $R = A[X]$ (Example 7.5 (i)). We already identified $\Gamma_A^d(R)$ with the A-algebra $A[\Sigma_1, \ldots, \Sigma_d]$, so any homogeneous multiplicative A-polynomial law $\mathrm{D} : R \longrightarrow B$ of degree d is uniquely determined by the image $\Sigma_i(\mathrm{D})$ of Σ_i in B. Unravelling the definitions, we see that $\Sigma_i(\mathrm{D}) = \Lambda_i(X)$, hence the claim in Example 7.5 (i).

(ii) If $\mathrm{D} : R \longrightarrow B$ is a homogeneous multiplicative A-polynomial law of degree d, and $r \in R$, we can restrict it to $A[X]$ via the A-algebra homomorphism $A[X] \longrightarrow R$, $X \mapsto r$. We get this way, and by the previous example, all the possible identities satisfied by determinants of polynomials over a single element of R. For example, the *Newton relations* hold, *i.e.* for all $r \in R$ we have the following equality in $B[[t]]$:

$$-t \frac{\frac{\partial}{\partial t}\mathrm{D}(1-tr)}{\mathrm{D}(1-tr)} = \sum_{n \geq 1} \mathrm{Tr}(r^n) t^n. \tag{7.3}$$

All the functions defined above satisfy a number of polynomial identities, we collect some of them in the following lemma.

Lemma 7.9. (i) *For all $r, r' \in R$, $\mathrm{D}(1 + rr') = \mathrm{D}(1 + r'r)$.*

(ii) *For all $r_1, \ldots, r_n \in R$ and $i \geq 0$, $\Lambda_i(r_1 + r_2 + \cdots + r_n)$ satisfies Amitsur's formula.*[16]

[16] See formula (7.5).

(iii) Tr *satisfies the* d*-dimensional* (B*-valued) pseudocharacter identity.*
(iv) *If* $B = A$*, then for all* $r, r_1, \ldots, r_n \in R$ *and all* $\alpha \in I_{n,d}$*,* $\mathrm{D}(1 + \chi_\alpha(r_1, \ldots, r_n)r) = 1$. •

Let $r, r' \in R$. We want to check that $\mathrm{D}(1 + rr') = \mathrm{D}(1 + r'r)$. Note that if r is invertible[17] in R, then this follows from the multiplicativity of D and the commutativity of B:

$$\mathrm{D}(1 + rr') = \mathrm{D}(r)\mathrm{D}(r^{-1} + r') = \mathrm{D}(r^{-1} + r')\mathrm{D}(r) = \mathrm{D}(1 + r'r).$$

We reduce to this case as follows. Set $r' = 1 + u$ and let us work in $R[t]$. We claim that

$$\mathrm{D}(1 + (1 + tu)r) = \mathrm{D}(1 + r(1 + tu)) \in B[t],$$

which will conclude the proof by evaluating t at 1. But this is an equality of polynomials in t with degree less than d, so it is enough to show that they coincide in $B[t]/(t^{d+1})$. But $1 + tu$ is invertible in $R \otimes_A A[t]/(t^{d+1})$ hence we are reduced to the previous argument.

Let us now prove Amitsur's formula. We mimic here (and actually for (iii) and (iv) below also) the beautiful argument of [RS].[18]

Let $n \geq 1$ be any positive integer, $X = \{x_1 < x_2 < \cdots < x_n\}$ a totally ordered alphabet, and X^+ the monoid of words in X equipped with the induced (total) lexicographic ordering $\leq$, with the convention that $\emptyset < x_i$ for each i. Recall that a word $w \in X^+$ is a *Lyndon word* if $w \leq w'$ for any suffix[19] w' of w (see [Lo, Ch. 5]). Denote by $\mathcal{L}$ the set of Lyndon words. By Lyndon's theorem, any word w writes uniquely as a product of Lyndon words $w = w_1 w_2 \ldots w_m$ where $w_1 \geq w_2 \geq \cdots \geq w_m$ (Lyndon factorization of w). This allows us to define a *sign* map

$$\varepsilon : X^+ \longrightarrow \{\pm 1\}$$

as follows. If $w \in X^+$ is a Lyndon word, set $\varepsilon(w) = (-1)^{\ell(w)-1}$ where $\ell(w)$ is the length of the word w. If $w \in X^+$ is any word, with Lyndon factorization $w = w_1 w_2 \ldots w_m$, we set $\varepsilon(w) = \prod_{i=1}^m \varepsilon(w_i)$.

We fix now some elements $r_1, \ldots, r_n$ in R, and consider the A-algebra

$$A_m - A[t_1, \ldots, t_n]/(t_1, \ldots, t_n)^m.$$

Lyndon's theorem writes then as the following equality in $R \otimes_A A_m$

$$\frac{1}{1 - (t_1 r_1 + \cdots + t_n r_n)} = \prod_w \frac{1}{1 - w},$$

17 By invertible we shall always mean on both sides.
18 We are grateful to Emmanuel Breuillard for pointing out this reference to us.
19 Recall that w' is a suffix of w if $w = mw'$ for some word m.

where the product on the right-hand side is taken over the finite set of Lyndon words of length $< m$ on the alphabet $\{t_1r_1 < \cdots < t_nr_n\}$, chosen in the decreasing order. Applying D and inverting, we get the following equality in $B \otimes_A A_m$

$$\mathrm{D}\left(1 - \left(\sum_{j=1}^{n} t_j r_j\right)\right) = \prod_{w \in \mathcal{L}} \left(\sum_{i=0}^{d} (-1)^i \Lambda_i(w)\right) \tag{7.4}$$

where the product on the right-hand side is now taken over all the Lyndon words on the $t_i r_i$, which is a well-defined element in $B[[t_1, \ldots, t_n]]$. Moreover, the term on the left is the image via $B[t_1, \ldots, t_n] \longrightarrow B \otimes_A A_m$ of the polynomial $\mathrm{D}(1 - (\sum_{j=1}^n t_j r_j)) \in B[t_1, \ldots, t_n]$ which does not depend on m. As a consequence, the formula (7.4), also called *Amitsur's formula*, holds in $B[[t_1, \ldots, t_n]]$, for any integer n. If $i \geq 0$ is any integer, the homogeneous part of degree i of this equality is (for any n)

$$\Lambda_i(t_1r_1 + \cdots + t_nr_n) = \sum_{\ell(w)=i} \epsilon(w)\Lambda(w), \tag{7.5}$$

where the sum is extended over the n^i words w on the $t_j r_j$ with length i, and if $w = w_1^{l_1} \ldots w_q^{l_q}$ is the Lyndon factorization of w with $w_1 > w_2 > \cdots > w_q$, where

$$\Lambda(w) = \Lambda_{l_q}(w_q) \cdots \Lambda_{l_2}(w_2)\Lambda_{l_1}(w_1).$$

Indeed, observe that for each such word, we have $\ell(w) = i = \sum_{k=1}^q l_k \ell(w_k)$, thus $\varepsilon(w) = (-1)^{(\sum_{k=1}^q l_k)-i}$. Equality (7.5) holds a priori in $B[[t_1, \ldots, t_n]]$ but both sides belong to $B[t_1, \ldots, t_n]$, hence it obviously holds in $B[t_1, \ldots, t_n]$. Sending each t_i to 1, we finally get Amistur's formula for $\Lambda_i(r_1 + \cdots + r_n)$ (in B).

Let us check now part (iii) of the Lemma. Let us look at Amitsur's formula (7.5) with $i = n = d+1$, and consider its homogeneous component with degree 1 in each t_j. We see at once that it is exactly the $d+1$-dimensional pseudocharacter identity for $\Lambda_1 = \mathrm{Tr}$.

Remark 7.10. Assume more generally that B is any associative A-algebra (not necessarily commutative) and $\mathrm{D} \in \mathcal{M}_d(R, B)$. Then the definition of the Λ_i also makes sense in this extended context and the same proof as above shows that Amitsur's formula 7.5 still holds (the *increasing ordering* chosen in the definition of $\Lambda(w)$ is important in this case). However, assertion (iv) only makes sense when $A = B$.

To prove assertion (iv), it amounts to show that $\Lambda_i(\chi_\alpha(r_1, \ldots, r_n)r) = 0$ for all $r, r_1, \ldots, r_n \in R$ and $i \geq 1$. We will prove it now only for $i = 1$. As Λ_1 is A-linear, and replacing R by $R[t_1, \ldots, t_n]$, it is enough to show that

$\Lambda_1(\chi(r)r') = 0$ for all $r, r' \in R$. Let us look at Amitsur's formula (7.5) with $i = d+1$, $n = 2$ and $(r_1, r_2) = (r, r')$, consider its homogeneous component with degree d in t_1 and 1 in t_2. Each word in the sum has the form $r^a r' r^b$ with $a+b = d$, whose Lyndon factorization is $(r^a r')(r)^b$, and whose sign is $(-1)^a$. As $\Lambda_{d+1} = 0$ we get an equality

$$0 = \sum_{a+b=d} (-1)^a \Lambda_1(r^a r')\Lambda_b(r) = \Lambda_1\left(\left(\sum_{a=0}^{d}(-1)^a \Lambda_{d-a}(r) r^a\right) r'\right)$$

what we wanted to show.

We still have to complete the proof of identity (iv), but before let us give a simple consequence of what we already proved.

Corollary 7.11. *Let* D *be an* A*-valued determinant on* G *(a monoid) of dimension* d *and* $B \subset A$ *the subring generated by the coefficients* $\Lambda_i(g)$ *of* $\chi(g,t)$ *for all* $g \in G$. *Then* D *factors through a (unique)* B*-valued determinant on* G *of dimension* d.

Proof. We have to show that for all $g_1, \ldots, g_n \in G$, $\mathrm{D}(g_1 t_1 + \cdots + g_n t_n) \in B[t_1, \cdots, t_n]$. By Amitsur's formula (7.5) such a determinant is a signed sum of monomials in $\Lambda_i(w)$ where w is a word on the g_i, in particular $w \in G$, and we are done. □

We now come back to the proof of part (iv). Although it might be possible to prove it in the style above, we will rather deduce it from a general theorem of Vaccarino. Actually, as we shall see, the multiplicativity assumption on D is strong enough to imply that all the polynomial identities between the $\Lambda_i(w)$ (where w is a word in elements of R) which hold for the determinant of matrix algebras also hold for D. These identities have to hold in principle in the universal ring $\Gamma^n_{\mathbb{Z}}(\mathbb{Z}\{X\})^{\mathrm{ab}}$, where $\mathbb{Z}\{X\} = \mathbb{Z}\{x, x \in X\}$ is the free ring over a set X (e.g. $X = \mathbb{N}$), but it might be a bit tedious in practice to compute in this ring. All we will need to know is actually contained in the aforementioned beautiful result of Vaccarino (relying on results of Donkin [D] and Zubkov) that we explain now.

Vaccarino's result ([V1, Thm 6.1], [V2, Thm 28]). Let X be any set, $\mathbb{Z}\{X\}$ as above, and $F_X(d) = \mathbb{Z}[x_{i,j}]$ the ring of polynomials on the variables $x_{i,j}$ for all $x \in X$ and $1 \leq i, j \leq d$. We have the natural *generic matrices* representation

$$\rho^{\mathrm{univ}} : \mathbb{Z}\{X\} \longrightarrow M_d(F_X(d))$$

defined by $x \mapsto (x_{i,j})_{i,j}$, hence we get by the usual Amitsur's formula (or by Corollary 7.11) a natural homogeneous multiplicative polynomial law of degree d given by

$$\det \circ \rho^{\mathrm{univ}} : \mathbb{Z}\{X\} \longrightarrow E_X(d),$$

where $E_X(d) \subset F_X(d)$ is the subring generated by the coefficients of the characteristic polynomials of the $\rho^{\mathrm{univ}}(w)$, $w \in \mathbb{Z}\{X\}$.

Theorem 7.12. *(Vaccarino)* $\det \circ \rho^{\mathrm{univ}}$ *induces an isomorphism* $\Gamma^d_{\mathbb{Z}}(\mathbb{Z}\{X\})^{\mathrm{ab}} \xrightarrow{\sim} E_X(d)$. *In particular,* $\Gamma^d_{\mathbb{Z}}(\mathbb{Z}\{X\})^{\mathrm{ab}}$ *is a free* $\mathbb{Z}$*-module.*

See [V1, Thm. 6.1] and [V2, Thm 28]. Set $X = R$ and consider the canonical map $\pi : \mathbb{Z}\{X\} \longrightarrow R$. Vaccarino's theorem shows that for any determinant $\mathrm{D} : R \longrightarrow A$ of dimension d, there is a unique ring homomorphism $\varphi_X : E_X(d) \longrightarrow A$ such that for all $w \in \mathbb{Z}\{X\}$,

$$\varphi_X(\det(\rho^{\mathrm{univ}}(w))) = \mathrm{D}(\pi(w)). \tag{7.6}$$

More generally, it asserts that $\varphi_X \circ (\det \circ \rho^{\mathrm{univ}}) = \mathrm{D} \circ \pi$ is an equality of $\mathbb{Z}$-polynomial laws. Via φ_X, we may view the A-module R as an $E_X(d)$-module and D becomes an $E_X(d)$-polynomial law. We get then a commutative square of $E_X(d)$-polynomial laws:

$$\begin{array}{ccc} R & \xrightarrow{\mathrm{D}} & A \\ \uparrow{\scriptstyle \pi\otimes\varphi_X} & & \uparrow{\scriptstyle \varphi_X} \\ \mathbb{Z}\{X\} \otimes_{\mathbb{Z}} E_X(d) & \xrightarrow{\det \circ \rho^{\mathrm{univ}}} & E_X(d) \end{array}$$

Now, all the assertions of Lemma 7.9 follow at once from this diagram and the classical formulae in matrix rings (here $M_d(F_X(d))$). For example for part (iv), the Cayley–Hamilton theorem shows that for $r_1, \ldots, r_n \in X = R$, $\rho^{\mathrm{univ}}(\chi_\alpha(r_1, \ldots, r_n)) = 0$, so

$$\mathrm{D}(1 + r\chi_\alpha(r_1, \ldots, r_n)) = \det(\rho^{\mathrm{univ}}(1 + r\chi_\alpha(r_1, \ldots, r_n))) = 1.$$

Remark 7.13. Actually, part (iv) would follow from an apparently weaker version of Theorem 7.12: For any (finite) set X, $\Gamma^d_{\mathbb{Z}}(\mathbb{Z}\{X\})^{\mathrm{ab}} = \mathrm{TS}^d_{\mathbb{Z}}(\mathbb{Z}\{X\})^{\mathrm{ab}}$ is torsion free as abelian group[20]. Unfortunately, as pointed out to us by Vaccarino, this is actually equivalent to Theorem 7.12 in view of Procesi's results.

1.4. Faithful and Cayley–Hamilton determinants

Let us first introduce the notion of *Kernel* of a polynomial law. Let M and N be two A-modules and $P \in \mathcal{P}_A(M, N)$. Define $\ker(P) \subset M$, as the subset whose elements are the $x \in M$ such that

[20] Indeed, it is enough to show (iv) when $A = \Gamma^d_{\mathbb{Z}}(\mathbb{Z}\{X\})^{\mathrm{ab}}$, $R = A\{X\}$, and $\mathrm{D} : R \longrightarrow A$ is the universal determinant. Fix $\alpha, r, r_1, \ldots, r_n$ as in the statement and set $x = \chi_\alpha(r_1, \ldots, r_n)r$. We showed in the proof above that $\Lambda_1(xy) = 0$ for all $y \in R$, and in particular that $\Lambda_1(x^m) = 0$ for all $m \geq 1$. By the Newton relations (7.3), this implies that $i\Lambda_i(x) = 0$ for $i \geq 1$, hence $\Lambda_i(x) = 0$ as A is torsion free.

$$\forall B \in \mathrm{Ob}(\mathcal{C}_A), \ \ \forall b \in B, \ \ \forall m \in M \otimes_A B, \ \ P(x \otimes b + m) = P(m).$$

Equivalently, $x \in \ker(P)$ if and only if for any integer n and any $m_1, \ldots, m_n \in M$, the element $P(tx + t_1m_1 + \cdots + t_nm_n) \in N[t, t_1, \ldots, t_n]$ is independent of t (i.e. lies in $N[t_1, \ldots, t_n]$). By definition, $\ker(P)$ is an A-submodule of M. We say that P is *faithful* if $\ker(P) = 0$.

Lemma 7.14. (i) $\ker(P)$ *is the biggest* A*-submodule* $K \subset M$ *such that* P *admits a factorization* $P = \widetilde{P} \circ \pi$ *where* π *is the canonical* A*-linear surjection* $M \longrightarrow M/K$ *and* $\widetilde{P} \in \mathcal{P}_A(M/K, N)$.

(ii) $\widetilde{P} : R/\ker(P) \longrightarrow S$ *is faithful.*

(iii) *If* B *is a commutative* A*-algebra, then*

$$\mathrm{Im}(\ker(P) \otimes_A B \to M \otimes_A B) \subset \ker(P \otimes_A B).$$

Proof. Assertion (iii) follows from the transitivity of tensor product. Moreover, it is clear that if $P = \widetilde{P} \circ \pi$ for some A-submodule $K \subset M$ as in the statement, then $K \subset \ker(P)$. We need to check that for any A-submodule $K \subset \ker(P)$, P factors through a polynomial map $\widetilde{P} : M/K \longrightarrow N$.

Let B be a commutative A-algebra and consider $K_B := \mathrm{Im}(K \otimes_A B \longrightarrow M \otimes_A B)$. Then $(M/K) \otimes_A B \overset{\sim}{\to} (M \otimes_A B)/K_B$ and $K_B \subset \ker(P \otimes_A B)$ by part (iii). In particular, the map $P_B : M \otimes_A B \longrightarrow N \otimes_A B$ satisfies $P_B(k + m) = P_B(m)$ for any $M \in M \otimes_A B$ and $k \in K_B$, hence we obtain a well-defined map $\widetilde{P}_B : (M/K) \otimes_A B \longrightarrow N \otimes_A B$ via the formula

$$\widetilde{P}_B((\pi \otimes_A B)(m)) = P_B(m), \ \forall m \in M \otimes_A B. \tag{7.7}$$

We check at once that the collection of maps $\widetilde{P}_B$ with B varying defines an element $\widetilde{P} \in \mathcal{P}_A(M/K, N)$.

If $K \subset \ker(P)$ and $P = \widetilde{P} \circ \pi$, it follows from formula (7.7) that $\ker(\widetilde{P}) = \ker(P)/K$, hence (ii). □

Of course, (7.7) shows that if P is homogeneous of some degree n and $P = \widetilde{P} \circ \pi$ as in the lemma, then so is $\widetilde{P}$.

Lemma 7.15. *Let* R *and* S *be two* A *algebras and* $P \in \mathcal{M}_A^d(R, S)$.

(i) $\ker(P) = \{r \in R, \ \forall B, \forall r' \in R \otimes_A B, \ P(1 + rr') = 1\} = \{r \in R, \ \forall B, \forall r' \in R \otimes_A B, \ P(1 + r'r) = 1\}$.

(ii) $\ker(P)$ *is a two-sided ideal of* R*, it is proper if* $d > 0$ *and* $R \neq 0$*. It is the biggest two-sided ideal* $K \subset R$ *such that* P *admits a factorization* $P = \widetilde{P} \circ \pi$ *where* π *is the canonical surjection* $R \longrightarrow R/K$ *and* $\widetilde{P} \in \mathcal{M}_A^d(R/K, S)$.

Proof. Denote by $J_1(P)$ and $J_2(P)$ the two sets on the right in the two equalities in part (i). Let $r \in \ker(P)$, B a commutative A-algebra, and $m = 1+h \in R \otimes_A B$. We want to show that the elements $P(1+r(1+th))$ and $P(1+(1+th)r)$ of $S \otimes_A B[t]$ are the unit element. As they are polynomial of degree d in t, it is enough to check that this holds in $S \otimes_A B[t]/(t^{d+1})$. But $1+th$ is invertible in $R \otimes_A B[t]/(t^{d+1})$ thus the multiplicativity assumption implies that

$$\begin{aligned} P(1+r(1+th)) &= P((1+th)^{-1}+r)P(1+th) \\ &= P((1+th)^{-1})P(1+th) = P(1) = 1, \end{aligned}$$

and for the same reason $P(1+(1+th)r) = 1$, so $\ker(P) \subset J_1(P), J_2(P)$. The same argument shows conversely that $J_i(P) \subset \ker(P)$, hence $\ker(P) = J_1(P) = J_2(P)$.

By (i), $\ker(P)$ is a two-sided ideal of R. As $P(1) = 1$ we have $P(1-t) = (1-t)^d$, thus $1 \notin \ker(P)$ if $d > 0$. Part (ii) follows from formula (7.7) as in the proof of Lemma 7.14 (i). □

Observe that Lemma 7.15 (i) shows that

$$\ker(P) = \{r \in R,\ \forall B\ , \forall r' \in R \otimes_A B,\ \forall i \geq 1,\ \Lambda_i(rr') = 0\}.$$

Moreover, $r \in \ker(P)$ if for any $r_1, \ldots, r_n \in R$, we have

$$P(1 + r(t_1r_1 + t_2r_2 + \ldots, +t_nr_n)) = 1.$$

When $S = A$ is an infinite domain, then $\ker(P) = \{r \in R,\ \forall r' \in R,\ P(1 + rr') = 1\}$.

Assume now that $S = A$, i.e. that $\mathrm{D} : R \longrightarrow A$ is a determinant of dimension d. We denote by $\mathrm{CH}(\mathrm{D}) \subset R$ the two-sided ideal of R generated by the coefficients of

$$\chi(t_1r_1 + \cdots + t_nr_n) \in R[t_1, \ldots, t_n],$$

with $r_1, \ldots, r_n \in R$, $n \geq 1$ (i.e. by the elements $\chi_\alpha(r_1, \ldots, r_n)$ defined in § 1.3). We say that D is *Cayley–Hamilton* if $\mathrm{CH}(\mathrm{D}) = 0$. Equivalently, D is Cayley–Hamilton if the polynomial law $\chi : R \longrightarrow R$ is identically zero. In this case, we will say also that (R, D) is a *Cayley–Hamilton A-algebra of degree d*. Note that by definition, if $\mathrm{D} : R \longrightarrow A$ is Cayley–Hamilton and if B is any commutative A-algebra, then $\mathrm{D} \otimes_A B : R \otimes_A B \longrightarrow B$ is also Cayley–Hamilton.

The Cayley–Hamilton property behaves rather well under several operations, which is in general not the case of the faithful property.

Example 7.16. (i) If R is an Azumaya algebra of rank d^2 over A and D is the reduced norm, then D is Cayley–Hamilton and faithful.

(ii) If D is Cayley–Hamilton and $S \subset R$ is any A-subalgebra, then the restriction of D to S is obviously Cayley–Hamilton. However, the analogous assertion with Cayley–Hamilton replaced by faithful does not hold. For example, if $T_d(A) \subset M_d(A)$ is the A-subalgebra of upper triangular matrices, then $\det : T_d(A) \longrightarrow A$ is Cayley–Hamilton, but not faithful. An easy computation shows that ker(det) is the kernel of the natural diagonal projection $T_d(A) \longrightarrow A^d$ in this case.

Lemma 7.17. ker(D) *contains* CH(D). *In particular, if* D *is faithful then R is Cayley–Hamilton.*

Proof. As ker(D) is a two-sided ideal by Lemma 7.15 (ii), the first assertion follows from the description of ker(D) given in Lemma 7.15 (i) and from Lemma 7.9 (iv). The second assertion follows from the first one. □

The next paragraph is a digression about the notion of Cayley–Hamilton representations; the reader urgently interested in the proofs of the results stated in the introduction may directly skip to section 2.

1.5. The $\mathcal{CH}_d(G)$ category of Cayley–Hamilon representations

Let us consider the counterpart of these notions on the space $X(G,d) = \mathrm{Spec}(\mathbb{Z}(G,d))$ defined in § 1.2. Consider the tautological (universal) determinant of dimension d

$$\mathrm{D}^u : \mathbb{Z}(G,d)[G] \longrightarrow \mathbb{Z}(G,d).$$

The *universal Cayley–Hamilton algebra*

$$R(G,d) := \mathbb{Z}(G,d)[G]/\mathrm{CH}(\mathrm{D}^u)$$

is equipped with a natural group homomorphism $\rho^u : G \longrightarrow R(G,d)^*$. This morphism has the following nice universal property.

Define a *Cayley–Hamilton A-representation (or* CH*-representation for short) of G of dimension d* as a triple $(A,(R,D),\rho)$ where A is a commutative ring, (R,D) is a Cayley–Hamilton A-algebra for the determinant $D : R \longrightarrow A$ of dimension d, and $\rho : G \longrightarrow R^*$ is a group homomorphism. Of course, usual representations give rise to CH-representations, but there are many more in general.

Consider the category $\mathcal{CH}_d(G)$ whose objects are the CH-representations of G of dimension d, and with arrows

$$(A_1,(R_1,D_1),\rho_1) \longrightarrow (A_2,(R_2,D_2),\rho_2)$$

the pairs (f, g) where $f : A_1 \longrightarrow A_2$ and $g : R_1 \longrightarrow R_2$ are ring homomorphisms such that if $\iota_i : A_i \longrightarrow R_i$ is the A_i-algebra structure on R_i, then $g \circ \iota_1 = \iota_2 \circ f$, $f \circ D_1 = D_2 \circ g$, and $\rho_2 = g \circ \rho_1$.

Proposition 7.18. $(\mathbb{Z}(G, d), (R(G, d), \mathrm{D}^u), \rho^u)$ *is the initial object of* $\mathcal{CH}_d(G)$.

Proof. Let $(A, (R, D), \rho)$ be a CH-representation of G of dimension d. The group homomorphism $\rho : G \longrightarrow R^*$ is induced by a unique A-algebra homomorphism $\widetilde{\rho} : A[G] \longrightarrow R$ and $\mathrm{D} \circ \widetilde{\rho}$ is then an A-valued determinant on G of dimension d. We get this way a unique ring homomorphism $f : \mathbb{Z}(G, d) \longrightarrow A$, hence a ring homomorphism $\mathbb{Z}(G, d)[G] \longrightarrow A[G] \longrightarrow R$. As (R, D) is Cayley–Hamilton, it factors through a ring homomorphism $g : R(G, d) \longrightarrow R$, and we check at once that (f, g) has all the required properties. □

The Cayley–Hamilton $\mathbb{Z}(G, d)$-algebra $R(G, d)$ is the global section of a quasi-coherent sheaf of Cayley–Hamilton algebras $\mathcal{R}(G, d)$ on $X(G, d)$. Its formation commutes with arbitrary base changes (contrary to the faithful quotient in general): for any morphism $\mathrm{Spec}(A) \longrightarrow X(G, d)$, corresponding to a determinant $D : A[G] \longrightarrow A$, then the natural surjective map

$$A[G] \longrightarrow \mathcal{R}(G, d) \otimes_{\mathbb{Z}(G,d)} A \tag{7.8}$$

induces an isomorphism $A[G]/\mathrm{CH(D)} \xrightarrow{\sim} \mathcal{R}(G, d) \otimes_{\mathbb{Z}(G,d)} A$.

Remark 7.19. (*CH-representations versus representations*) In general, given a point $\mathrm{Spec}(A) \longrightarrow X(G, d)$, i.e. a determinant $D : A[G] \longrightarrow A$, there is no representation $\rho : A[G] \longrightarrow M_d(A)$ such that $D = \det \circ \rho$ (see e.g. [BC, Thm. 1.6.3]). However, we have a natural candidate for a substitute which is the CH-representation (7.8), i.e.

$$G \longrightarrow (A[G]/\mathrm{CH}(D))^*.$$

Thus it is an important task to study the sheaf $\mathcal{R}(G, d)$ of CH-algebras. It turns out to be extremely nice over some specific loci of $X(G, d)$. For instance, we will show in Corollary 7.44 that it is a sheaf of Azumaya algebras of rank d^2 over the *absolute irreducibility locus* of $X(G, d)$; in particular, étale-locally on this (open) subspace D^{univ} is the determinant of a true representation (unique, surjective).

The situation is more complicated over the reducible locus. In Theorem 7.43 we will study more generally the algebra $\mathcal{R}(G, d) \otimes \mathcal{O}_x^{\mathrm{hens}}$ when $x \in X(G, d)$ is reducible but *multiplicity free*: it is a *generalized matrix algebra* in the sense of [BC, §1.3] (and all such algebras occur somehow this way; when $d! \in A^*$, this result follows from [BC, Thm. 1.4.4]).

Remark 7.20. *(The embedding problem)* The embeding problem is to decide whether the CH-algebra $(R(G,d), D^{\mathrm{univ}})$ admits a CH-embedding in $(M_d(B), \det)$ for some commutative ring B. A result of Procesi [P3] asserts that it holds after tensoring by $\mathbb{Q}$, but the result over $\mathbb{Z}$ still seems to be open (see [V3]). The problem is local on $X(G,d)$, and there are some partial known results. For instance, we will show in Theorem 7.43 that it holds at $x \in X(G,d)$ (i.e for $R(G,d) \otimes \mathcal{O}_x$) whenever x is multiplicity free (compare with [BC, Prop. 1.3.13]).

1.6. Determinants and pseudocharacters

We end this section by a comparison between determinants and pseudocharacters. Let us start with the following result, whose conclusion will be actually sharpened below.

Proposition 7.21. *The map* $\mathrm{D} \mapsto \mathrm{Tr}$ *defined in § 1.3 induces an injection between the set of d-dimensional A-valued determinants on R and the set of d-dimensional A-valued pseudocharacters on R. When A is a $\mathbb{Q}$-algebra, it is a bijection.*

Proof. Let $\mathrm{D} : R \longrightarrow A$ be a determinant of dimension d. By Lemma 7.9 (iii), Tr is a d-dimensional pseudocharacter on R (note that $\mathrm{Tr}(1) = d$). Moreover, the Newton relations (7.3) show that for each commutative A-algebra B, each $r \in R \otimes_A B$ and $i \leq d$, $\Lambda_i(r)$ lies in the $\mathbb{Z}[1/d!]$-algebra generated by $\mathrm{Tr}(r^j)$ for $j \leq i$, hence Tr determines D.

Let T be a d-dimensional A-valued pseudocharacter, it remains to show that it has the form Tr for some D. By the Newton relations again, there is a unique element

$$P \in \mathbb{Z}[1/d!][S_1, \ldots, S_d]$$

such that for any commutative ring B and $r \in M_d(B)$, we have $P(\ldots, \mathrm{tr}(r^i), \ldots) = \det(r)$. Of course, if we ask S_i to have degree i, then P is homogeneous of degree d. Moreover, $P(d, d(d-1)/2, \ldots, d, 1) = 1$. We consider then the A-polynomial law $\mathrm{D} : R \longrightarrow A$ defined by $\mathrm{D} = P(\ldots, T(r^i), \ldots)$. It is homogenenous of degree d and satisfies $\mathrm{D}(1) = 1$. By construction, it is enough to check that $\mathrm{D}(rr') = \mathrm{D}(r)\mathrm{D}(r')$ for all commutative A-algebra B and $r, r' \in R \otimes_A B$. By construction, $\mathrm{D}_B(r) = P(\ldots, (T \otimes_A B)(r^i), \ldots)$ for all $r \in R \otimes_A B$, so we may assume that $A = B$. By a result of Procesi [P3], there is a commutative A-algebra C with $A \longrightarrow C$ injective and an A-algebra homomorphism

$$\rho : R \longrightarrow M_d(C)$$

such that $\mathrm{tr} \circ \rho = T$. But then $\det(\rho(x)) = \mathrm{D}(x)$ is multiplicative, and we are done. □

Remark 7.22. The proposition might hold under the weaker assumption $d! \in A^*$ but we don't know how to prove it in general, namely: we don't know how to show that the obvious A-polynomial law of degree d attached to a pseudocharacter $T : R \longrightarrow A$ is multiplicative (compare with [BC, Remark 1.2.9]). However, using structure theorems for pseudocharacters over fields and over local rings instead of [P3], we know that this holds in either of the following situations:

(i) A is reduced,
(ii) For all $x \in \mathrm{Specmax}(A)$, and k an algebraic closure of the residue field at x, the induced pseudocharacter $T \otimes_A k$ is multiplicity free (use [BC, Prop. 1.3.13] and [BC, Thm.1.4.4]).

In general, it would be enough (actually equivalent) to know that if G is the free monoid over two letters $\{a, b\}$, and $T : \mathbb{Z}[G] \longrightarrow A$ is the universal pseudocharacter on G of dimension d (with $d! \in A^*$), then A (which is easy to describe by generators and relations) is torsion free over $\mathbb{Z}$. The next result is an evidence for the general case.

Proposition 7.23. *Assume:*

(i) *either that* 2 *is invertible in* A *and* $d = 2$,
(ii) *or that* $(2d)!$ *is invertible in* A,

then the map $\mathrm{D} \mapsto \mathrm{Tr}$ *defined in § 1.3 induces a bijection between the set of* d*-dimensional* A*-valued determinants on* R *and the set of* d*-dimensional* A*-valued pseudocharacters on* R.

Proof. We first show (i). For $x, y \in R$ set $f(x, y) = T(x)T(y) - T(xy)$ and $D(x) = f(x, x)/2$. Then $f : R \times R \longrightarrow A$ is an A-bilinear map and $D : R \longrightarrow A$ is a quadratic A-map with associated bilinear map f. In particular, D defines a quadratic A-polynomial law $R \longrightarrow A$ which satisfies $D(1) = 1$ (see Example 7.1 (ii)). We have to check that $D(xy) = D(x)D(y)$ for all $x, y \in R$. We check at once as in Example 7.6 that it suffices to show that for all $x, x', y, y' \in R$, we have

$$f(xy, x'y') + f(xy', x'y) = f(x, x')f(y, y'). \tag{7.9}$$

For $m \geq 1, \sigma \in \mathfrak{S}_m$ and $x = (x_1, \ldots, x_m) \in X^m$, set $T^\sigma(x) = T(x_{i_1} \ldots x_{i_r})$ if x is the cycle $(i_1, \ldots, i_r)$, and $T^\sigma(x) = \prod_i T^{c_i}(x)$ if $\sigma = \prod_i c_i$ is the cycle decomposition of σ. For example for $m = 3$, the 2-dimensional pseudocharacter relation reads $s_3(T) := \sum_{\sigma \in \mathfrak{S}_3} \varepsilon(\sigma) T^\sigma = 0$ on R^3, where ϵ is the signature on $\mathfrak{S}_m$. We have to show that this relation implies (7.9) if 2 is invertible in A.

Let us fix now $m = 4$ and consider the order 8 subgroup $H \subset \mathfrak{S}_4$ generated by $H_0 = \langle (1,2), (3,4) \rangle$ and (1324). Let $s : H \to \{\pm 1\}$ denote the unique character which coincides with the signature ε on H_0 and such that $s((1324)) = 1$. Condition (7.9) reads

$$\forall x \in R^4, \sum_{h \in H} s(h) T^h(x) = 0. \tag{7.10}$$

Let $B = \mathbb{Z}[1/2][\mathfrak{S}_4]$ be the group ring of $\mathfrak{S}_4$ over $\mathbb{Z}[1/2]$ and consider the two elements of B

$$p := \frac{1}{8} \sum_{h \in H} s(h) h, \quad q = \sum_{g \in \mathfrak{S}_3} \varepsilon(g) g$$

where $\mathfrak{S}_3$ is viewed as the subgroup of $\mathfrak{S}_4$ fixing $\{4\}$. Note that p is an idempotent of B. To prove that the pseudocharacter relation implies (7.10), it is enough to show that $p \in BqB$ (see e.g. remarks (i) to (v) following Thm. 4.5 of [P2]). For that it is actually enough to show that for any field k in which 2 is invertible then $p \in B_k q B_k$, where $B_k := B \otimes_{\mathbb{Z}} k = k[\mathfrak{S}_4]$. We fix such a field.

Let k^4 be the natural permutation representation of $\mathfrak{S}_4$. As $2 \in k^*$ we have $k^4 = 1 \oplus \mathrm{St}$, and we check at once that St absolutely irreducible. Let $V = \mathrm{St} \otimes \varepsilon$. We have[21]

$$\mathrm{Ind}_H^{\mathfrak{S}_4} s = V, \quad \mathrm{Ind}_{\mathfrak{S}_3}^{\mathfrak{S}_4} \varepsilon = \varepsilon \oplus V. \tag{7.11}$$

As $|\mathfrak{S}_4| / \dim(V) = 8 \in k^*$, V is a projective B_k-module and we may find a central idempotent $e \in B_k$ acting on V as the identity, and as 0 on the k-representations of $\mathfrak{S}_4$ not containing V. Moreover, the k-algebra eB_k (with unit e) is isomorphic to $\mathrm{End}_k(V) \simeq M_3(k)$ as V is absolutely irreducible. As $\mathrm{Ind}_H^{\mathfrak{S}_4} s = V$, the idempotent p acts non-trivially in a k-representation U of $\mathfrak{S}_4$ if and only if U contains V. Applying this to $U = B_k(1-e)$ we obtain $p(1-e) = 0$, so $p \in eB_k$. But one easily sees that $q(V) \neq 0$, for instance $q \cdot (1,0,0,-1) = (2,2,2,-6)$. It follows that $eq \neq 0$ so $B_k q B_k \supset B_k eq B_k = eB_k$ by simplicity of eB_k, thus $p \in B_k q B_k$.

The second statement is actually a formal consequence of Procesi's results [P2]. Let us consider the full polarization of the polynomial map

[21] Note that the vector $(1,1,-1,-1)$ (resp. $(1,1,1,-3)$) generates the representation $s\varepsilon$ under the action of H (resp. is invariant under the action of $\mathfrak{S}_3$).

$\det(g)\det(h) - \det(gh)$ on M_{2d}^2; it is given by some element $p \in \mathbb{Z}[\mathfrak{S}_{2d}]$ (see below for an explicit formula of a partial polarization), and as above we have to show that $p \in BqB$ where $q = \sum_{\sigma\in\mathfrak{S}_{d+1}} \varepsilon(\sigma)\sigma$ and $B = \mathbb{Z}[1/(2d!)][\mathfrak{S}_{2d}]$. By the second fundamental theorem of invariants of set of matrices [P2], we know that this holds over $\mathbb{Q}$, so $mp \in BqB$ for some $m \in \mathbb{Z}$. As B is isomorphic to a direct product of matrix rings over $\mathbb{Z}[1/(2d)!]$, and as $\frac{q}{(d+1)!}$ is an idempotent of B, it turns out that B/BqB is torsion free, and we are done. □

We end this section by giving an explicit (d,d)-*partial* polarization of the homogeneous (of degree $2d$) polynomial map

$$(g,h) \mapsto \det(g)\det(h) - \det(gh),\ \ M_d(A)^2 \to A \tag{7.12}$$

when $d! \in A$, which extends the relation (7.10) obtained in dimension 2. By this we mean an A-multilinear map $\varphi : M_d(A)^{2d} \to A$ which is symmetric only in the first d (resp. last d) variables, and such that $\varphi(g,g,\ldots,g,h,h,\ldots,h) = (d!)^2(\det(g)\det(h) - \det(gh))$ for any $(g,h) \in M_d(A)^2$.

Let $H_0 \subset \mathfrak{S}_{2d}$ be the subgroup preserving $\{1,\ldots,d\}$ (thus $H_0 \simeq \mathfrak{S}_d^2$); the element

$$\tau = \prod_{i=1}^{d} (i\ d+i)$$

has order 2 and normalizes H_0, thus $H = \langle H_0, \tau\rangle$ is a subgroup of order $2(d!)^2 \in A^*$. The signature on H_0 being τ-invariant, there exists a unique character $s : H \to \{\pm 1\}$ such that $s(\tau) = -1$ and such that s coincides with the signature on H_0. We define an A-multilinear map φ: $M_d(A)^{2d} \to A$ by

$$\varphi_T = \sum_{\sigma\in H} s(\sigma) T^{\sigma},$$

where T is the trace.

Proposition 7.24. *φ_T is a (d,d)-partial polarization of (7.12).*

Proof. Note that φ_T is H_0-invariant by construction. Let us first consider the multilinear invariant map $\psi : M_d(A) \to A$ associated to the element $u = \sum_{\sigma\in\mathfrak{S}_d} \epsilon(\sigma)\sigma \in \mathbb{Z}[\mathfrak{S}_d]$, that is $\psi = \sum_{\sigma\in\mathfrak{S}_d} \epsilon(\sigma)T^{\sigma}$. As is well-known, ψ is the full polarization of det, being the trace of $(g_1,\ldots,g_n)$ on $\frac{u}{d!}(V^{\otimes_A d}) = \Lambda^d(V)$, where $V = A^d$. We deduce from this an expression for a partial polarization of $(g,h) \mapsto \det(gh)$ by (d,d)-polarizing each term of the form $T^{\sigma}(g_1h, g_2h,\ldots,g_dh)$ as

$$\sum_{\sigma'\in\mathfrak{S}_d} T^{\sigma}(g_1h_{\sigma'(1)}, g_2h_{\sigma'(2)},\ldots,g_dh_{\sigma'(d)}). \tag{7.13}$$

It only remains to identify the associated elements of $\mathbb{Z}[\mathfrak{S}_{2d}]$.

Write $H_0 = H_1.H_2$ where H_1 is the subgroup fixing $d+1$ to $2d$ and H_2 fixes 1 to d, and identify H_2 with $\mathfrak{S}_d$ under the bijection $\{1,\ldots,d\} \overset{\sim}{\to} \{d+1,\ldots,2d\}$, $i \mapsto i+d$. A simple cycle computation shows that

$$T^{\sigma}(g_1 h_{\sigma'(1)}, g_2 h_{\sigma'(2)}, \ldots, g_d h_{\sigma'(d)}) = T^{\sigma''}(g_1, g_2, \ldots, g_d, h_1, h_2, \ldots, h_d)$$

where $\sigma'' = \sigma\sigma'\tau\sigma'^{-1} \in \mathfrak{S}_{2d}$. The key fact is that for $(i_1\, i_2 \cdots i_r) \in H_1$ any cycle, and for $j_1, j_2, \ldots, j_r$ any distinct elements in $\{d+1,\ldots,2d\}$, then

$$(i_1\, i_2 \cdots i_r)\,(i_1\, j_1)\,(i_2,\, j_2) \cdots (i_r,\, j_r) = (i_1\, j_1\, i_2\, j_2 \cdots i_r\, j_r).$$

As a consequence, we get a (d,d)-polarization of (7.12) as the multilinear invariant associated to the element

$$p = \sum_{\sigma \in H_0} \varepsilon(\sigma)\sigma - \sum_{(\sigma,\sigma') \in H_1 \times H_2} \epsilon(\sigma)\sigma\sigma'\tau\sigma'^{-1} \in \mathbb{Z}[\mathfrak{S}_{2d}].$$

A simple change of variables $(\sigma_1, \sigma_2) = (\sigma\tau\sigma'\tau, \sigma'^{-1})$ identifies this map with φ_T. □

2. Structure and finiteness theorems

In this section, we will give some necessary conditions ensuring that a determinant $D : R \longrightarrow A$ of dimension d is the determinant of a true representation $R \longrightarrow M_d(A)$. As explained in Remark 7.19, we will get these results by first proving some structure theorems for certain Cayley–Hamilton algebras.

In an independent last paragraph, we will also state and discuss a result of Vaccarino and Donkin asserting in particular that $\mathbb{Z}(G,d)$ is finite type over $\mathbb{Z}$ when the group (or monoid) G is finitely generated.

2.1. Some preliminary lemmas

Let S_1 and S_2 be two A-algebras, B a commutative A-algebra, d an integer, and let $p_j : S_j \longrightarrow B$ be a multiplicative A-polynomial law which is homogeneous of degree d_i, with $d_1 + d_2 = d$. Then we check at once that the A-polynomial map

$$p_1 p_2 : (x_1, x_2) \mapsto p_1(x_1)p_2(x_2), \; S_1 \times S_2 \longrightarrow B,$$

which is homogeneous of degree $d_1 + d_2$, is again multiplicative. We will call $p_1 p_2$ the product of p_1 and p_2. This operation induces a natural A-algebra homomorphism

$$\Gamma_A^d(S)^{\mathrm{ab}} \longrightarrow \prod_{i=0}^{d} \Gamma_A^i(S_1)^{\mathrm{ab}} \otimes_A \Gamma_A^{d-i}(S_2)^{\mathrm{ab}}. \tag{7.14}$$

Recall that an A-algebra is called finite diagonal if it is isomorphic to A^n (with coordinate-wise addition and multiplication) for some integer $n \geq 1$.

Lemma 7.25. (i) *The map (7.14) is an A-algebra isomorphism.*

(ii) *If S is a finite diagonal A-algebra, then so is $\Gamma^d_A(S) = \Gamma^d_A(S)^{\mathrm{ab}}$.*

(iii) *Assume* $\mathrm{Spec}(B)$ *is connected and* $B \neq 0$. *Then any multiplicative homogeneous A-polynomial law $S_1 \times S_2 \longrightarrow B$ of degree d is the product $p_1 p_2$ of two unique multiplicative homogeneous A-polynomial laws $p_i : S_i \longrightarrow B$ with degree d_i, and we have $d_1 + d_2 = d$.*

Proof. Note that if B is any A-algebra (non necessarily commutative), and if $p_j : S_j \longrightarrow B$ are multiplicative A-polynomial laws of degree d_j such that the images of p_1 and p_2 commute in the obvious sense, then $(p_1 p_2)(x_1, x_2) := p_1(x_1) p_2(x_2)$ still defines an A-multiplicative polynomial law of degree $d_1 + d_2$. This defines a natural A-algebra homomorphism

$$\Gamma^d_A(S) \longrightarrow \prod_{i=0}^{d} \Gamma^i_A(S_1) \otimes_A \Gamma^{d-i}_A(S_2) \tag{7.15}$$

of which (7.14) results by abelianization, thus it is enough to check that (7.15) is an isomorphism. By definition, the projection of (7.15) to the i-th factor corresponds to the homogeneous A-polynomial law of degree d

$$S_1 \times S_2 \longrightarrow \Gamma^i_A(S_1) \otimes_A \Gamma^{d-i}_A(S_2), \quad (s_1, s_2) \mapsto {s_1}^{[i]} \otimes s_2^{[d-i]},$$

(which is incidentally obviously multiplicative) hence is exactly the map defined more generally by Roby in [Ro1, §9] for any pair of A-modules (S_1, S_2), and which is an A-linear isomorphism by [Ro1, Thm. III.4], which proves (i). More precisely, we showed that as A-algebras there is an isomorphism

$$\Gamma^d_A(S_1 \times S_2) \overset{\sim}{\to} \prod_{i=0}^{d} \Gamma^i_A(S_1) \otimes_A \Gamma^{d-i}_A(S_2). \tag{7.16}$$

Note that $\Gamma^i_A(A) = A \cdot 1^{[i]} \simeq A$ for each $i \geq 0$. In particular, if $S_1 = A$ then (7.16) shows that $\Gamma^d_A(S) \overset{\sim}{\to} \prod_{i=0}^{d} \Gamma^i_A(S_2)$ as A-algebras. Part (ii) follows then by induction.

We now show assertion (iii). It follows from (i) and the following general fact. Consider a finite number of rings with unit $C_1, \ldots, C_m$, and set $C = \prod_{i=1}^{m} C_i$. Let B be a nonzero commutative ring with unit, with connected spectrum, and let $f : C \to B$ be a ring homomorphism. Then f factors through the projection $C \to C_j$ for a unique j. Indeed, let c_i be the central idempotent of C whose j-th component is 0 if $i \neq j$, and the unit of C_i if

$j = i$. Set $e_i = f(c_i)$. Then $\{e_i, 1 \leq i \leq m\} \subset B$ is a set of m idempotents of B such that $\sum_i e_i = 1$ and $e_i e_j = 0$ if $i \neq j$. As Spec(B) is connected, it follows that $e_i = 1$ or 0 for each i, exclusively as $B \neq 0$, and that there is a necessarily unique j such that $e_j = 1$, and the claim follows. □

Remark 7.26. We actually proved that the morphism (7.16) is an isomorphism, which is stronger than assertion (i) of the lemma.

Let R be an A-algebra. Recall that an element $e \in R$ is said to be idempotent if $e^2 = e$, in which case $1 - e$ is also idempotent. The subset $eRe \subset R$ is then an A-algebra whose unit element is e and $eRe \oplus (1-e)R(1-e)$ is an A-subalgebra of R isomorphic to $eRe \times (1-e)R(1-e)$. We say that a family of idempotents $\{e_i\}$ is orthogonal if $e_i e_j = 0$ if $i \neq j$. Let $\mathrm{D} : R \longrightarrow A$ be a determinant of dimension d and assume $A \neq 0$.

Lemma 7.27. *Assume that* Spec(A) *is connected and let* $e \in R$ *be an idempotent.*

(1) *The polynomial map* $\mathrm{D}_e : eRe \longrightarrow A$, $x \mapsto \mathrm{D}(x+1-e)$, *is a determinant whose dimension* $r(e)$ *is* $\leq d$.
(2) *We have* $r(1-e) + r(e) = d$. *Moreover, the restriction of* D *to the* A-*subalgebra* $eRe \oplus (1-e)R(1-e)$ *is the product determinant* $\mathrm{D}_e \mathrm{D}_{1-e}$.
(3) *If* D *is Cayley–Hamilton (resp. faithful), then so is* D_e.
(4) *Assume that* D *is Cayley–Hamilton. Then* $e = 1$ *(resp.* $e = 0$*) if and only if* $\mathrm{D}(e) = 1$ *(resp.* $r(e) = 0$*). Let* $e_1, \ldots, e_s$ *be a family of (nonzero) orthogonal idempotents of* R. *Then* $s \leq d$, *and we have the inequality* $\sum_{i=1}^{s} r(e_i) \leq d$, *which is an equality if and only if* $e_1 + e_2 + \cdots + e_s = 1$.

Proof. Set $S_1 = eRe$, $S_2 = (1-e)R(1-e)$, and consider the A-subalgebra $S = S_1 \oplus S_2 \subset R$. Then e is a central idempotent in S, hence the map $x \mapsto (ex, (1-e)x)$ is an A-algebra isomorphism $S \xrightarrow{\sim} S_1 \times S_2$. Lemma 7.25 (iii) applied to the restriction of D to S shows parts (1) and (2).

Assume that D is faithful. Let $x \in \ker(\mathrm{D}_e)$, B a commutative A-algebra and $y \in R \otimes_A B$. Note that

$$eRe \otimes_A B = e(R \otimes_A B)e \tag{7.17}$$

is a direct summand of $R \otimes_A B$. We have (using Lemma 7.9)

$$\begin{aligned}\mathrm{D}(1+xy) &= \mathrm{D}(1+exey) = \mathrm{D}(1+eyex)\\ &= \mathrm{D}(1-e+e+eyex) = \mathrm{D}_e(e+eyex) = 1,\end{aligned}$$

so $x \in \ker(\mathrm{D})$, and $x = 0$.

Assume that D is Cayley–Hamilton. If $x \in R$, then $\chi^{\mathrm{D}}(x, x) = 0$. For $x \in eRe \oplus (1-e)R(1-e)$, we know from part (2) that

$$\chi^{\mathrm{D}}(x, t) = \chi^{\mathrm{D}_e}(ex, t)\chi^{\mathrm{D}_{1-e}}((1-e)x, t). \tag{7.18}$$

For $x \in eRe$, apply the Cayley–Hamilton identity to x and $x + 1 - e$. We get that

$$P(x)x^{d_2} = P(x)(x-1)^{d_2} = 0$$

in R, where $P = \chi^{\mathrm{D}_e}(ex, t) \in A[t]$ is the characteristic polynomial of x in eRe with respect to D_e. But the ideal of $A[t]$ generated by t^{d_2} and $(t-1)^{d_2}$ is $A[t]$, so $P(x) = 0$. Applying this argument to $R \otimes_A B$ for all commutative A-algebras B, we get that D_e is Cayley–Hamilton.

Let us show assertion (4). If $e^2 = e$, then $\chi(e, e) - \mathrm{D}(e) \in Ae \subset R$. If moreover D is Cayley–Hamilton and $\mathrm{D}(e) = 1$, then $ae = 1$ in R for some $a \in A$, thus $1 = ae = ae^2 = e$. If $r(e) = 0$, then $\mathrm{D}((1-e)+x)$ is a determinant of degree 0 on eRe, so it is constant and equal to 1. But then $\mathrm{D}(1-e) = 1$ and $e = 0$ by the previous case. For the last property, set $e_{s+1} := 1-(e_1+\cdots+e_s)$. Note that $r(e_i) \leq d$ for each $i \leq s+1$ and $\sum_{i=1}^{s+1} r(e_i) = d$ applying part (2) s times. We conclude as $r(e_j) = 0$ implies $e_j = 0$. □

Exercise 7.28. *(Another proof or Ziplies's result* [Z1]*)* Let R be an Azumaya algebra of rank n^2 over A, show as follows that the reduced norm N induces an isomorphism

$$\varphi : \Gamma_A^d(R)^{\mathrm{ab}} \xrightarrow{\sim} \Gamma_A^{d/n}(A) = A$$

if n divides d, and that $\Gamma_A^d(R)^{\mathrm{ab}} = 0$ otherwise, using only Lemmas 7.25 and 7.9 (i) and (ii). Using a faithfully flat commutative A-algebra C such that $R \otimes_A C \xrightarrow{\sim} M_n(C)$, we may assume $R = M_n(A)$, in which case $N = \det$. Let $E_{i,j}$ be the usual A-basis of R and $\mathrm{D} : R \longrightarrow B$ be any homogeneous multiplicative $\mathbb{Z}$-polynomial law of degree d (B a commutative A-algebra). Using Lemma 7.25 and the fact that the $E_{i,i}$ are conjugate under $\mathrm{GL}_n(A)$, show that n divides d and that $\mathrm{D}_e : E_{1,1}A \longrightarrow B$ is a homogeneous multiplicative A-polynomial law of degree d/n. This provides an A-algebra morphism

$$\iota : A = \Gamma_A^{d/n}(A) \longrightarrow \Gamma_A^d(M_n(A))^{\mathrm{ab}}$$

such that $\varphi \circ \iota = \mathrm{id}$. To conclude, it is enough to check that that D_e determines D uniquely. Note that $\mathrm{D}(1 - tE_{i,i}) = \mathrm{D}(1 - tE_{1,1}) = (1-t)^{d/n}$ and for $i \neq j$,

$$\mathrm{D}(1 - tE_{i,j}) = \mathrm{D}(1 - tE_{i,i}E_{i,j}) = \mathrm{D}(1 - tE_{i,j}E_{i,i}) = 1$$

and conclude using Amitsur's formula.

Lemma 7.29. *Let* $\mathrm{D} : R \longrightarrow A$ *be a Cayley–Hamilton determinant of dimension* 1*. Then* $R = A$ *and* D *is the identity.*

Proof. By assumption $x = \mathrm{Tr}(x) = \mathrm{D}(x)$ for all $x \in R$, so the A-linear map $\mathrm{Tr} = \mathrm{D} : R \longrightarrow A$ is an A-algebra isomorphism. □

We now study the Jacobson radical (denoted by Rad) of an algebra with a determinant. We shall need the Nagata–Higman theorem [Hi], that we recall now. Let d be an integer and let k be a field such that either $\mathrm{char}(k) = 0$ or $\mathrm{char}(k) > d$. Let R be an algebra without unit over k, and assume that $x^d = 0$ for all $x \in R$. Then there is an integer $N(d) \leq 2^d - 1$ (independent of R) such that for all $x_1, \ldots, x_{N(d)}$ in R, the product $x_1 \ldots x_{N(d)}$ vanishes.

Lemma 7.30. *Assume that* $\mathrm{D} : R \longrightarrow A$ *is Cayley–Hamilton of dimension* d*.*

(i) $\mathrm{Rad}(R)$ *is the largest two-sided ideal* $J \subset R$ *such that* $\mathrm{D}(1 + J) \subset A^*$*.*
(ii) $\ker(\mathrm{D}) \subset \mathrm{Rad}(R)$.
Assume from now on that A is a field.
(iii) *For all* $x \in \ker(\mathrm{D})$ *we have* $x^d = 0$*. In particular, if* $d!$ *is invertible in* A *then* $\ker(\mathrm{D})^{N(d)} = 0$.
(iv) $x \in R$ *is nilpotent if and only if* $\mathrm{D}(t - x) = t^d$*. Morever,* $\mathrm{Rad}(R)$ *consists of nilpotent elements.*
(v) *If* $J \subset R$ *is a two-sided ideal such that* $J^n = 0$ *for some* $n \geq 1$*, then* $J \subset \ker(\mathrm{D})$ *(here it is not necessary to assume that* D *is Cayley–Hamilton).*

Proof. By the Cayley–Hamilton identity, if $x \in R$, then x is invertible in R if and only if $\mathrm{D}(x)$ is invertible in A, hence (i). Assertion (ii) follows as $\mathrm{D}(1 + \ker(\mathrm{D})) = 1$.

Assume that $A{=}k$ is a field. If $x \in \ker(\mathrm{D})$, then $\chi(x, t){=}t^d$ thus $x^d = 0$ as D is Cayley–Hamilton. When $d!$ is invertible in k, the Nagata–Higman theorem applies and proves (iii). If $x \in R$ is nilpotent, then $1 + tx$ is invertible in R, hence $\mathrm{D}(1 + tx)$ is invertible in $k[t]$, so $\mathrm{D}(1 + tx) = 1$. The converse follows from the Cayley–Hamilton identity, which even shows that $x^d = 0$. Assume that $x \in \mathrm{Rad}(R)$. For all $y \in k[x]$, $1 + yx$ is invertible in R, so $\mathrm{D}(1 + yx) \in k^*$ and the Cayley–Hamilton identity implies that $1 + yx$ is actually invertible in $k[x]$. In particular, $x \in \mathrm{Rad}(k[x])$. This implies that x is nilpotent as $k[x]$ is a finite dimensional k-algebra, hence (iv) follows.

Let $J \subset R$ be as in (v) and $x \in J$. If $y \in R[t_1, \ldots, t_n]$, we see that xy is nilpotent, so $\mathrm{D}(1 + txy) \in k[t_1, \ldots, t_n, t]$ is invertible, hence constant equal to 1, and $x \in \ker(\mathrm{D})$. □

The following lemma strengthens part (iv) of the previous one.

Lemma 7.31. *Assume that k is a field and let* $\mathrm{D} : R \longrightarrow k$ *be a determinant of dimension d.*

(i) *If K/k is a separable algebraic extension, then the natural injection $R \longrightarrow R \otimes_k K$ induces isomorphisms* $\mathrm{Rad}(R) \otimes_k K \xrightarrow{\sim} \mathrm{Rad}(R \otimes_k K)$ *and* $\ker(\mathrm{D}) \otimes_k K \xrightarrow{\sim} \ker(\mathrm{D} \otimes_k K)$.

(ii) *Assume that* D *is Cayley–Hamilton. Then* $\ker(\mathrm{D}) = \mathrm{Rad}(R)$.

Proof. We first check (i). The assertion concerning the Jacobson radical is well-known. Moreover, the injection of the statement induces an injection (Lemma 7.14)

$$\ker(\mathrm{D}) \otimes_k K \longrightarrow \ker(\mathrm{D} \otimes_k K),$$

and it only remains to check its surjectivity. Enlarging K if necessary, we may assume that K/k is normal. Let $G := \mathrm{Gal}(K/k)$ act semilinearily on $R \otimes_k K$. By Galois descent, each G-stable K-subvector space V of $R \otimes_k K$, has the form $V^G \otimes_k K$ where $V^G \subset R$ is k-vector space of fixed points. We claim that $\ker(\mathrm{D} \otimes_k K)$ is G-stable. Indeed, if we let G act on $K[t_1, \ldots, t_s]$ by $\sigma(\sum_\alpha a_\alpha t^\alpha) = \sum_\alpha \sigma(a_\alpha) t^\alpha$, and then $K[t_1, \ldots, t_s]$-semilinearily on $R \otimes_k K[t_1, \ldots, t_s]$, then for any $r \in R \otimes_k K[t_1, \ldots, t_s]$ we have

$$\mathrm{D}(\sigma(r)) = \sigma(D(r)),$$

from which the claim follows at once. For the same reason, we see that $\ker(\mathrm{D} \otimes_k K)^G \subset \ker(\mathrm{D})$, which concludes the proof.

We now prove assertion (ii). We already know from Lemma 7.30 (ii) that $\ker(\mathrm{D}) \subset \mathrm{Rad}(R)$. By extending the scalars to a separable algebraic closure of k and part (i), we may assume that k is infinite. In this case (see § 1.4),

$$\ker(\mathrm{D}) = \{x \in R,\ \forall y \in R, \mathrm{D}(1 + xy) = 1\}. \tag{7.19}$$

By Lemma 7.30, $\mathrm{Rad}(R)$ is a two-sided ideal of R consisting of nilpotent elements x, for which $\mathrm{D}(1 + x) = 1$, hence (7.19) implies that $\mathrm{Rad}(R) \subset \ker(\mathrm{D})$. □

Example 7.32. In part (i) above it is necessary to assume that K/k is separable. Indeed, let k be a field of characteristic $p > 0$, K/k a purely inseparable extension of k such that for some pth-power $q \geq 1$, $x^q \in k$ for all $x \in K$. Then the k-polynomial map $F^q : K \longrightarrow k$, defined by

$$x \mapsto x^q$$

for any $x \in K \otimes_k B$ with B a commutative k-algebra, is a determinant of dimension q, necessarily faithful as K is a field. However, $K \otimes_k K$ is not reduced when $q > 1$, in which case $F^q \otimes_k K$ is not faithful by Lemma 7.30 (v).

In what follows, A is a local ring with maximal ideal m and residue field $k := A/m$. We will denote by $\overline{\mathrm{R}}$ the k-algebra $R \otimes_A k = R/mR$, and by $\overline{\mathrm{D}}$ the induced determinant $\mathrm{D} \otimes k : \overline{\mathrm{R}} \longrightarrow k$.

Lemma 7.33. *Assume that* D *is Cayley–Hamilton.*

(i) *The kernel of the canonical surjection* $R \longrightarrow \overline{\mathrm{R}}/\ker\overline{\mathrm{D}}$ *is* $\mathrm{Rad}(R)$.
(ii) *If* $m^s = 0$ *for* $s \geq 1$ *an integer, and if* $d!$ *is invertible in* A*, then* $\mathrm{Rad}(R)^{N(d)s} = 0$.

Proof. Let J be the two-sided ideal of the statement (i); we check first that $J \subset \mathrm{rad}(R)$. It is enough to check that $1 + J \subset R^*$, i.e. $\mathrm{D}(1 + J) \in A^*$, but this is obvious as $\mathrm{D}(1+J) \in 1+m$ by definition. In particular, $mR \subset \mathrm{rad}(R)$, hence to check the converse we may (and do) assume that $A = k$ and even that D is faithful. But then $\mathrm{rad}(R) = 0$ by Lemma 7.31 (ii).

Assume that $m^s = 0$. Replacing R by R/mR if necessary, we may assume that $A = k$ is a field and we have to show that $\mathrm{Rad}(R)^{N(d)} = 0$. Here $N(d)$ is the integer coming from the Nagata–Higman theorem, in particular $N(d) \leq 2^d - 1$. We conclude by the equality $\mathrm{Rad}(R) = \ker(\mathrm{D})$ and Lemma 7.30. □

2.2. Determinants over a field

In all this section, k is a field and R a k-algebra. We fix $\overline{k}$ an algebraic closure of k, and by $k^{\mathrm{sep}} \subset \overline{k}$ a separable algebraic closure of k.

Theorem 7.34. *Assume that* k *is algebraically closed. For any* d*-dimensional determinant* $\mathrm{D} : R \longrightarrow k$*, there exists a semisimple representation* $\rho : R \longrightarrow M_d(k)$ *such that* $\mathrm{D} = \det \circ \rho$.

Such a ρ *is unique up to isomorphism, and* $\ker\rho = \ker(\mathrm{D})$.

Corollary 7.35. *Let* G *be a group (or a unital monoid), then for any algebraically closed field* k*,* $X(G, d)(k)$ *is in natural bijection with the isomorphism classes of* d*-dimensional semisimple* k*-linear representations of* G.

Let us prove the first part of the theorem. By replacing R by $R/\ker(\mathrm{D})$ if necessary, we may assume that D is faithful. By Lemmas 7.27 and 7.33, R satisfies the assumptions of the following general fact from classical noncommutative ring theory.[22]

Lemma 7.36. *Let* k *be a field,* R *a* k*-algebra with trivial Jacobson radical, and* $n \geq 1$ *an integer. Assume that each element of* R *(resp. of* $R \otimes_k k^{\mathrm{sep}}$*)*

[22] This result is presumably well-known; it is close to some old results of Kaplanski (see [H, Chap. 6.3], as well as [P1] for a related use). We have learnt most of it from Rouquier [Rou, Lemme 4.1].

is algebraic over k (resp. $k^{\rm sep}$) of degree less than n, and that the length of families of orthogonal idempotents of $R \otimes_k k^{\rm sep}$ is also bounded by n. Then

$$R \overset{\sim}{\to} \prod_{i=1}^{s} M_{n_i}(E_i)$$

where E_i is a division k-algebra which is finite dimensional over its center k_i. Moreover, each k_i is a finite separable extension of k, unless maybe k has characteristic $p > 0$, in which case $k[k_i^q]$ is separable over k where q is the biggest power of p less than n.

In particular, R is semisimple. It is finite dimensional over k in each of the following three cases: k is a perfect field, or k has characteristic $p > 0$ and $[k : k^p] < \infty$, or $p > n$.

Proof. Let A be a commutative k-algebra such that each element of A is algebraic over k of degree less than n, and that the length of families of orthogonal idempotents of A is also bounded by n. If k has characteristic $p > 0$ we define q as in the statement, and we set $q = 1$ else. Then we check at once that there is a k-algebra isomorphism

$$A \overset{\sim}{\to} \prod_{i=1}^{r} A_i$$

where $r \leq n$ and where A_i is a field whose maximal separable k-subextension $A_i^{\rm et}$ is finite dimensional over k (with dimension $\leq n$), and satisfies $A_i^q \subset A_i^{\rm et}$. These facts apply in particular to the center Z of R. We get moreover that $\dim_k(Z) < \infty$ in the three cases discussed in the last assertion on the statement.

We prove now that R is semisimple. Let M be a simple R-module, E the division k-algebra $\mathrm{End}_R(M)$. We claim first that M is finite dimensional over E. Indeed, by Jacobson's density theorem, we know that either M is finite dimensional over E and $R \longrightarrow \mathrm{End}_E(M) \simeq M_s(E^{\rm opp})$ is surjective, or for each $r \geq 1$ there is a k-subalgebra $S_r \subset R$ and a surjective k-algebra homomorphism $S_r \longrightarrow M_r(E^{\rm opp})$, but this second possibility is impossible as elements of R (hence of S_r) are algebraic over k of bounded degree by assumption.

We claim now that there are only finitely many (pairwise non isomorphic) simple R-modules $M_1, \ldots, M_s$, which will conclude the proof. Indeed, assuming this claim and as $\mathrm{Rad}(R) = 0$, there is a natural injective homomorphism

$$R \longrightarrow \prod_{i=1}^{s} \mathrm{M}_{n_i}(E_i^{\rm opp}), \tag{7.20}$$

$E_i = \mathrm{End}_R(M_i)$, which is surjective as the M_i are pairwise non isomorphic and simple, hence (7.20) is an isomorphism. It remains to check the claim. If $M_1, \dots, M_s$ are any pairwise non isomorphic simple R-modules, and $E_i = \mathrm{End}_R(M_i)$, the morphism (7.20) is still surjective. As $\mathrm{Rad}(R) = 0$ and R is algebraic over k, the family of orthogonal idempotents of the right-hand side lifts in R ([Bki, §4, ex.5(b)]), hence $s \leq n$ by assumption, and we are done.

It only remains to show that R is a finite type Z-module. As $Z \otimes_k k^{\mathrm{sep}}$ is faithfully flat over Z, it is enough to check that $R \otimes_k k^{\mathrm{sep}}$ is a finite type $Z \otimes_k k^{\mathrm{sep}}$-module. Note that the k^{sep}-algebra $R \otimes_k k^{\mathrm{sep}}$ satisfies also the assumptions of the lemma, hence is semisimple by what we proved so far. Moreover, its center is easily seen to be $Z \otimes_k k^{\mathrm{sep}}$. By the Wedderburn–Artin theorem,

$$R \otimes_k k^{\mathrm{sep}} \tilde{\to} \prod_{j=1}^{t} M_{d_j}(k_j)$$

where k_j is a division k^{sep}-algebra, which is moreover algebraic over k^{sep} here. The Jacobson–Noether theorem implies that such a division algebra is commutative, hence each k_j is a field extension of k^{sep}, which concludes the proof. □

Going back to the proof of Theorem 7.34, we get that R is isomorphic to a finite product of matrix k-algebras,

$$R \tilde{\to} \prod_{i=1}^{s} M_{n_i}(k).$$

In particular, fixing such a k-algebra isomorphism, D appears as a determinant of such an algebra. By Lemma 7.25 (Spec(k) is certainly connected), there are unique determinants $\mathrm{D}_i : M_{n_i}(k) \longrightarrow k$, say of dimension d_i, such that D is the product of the D_i and $d = \sum_i d_i$.

Lemma 7.37. *If* $\mathrm{D} : M_n(k) \longrightarrow k$ *is a determinant of dimension d, then $d = mn$ is divisible by n and and* D *is the mth-power of the usual determinant (here k is actually any commutative ring, and* $(M_n(k), \det)$ *can be replaced by any Azumaya algebra equipped with its reduced norm).*

Proof. Indeed, by Ziplies theorem [Z1, Thm. 3.17] (or Ex. 7.28), any such determinant is a composition of the usual determinant with a multiplicative k-polynomial law $k \longrightarrow k$. It is clear that any such law is of the form $x \mapsto x^m$ for some integer $m \geq 0$. □

As a conclusion, we may write $d_i = m_i n_i$, and if M_i is the simple module of R corresponding to $M_{n_i}(k)$, then D is the determinant of the semisimple representation $\oplus_{i=1}^{s} M_i^{m_i}$. As a semisimple representation is well known to be uniquely determined by its characteristic polynomials (Brauer–Nesbitt's theorem), this representation is unique up to isomorphism. As ρ is obviously injective, the second assertion on $\ker\rho$ follows. This concludes the proof of Theorem 7.34.

We now investigate the case of a general field k, starting with the following useful observations.

Let K be a field extension of k and denote by $k' \subset K$ the maximal separable k-subextension of K. Assume that k' is finite over k. If $p := \mathrm{char}(k) > 0$ assume also that there exists a finite power q of p such that $K^q \subset k'$. We define the *exponent* $(f, q) \in \mathbb{N}^2$ of K/k by $f = [k' : k]$, $q = 1$ if $K = k'$, and q is the smallest power of $p = \mathrm{char}(k) > 0$ as above if $K \neq k'$.

Let S be a central simple K-algebra with rank n^2 over K and reduced norm $N : S \longrightarrow K$, let $N_{k'/k} : k' \longrightarrow k$ be the usual norm (i.e. the determinant of the regular k-representation) and $F^q : K \longrightarrow k$ the qth-Frobenius law (see Ex. 7.32). Then we have a natural determinant

$$\det_S : S \longrightarrow k$$

of dimension nqf defined by $\det_S = N_{k'/k} \circ F^q \circ N$.

Theorem 7.38. *Let* $\mathrm{D} : R \longrightarrow k$ *be a determinant of dimension* d*. Then as a* k*-algebra*

$$R/\ker(\mathrm{D}) \overset{\sim}{\to} \prod_{i=1}^{s} S_i$$

where S_i *is a simple* k*-algebra which is of finite dimension* n_i^2 *over its center* k_i*, and where* k_i/k *has a finite exponent* (f_i, q_i)*.*

Moreover, under such an isomorphism, D *coincides with the product determinant*

$$\mathrm{D} = \prod_{i=1}^{s} \det_{S_i}{}^{m_i}, \quad d = \sum_i m_i n_i q_i f_i,$$

where m_i *are some uniquely determined integers.*

In particular, $R/\ker(\mathrm{D})$ *is semisimple. It is finite dimensional over* k *if and only if each* k_i *is. This always occurs in each of the following three cases:* k *is perfect, or* k *has characteristic* $p > 0$ *and* $[k : k^p] < \infty$*, or* $d < p$*.*

By Lemmas 7.31 (i) and 7.36, it only remains to show the following lemma.

Lemma 7.39. *Let* K/k *be a field extension with finite exponent* (f, q) *and* S *a central simple* K*-algebra which is finite dimensional over* K*. Then any determinant* $S \longrightarrow k$ *has the form* $\det_S^m$ *for some unique integer* $m \geq 0$*.*

Proof. Let $\mathrm{D} : S \longrightarrow k$ be a determinant of dimension d and $n^2 := \dim_K(S)$. Note that if $\mathrm{D} = \det_S^m$, then we have by homogeneity $d = fmnq$ thus m is unique if it exists. Moreover, note that by Proposition 7.4, if two determinants $D_1, D_2 : R \longrightarrow A$ of dimension d are such that $D_1 \otimes_A B = D_2 \otimes_A B$ for some commutative A-algebra B with $A \to B$ injective, then $D_1 = D_2$. We will apply this below when B is a field extension of a field A.

Assume first that k is separably closed (hence so is K); by the Noether–Jacobson theorem $S \xrightarrow{\sim} M_n(K)$ for some $n \geq 1$. Set $A := K \otimes_k K$ and consider the kernel I of natural split surjection $A \longrightarrow K$; I is generated as A-module by the $x \otimes 1 - 1 \otimes x$, which are nilpotent of index $\leq q$, thus any finite type A-submodule of I is nilpotent. Lemma 7.30 (iv) implies then that any determinant $M_n(A) \longrightarrow K$ factors through $\pi : M_n(A) \to M_n(A/I) = M_n(K)$. Applying this to $\mathrm{D} \otimes_k K$, we get a determinant $M_n(K) \longrightarrow K$, which is an integral power of the usual determinant by Lemma 7.28, say $\mathrm{D} \otimes_k K = \det^s \circ \pi$ and $d = ns$. A necessary condition is that $\det^s(M_n(K)) \subset k$, which implies that q divides s. In particular, there is a unique possibility for $\mathrm{D} \otimes_k K$, hence applying this again to $\mathrm{D}' = \det_S^{s/q}$, the remark above shows that $\mathrm{D} = \det_S^{s/q}$.

We now reduce to the previous case. We have

$$K \otimes_k k^{\mathrm{sep}} \xrightarrow{\sim} \prod_{i=1}^{f} K_i$$

where $K_i = K.k^{\mathrm{sep}}$ is a separable algebraic closure of K such that $K_i^q \subset k^{\mathrm{sep}}$ (and q is still minimal for that property), and where $\mathrm{Gal}(k^{\mathrm{sep}}/k)$ permutes transitively the K_i. Moreover,

$$S \otimes_k k^{\mathrm{sep}} = S \otimes_K (K \otimes_k k^{\mathrm{sep}}) \xrightarrow{\sim} \prod_{i=1}^{f} S_i,$$

and $S_i = S \otimes_K K_i$ is central simple of rank n^2 over K_i. By Lemma 7.25 (iii), each $\mathrm{D} \otimes_k k^{\mathrm{sep}}$ is a product of determinants $S_i \xrightarrow{\sim} M_n(K_i) \longrightarrow k^{\mathrm{sep}}$, which have the form $\det_{S_i}^{m_i}$ by the previous step and $d = n(\sum_{i=1}^{f} m_i)$. As $\mathrm{D} \otimes_k k^{\mathrm{sep}}$ is $\mathrm{Gal}(k^{\mathrm{sep}}/k)$-equivariant, this implies that $m := m_i$ is independent of i, thus $m = d/nf$. In particular, there is a unique possibility for $\mathrm{D} \otimes_k k^{\mathrm{sep}}$ and thus $\mathrm{D} = \det_S^m$. □

Definition-Proposition 7.40. Let $\mathrm{D} : R \longrightarrow k$ be a determinant of dimension d. We say that D is *absolutely irreducible* if one of the following equivalent properties is satisfied:

(i) The unique semisimple representation $R \longrightarrow M_d(\overline{k})$ with determinant D (which exists by Theorem 7.34) is irreducible,
(ii) $(R \otimes_k \overline{k})/\ker(\mathrm{D} \otimes_k \overline{k}) \simeq M_d(\overline{k})$,
(iii) $R/\ker(\mathrm{D})$ is a central simple k-algebra of rank d^2,
(iv) $R/\mathrm{CH}(\mathrm{D})$ is a central simple k-algebra of rank d^2,
(v) for some (resp. all) subset $X \subset R$ generating R as a k-vector space, there exists $x_1, x_2, \ldots, x_{d^2} \in X$ such that the abstract $d^2 \times d^2$ matrix $(\Lambda_1(x_i x_j))_{i,j}$ belongs to $\mathrm{GL}_{d^2}(k)$.

If they are satisfied, then $\mathrm{CH}(\mathrm{D}) = \ker(\mathrm{D}) = \{x \in R, \forall y \in R, \Lambda_1(xy) = 0\}$.

Proof. It is clear that (ii) implies (i). If $\rho : R \otimes_k \overline{k} \longrightarrow M_d(\overline{k})$ is as in (i), then a standard result of Wedderburn asserts that ρ surjective, and we check at once that $\ker(\rho) \subset \ker(\mathrm{D} \otimes_k \overline{k})$, hence (ii) follows by Theorem 7.38, and (v) (for any X) is the nondegeneracy of the trace on $M_d(\overline{k})$. If (v) holds for some X, we see that

$$\dim_{\overline{k}}\left((R \otimes_k \overline{k})/\ker(\mathrm{D} \otimes_k \overline{k})\right) \geq d^2,$$

hence (v) implies (ii) by Theorem 7.38. So far, we showed that (i), (ii) and (v) are equivalent.

By Lemma 7.30, the kernel of the natural surjective map $R/\mathrm{CH}(\mathrm{D}) \longrightarrow R/\ker(\mathrm{D})$ is a nilideal (and Lemma 7.31 shows that it is the Jacobson radical), hence a standard argument shows that (iii) $\Leftrightarrow$ (iv) and that if they hold this map is an isomorphism. As the formation of $R/\mathrm{CH}(D)$ commutes with arbitrary base changes (hence with $k \rightarrow \overline{k}$), and as a k-algebra E is central simple of rank d^2 if and only if $E \otimes_k \overline{k}$ has this property over $\overline{k}$, then (iv) $\Leftrightarrow$ (ii). □

Let us give some more definitions.

Definition 7.41. We say that $\mathrm{D} : R \longrightarrow k$ is *multiplicity free* if $\mathrm{D} \otimes_k \overline{k}$ is the determinant of a direct sum of pairwise non-isomorphic absolutely irreducible $\overline{k}$-linear representations of R. In the notation of Theorem 7.38, it means that $m_i = q_i = 1$ for each i.

We say that D is *split* if it is the determinant of a representation $R \longrightarrow M_d(k)$. Equivalenty, D is split if and only if $R/\ker(\mathrm{D})$ is a finite product of matrix algebras over k.

We leave as an exercise to the reader to check the equivalences in the definition above. Moreover, we see easily that $\mathrm{D} : R \longrightarrow k$ is split and absolutely irreducible (resp. multiplicity free) if, and only if, D is the determinant of a surjective k-representation $R \longrightarrow M_d(k)$ (resp. of a direct sum of pair-wise non isomorphic absolutely irreducible representations of R defined over k).

Example 7.42. *(The absolute irreducible locus)* Let G be group (or a unital monoid) and $d \geq 1$ an integer. If $x \in X(G,d)$, we say that x is *absolutely irreducible* if the induced determinant $k(x)[G] \longrightarrow k(x)$ has this property, where $k(x)$ is the residue field at x. Let

$$X(G,d)^{\mathrm{irr}} \subset X(G,d)$$

be the subset of absolutely irreducible points. It is a Zariski open subset. Indeed, for each sequence of elements $\underline{g} = (g_1, \ldots, g_{d^2}) \in G^{d^2}$, consider the abstract $d^2 \times d^2$ matrix

$$m_{\underline{g}} = (\mathrm{Tr}(g_i g_j)) \in M_{d^2}(\mathbb{Z}(G,d)),$$

where $\mathrm{Tr} = \Lambda_1$ is the trace of the universal determinant of G of dimension d, and define $I \subset \mathbb{Z}(G,d)$ as the ideal generated by the $\det(m_{\underline{g}})$ when $\underline{g}$ varies in all the sequences as above. Then $X(G,d)^{\mathrm{irr}} = X(G,d) - V(I)$ by Definition-Proposition 7.40, hence the claim.

2.3. Determinants over henselian local rings

We now study the determinants $\mathrm{D} : R \longrightarrow A$ where A is a local ring. We shall use the notation of Lemma 7.33. Let $\mathrm{D} : R \longrightarrow A$ be a determinant of dimension d, we call $\overline{\mathrm{D}} : \overline{\mathrm{R}} \longrightarrow \overline{k}$ the residual determinant.

Theorem 7.43. *Assume that* D *is Cayley–Hamilton and that A is henselian.*

(i) *If $\overline{\mathrm{D}}$ is split and absolutely irreducible, then there is an A-algebra isomorphism*

$$\rho : R \longrightarrow M_d(A)$$

such that $\mathrm{D} = \det \circ \rho$.

(ii) *More generally, if $\overline{\mathrm{D}}$ is split and multiplicity free, then* (R, Tr) *is a generalized matrix algebra in the sense of* [BC, §1.3].

Proof. The proof is almost verbatim the same as in [BC, Lemma 1.4.3], replacing the appeals to [BC, Lemma 1.2.5] and [BC, Lemma 1.2.7] by the ones of § 2.1, so we will be a bit sketchy.

By assumption, we have an isomorphism

$$\varphi : R/\ker(\overline{\mathrm{D}}) \xrightarrow{\sim} \prod_{i=1}^{s} M_{n_i}(k)$$

such that $\overline{\mathrm{D}} = \det \circ \varphi$ and $\sum_{i=1}^{s} n_i = d$. Call $\overline{\mathrm{D}}_i$ the determinant of the representation $R \longrightarrow M_{n_i}(k)$ on the ith factor, so $\overline{\mathrm{D}} = \prod_{i=1}^{s} \overline{\mathrm{D}}_i$.

Assume first that we are in case (i), i.e. $s = 1$. As R is integral over A, A is henselian, and $\mathrm{Rad}(R) = \ker\overline{\mathrm{D}}$ by Lemma 7.33, we may find some elements $E_{i,j} \in R$ such that $E_{i,j}E_{k,l} = \delta_{j,k}E_{i,l}$ lifting the usual basis of $M_d(k)$ (see [Bki, chap. III, §4, exercice 5]). Set $e_i = E_{i,i}$. Lemma 7.27 shows that $\mathrm{D}_{e_i} : e_iRe_i \longrightarrow A$ is a Cayley–Hamilton determinant. Its dimension is the integer $r(e_i)$ such that $t^{r(e_i)} = \mathrm{D}_{e_i}(te_i) = \mathrm{D}(1 - e_i + te_i) \in A[t]$. Projecting this equality in $k[t]$ we get that

$$t^{r(e_i)} = \overline{\mathrm{D}}(1 - \overline{e_i} + t\overline{e_i}) = t \in k[t]$$

so $r(e_i) = 1$. By Lemma 7.27 again, $e_1 + \cdots + e_s = 1$ and $\mathrm{D}_{e_i} : e_iRe_i \longrightarrow A$ is Cayley–Hamilton of dimension 1, so $e_iRe_i = Ae_i$ is free of rank 1 over A by Lemma 7.29. But if $x \in e_iRe_j$, then $x = E_{i,j}(e_jE_{j,i}x) \in AE_{i,j}$ and we check at once that $R = \oplus_{i,j}AE_{i,j} \simeq M_d(A)$, in which case D necessarily coincides with the usual determinant by Ziplies' theorem.

Assume now that we are in case (ii). Let us lift the family of central orthogonal idempotents of length s of $R/\ker(\overline{\mathrm{D}})$ to a family of orthogonal idempotents $e_1+\cdots+e_s = 1$ in R. Arguing as above we see again that $\mathrm{D}_{e_i} : e_iRe_i \longrightarrow A$ is a Cayley–Hamilton determinant of dimension n_i, which is residually split and absolutely irreducible. By (i) we get that $e_iRe_i \simeq M_{d_i}(A)$, and it is immediate to check that R is a generalized matrix algebra whose trace coincides with the trace of D. □

We get in particular the following nice corollary (see §1.5).

Corollary 7.44. *Let G be a group (or a unital monoid).*

(i) *Over $X(G,d)^{\mathrm{irr}}$, the Cayley–Hamilton $\mathcal{O}_X$-algebra $\mathcal{R}(G,d)$ is an Azumaya $\mathcal{O}_X$-algebra of rank d^2 equipped with its reduced norm.*

(ii) *For each split $x \in X(G,r)^{\mathrm{irr}}$, the pro-artinian completion of $\mathcal{O}_x$ is canonically isomorphic to the usual deformation ring of the associated absolutely irreducible representation $G \longrightarrow \mathrm{GL}_d(k(x))$ (see e.g. [M]).*

Proof. Let $x \in X(G,d)^{\mathrm{irr}}$ and A the strict henselianization of $\mathcal{O}_x$. Recall that the formation of the Cayley–Hamilton quotient commutes with arbitrary base change. In particular,

$$R(G,d) \otimes_{\mathbb{Z}(G,d)} A \overset{\sim}{\rightarrow} A[G]/\mathrm{CH}(\mathrm{D}^u \otimes A).$$

Theorem 7.43 (i) shows that the A-algebra on the right side is isomorphic to $M_{d^2}(A)$, thus $R(G,d) \otimes_{\mathbb{Z}(G,d)} \mathcal{O}_x$ is an Azumaya algebra of rank d^2 as $\mathcal{O}_x \longrightarrow A$ is faithfully flat. Part (i) follows then from the following abstract result, a variant of which is implicitly used in [Rou, Thm. 5.1][23]: Let C be a

[23] We are grateful to R. Rouquier for providing us with a proof of this result.

commutative ring, $d \geq 1$ an integer, and R a C-algebra. Assume that for all $x \in \mathrm{Spec}(C)$, then R_x is Azumaya of rank d^2 over C_x, then R is an Azumaya C-algebra (locally free) of rank d^2.

Part (ii) follows at once from Theorem 7.43 (i), which moreover identifies canonically the universal representation to the natural map $G \longrightarrow (R(G,d) \otimes_{\mathbb{Z}(G,d)} \mathcal{O}_x)^*$. □

2.4. Determinants over $A[\epsilon]$

Let us fix a commutative ring A and a determinant $D_0 : R \longrightarrow A$ of dimension d. Consider the A-algebra $A[\epsilon]$ with $\epsilon^2 = 0$; if M is an A-module we will write more generally $M[\epsilon]$ for $M \otimes_A A[\epsilon]$.

We are interested in the set of determinants $D : R[\epsilon] \longrightarrow A[\epsilon]$ *lifting* D_0, i.e. such that $D \otimes_A A[\epsilon] = D_0$. Via the identification $\mathcal{M}^d_A(R, A[\epsilon]) \xrightarrow{\sim} \mathcal{M}^d_{A[\epsilon]}(R[\epsilon], A[\epsilon])$, it coincides with the set $\mathcal{T}$ of d-homogeneous multiplicative A-polynomial laws $P : R \longrightarrow A[\epsilon]$ which map to D_0 via the A-algebra homomorphism $\pi : A[\epsilon] \xrightarrow{\epsilon \mapsto 0} A$. In other words, $\mathcal{T} = (\pi^*)^{-1}(D_0)$ where

$$\pi^* : \mathrm{Hom}_{A-\mathrm{alg}}(\Gamma^d_A(R)^{\mathrm{ab}}, A[\epsilon]) \longrightarrow \mathrm{Hom}_{A-\mathrm{alg}}(\Gamma^d_A(R)^{\mathrm{ab}}, A), \quad f \mapsto \pi \circ f.$$

This expression makes $\mathcal{T}$ appear as a relative tangent space, thus $\mathcal{T}$ carries a natural structure of A-module in the usual way.

Recall that we have a natural A-module isomorphism

$$\mathcal{P}^d_A(R, A[\epsilon]) \xrightarrow{\sim} \mathcal{P}^d_A(R, A)^2, \; P \mapsto (P_0, P_1), \; P = P_0 + \epsilon P_1,$$

and any $P \in \mathcal{T}$ writes by definition as

$$P = D_0 + \epsilon \Delta$$

for some $\Delta \in \mathcal{P}^d_A(R, A)$.

Proposition 7.45. *The map $P \mapsto \Delta$ above induces an A-module isomorphism between $\mathcal{T}$ and the A-submodule of elements δ of $\mathcal{P}^d_A(R, A)$ such that for any commutative A-algebra B and any $x, y \in R \otimes_A B$,*

$$\delta(xy) = D_0(x)\delta(y) + D_0(y)\delta(x).$$

Proof. Immediate from the definitions. □

As in the case of determinants, the polynomial map Δ (associated to some $P \in \mathcal{T}$) satisfies a number of polynomial identities. For example $\Delta(1) = 0$, $\Delta(xy) = \Delta(yx)$, and Δ satisfies a variant of Amitsur's formula.

In what follows, an important role will be played by the two-sided ideal

$$I := \ker(D_0) \subset R.$$

The main reason for this are the following lemmas.

Lemma 7.46. *Assume that A is a field of characteristic 0 or $> d$. For any $P \in \mathcal{T}$, $I^{2N(d)} \subset \ker(P)$. In other words, $\mathcal{T} \subset \mathcal{P}_A^d(R/I^{2N(d)}, A)$.*

Here, $N(d)$ is the integer coming from the Nagata–Higman theorem, in particular $N(d) \leq 2^d - 1$.

Proof. Let $P \in \mathcal{T}$ and $D : R[\epsilon] \longrightarrow A[\epsilon]$ the associated determinant. We check at once that via the natural injection $R \longrightarrow R[\epsilon]$, we have

$$\ker(P) = R \cap \ker(D),$$

so it suffices to show that $I^{2N(d)} \subset \ker(D)$.

Remark that $I \supset \mathrm{CH(D)}$ and consider the Cayley–Hamilton quotient $S = R[\epsilon]/\mathrm{CH(D)}$. For $r \in I[\epsilon] \subset R[\epsilon]$, we have by assumption $\Lambda_i(r) \in \epsilon A$ for all $i \geq 1$, thus $s^d \in \epsilon A[s]$ for all $s \in J = I[\epsilon]/\mathrm{CH(D)}$. The Nagata–Higman theorem implies that $(J/\epsilon J)^{N(d)} = 0$ and then that $J^{2N(d)} = 0$. As a consequence, $I^{2N(d)} \subset \mathrm{CH}(D) \subset \ker(D)$, and we are done. □

The next lemma is well-known.

Lemma 7.47. *There is a natural A-module isomorphism*

$$\mathrm{Hom}_R(I/I^2, R/I) \xrightarrow{\sim} \mathrm{Ext}^1_R(R/I, R/I).$$

(The Hom and Ext above are understood in the category of (left) R-modules.)

Proof. Apply $\mathrm{Hom}_R(-, R/I)$ to the exact sequence of R-modules

$$0 \longrightarrow I \longrightarrow R \longrightarrow R/I \longrightarrow 0,$$

and use that $\mathrm{Ext}^1_R(R, -) = 0$. □

Let us study now a more specific example where those concepts apply. Assume that $A = k$ is a field, and that $S := R/I$ is a finite dimensional semisimple k-algebra. Recall that by Theorem 7.38, this is always the case if k is perfect, or if $\mathrm{char}(k) = p > 0$ and $[k : k^p] < \infty$, or if $d < p$. Let $M_1, \ldots, M_r$ denote the simple S-modules and $M := \oplus_{i=1}^r M_r$.

Proposition 7.48. *Assume either* $\mathrm{char}(k) = 0$ *or* $\mathrm{char}(k) > d$. *If* $\mathrm{Ext}^1_R(M, M)$ *is finite dimensional over k, then so is $\mathcal{T}$.*

Proof. For any semisimple ring S, the left S-module S is a finite direct sum of simple modules, hence $\mathrm{Ext}^1_R(S, S)$ is a finite dimensional k-vector space by assumption. As a consequence, the S-module $\mathrm{Hom}_S(I/I^2, S)$ is also finite dimensional over k by Lemma 7.47, which implies that I/I^2 has a finite length as S-module as S is semisimple, hence $\dim_k I/I^2 < \infty$. But then $R/I^{2N(d)}$ is also finite dimensional over k and we are done by Lemma 7.46. □

Assume moreover that $R = k[G]$ and say that k is perfect. Let $\rho : G \longrightarrow \mathrm{GL}_d(\overline{k})$ denote the unique semisimple representation of G such that $\det(1 - t\rho(g)) = D_0(1 - tg)$ for all $g \in G$ (Thm. 7.34), the assumption in the proposition is equivalent to

$$\dim_{\overline{k}} H^1(G, \mathrm{ad}(\rho)) < \infty,$$

which generalizes a well-known result in the case ρ is irreducible (see the remark below).

Remark 7.49. It would be interesting to know whether the known improvements of the Nagata–Higman theorem (as Shirshov's height theorem) lead to a generalization of this proposition to fields of characteristic $\leq d$. The arguments above actually give an explicit upper bound for $\dim_k \mathcal{T}$, which is however very bad in general.[24] For example, when ρ is defined over k (say) and irreducible, Theorem 7.43 and a standard argument give a natural identification $\mathcal{T} \xrightarrow{\sim} H^1(G, \mathrm{ad}(\rho))$, which is much finer than what we got by the previous analysis. When ρ is defined over k and multiplicity free (and in the context of pseudocharacters), this space $\mathcal{T}$ has recently been studied by Bellaïche [B].

2.5. Continuous determinants

For later use we shall need to study a variant of the notions we have studied so far taking care of some topology.

Assume that G is a topological group and that A is a topological ring. Let $\mathrm{D} : A[G] \longrightarrow A$ be a determinant of dimension n, or which is the same, a homogeneous multiplicative C-polynomial law $C[G] \longrightarrow A$ of degree n for any subring $C \subset A$. We say that D is *continuous* if for each $\alpha \in I_n$, the map $\mathrm{D}^{[\alpha]} : G^n \longrightarrow A$ defined in §1.1 is continuous. By Amitsur's formula, D is continuous if, and only if, $\Lambda_i : G \longrightarrow A$ is continuous for all $i \leq n$ (same argument as in the proof of Lemma 7.11).

Example 7.50. *(Restriction to a dense subgroup)* Assume that $H \subset G$ is a dense subgroup, then a continuous determinant on G is uniquely determined

[24] It could actually be improved by studing more carefully the successive restrictions of elements of $\mathcal{T}$ to the subspaces $I^k/I^{2N(d)}$ but still the general bound would be rather bad.

by its restriction to H. Indeed, if two such determinants D_1 and D_2 coincide on $\mathbb{Z}[H]$, and if n denotes their common dimension, then for each $\alpha \in I_n$ the continuous maps $\mathrm{D}_1^{[\alpha]}, \mathrm{D}_2^{[\alpha]} : G^n \longrightarrow A$ coincide on H^n, hence on the whole of G^n, so $\mathrm{D}_1 = \mathrm{D}_2$.

Example 7.51. *(Glueing determinants)* In some applications to number theory, we are in the following situation. Let G be a compact topological group, A and $\{A_i, i \in I\}$ topological rings with A compact, $\iota : A \longrightarrow \prod_i A_i$ a continuous injective map, $\mathrm{D}_i : A_i[G] \longrightarrow A_i$ a continuous determinant on G of dimension d. We assume that for each g in a dense subset $X \subset G$, $(\chi_i(g,t)) \in A[t]$ (of course $\chi_i(g,t)$ denotes here the characteristic polynomial of g with respect to D_i). We claim that there is a continuous determinant $\mathrm{D} : A[G] \longrightarrow A$ such that $\mathrm{D}_i = \mathrm{D} \otimes_A A_i$ for each i. Indeed, set $C = \prod_i A_i$ and consider the map $\psi : G \longrightarrow C[t]$, $g \mapsto (\chi_i(g,t))$. By assumption, $\psi(X) \subset A[t]$. As A is compact, $\iota(A)$ is a closed subspace of C homeomorphic to A, hence $\psi(G) \subset A[t]$ for X is dense in G and the D_i are continuous. The claim follows then from Corollary 7.11 and the dicussion above.

From now on, we equip A, as well as all the commutative A-algebras B, with the discrete topology, and we assume that G is a profinite group. In this context, a B-valued determinant D on G is continuous if, and only if, the characteristic polynomal map

$$G \longrightarrow B[t],\ g \mapsto D(1+tg)$$

factors through $G \mapsto G/H$ for some normal open subgroup H of G.

This leads us to define for each normal open subgroup $H \subset G$ the two-sided ideal of $A[G]$

$$J(H) := \ker\left(A[G] \xrightarrow{\text{can}} A[G/H]\right)$$

and to equip $A[G]$ with the topology defined by this filtered set of ideals.

Lemma 7.52. *A B-valued determinant D on G, viewed as an element $P \in \mathcal{M}_A^d(A[G], B)$, is continuous if, and only if,* $\ker(P) \subset A[G]$ *is open (that is, contains some $J(H)$).*

If it is the case, then the natural representation

$$G \longrightarrow (B[G]/\ker(D))^*$$

factors through a finite quotient G/H of G for some open subgroup H.

Proof. If $\ker(P) \supset J(H)$, then P factors through an element of $\mathcal{M}_A^d(A[G/H], B)$ hence D is obviously continuous. Assume conversely that D

is continuous. As B is discrete and G profinite, there is an open normal subgroup $H \subset G$ such that all the $\Lambda_i : G \longrightarrow B$ factor through G/H. As a consequence, Amitsur's formula shows that for $g \in G$ and $h \in H$,

$$D(t(g-gh)+\sum_i t_i g_i) = D(\sum_i t_i g_i)$$

so $g - gh \in \ker(P)$, and $J(H) = \sum_{g\in G, h\in H} Ag(h-1) \subset \ker(P)$. The last assertion is obvious. □

Example 7.53. Assume that $A = k$ is a field and consider the (unique) semisimple representation

$$\rho : G \longrightarrow \mathrm{GL}_d(\overline{k})$$

such that $\det(1+t\rho(g)) = D(1+tg)$ for all $g \in G$ (see Theorem 7.34). Equip $\mathrm{GL}_d(\overline{k})$ with the discrete topology. Then ρ is continuous if, and only if, D is continuous.

We end by discussing continuous deformations of a continuous determinant. We adopt the notation of § 2.4 with $R = A[G]$ as above and with D_0 a continuous determinant $A[G] \longrightarrow A$ of dimension d. Consider the A-submodule

$$\mathcal{T}^c \subset \mathcal{T}$$

of continuous liftings of D_0. This A-module writes

$$\mathcal{T}^c = \bigcup_H \mathcal{T}^H$$

where H varies in the set of all normal open subgroups of G such that $\ker(D_0) \supset J(H)$, and where $\mathcal{T}^H$ is defined as the subset of liftings P such that $\ker(P) \supset J(H)$.

Assume now that $A = k$ and that $k[G]/\ker(D_0)$ is finite dimensional over k (see 7.38), and let $\rho : G \longrightarrow \mathrm{GL}_d(\overline{k})$ be the continuous representation associated to D_0 as in Example 7.53 above. The following result is a variant of Prop. 7.48.

Proposition 7.54. *Assume* $\mathrm{char}(k) = 0$ *or* $> d$. *If the continuous cohomology group* $H^1_c(G, \mathrm{ad}(\rho))$ *is finite dimensional over* $\overline{k}$, *then* $\mathcal{T}^c$ *is finite dimensional over* k.

Proof. It is enough to show that $\dim_k \mathcal{T}^H$ is bounded independently of the normal open subgroup H such that $J(H) \subset I := \ker(D_0)$. Fix such an H. By Lemma 7.46

$$\mathcal{T}^H \subset \mathcal{P}^d_k(k[G]/(I^{2N(d)} + J(H)), k),$$

so it is enough to show that $\dim_k(k[G]/(I^{2N(d)} + J(H)))$ is bounded independently of H.

As $J(H) \subset I$, we have for each $n \geq 1$ a natural k-linear surjection

$$\left(I/(I^2 + J(H))\right)^{\otimes_k n} \longrightarrow (I^n + J(H))/(I^{n+1} + J(H)),$$

hence it is enough to show that $\dim_k(I/(I^2+J(H)))$ is bounded independently of H (recall that $k[G]/I$ is finite dimensional). As $k[G]/I$ is a semisimple k-algebra, it is then enough to show that

$$\dim_k\left(\mathrm{Hom}_{k[G]}(I/(I^2 + J(H)), k[G]/I)\right)$$

is bounded independently of H. But by Lemma 7.47,

$$\mathrm{Hom}_{k[G]}(I/(I^2 + J(H)), k[G]/I) \xrightarrow{\sim} \mathrm{Ext}^1_{k[G/H]}(k[G]/I, k[G]/I)$$

and this latter space is naturally a subvector space of the space of continuous G-extensions of $k[G]/I$ by itself, which does not depend on H, and which is finite dimensional by assumption. □

Remark 7.55. The proof above shows moreover that if $H^1_c(G, \mathrm{ad}(\rho)) = 0$, then $\mathcal{T}^c = 0$.

2.6. A finiteness result

We end this section by the following important finiteness result, which follows from works of Donkin [D], Seshadri [S] and Vaccarino [V1].

Proposition 7.56. *Assume that R is finitely generated as A-algebra and let $d \geq 1$ be an integer, then $\Gamma^d_A(R)^{\mathrm{ab}}$ is a finite type A-algebra.*

Proof. Let X be a finite set and $\mathbb{Z}\{X\}$ the free ring on X, and set $m = |X|$. By assumption there is a surjective A-algebra homomorphism

$$A\{X\} := A \otimes_{\mathbb{Z}} \mathbb{Z}\{X\} \longrightarrow R,$$

hence a surjective A-algebra homomorphism $\Gamma^d_A(A\{X\})^{\mathrm{ab}} \longrightarrow \Gamma^d_A(R)^{\mathrm{ab}}$, so we may assume that $R = A\{X\}$. As $\Gamma^d_A(A\{X\})^{\mathrm{ab}}$ is canonically isomorphic to $A \otimes \Gamma^d_{\mathbb{Z}}(\mathbb{Z}\{X\})^{\mathrm{ab}}$, we may also assume that $A = \mathbb{Z}$.

Let $B = F_X(d)$ as in § 1.3 be the coordinate ring of M^m_d over $\mathbb{Z}$ and $B^H \subset B$ the ring of invariant elements under the componentwise conjugacy of $H := \mathrm{GL}_d(\mathbb{Z})$. Recall that we have a natural ring homomorphism

$$\rho^{\mathrm{univ}} : \mathbb{Z}\{X\} \longrightarrow M_d(B)$$

sending $x \in X$ to the matrix $(x_{i,j})_{i,j}$, and that $E_X(d) \subset B$ is the subring generated by the coefficients of the characteristic polynomials of the elements of $\rho^{\mathrm{univ}}(\mathbb{Z}\{X\})$. Clearly we have $E_X(d) \subset B^H$ and a theorem of S. Donkin ([D, Thm. 1 and §3]) shows that $E_X(d) = B^H$. As $\mathrm{GL}_d/\mathbb{Z}$ is reductive (and in particular reduced), and as $H \longrightarrow \mathrm{PGL_n}(\mathbb{C})$ has a Zariski-dense image, a general result of Seshadri [S, Thm. 2] implies that $E_X(d) = B^H$ is a finite type $\mathbb{Z}$-algebra. By the result of Vaccarino (Theorem 7.12) recalled in § 1.3, $\Gamma^d_{\mathbb{Z}}(\mathbb{Z}\{X\}) \simeq E_X(d)$, and we are done. □

Corollary 7.57. *Assume that G is finitely generated and fix $d \geq 1$.*

(i) *$\mathbb{Z}(G, d)$ is a finite type $\mathbb{Z}$-algebra,*
(ii) *There exists a finite set $X \subset G$ such that for each commutative ring A and any two d-dimensional determinants $\mathrm{D}_1, \mathrm{D}_2 : A[G] \longrightarrow A$, then $\mathrm{D}_1 = \mathrm{D}_2$ if and only if $\chi^{\mathrm{D}_1}(x,t) = \chi^{\mathrm{D}_2}(x,t)$ for all $x \in X$.*[25]

Proof. Part (i) is a special case of the proposition and part (ii) follows from part (i) and Amitsur's formula (see Lemma 7.11): $E_X(d)$ is generated as $\mathbb{Z}$-algebra by the coefficients of the $\chi(g,t)$ for $g \in G$. □

3. The universal rigid-analytic families of pseudocharacters of a profinite group

In this section, G is a profinite group,[26] p is a fixed prime number and $d \geq 1$ is an integer.

3.1. The deformation space of a given residual determinant

Let k be a finite field of characteristic p equipped with its discrete topology and

$$\overline{\mathrm{D}} : k[G] \longrightarrow k$$

a continuous determinant of dimension d. Recall that by Theorem 7.34 (and Example 7.53), it is equivalent to give such a determinant and (the isomorphism class of) a continuous semisimple representation

$$\bar{\rho} : G \longrightarrow \mathrm{GL}_d(\bar{k})$$

such that $\det(1 + t\bar{\rho}(g)) \in k[t]$ for all $g \in G$, the relation being then $\overline{\mathrm{D}}(x) = \det(\bar{\rho}(x))$ for all $x \in k[G]$.

[25] Of course, $\chi^{\mathrm{D}}(x, T)$ denotes here the characteristic polynomial of x with respect to D.
[26] The methods of this section could easily be extended to study more general topological groups, as locally profinite ones.

Let $W(k)$ be the ring of Witt vectors of k. Let $\mathcal{C}$ be the category whose objects are the local $W(k)$-algebras which are finite (as a set) and with residue field isomorphic to k, and whose morphisms are $W(k)$-algebra homomorphisms. If $A \in \mathrm{Ob}(\mathcal{C})$, we will denote by m_A its maximal ideal. The given map $W(k) \to A$ induces then a canonical isomorphism $k \xrightarrow{\sim} A/m_A$, and we shall always identify A/m_A with k using this isomorphism. We shall always equip such an A with the discrete topology. Moreover any arrow $A \longrightarrow A'$ is local, i.e. sends m_A into $m_{A'}$, and is continuous. We define a covariant functor

$$F : \mathcal{C} \longrightarrow \mathrm{Ens}$$

as follows. For an object A, define $F(A)$ as the set of continuous homogeneous multiplicative $W(k)$-polynomial laws $P : W(k)[G] \longrightarrow A$ of degree d (or equivalently, of continuous A-valued determinant $D : A[G] \longrightarrow A$ of dimension d), such that $P \otimes_A k = \overline{\mathrm{D}}$ (see §2.5). If $\iota : A \longrightarrow A'$ is an arrow in $\mathcal{C}$, and $P \in F(A)$, then we check immediately that $\iota \circ P \in F(A')$, which makes F a functor.

Let us extend the functor F a little bit. Consider more generally the category $\mathcal{C}'$ whose objects are the profinite[27] local $W(k)$-algebras A with residue field k, and whose morphisms are the local continuous $W(k)$-algebra homomorphisms. Denote by $F'(A)$ the set of continuous homogeneous multiplicative $W(k)$-polynomial laws $P : W(k)[G] \longrightarrow A$ of degree d such that $P \otimes_A k = \overline{\mathrm{D}}$ (here $A \to k$ is the natural $W(k)$-algebra morphism). As before, $F' : \mathcal{C}' \to \mathrm{Ens}$ is a covariant functor; it coincides by definition with F over the full subcategory $\mathcal{C}$ of $\mathcal{C}'$. It turns out that F' coincides with the natural pro-extension of F.

Lemma 7.58. *If $A \xrightarrow{\sim} \varprojlim_i A_i$ is a projective limit in $\mathcal{C}'$, then the natural map $F'(A) \longrightarrow \varprojlim_i F'(A_i)$ is a bijection.*

Proof. If R is any A-algebra, the functor $\mathcal{M}^d_A(R, -)$ from A-algebras to sets is representable, hence commutes with any projective limit. As a map $G \longrightarrow \varprojlim_i A_i$ is continuous if and only if each coordinate map $G \to A_i$ is continuous, we get the lemma. □

Proposition 7.59. *The functor F' is representable.*

This means that there is a profinite local $W(k)$-algebra $A(\bar{\rho})$ with residue field k, and a determinant

$$\mathrm{D}(\bar{\rho}) : W(k)[G] \longrightarrow A(\bar{\rho}),$$

[27] By this we shall always mean a directed projective limit of finite rings with surjective transition maps.

such that for any $A \in \mathrm{Ob}(\mathcal{C})$ and $D \in F(A)$, there is a unique continuous $W(k)$-algebra homomorphism $\varphi_D : A(\bar{\rho}) \longrightarrow A$ such that $\mathrm{D}(\bar{\rho}) \otimes_{\varphi_D} A = D$. Such a pair $(A(\bar{\rho}), \mathrm{D}(\bar{\rho}))$ is unique, if it exists.

Proof. Let us show the existence. Consider the $W(k)$-algebra

$$B = \Gamma^d_{W(k)}(W(k)[G])^{\mathrm{ab}} = W(k) \otimes_{\mathbb{Z}} \Gamma^d_{\mathbb{Z}}(\mathbb{Z}[G])^{\mathrm{ab}},$$

the universal multiplicative polynomial law $P^u : W(k)[G] \longrightarrow B$, and let $\psi : B \longrightarrow k$ be the $W(k)$-algebra homomorphism corresponding to $\overline{\mathrm{D}}$. Say that an ideal $I \subset B$ is open if $I \subset \ker(\psi)$, B/I is a finite local ring and if the induced multiplicative polynomial law $P_I : W(k)[G] \longrightarrow B/I$ obtained as the composition of P with $B \to B/I$ is continuous. If I and J are open, then so is $I \cap J$, as $B/(I \cap J) \longrightarrow B/I \times B/J$ is injective, and a homeomorphism onto its image (!), so those ideals define a topology on B. Set

$$A(\bar{\rho}) := \varprojlim_{I \text{ open}} B/I$$

and consider the law $P(\bar{\rho}) = \iota \circ P : W(k)[G] \longrightarrow A(\bar{\rho})$ where $\iota : B \longrightarrow A(\bar{\rho})$ is the canonical map. Then $A(\bar{\rho})$ is an object of $\mathcal{C}'$ and

$$P(\bar{\rho}) = (P_I) \in F'(A(\bar{\rho})) = \varprojlim_I F(B/I)$$

by the previous lemma.

If A is an object of $\mathcal{C}$ and $P \in F(A)$, then by Proposition 7.4 there is a unique $W(k)$-algebra homomorphism $\phi : B \longrightarrow A$ such that $P = \phi \circ P^u$ and $\phi \bmod m_A$ is ψ, hence $\ker(\phi) \subset \ker(\psi)$. But $B/\ker(\phi) \subset A$ is necessarily finite local, and the continuity of P implies that $\ker(\phi)$ is open, and we are done by Lemma 7.58. □

Example 7.60. If we assume that $\bar{\rho}$ is absolutely irreducible and (say) defined over k, then F is canonically isomorphic with the usual deformation functor of $\bar{\rho}$ defined by Mazur in [M], by Theorem 7.43.

Remark 7.61. By construction, $A(\bar{\rho})$ is topologically generated by the $\Lambda_i(g)$ for $g \in G$ and $i \geq 1$.

Recall that for a profinite local $W(k)$-algebra B, say with maximal ideal m and residue field k, the following properties are equivalent:

(i) there is a continuous $W(k)$-algebra surjection $W(k)[[t_1, \cdots, t_h]] \longrightarrow B$,
(ii) B is noetherian,
(iii) $\dim_k(m/m^2) < \infty$.

As is well known, the *tangent space* $F(k[\epsilon])$ has a natural structure of k-vector space, isomorphic to the space of continuous k-linear forms on $m_{A(\bar{\rho})}/\overline{m^2_{A(\bar{\rho})}}$, where $\overline{m^2_{A(\bar{\rho})}}$ denotes the closure of $m^2_{A(\bar{\rho})}$ in $A(\bar{\rho})$. This leads us to consider the following equivalent hypotheses, that we will denote by $C(\bar{\rho})$:

(a) $\dim_k F(k[\epsilon]) < \infty$,
(b) $A(\bar{\rho})$ is topologically of finite type as $W(k)$-algebra.

As an immediate consequence of Corollary 7.57 (ii) and Example 2.5, $C(\bar{\rho})$ holds if G is topologically of finite type. Following Mazur, consider the following weaker condition:

(F) For any open subgroup $H \subset G$, there are only finitely many continuous group homomorphisms $H \longrightarrow \mathbb{Z}/p\mathbb{Z}$.

Example 7.62. (F) is satisfied if G is the absolute Galois group of a local field of characteristic 0 or if $G = \mathrm{Gal}(K_S/K)$ with K a number field, S a finite set of places of K and K_S a maximal algebraic extension of K which is unramified outside S (by class field theory and a result of Hermite). The condition (F) is not satisfied when $G = (\mathbb{Z}/p\mathbb{Z})^{\mathbb{N}}$. We leave as an exercise to the reader to check that for a given G, $H^1_c(G, \mathrm{ad}(\bar{\rho}))$ is finite dimensional for any continuous semisimple representation $G \longrightarrow \mathrm{GL}_m(\overline{\mathbb{F}}_p)$ (for any m) if and only if (F) holds.

Proposition 7.63. *Assume either that (F) is satisfied, or that $p > d$ and $H^1_c(G, \mathrm{ad}(\bar{\rho}))$ is finite dimensional. Then $C(\bar{\rho})$ holds.*

Proof. In the second case, it follows from Proposition 7.54. When G is topologically of finite type we already explained that $C(\bar{\rho})$ holds. When we only assume (F), we are reduced to this case by the following lemma. Indeed, if $F^* : \mathcal{C} \to \mathrm{Ens}$ is the determinantal deformations of $\bar{\rho}$ viewed as a representation of G/H, the lemma shows that the natural transformation $F^* \longrightarrow F$ is an equivalence. □

Lemma 7.64. *Let A be a commutative, profinite, local $W(k)$-algebra with residue field k and let* $\mathrm{D} : A[G] \longrightarrow A$ *be a continuous determinant deforming* $\det(\bar{\rho})$. *Then* D *factors through* $A[G] \to A[G/H]$ *where* $H \subset \ker(\bar{\rho})$ *is the smallest closed normal subgroup such that* $\ker(\bar{\rho})/H$ *is pro-p.*

Proof. We have to check that $\mathrm{D}(T - gh) = \mathrm{D}(T - g)$ for all $g \in G$ and $h \in H$. We may assume that A is a finite ring. By Lemma 7.52, we may assume that

D factors through a finite quotient G'. By Lemma 7.33 and Theorem 7.38, the radical of the finite ring

$$B := A[G']/\ker(\mathrm{D})$$

is the kernel of the natural extension of $\bar{\rho} : k[G] \to M_d(\bar{k})$. In particular, the image of the natural continuous group homomorphism $\ker(\bar{\rho}) \to B^*$ falls into the p-group $1 + \mathrm{Rad}(B)$, what we had to prove. □

Assume that $C(\bar{\rho})$ is satisfied, and consider the affine formal scheme

$$\mathcal{X}(\bar{\rho}) := \mathrm{Spf}(A(\bar{\rho}))$$

over $\mathrm{Spf}(W(k))$, as well as the rigid analytic space

$$X(\bar{\rho}) := \mathcal{X}(\bar{\rho})[1/p]$$

attached to $\mathcal{X}(\bar{\rho})$ by Berthelot. Our next aim is to describe the functors that those two spaces represent. More generally, we will identify them as component parts of the universal formal (resp. rigid analytic) determinant of dimension d.

3.2. Formal and rigid analytic determinants

3.2.1. The formal scheme of continuous determinants

We refer to [EGA, Ch. 0 §7, Ch. 1 §10] for the basics of topological rings and formal schemes.

Let us consider $\mathbb{Z}_p$ as a topological ring, equipped with the p-adic topology. We denote by $\mathcal{F}$ the category whose objects are the admissible topological rings A equipped with a continuous homomorphism $\mathbb{Z}_p \longrightarrow A$, and whose morphisms are continuous ring homomorphisms. Recall that the admissibility of A means that there is a topological isomorphism

$$A \overset{\sim}{\to} \lim_{\leftarrow} A_\lambda$$

where the limit is taken over a directed ordered set S with minimal element 0, A_λ is a discrete ring, and each $A_\lambda \to A_0$ is surjective with nilpotent kernel.

An object A is said *topologically of finite type over* $\mathbb{Z}_p$, if it is a quotient of the topological ring[28]

$$\mathbb{Z}_p[[t_1, \ldots, t_s]]\langle x_1, \ldots, x_r\rangle$$

[28] Recall that $A\langle t\rangle$ is the A-subalgebra of $A[[t]]$ of power series $\sum_n a_n t^n$ with $a_n \to 0$ (say A is admissible here).

(for some s and r) equipped with its I-adic topology defined by $I = (t_1, \ldots, t_s, p)$. Actually, we would not lose much in restricting to the full subcategory of such objects of $\mathcal{F}$ but it is unnecessary.

Lemma 7.65. *Let A be an object of $\mathcal{F}$ and let $D : A[G] \longrightarrow A$ be a continuous determinant. Denote by $B \subset A$ the closure of the $\mathbb{Z}_p$-algebra generated by the $\Lambda_i(g)$ for $g \in G$ and $i \geq 1$.*

(i) *B is an admissible profinite subring of A. In particular, it is finite product of local $\mathbb{Z}_p$-algebras.*
(ii) *Assume moreover that $\iota : A \longrightarrow A'$ is a continuous $\mathbb{Z}_p$-algebra homomorphism and let $D' : A'[G] \longrightarrow A'$ be the induced determinant and $B' \subset A'$ the ring associated as above. Then ι induces a continuous surjection $B \longrightarrow B'$.*

Proof. Assume first that A is discrete, in which case the assumption reads $p^n A = 0$ for some integer $n \geq 1$. Let $P : (\mathbb{Z}/p^n\mathbb{Z})[G] \longrightarrow A$ the continuous multiplicative polynomial law associated to D. By Lemma 7.52, P factors through $(\mathbb{Z}/p^n\mathbb{Z})[G/H]$ for some normal open subgroup $H \subset G$. But $\Gamma^d_{\mathbb{Z}/p^n\mathbb{Z}}((\mathbb{Z}/p^n\mathbb{Z})[G/H])$ is a finite ring as G/H is finite, hence so is the ring of the statement which is (by Amitsur's relations) the image of the natural ring homomorphism

$$\Gamma^d_{\mathbb{Z}/p^n\mathbb{Z}}((\mathbb{Z}/p^n\mathbb{Z})[G/H]) \longrightarrow A$$

attached to P (and D).

Consider now the general case. Write $A \overset{\sim}{\to} \varprojlim A_\lambda$ as above and denote by $\pi_\lambda : A \to A_\lambda$ the natural projection. Let $P : \mathbb{Z}_p[G] \longrightarrow A$ denote the continuous multiplicative polynomial law associated to D, and $P_\lambda = \pi_\lambda \circ P$. By the discrete case, the image B_λ of B in A_λ is a finite ring, hence

$$B \overset{\sim}{\to} \varprojlim B_\lambda$$

is a profinite admissible $\mathbb{Z}_p$-subalgebra. The last part of the first statement holds obviously for any profinite admissible ring B: the radical of B contains any ideal of definition of B by admissibility, hence $B/\mathrm{Rad}(B)$ is finite as B is profinite.

By Amitsur's relations, the ring B is the closure of the image of the natural map

$$\Gamma^d_{\mathbb{Z}_p}(\mathbb{Z}_p[G])^{\mathrm{ab}} \longrightarrow A$$

given by D, so the last assertion follows. □

Definition 7.66. We denote by $|G(d)| \subset \mathrm{Spec}(\Gamma^d_{\mathbb{Z}_p}(\mathbb{Z}_p[G])^{\mathrm{ab}})$ the subset of closed points z with finite residue field, that we shall denote by $k(z)$.

For each $z \in |G(d)|$, there is a canonical determinant

$$D_z : k(z)[G] \longrightarrow k(z).$$

By Theorem 7.38, and Example 7.53, $|G(d)|$ is in bijection with the set of continuous semisimple representations $G \longrightarrow \mathrm{GL}_d(\overline{\mathbb{F}}_p)$ taken up to isomorphism and Frobenius actions on coefficients.[29]

Definition 7.67. Let A, D and $B \subset A$ be as in the statement of Lemma 7.65. If B is local, we will say that D is *residually constant.*

If it is so, the radical of the kernel of the natural surjective ring homomorphism

$$\Gamma^d_{\mathbb{Z}_p}(\mathbb{Z}_p[G])^{\mathrm{ab}} \longrightarrow B_0$$

defines a point $z \in |G(d)|$ which is independent of the ideal of definition I of B chosen such that $B_0 = B/I$. The field $k(z)$ is canonically isomorphic to the residue field of B and the determinant $\overline{\mathrm{D}}$ obtained by reduction of D via $B \to k(z)$ coincides by definition with D_z: we will say that D *is residually equal to* D_z.

For instance, if $\mathrm{Spec}(A)$ is connected then any $D \in E(A)$ is residually constant. In general, if $D \in E(A)$ then there is a unique finite set $|D| \subset |G(d)|$, a unique decomposition $A = \prod_{i \in |D|} A_i$ in $\mathcal{F}$, and a unique collection of determinants $D^i : A_i[G] \longrightarrow A_i$ with D^i residually constant equal to D_i, such that $D = (D^i)_i : A[G] \longrightarrow A$.

Let us define a covariant functor

$$E : \mathcal{F} \longrightarrow \mathrm{Ens}$$

as follows. For an object A of $\mathcal{F}$, define $E(A)$ as the set of continuous determinants $A[G] \longrightarrow A$ of (fixed) dimension d, or which is the same, of continuous homogeneous multiplicative $\mathbb{Z}_p$-polynomial laws $\mathbb{Z}_p[G] \longrightarrow A$ of degree d. If $\iota : A \to A'$ is a morphism in $\mathcal{F}$ and $P \in E(A)$ is such a law, then $E(\iota)(P) := \iota \circ P \in E(A')$, which makes E a covariant functor. For each $z \in |G(d)|$, define $E_z(A) \subset E(A)$ as the subset of determinants which are residually constant and equal to D_z. As the formation of the ring B of Lemma 7.65 is functorial, E_z is a subfunctor of E.

[29] This means that we identify such a representation ρ (whose image actually falls into some $\mathrm{GL}_d(F)$ with F a finite subfield) exactly with the representations $Q(\mathrm{Frob}^m \circ \rho)Q^{-1}$ for any $Q \in \mathrm{GL}_d(\overline{\mathbb{F}}_p)$ and $m \geq 1$.

As a start, let us fix some $z \in |G(d)|$ and let $\bar{\rho}_z : G \longrightarrow \mathrm{GL}_d(\overline{k(z)})$ be "the" continuous semisimple representation such that $\det(1 + t\bar{\rho}_z(g)) = D_z(1 + gt)$ for all $g \in G$ (see Example 7.53).

Proposition 7.68. *Assume that $C(\bar{\rho}_z)$ holds. Then E_z is representable by an object $A(z)$ of $\mathcal{F}$. This object $A(z)$ is a local ring whose residue field is canonically isomorphic to $k(z)$, moreover it is topologically of finite type over $\mathbb{Z}_p$. Actually, the $W(k(z))$-algebra $A(z)$ is canonically topologically isomorphic to $A(\bar{\rho}_z)$ of Proposition 7.59.*

Proof. By Lemma 7.65, for any object A and any $P : \mathbb{Z}_p[G] \longrightarrow A$ in $E(A)$, P is the composite of a continuous multiplicative polynomial law $P' : \mathbb{Z}_p[G] \longrightarrow B$ with $B \to A$. If $P \in E_z(A)$, then B is a $W(k(z))$-algebra in a natural way and P' extends to a continuous multiplicative polynomial law $P'' : W(k(z))[G] \longrightarrow B$ which reduces to D_z, thus $P'' \in F'(B)$ where F' is the functor defined in Section 3.1. As a consequence, there is a unique continuous $W(k(z))$-algebra homomorphism $A(\bar{\rho}_z) \longrightarrow B$ corresponding to P''. As $C(\rho_z)$ holds, $A(\bar{\rho}_z)$ is an object of $\mathcal{F}$ which is moreover local and topologically of finite type over $\mathbb{Z}_p$. Unravelling the definitions we get the result. □

It is then essentially formal to deal with E rather than a given E_z. For that we need to extend E and the E_z to the category $\mathcal{FS}/\mathbb{Z}_p$ of all formal schemes over $\mathrm{Spf}(\mathbb{Z}_p)$.

For an object $\mathcal{X}$ of $\mathcal{FS}/\mathbb{Z}_p$, let $\widetilde{E}(\mathcal{X})$ be the set of continuous determinants $\mathcal{O}(\mathcal{X})[G] \longrightarrow \mathcal{O}(\mathcal{X})$ of dimension d, which makes

$$\widetilde{E} : \mathcal{FS}/\mathbb{Z}_p \longrightarrow \mathrm{Ens}$$

a contravariant functor in the obvious way. The restriction of $\widetilde{E}$ to the full subcategory of affine formal scheme coincides with E^{opp}. In the same way, define a subfunctor $\widetilde{E}_z \subset \widetilde{E}$ where $\widetilde{E}_z(\mathcal{X}) \subset \widetilde{E}(\mathcal{X})$ is the subset of elements D such that for any open affine $\mathcal{U} \subset \mathcal{X}$, the image of D in $\widetilde{E}(\mathcal{U}) = E(\mathcal{O}(\mathcal{U}))$ belongs to $E_z(\mathcal{O}(\mathcal{U}))$. If $\mathrm{Spf}(A) = \bigcup_i \mathcal{U}_i$ is an affine covering, note that an element $D \in E(A)$ belongs to $E_z(A)$ if, and only if, its image in each D_i belongs to $E_z(\mathcal{O}(\mathcal{U}_i))$, by Lemma 7.65. In particular, the restriction of $\widetilde{E}_z$ to the full subcategory of affine formal scheme coincides with E_z^{opp}.

Corollary 7.69. *Assume that condition (F) holds for G (see 3.1). Then $\widetilde{E}$ (resp. $\widetilde{E}_z$) is representable by the formal scheme $\coprod_{z \in |G(d)|} \mathrm{Spf}(A(z))$ (resp. by $\mathrm{Spf}(A(z))$).*

Proof. By definition, if $\mathcal{X}$ is a formal scheme then the topology on $\mathcal{O}(\mathcal{X})$ is the weakest topology such that the $\mathcal{O}(\mathcal{X}) \longrightarrow \mathcal{O}(\mathcal{U})$ are continuous for each open affine $\mathcal{U}$. From this we check at once as in Lemma 7.58 that $\widetilde{E}$ and $\widetilde{E}_z$ are sheaves for the Zariski topology on $\mathcal{FS}/\mathbb{Z}_p$. As $\widetilde{E}_z$ coincides with E_z^{opp} on $\mathcal{F}^{\text{opp}}$, we have (by Proposition7.68) a canonical isomorphism $\widetilde{E}_z \overset{\sim}{\to} \mathrm{Spf}(A(z))$. The assertion on $\widetilde{E}$ follows then from Lemma 7.65 (i). □

3.2.2. Rigid analytic determinants

Let Aff be the category of affinoid $\mathbb{Q}_p$-algebras ([BGR, Ch. 6]). We define again an obvious covariant functor

$$E^{\text{an}} : \text{Aff} \to \text{Ens}$$

as follows. If A is an affinoid algebra, $E^{\text{an}}(A)$ is the set of continuous determinants $A[G] \longrightarrow A$ of dimension d, and if $\varphi : A \longrightarrow B$ is a $\mathbb{Q}_p$-algebra homomorphism (necessarily continuous) and $P : \mathbb{Z}_p[G] \longrightarrow A$ is in $E^{\text{an}}(A)$, then we set $E^{\text{an}}(\varphi)(P) = \varphi \circ P$. Remark that by Proposition7.21, $E^{\text{an}}(A)$ also coincides with the set of continuous pseudocharacters $G \longrightarrow A$ of dimension d.

Recall that for any object $\mathcal{A}$ of $\mathcal{F}$ which is topologically of finite type over $\mathbb{Z}_p$, the algebra $A := \mathcal{A}[1/p]$ is an affinoid algebra and the map $\mathcal{A} \longrightarrow A$ is continuous and open. We say that $\mathcal{A}$ is a model of A. Any affinoid algebra admits at least one (and in general many) such model, as $\mathbb{Q}_p\langle t_1, \ldots, t_n\rangle$ does. If $\mathcal{A}$ is a model of A we have a natural map

$$\iota_{\mathcal{A}} : E(\mathcal{A}) \longrightarrow E^{\text{an}}(A),$$

which is moreover injective if $\mathcal{A}$ is torsion free over $\mathbb{Z}_p$. If $\mathcal{A}'$ is another model of A and if we have a continuous ring homomorphism $\mathcal{A} \longrightarrow \mathcal{A}'$, we get a natural map $E(\mathcal{A}) \longrightarrow E(\mathcal{A}')$, whose composite with $\iota'_{\mathcal{A}}$ is $\iota_{\mathcal{A}}$, so we get a natural injective map

$$\iota : \varinjlim E(\mathcal{A}) \longrightarrow E^{\text{an}}(A),$$

the colimit being over the (directed set of) models $\mathcal{A}$ of A.

If A is affinoid, we denote by $A^0 \subset A$ the subset of elements a with bounded powers (i.e. such that the sequence a^n, $n \geq 1$ is bounded in A). It is an open $\mathbb{Z}_p$-subalgebra, such that $A^0[1/p] = A$. When A is reduced, A^0 is a model of A (actually, the biggest torsion free model), but not in general (think about $A = \mathbb{Q}_p[\epsilon]/(\epsilon^2)$).

Lemma 7.70. *Let A be an affinoid algebra and $D \in E^{\text{an}}(A)$.*

(i) *For any $g \in G$ and any $i \geq 1$, $\Lambda_i(g) \in A^0$.*
(ii) *The map ι is bijective.*
(iii) *When A is reduced, then $E(A^0) = E^{\text{an}}(A)$.*

Proof. We first check (i). Recall that an element of an affinoid algebra A has bounded powers if and only if its image in all the residue fields A/m has norm ≤ 1: we may assume that A is a finite extension of $\mathbb{Q}_p$. Fix $g \in G$; up to replacing A by a finite extension, we may assume that $D(t-g)=\prod_{i=1}^{d}(t-x_i) \in A[t]$ splits in A, and we have to show that each x_i has norm 1. As $D(g) \in A^*$, each x_i is in A^*. By Newton's relations (or by Theorem 7.34), $D(t-g^n) = \prod_{i=1}^{d}(t-x_i^n)$ for each $n \in \mathbb{Z}$. By the continuity assumption, $D(t-g^n) = \prod_i(t-x_i^n)$ goes to $(t-1)^d$ (in A^d) when n tends to 0 in $\widehat{\mathbb{Z}}$, and it is a simple exercise to conclude.

We check (ii); it only remains to see the surjectivity of ι. Let $D \in E^{\mathrm{an}}(A)$ and $\mathcal{A} \subset A$ a model of A. Consider the compact subset $K = \cup_{i=1}^{d}\Lambda_i(G) \subset A$. As $\mathcal{A}$ is open in A, K meets only finitely many of the translates of $\mathcal{A}$. In particular, there exists a finite number of elements $k_1, \ldots, k_s \in K$ such that

$$K \subset \sum_{i=1}^{s}(k_i + \mathcal{A}).$$

By part (i), those k_i have bounded powers, thus

$$\mathcal{A}' = \mathcal{A}\langle k_1, \ldots, k_s\rangle \subset A$$

is a model of A containing K. By Amitsur's relations, we obtain that $D \in \mathrm{Im}(\iota_{\mathcal{A}'})$, hence (ii). Part (iii) is a consequence of (ii) and the fact that A^0 is the biggest model of A included in A. $\square$

For $z \in |G(d)|$ and an affinoid algebra A, let us define $E_z^{\mathrm{an}}(A)$ as the colimit of the $E_z(\mathcal{A})$ with $\mathcal{A}$ a model of A. Equivalently, a $D \in E^{\mathrm{an}}(A)$ belongs to $E_z^{\mathrm{an}}(A)$ if and only if $D = \iota_{\mathcal{A}}(D')$ for *some* model $\mathcal{A}$ and some $D' \in E_z(\mathcal{A})$. Obviously, this defines a subfunctor

$$E_z^{\mathrm{an}} : \mathrm{Aff} \to \mathrm{Ens}$$

of E^{an}. Let us first give a useful alternative description of this functor. Fix an affinoid A and consider $x \in \mathrm{Specmax}(A)$, L its residue field (a finite extension of $\mathbb{Q}_p$), $\mathcal{O}_L = L^0$ its ring of integers and k the residue field of $\mathcal{O}_L$. We have natural maps

$$E^{\mathrm{an}}(A) \longrightarrow E^{\mathrm{an}}(L) = E(\mathcal{O}_L) \longrightarrow E(k),$$

hence a natural *reduction map*

$$\mathrm{Red}_x : E^{\mathrm{an}}(A) \longrightarrow |G(d)|. \tag{7.21}$$

We check at once the following characterization of E_z^{an}:

Lemma 7.71. $E_z^{\mathrm{an}}(A) = \{D \in E^{\mathrm{an}}(A),\ \forall x \in \mathrm{Specmax}(A),\ \mathrm{Red}_x(D) = z\}$.

Proof. The inclusion $\subset$ is immediate as E_z is a functor. Conversely, let $D \in E^{\text{an}}(A)$ belong to the set on the right. By Lemma 7.70, it comes from an element $D' \in E(\mathcal{A})$ for some model $\mathcal{A} \subset A$. Consider the ring $B \subset \mathcal{A}$ associated to D' as in Lemma 7.65, and write it as a product of local rings $B = \prod_{i=1}^n B_i$. In particular, $A = \prod_{i=1}^n A_i$ itself is a product of affinoid algebras. If x_i is a closed point of Specmax(A_i), with residue field L_i, then the kernel of the natural continuous map $B_i \longrightarrow \mathcal{O}_{L_i}/m_{\mathcal{O}_{L_i}}$ corresponds to z by assumption on D'. As the natural map $\Gamma^d_{\mathbb{Z}_p}(\mathbb{Z}_p[G])^{\text{ab}} \to B/\text{Rad}(B)$ is surjective by construction, B is local and $D' \in E_z(\mathcal{A})$. □

Let us denote by An the category of rigid analytic spaces over $\mathbb{Q}_p$ ([BGR]). For any rigid space X, we endow the $\mathbb{Q}_p$-algebra $\mathcal{O}(X)$ with the weakest topology such that all the $\mathbb{Q}_p$-algebra homomorphisms $\mathcal{O}(X) \longrightarrow \mathcal{O}(U)$, with $U \subset X$ open affinoid, are continuous. Of course, such an $\mathcal{O}(U)$ is equipped here with its usual Banach topology; if X itself is affinoid then this weak topology on $\mathcal{O}(X)$ coincides with its Banach topology. For a general X, we check at once that $\mathcal{O}(X)$ is a complete topological $\mathbb{Q}_p$-algebra (it is even a Fréchet space if X is separable), and that the sheaf $\mathcal{O}_X$ becomes a sheaf of topological $\mathbb{Q}_p$-algebras.

Define a contravariant functor of continuous determinants

$$\widetilde{E}^{\text{an}} : \text{An} \longrightarrow \text{Ens}$$

as usual: for any rigid space X, let $\widetilde{E}^{\text{an}}(X)$ be the set of continuous determinants $\mathcal{O}(X)[G] \longrightarrow \mathcal{O}(X)$ of (fixed) dimension d. Of course, over the full subcategory of affinoids, $\widetilde{E}^{\text{an}}$ coincides by definition with the opposite of E^{an}.

For $z \in |G(d)|$, define $\widetilde{E}^{\text{an}}_z : \text{An} \longrightarrow \text{Ens}$ as the following subfunctor of $\widetilde{E}^{\text{an}}$: $\widetilde{E}^{\text{an}}_z(X)$ is the set of determinants such that for all closed points $x \in X$ (with residue field k_x) the induced determinant in $\widetilde{E}^{\text{an}}(\{x\}) = E^{\text{an}}(k_x) = E(\mathcal{O}_{k_x})$ is residually equal to z. By Lemma 7.71, $\widetilde{E}^{\text{an}}_z$ is the opposite functor of E^{an}_z over the full subcategory of affinoids.

Theorem 7.72. *Assume that condition (F) holds. The functor $\widetilde{E}^{\text{an}}$ (resp. $\widetilde{E}^{\text{an}}_z$) is representable by a rigid analytic space X (resp X_z). It is canonically isomorphic to the generic fiber of the formal scheme $\widetilde{E}$ (resp. $\widetilde{E}_z$).*

Moreover, X is the disjoint union of the X_z, $z \in |G(d)|$, and each X_z is isomorphic to a closed subspace of some h_z-dimensional open unit ball $\mathbb{B}^{h_z}_{[0,1[}$, $h_z \in \mathbb{N}$. In particular, X is a quasi-Stein space.

Proof. Recall that Berthelot [Ber, 0.2.6] constructed a functor $\mathcal{F}S'/\mathbb{Z}_p \longrightarrow$ An,

$$\mathcal{X} \mapsto \mathcal{X}^{\text{rig}},$$

extending Raynaud's one, where $\mathcal{FS}'/\mathbb{Z}_p$ is the full subcategory of $\mathcal{FS}/\mathbb{Z}_p$ whose objects are locally topologically of finite type (see also [DJ, Ch. 7]). The universal property of $\mathcal{X}^{\mathrm{rig}}$ is given by ([DJ, §7.1.7.1])

$$\lim_{\mathcal{Y} \text{ model of } Y} \mathrm{Hom}_{\mathcal{FS}'/\mathbb{Z}_p}(\mathcal{Y}, \mathcal{X}) = \mathrm{Hom}_{\mathrm{An}}(Y, \mathcal{X}^{\mathrm{rig}}), \tag{7.22}$$

where Y is any affinoid. In the case we are interested in of a $\mathrm{Spf}(A)$ with

$$A \xrightarrow{\sim} W(k)[[t_1, \dots, t_h]]/I,$$

then $\mathrm{Spf}(A)^{\mathrm{rig}}$ is isomorphic to the closed subspace of the open unit ball of dimension h

$$\mathrm{Spf}(A)^{\mathrm{rig}} \subset \mathbb{B}^h_{[0,1[}$$

defined by $I = 0$. In particular, by Corollary 7.69, it is enough to prove the theorem to show that $\widetilde{E}^{\mathrm{an}}$ (resp. $\widetilde{E}^{\mathrm{an}}_z$) represents the generic fiber of $\widetilde{E}$ (resp. $\widetilde{E}_z$). As those functors are sheaves for the rigid-analytic Grothendieck topology on An, it is enough to check the universal property over affinoids, in other words (7.22). But that follows from Lemma 7.70 (ii). □

Remark 7.73. Of course, if we are only interested in $\widetilde{E}^{\mathrm{an}}_z$ for some z, and if $C(\bar{\rho}_z)$ holds, then the same argument and Proposition 7.68 shows that $\widetilde{E}^{\mathrm{an}}_z$ is representable by $\mathrm{Spf}(A(z))^{\mathrm{rig}}$.

4. Complements

We keep the notation of § 3. Let us assume that condition (F) holds for G and denote by $\mathcal{X}$ the formal scheme $\widetilde{E} = \coprod_{z \in |G(d)|} \mathrm{Spf}(A(z))$ and $X = \mathcal{X}^{\mathrm{rig}} = \widetilde{E}^{\mathrm{an}}$. We shall also denote by $\mathcal{D}$ and D the respective universal determinants of G over $\mathcal{X}$ and X.

Alternatively, we might fix some $z \in |G(d)|$ and assume only that $C(\bar{\rho}_z)$ holds, in which case all that we say below would also apply to the restricted spaces $\mathcal{X} = \widetilde{E}_z$ and $X = \mathcal{X}^{\mathrm{rig}} = \widetilde{E}^{\mathrm{an}}_z$.

4.1. Completion at a point

Let us fix some (closed) point $x \in X$, with residue field k_x (a finite extension of $\mathbb{Q}_p$), and associated continuous determinant $D(x) : k_x[G] \longrightarrow k_x$. As X represents a functor, we get a natural interpretation for the completed local ring $\widehat{\mathcal{O}}_x$, viewed as a k_x-algebra, as pro-representing the functor $F(x)$ of continuous deformations of $D(x)$ to the category local artinian k_x-algebras with residue field k_x.

This applies in particular when $D(x) = \det \circ \rho(x)$ is absolutely irreducible and split, in which case this functor $F(x)$ is canonically isomorphic to the usual deformation functor of $\rho(x)$ in the sense of Mazur by 7.43 (i).[30]

4.2. The absolutely irreducible locus

For the same reason as in Example 7.42, the locus

$$X^{\mathrm{irr}} \subset X$$

whose points x parameterize the absolutely irreducible $D(x)$ is an admissible (Zariski) open subset. In particular, the subfunctor $\widetilde{E}^{\mathrm{an,irr}} \subset \widetilde{E}^{\mathrm{an}}$, parameterizing determinants $D \in \widetilde{E}^{\mathrm{an}}(Y)$ whose evaluation at each closed point of Y is absolutely irreducible, is representable by the rigid analytic space X^{irr}.

The universal Cayley–Hamilton algebra on X is the sheaf

$$U \mapsto R(U) = \mathcal{O}(U)[G]/\mathrm{CH}(T(U)),$$

where $T(U) : G \to \mathcal{O}(U)$ is the tautological pseudocharacter on the open affinoid U. It defines a sheaf on X as the formation of the biggest Cayley–Hamilton quotient commutes with any base change.

Let us now prove Proposition G of the introduction.[31] We have to show that E^{irr} is represented by X^{irr} and $\rho : G \to R^*_{|X^{\mathrm{irr}}}$. First, the reduced trace of Azumaya algebras induces a natural transformation $E^{\mathrm{irr}} \to E^{\mathrm{an}}$ which factors by definition through $X^{\mathrm{irr}} \subset X$. To show that $E^{\mathrm{irr}} \to X^{\mathrm{irr}}$ is an isomorphism it is enough to show that for any affinoid $\mathbb{Q}_p$-algebra A, and any $T \in E^{\mathrm{an}}(A)$ such that all the evaluations T_x, for all closed points x, are absolutely irreducible, there is a unique isomorphism class of continuous representations $\rho : G \to B^*$ where B is an Azumaya A-algebra of rank d^2, namely: the canonical map $\rho^u : G \to (A[G]/\mathrm{CH}(T))^*$. But this follows from Theorem 7.43 as in Corollary 7.44 (i) (ρ^u is continuous as T is and the reduced trace of an Azumaya algebra is nondegenerate).

30 Of course, when an absolutely irreducible $D(x)$ is not split, but splits over L/k_x, we get such an interpretation for the L-algebra $\widehat{\mathcal{O}}_x \otimes_{k_x} L$.

31 Recall that if L is a field and if R is an Azumaya algebra over L of rank d^2, that is a central simple L-algebra of dimension d^2, and if $\rho : G \to R^*$ is a group homomorphism, then ρ is said to be absolutely irreducible if

$$\rho \otimes_L \overline{L} : G \to (R \otimes_L \overline{L})^* \simeq \mathrm{GL}_d(\overline{L})$$

is irreducible. If $\det : R \to L$ is the reduced norm of R, then $\det(\rho)$ is absolutely irreducible if, and only if, ρ is.

5. An application to Galois deformations

Let G be the Galois group of a maximal algebraic extension of $\mathbb{Q}$ unramified outside $\{2, \infty\}$ and consider

$$\bar{\rho} : G \longrightarrow \mathrm{GL}_2(\mathbb{F}_2)$$

the *trivial* representation. Our main aim here is to study the generic fiber $X(\bar{\rho})$ of the universal deformation of $\det(\bar{\rho}) : \mathbb{F}_2[G] \to \mathbb{F}_2$ as in §3.1, and more precisely its *odd* locus, i.e. the closed and open subspace

$$X(\bar{\rho})^{\mathrm{odd}} \subset X(\bar{\rho})$$

where a complex conjugation $c \in G$ has determinant -1. By class field theory, the (separated) abelianization G^{ab} of G is isomorphic to $\mathbb{Z}_2^*$, thus condition $C(\bar{\rho})$ is satisfied and $X(\bar{\rho})$ makes sense.

Theorem 7.74. *$X(\bar{\rho})^{\mathrm{odd}}$ is the open unit ball of dimension 3 over $\mathbb{Q}_2$.*

Remark 7.75. By a well-known result of Tate [Ta], $\bar{\rho}$ is the unique continuous semisimple representation $G \longrightarrow \mathrm{GL}_2(\overline{\mathbb{F}}_2)$, so $|G(2)| = \{\bar{\rho}\}$ and $X(\bar{\rho})$ is actually the universal 2-dimensional 2-adic analytic pseudocharacter of G.

In order to prove Theorem 7.74, we shall consider the subfunctor

$$F^{\mathrm{odd}} \subset F$$

of the deformation functor F of $\det(\bar{\rho})$ which is defined by the condition that the characteristic polynomial of c be $T^2 - 1$ (here and below we shall use the notation of §3.1). This subfunctor F^{odd} is pro-representable by the quotient of

$$A(\bar{\rho})^{\mathrm{odd}} = A(\bar{\rho})/(f, g-1),$$

where $\mathrm{D}(\bar{\rho})(T - c) = T^2 - fT + g \in A(\bar{\rho})[T]$. Moreover, we check at once (following the proofs of Lemmas 7.70 (ii) and 7.65) that the generic fiber of F^{odd} is $X(\bar{\rho})^{\mathrm{odd}}$. As a consequence, it is enough to show that

$$A(\bar{\rho})^{\mathrm{odd}} \simeq \mathbb{Z}_2[[x, y, z]]. \tag{7.23}$$

We start with a tangent space computation that we explain in its natural generality. In the following lemma, G is any profinite group, A is a discrete commutative ring such that $2A = 0$, and $\mathrm{D}_0 : A[G] \longrightarrow A$ is *the trivial determinant of dimension* 2, so

$$D_0(T - g) = (T-1)^2 \in A[T], \quad \forall g \in G.$$

We denote by G^2 the closed subgroup of G generated by the squares of the elements of G. This is a normal subgroup containing[32] the commutators of G, so G/G^2 is a profinite $\mathbb{F}_2$-vector space. We shall be interested in the A-module $\mathcal{T}$ of continuous deformations of D_0 to $A[\epsilon]$ (see §2.4). By Lemma 7.7, any $\mathrm{D} \in \mathcal{T}$ may be written uniquely as a pair

$$(2 + \epsilon\tau, 1 + \epsilon\delta)$$

for some maps $\tau, \delta : G \longrightarrow A$.

Lemma 7.76. *The map $\mathrm{D} \in \mathcal{T} \mapsto (\tau, \delta)$ is an A-linear isomorphism onto the A-module of pairs of continuous maps $(t, d) : G/G^2 \longrightarrow A$ where $t(1) = 0$ and d is a group homomorphism.*

Proof. Let $\mathrm{D} = (2 + \epsilon\tau, 1 + \epsilon\delta) \in \mathcal{T}$ be an $A[\epsilon]$-valued determinant. As $2A = 0$, condition (b) in Lemma 7.7 is reduced to

$$\tau(g^{-1}h) = \tau(gh), \ \forall g, h \in G,$$

or which is the same $\tau(h) = \tau(g^2h)$ for all $g, h \in G$. The lemma follows then from Lemma 7.7. □

Let us go back now to the case of the Galois group G, for which $G/G^2 \simeq \mathbb{F}_2 \times \mathbb{F}_2$. By Lemma 7.76, and taking into account the *odd* condition, the tangent space $F^{\mathrm{odd}}(\mathbb{F}_2[\epsilon])$ is isomorphic to the $\mathbb{F}_2$-vector space of pairs of maps (τ, δ) with $\tau(1) = \tau(c) = 0$ and $\delta : G/G^2 \to \mathbb{F}_2$ a group homomorphism with $\delta(c) = 0$, so

$$\dim_{\mathbb{F}_2}(F^{\mathrm{odd}}(\mathbb{F}_2[\epsilon])) = 3.$$

In particular, if m is the maximal ideal of $A(\bar{\rho})^{\mathrm{odd}}$, then

$$\dim_{\mathbb{F}_2}(m/m^2) \leq 4,$$

and it only remains to show that the Krull dimension of $A(\bar{\rho})^{\mathrm{odd}}$ is at least 4, or better that the Krull dimension of $A(\bar{\rho})^{\mathrm{odd}}[\frac{1}{2}]$ is at least 3. For that it is enough to show that for some (closed) point $x \in X(\bar{\rho})^{\mathrm{odd}}$, the completed local ring $\widehat{\mathcal{O}}_{X,x}$ has Krull dimension at least 3. Indeed, the Krull dimension of a local noetherian ring does not change after completion, and $\widehat{\mathcal{O}}_{X,x}$ is (canonically) isomorphic to the completion of $A(\bar{\rho})^{\mathrm{odd}}[\frac{1}{2}]$ at its maximal ideal defined by x (see [DJ, Lemma 7.1.9]).

Consider for instance the point x parameterizing the Galois representation

$$\rho_\Delta : G \longrightarrow \mathrm{GL}_2(\mathbb{Q}_2)$$

32 Indeed, $xyx^{-1}y^{-1} = (xy)^2(y^{-1}x^{-1}y)^2y^{-2}$.

attached by Deligne to Ramanujan's Δ modular form. This representation is irreducible, odd, with trivial residual associated determinant (actually, any such representation would allow us to conclude below). By §4.1, $\widehat{\mathcal{O}}_{X,x}$ is the universal deformation ring (in the sense of Mazur) of ρ_Δ. But it is a well-known observation of Mazur [M] that the Krull dimension of such a deformation ring is at least

$$\dim_{\mathbb{Q}_2} H^1(G, \mathrm{ad}(\rho_\Delta)) - \dim_{\mathbb{Q}_2} H^2(G, \mathrm{ad}(\rho_\Delta)) = 3$$

by the global Euler characteristic formula of Tate (and as ρ_Δ is odd). This concludes the proof of (7.23), and of Theorem 7.74.

Remark 7.77. (i) If $g \in G$ is any element such that g and -1 generate topologically $G^{\mathrm{ab}} \xrightarrow{\sim} \mathbb{Z}_2^*$, we actually showed that $A(\bar{\rho})^{\mathrm{odd}} \doteq \mathbb{Z}_2[[\mathrm{Tr}(g) - 2, \mathrm{Tr}(cg) - 2, \mathrm{D}(g) - 1]]$, where Tr and D denote the universal trace and determinant.

(ii) A maybe more elementary method to show the smoothness of F^{odd} would have been to study abstractly the relations occuring in the process of lifting determinants.

To end the proof of Theorem H of the introduction, we still have to study the other (less interesting) components $X(\bar{\rho})^{\pm}$ over which the universal trace of c is ± 2. As there are continuous characters $\chi : G \longrightarrow \{\pm 1\}$ such that $\chi(c) = -1$, $X(\bar{\rho})^+$ and $X(\bar{\rho})^-$ are isomorphic, so we focus on $X(\bar{\rho})^+$. We claim that over $X(\bar{\rho})^+$, the universal pseudocharacter Tr factors through $G^{\mathrm{ab}}/\langle c\rangle = \mathbb{Z}_2^*/\{\pm 1\}$. It is enough to show that:

(a) Over the whole of $X(\bar{\rho})$, Tr factors through the maximal pro-2 quotient P of G.

(b) Over $X(\bar{\rho})^+$, Tr factors through G/H where H is the closed normal subgroup of G generated by c.

Indeed, assuming (a) and (b), Tr factors through the quotient of P by the image H' of H in P. But $(P/H')^{\mathrm{ab}} = \mathbb{Z}_2^*/\{\pm 1\}$ is monogenic, so $P/H' = \mathbb{Z}_2^*/\{\pm 1\}$ by Frattini's argument.

Part (b) above follows from the fact that

$$e := (1 - c)/2$$

is an idempotent of $\mathbb{Q}_2[G]$ such that $\mathrm{Tr}(e) = 0$, so $e \in \ker(\mathrm{Tr})$ by [BC, Lemme 1.2.5 (5)], thus $\mathrm{Tr}(cg) = \mathrm{Tr}(g)$ for all $g \in G$.

Part (a) is a consequence of Lemma 7.64 (recall that in this lemma, G is any profinite group, k a finite field of characteristic p, and $\bar{\rho} : G \to \mathrm{GL}_d(\overline{k})$ is any continuous semisimple representation such that $\det(T - \bar{\rho}(g)) \in k[t]$ for all $g \in G$).

As a consequence, we may replace G by its quotient $G' = \mathbb{Z}_2^*/\{\pm 1\} \simeq \mathbb{Z}_2$ to study $X(\bar{\rho})^+$, which is now a trivial exercise. Consider the (pro-representable) subfunctor

$$F^+ =: \mathrm{Spf}(A(\bar{\rho})^+) \subset F$$

of deformations of $\det(\bar{\rho})$ *as determinants on* G'. Its generic fiber is $X(\bar{\rho})^+$, and we claim that $F^+ \simeq \mathrm{Spf}(\mathbb{Z}_2[[x, y]])$. Indeed, as $G'/G'^2 \simeq \mathbb{F}_2$, Lemma 7.76 shows that

$$\dim_{\mathbb{F}_2} F^+(\mathbb{F}_2[\epsilon]) = 2.$$

It remains to show that the Krull dimension of $A(\bar{\rho})^+$ is at least three. Consider two copies $\chi_i : \mathbb{Z}_2 \longrightarrow \mathbb{Z}_2[[T_i]]^*$, $i = 1, 2$, of the universal 2-adic character of $\mathbb{Z}_2$ (1 being sent to $1 + T_i$), and set

$$D := \chi_1 \chi_2, \quad \mathbb{Z}_2[G] \longrightarrow \mathbb{Z}_2[[T_1, T_2]].$$

This 2-dimensional determinant takes its values in the subring[33] $\mathbb{Z}_2[[x, y]]$ where $x = T_1 + T_2$ and $y = T_1 T_2$. The induced map

$$A(\bar{\rho})^+ \longrightarrow \mathbb{Z}_2[[x, y]]$$

is clearly surjective, hence an isomorphism, which concludes the proof of Theorem H.

Remark 7.78. We showed that the universal pseudocharacter on $X(\bar{\rho})^{\pm}$ is everywhere absolutely reducible: precisely, it becomes a sum of two characters over a covering of $X(\bar{\rho})^{\pm}$ of degree 2 by the 2-dimensional open unit ball. In the same vein, it is easy to determine the reducible locus of $X(\bar{\rho})^{\mathrm{odd}}$: in terms of the coordinates $x = \mathrm{Tr}(g) - 2$, $y = \mathrm{Tr}(cg) - 2$ and $z = \det(g) - 1$ (see Remark 7.77); it is given by the relation

$$x^2 - y^2 = 4(1 - x + y + z).$$

Moreover, we could show that over $X(\bar{\rho})^{\mathrm{odd}}$ there exists a continuous representation $G \to \mathrm{GL}_2(\mathcal{O})$ whose trace is the universal pseudocharacter Tr.

References

[A] S. A. Amitsur, On the characteristic polynomial of a sum of matrices, *J. Linear and Multilinear Algebra* **8**, 177–182 (1980).

[33] Note that this ring coincides with $\mathbb{Z}_2[[T_1, T_2]]^{\mathfrak{S}_2}$.

[B] J. Bellaïche, Pseudodeformations, *Math. Z.* **270**, 1163–1180 (2012).

[BC] J. Bellaïche & G. Chenevier, Families of Galois representations and Selmer groups, *Astérisque* **324**, Soc. Math. France (2009).

[Ber] P. Berthelot, *Cohomologie rigide et cohomologie rigide à supports propres*, Première partie (version provisoire 1991), Prépublication de l'Inst. Rech. Math. Rennes (1996).

[BGR] S. Bosch, U. Güntzer & R. Remmert, *Non-Archimedean Analysis*, Springer Verlag, Grundlehren der math. wissenschaften 261 (1983).

[Bki] N. Bourbaki, *Eléments de mathématiques*, Algèbre Ch. III, Actualités Scientifiques et Industrielles, Hermann, Paris (1961).

[DCPRR] C. De Concini, C. Procesi, N. Reshetikhin & M. Rosso, Hopf algebras with trace and representations, *Invent. Math.* **161**, 1–44 (2005).

[DJ] A.J. De Jong, Crystalline and module Dieudonné theory via formal and rigid geometry, *Publ. Math. I.H.É.S.* **82**, 5–96 (1995).

[D] S. Donkin, Invariants of several matrices, *Inv. Math.* **110**, 389–410 (1992).

[EGA] A. Grothendieck, Éléments de géométrie algébrique, I. Le language des schémas, *Publ. Math. I.H.É.S.* **4** (1960).

[Fe] D. Ferrand, Un foncteur norme, *Bull. Soc. Math. France* **126**, 1–49 (1998).

[Fr] F. G. Frobenius, *Über die Primfactoren der Gruppendeterminante*, Ges. Abh. III (1968) (S'ber. Akad. Wiss. Berlin 1343–1382).

[H] I. N. Herstein, *Noncommuntative rings*, The Carus Mathematical Monographs 15, Math. Assoc. of America (1968).

[Hi] G. Higman, On a conjecture of Nagata, *Proc. Camb. Phil. Soc.* **52**, 1–4 (1956).

[Lo] M. Lothaire, *Combinatorics on words*, Encycl. of Math. and its applications, Addison Wesley (1983).

[M] B. Mazur, Deforming Galois representations, in *Galois groups over* $\mathbb{Q}$, Y. Ihara, K. Ribet, J.-P. Serre, eds., MSRI Publ. 16, 385–437 (1987).

[N] L. Nyssen, Pseudo-représentations, *Math. Annalen* **306**, 257–283 (1996).

[P1] C. Procesi, Finite dimensional representations of algebras, *Israel J. Math.* **19**, 169–182 (1974).

[P2] C. Procesi, Invariant theory of $n \times n$ matrices, *Adv. Math.* **19**, 306–381 (1976).

[P3] C. Procesi, A formal inverse to the Cayley Hamilton theorem, *J. Algebra* **107**, 63–74 (1987).

[P4] C. Procesi, Deformations of representations, *Methods in ring theory* (Levico Terme, 1997), Lecture Notes in Pure and Appl. Math. 198, Dekker, New York, p. 247–276 (1998).

[RS] C. Reutenauer & M.-P. Schützenberger, A formula for the determinant of a sum of matrices, *Lett. Math. Phys.* **13**, 299–302 (1987).

[Ro1] N. Roby, Lois polynômes et lois formelles en théorie des modules, *Ann. Sc. É.N.S.* **80**, 213–348 (1963).

[Ro2] N. Roby, Lois polynômes multiplicatives universelles, *C. R. Acad. Ac. Paris* **290**, 869–871 (1980).

[Rou] R. Rouquier, Caractérisation des caractères et pseudo-caractères, *J. Algebra* **180**, 571–586 (1996).

[Ru] K. Rubin, *Euler systems*, Princeton University Press, Annals of math. studies 147 (2000).

[Sen] S. Sen, An infinite dimensional Hodge-Tate theory, *Bull. Soc. Math. France* **121**, 13–34 (1993).

[S] C. S. Seshadri, Geometric reductivity over arbitrary bases, *Adv. Math.* **26**, 225–274 (1977).

[Ta] J. Tate, The non-existence of certain Galois extensions of $\mathbb{Q}$ unramified outside 2, in *Arithmetic geometry* (Tempe, AZ, 1993), Contemp. Math. 174, p. 153–156.

[T] R. Taylor, Galois representations associated to Siegel modular forms of low weight, *Duke Math. J.* **63**, 281–332 (1991).

[V1] F. Vaccarino, Generalized symmetric functions and invariants of matrices, *Math. Z.* **260**, 509–526 (2008).

[V2] F. Vaccarino, Homogeneous multiplicative polynomial laws are determinant, *J. Pure Appl. Algebra* **213**, 1283–1289 (2009).

[V3] F. Vaccarino, On the invariants of matrices and the embedding problem, preprint (2004), arXiv:math/0406203v1.

[W] A. Wiles, On ordinary λ-adic representations associated to modular forms, *Invent. Math.* **94**, 529–573 (1988).

[Z1] D. Ziplies, A characterization of the norm of an Azumaya algebra of constant rank through the divided powers algebra of an algebra, *Beiträge Algebra Geom.* **22**, 53–70 (1986).

[Z2] D. Ziplies, Generators for the divided powers algebra of an algebra and trace identities, *Beiträge Algebra Geom.* **24**, p. 9–27 (1987).

8

La série principale unitaire de $\mathbf{GL}_2(\mathbf{Q}_p)$: vecteurs localement analytiques

Pierre Colmez

Abstract

Résumé. Nous déterminons les vecteurs localement analytiques des représentations de la série principale unitaire de $\mathbf{GL}_2(\mathbf{Q}_p)$

Abstract We identify the locally analytic vectors of representations from the unitary principal series of $\mathbf{GL}_2(\mathbf{Q}_p)$

Contents

Introduction *page* 288
- 0.1 Notations 288
- 0.2 La correspondance de Langlands locale p-adique pour $\mathbf{GL}_2(\mathbf{Q}_p)$ 289
- 0.3 (φ, Γ)-modules triangulables de rang 2 291
- 0.4 La série principale localement analytique 293
- 0.5 Céoukonfaikoi 295
- 0.6 Remerciements 296

1 Cohomologie de $\mathscr{R}(\delta)$ 297
- 1.1 L'anneau de Robba $\mathscr{R}$ 297
- 1.2 Dictionnaire d'analyse fonctionnelle p-adique 298
  - 1.2.1 Quelques faisceaux P^+-équivariants sur $\mathbf{Z}_p$ 298
  - 1.2.2 Dualité 301
  - 1.2.3 Le théorème d'Amice 302
- 1.3 Extensions de (φ, Γ)-modules et cohomologie de A^+ 303
- 1.4 Cohomologie de Φ^+ et Φ^- 304

Automorphic Forms and Galois Representations, ed. Fred Diamond, Payman L. Kassaei and Minhyong Kim. Published by Cambridge University Press.

1.4.1 Les groupes $H^i(\Phi^\pm, \mathrm{LA}(\mathbf{Z}_p) \otimes \delta)$ et $H^i(\Phi^\pm, \mathscr{D}(\mathbf{Z}_p) \otimes \delta^{-1})$ 304
1.4.2 Un peu de topologie 305
1.4.3 Actions de φ et ψ sur $\mathrm{LA}(\mathbf{Z}_p)$ 306
1.4.4 Actions de φ et ψ sur $\mathscr{D}(\mathbf{Z}_p)$ 308
1.5 Cohomologie de A^0 309
1.5.1 Sur $\mathscr{R}(\delta)$ 309
1.5.2 Sur $\mathscr{R}(\delta) \boxtimes \mathbf{Z}_p^*$. 310
1.5.3 Cohomologie de A^1 311
1.6 Cohomologie de A^+ 312
1.6.1 A valeurs dans $\mathrm{LA}(\mathbf{Z}_p) \otimes \delta$ et $\mathscr{D}(\mathbf{Z}_p) \otimes \delta^{-1}$ 312
1.6.2 A valeurs dans $\mathscr{R}(\delta)$ 314
2 La série principale localement analytique 316
2.1 Construction de représentations localement analytiques 317
2.2 Composantes de Jordan-Hölder 318
2.3 Extensions de $W(\delta_1, \delta_2)$ par $\mathrm{St}^{\mathrm{an}}(\delta_1, \delta_2)$ 319
3 Le module $\Delta \boxtimes \{0\}$ 324
3.1 (φ, Γ)-module sur $L\{\{t\}\}$ 324
3.1.1 L'anneau $L\{\{t\}\}$ 324
3.1.2 φ-modules 325
3.1.3 (φ, Γ)-modules 327
3.2 L'action de φ sur $\mathrm{Fr}(\mathscr{R})$ 328
3.2.1 L'action de φ sur $\mathscr{R}$ 328
3.2.2 Le corps $\mathrm{Fr}(\mathscr{R})$ 328
3.2.3 L'action de φ sur $\mathrm{Fr}(\mathscr{R})$ 330
3.3 Le $L\{\{t\}\}$-module $\Delta \boxtimes \{0\}$ 331
3.3.1 Structure de $\Delta \boxtimes \{0\}$ 331
3.3.2 Le module $\Delta \boxtimes \{0\}$ dans le cas de rang 2 333
4 Les vecteurs localement analytiques de la série principale unitaire 335
4.1 La représentation localement analytique $\mathbf{\Pi}(\Delta)$ 335
4.1.1 Le faisceau $U \mapsto \Delta \boxtimes U$ 335
4.1.2 Dualité 337
4.2 Le foncteur de Jacquet dual 338
4.3 Dévissage de $\Delta(s) \boxtimes \mathbf{P}^1$ 340
4.4 Dévissage de $\mathbf{\Pi}(\Delta(s))$ 343
4.5 Vecteurs localement algébriques 345
4.6 Cas particuliers 346
4.6.1 Le cas cristallin 346
4.6.2 Le cas Hodge-Tate non de Rham 347

4.7 La transformée de Stieljes 348
4.7.1 Le cas générique 348
4.7.2 Le cas spécial 350
A Cohomologie des semi-groupes 351
A.1 Dévissage de la cohomologie de A^+ 352
A.2 Dualité 353
Références 356

Introduction

0.1. Notations

Soit p un nombre premier. On fixe une clôture algébrique $\overline{\mathbf{Q}}_p$ de $\mathbf{Q}_p$, et on note $\mathscr{G}_{\mathbf{Q}_p} = \mathrm{Gal}(\overline{\mathbf{Q}}_p/\mathbf{Q}_p)$ le groupe de Galois absolu de $\mathbf{Q}_p$ et $W_{\mathbf{Q}_p} \subset \mathscr{G}_{\mathbf{Q}_p}$ son groupe de Weil (qui est dense dans $\mathscr{G}_{\mathbf{Q}_p}$). Si $g \in W_{\mathbf{Q}_p}$, on note $\deg(g) \in \mathbf{Z}$ l'entier défini par $g(x) = x^{p^{\deg(g)}}$ si $x \in \overline{\mathbf{F}}_p$. Soit $\chi : \mathscr{G}_{\mathbf{Q}_p} \to \mathbf{Z}_p^*$ le caractère cyclotomique. Si $F_\infty = \mathbf{Q}_p(\boldsymbol{\mu}_{p^\infty})$, alors $\mathscr{H}_{\mathbf{Q}_p} = \ker\chi = \mathrm{Gal}(\overline{\mathbf{Q}}_p/F_\infty)$, ce qui permet de voir χ aussi comme un isomorphisme de $\Gamma = \mathscr{G}_{\mathbf{Q}_p}/H_{\mathbf{Q}_p} = \mathrm{Gal}(F_\infty/\mathbf{Q}_p)$ sur $\mathbf{Z}_p^*$; on note $a \mapsto \sigma_a$ son inverse.

On fixe une extension finie L de $\mathbf{Q}_p$, et on note $\mathscr{O}_L$ l'anneau de ses entiers. Soit $\widehat{\mathscr{T}}(L)$ l'ensemble des caractères continus $\delta : \mathbf{Q}_p^* \to L^*$. La notation est justifiée par le fait que $\widehat{\mathscr{T}}(L)$ est l'ensemble des points L-rationnels d'une variété analytique $\widehat{\mathscr{T}}$. On note juste $x \in \widehat{\mathscr{T}}(L)$ le caractère induit par l'inclusion de $\mathbf{Q}_p$ dans L, et $|x|$ le caractère envoyant $x \in \mathbf{Q}_p^*$ sur $p^{-v_p(x)}$. Si $\delta \in \widehat{\mathscr{T}}(L)$, on note $w(\delta)$ son *poids*, défini par $w(\delta) = \frac{\log\delta(u)}{\log u}$, où $u \in \mathbf{Z}_p^*$ n'est pas une racine de l'unité.

L'abélianisé $W_{\mathbf{Q}_p}^{\mathrm{ab}}$ de $W_{\mathbf{Q}_p}$ est isomorphe à $\mathbf{Q}_p^*$ d'après la théorie locale du corps de classes, ce qui permet de voir un élément de $\widehat{\mathscr{T}}(L)$ aussi comme un caractère continu de $W_{\mathbf{Q}_p}$. De manière explicite, si $g \in W_{\mathbf{Q}_p}$ et $\delta \in \widehat{\mathscr{T}}(L)$, alors $\delta(g)$ est défini par la formule

$$\delta(g) = \delta(p)^{-\deg(g)}\delta(\chi(g)).$$

Si δ est *unitaire* (i.e. si δ est à valeurs dans $\mathscr{O}_L^*$), alors δ se prolonge par continuité à $\mathscr{G}_{\mathbf{Q}_p}$, ce qui permet aussi de voir δ comme un caractère de $\mathscr{G}_{\mathbf{Q}_p}$, et $w(\delta)$ est alors le poids de Hodge-Tate généralisé de δ. Par exemple $x|x|$, qui est unitaire, correspond au caractère cyclotomique χ ; son poids est 1 et il sera noté χ dans l'article.

On note :

- G le groupe $\mathbf{GL}_2(\mathbf{Q}_p)$,
- $B = \left(\begin{smallmatrix} \mathbf{Q}_p^* & \mathbf{Q}_p \\ 0 & \mathbf{Q}_p^* \end{smallmatrix}\right)$ le *borel* supérieur,

- $Z = \{\left(\begin{smallmatrix} a & 0 \\ 0 & a \end{smallmatrix}\right),\ a \in \mathbf{Q}_p^*\}$ le centre de G,
- $P = \left(\begin{smallmatrix} \mathbf{Q}_p^* & \mathbf{Q}_p \\ 0 & 1 \end{smallmatrix}\right)$ le *mirabolique*,
- $N = \left(\begin{smallmatrix} 1 & \mathbf{Q}_p \\ 0 & 1 \end{smallmatrix}\right)$ l'*unipotent* supérieur,
- $T = \left(\begin{smallmatrix} \mathbf{Q}_p^* & 0 \\ 0 & \mathbf{Q}_p^* \end{smallmatrix}\right)$ le *tore maximal déployé*, et $A = \left(\begin{smallmatrix} \mathbf{Q}_p^* & 0 \\ 0 & 1 \end{smallmatrix}\right)$ son sous-groupe,
- $P^+ \subset P$ le semi-groupe $\left(\begin{smallmatrix} \mathbf{Z}_p - \{0\} & \mathbf{Z}_p \\ 0 & 1 \end{smallmatrix}\right)$,
- $A^+ = A \cap P^+$ le semi-groupe $\left(\begin{smallmatrix} \mathbf{Z}_p - \{0\} & 0 \\ 0 & 1 \end{smallmatrix}\right)$ que l'on décompose sous la forme $A^+ = \Phi^+ \times A^0$, avec $\Phi^+ = \left(\begin{smallmatrix} p^{\mathbf{N}} & 0 \\ 0 & 1 \end{smallmatrix}\right)$ et $A^0 = \left(\begin{smallmatrix} \mathbf{Z}_p^* & 0 \\ 0 & 1 \end{smallmatrix}\right)$ (si $n \geq 1$, on note A^n le sous-groupe $\left(\begin{smallmatrix} 1+2p^n\mathbf{Z}_p & 0 \\ 0 & 1 \end{smallmatrix}\right)$ de A^0),
- w la matrice $\left(\begin{smallmatrix} 0 & 1 \\ 1 & 0 \end{smallmatrix}\right)$.

0.2. La correspondance de Langlands locale p-adique pour $\mathbf{GL}_2(\mathbf{Q}_p)$

On dispose [14] à présent d'une correspondance $V \mapsto \mathbf{\Pi}(V)$ associant à une L-représentation V de $\mathscr{G}_{\mathbf{Q}_p}$, de dimension 2, une représentation unitaire admissible de $G = \mathbf{GL}_2(\mathbf{Q}_p)$. Cette correspondance satisfait aux espoirs de Breuil [5] dans le cas potentiellement semi-stable (en particulier, elle encode la correspondance locale classique).

- Si V est absolument irréductible, il en est de même de $\mathbf{\Pi}(V)$ (et même de la restriction de $\mathbf{\Pi}(V)$ à B) ;
- si $V^{\mathrm{ss}} = L(\delta_1) \oplus L(\delta_2)$, alors $\mathbf{\Pi}(V)^{\mathrm{ss}} = B(\delta_1, \delta_2)^{\mathrm{ss}} \oplus B(\delta_2, \delta_1)^{\mathrm{ss}}$, où la représentation $B(\delta_i, \delta_{3-i})$ est l'induite continue de B à G du caractère unitaire $\delta_{3-i} \otimes \delta_i \chi^{-1}$ de B [défini par $(\delta_{3-i} \otimes \delta_i \chi^{-1})\left(\begin{smallmatrix} a & b \\ 0 & d \end{smallmatrix}\right) = \delta_{3-i}(a)\delta_i(d)(d|d|)^{-1}$].

La construction de la correspondance $V \mapsto \mathbf{\Pi}(V)$ s'appuie sur l'équivalence de catégories $\mathrm{Rep}_L \mathscr{G}_{\mathbf{Q}_p} \cong \Phi\Gamma^{\mathrm{et}}(\mathscr{E})$ de Fontaine [20] entre représentations de $\mathscr{G}_{\mathbf{Q}_p}$ et (φ, Γ)-modules étales, et ses raffinements [9, 22] $\Phi\Gamma^{\mathrm{et}}(\mathscr{E}) \cong \Phi\Gamma^{\mathrm{et}}(\mathscr{E}^\dagger) \cong \Phi\Gamma^{\mathrm{et}}(\mathscr{R})$, où $\mathscr{R}$ désigne l'anneau de Robba (i.e. l'ensemble des fonctions analytiques f sur une couronne $0 < v_p(T) \leq r(f)$, où $r(f) > 0$ dépend de f), $\mathscr{E}^\dagger$ désigne l'ensemble des éléments bornés de $\mathscr{R}$ et $\mathscr{E}$ est le complété de $\mathscr{E}^\dagger$ pour la valuation p-adique. Si $V \in \mathrm{Rep}_L \mathscr{G}_{\mathbf{Q}_p}$, on dispose donc de (φ, Γ)-modules $D \in \Phi\Gamma^{\mathrm{et}}(\mathscr{E})$, $D^\dagger \in \Phi\Gamma^{\mathrm{et}}(\mathscr{E}^\dagger)$ et $D_{\mathrm{rig}} \in \Phi\Gamma^{\mathrm{et}}(\mathscr{R})$ (le module $D^\dagger$ est le moins intéressant, mais c'est lui qui permet de passer de D à D_{rig} et vice-versa car $D = \mathscr{E} \otimes_{\mathscr{E}^\dagger} D^\dagger$ et $D_{\mathrm{rig}} = \mathscr{R} \otimes_{\mathscr{E}^\dagger} D^\dagger$).

Si V est de dimension 2, on construit des faisceaux G-équivariants sur $\mathbf{P}^1 = \mathbf{P}^1(\mathbf{Q}_p)$ dont les sections globales $D \boxtimes \mathbf{P}^1$, $D^\dagger \boxtimes \mathbf{P}^1$ et $D_{\mathrm{rig}} \boxtimes \mathbf{P}^1$

fournissent une description de la représentation $\mathbf{\Pi}(V)$ et de ses vecteurs localement analytiques $\mathbf{\Pi}(V)^{\rm an}$: on a des suites exactes de G-espaces vectoriels topologiques

$$0 \to \mathbf{\Pi}(V)^* \otimes \omega_D \to D \boxtimes \mathbf{P}^1 \to \mathbf{\Pi}(V) \to 0,$$
$$0 \to (\mathbf{\Pi}(V)^{\rm an})^* \otimes \omega_D \to D_{\rm rig} \boxtimes \mathbf{P}^1 \to \mathbf{\Pi}(V)^{\rm an} \to 0,$$

où ω_D désigne le caractère $\chi^{-1} \det V$ de $\mathbf{Q}_p^*$ vu comme caractère de G en composant avec le déterminant (le module $D^\dagger \boxtimes \mathbf{P}^1$ permet de passer de $D \boxtimes \mathbf{P}^1$ à $D_{\rm rig} \boxtimes \mathbf{P}^1$, mais n'a pas vraiment d'intérêt en lui-même). Le caractère central de $\mathbf{\Pi}(V)$ est ω_D ; c'est donc le déterminant de V à multiplication près par l'inverse χ^{-1} du caractère cyclotomique.

Dans cet article, on s'intéresse de plus près à la représentation $\mathbf{\Pi}(V)^{\rm an}$. En particulier, on calcule son module de Jacquet ou plutôt son dual $J^*(\mathbf{\Pi}(V)^{\rm an}) = H^0(N, (\mathbf{\Pi}(V)^{\rm an})^*)$ (Il s'agit donc du module de Jacquet naïf, qui contrôle les morphismes vers la série principale, et pas de celui d'Emerton [18] qui contrôle les morphismes depuis la série principale). Cela nous permet de répondre à des questions de Breuil et Emerton concernant les vecteurs localement analytiques des représentations de la *série principale unitaire* qui correspondent [13, 2, 28] aux représentations *trianguline*s [ce qui signifie [10] que le (φ, Γ)-module $D_{\rm rig}$ est *triangulable* c'est-à-dire est une extension de deux (φ, Γ)-modules de rang 1 sur $\mathscr{R}$ (non étales si V est irréductible)]. On peut aussi obtenir ces représentations de la série principale unitaire par complétion de certaines induites analytiques de caractères (non unitaires) du borel ; cette description joue d'ailleurs un grand rôle dans la théorie (au moins pour le moment) : c'est grâce à elle que l'on peut vérifier que le module $D \boxtimes \mathbf{P}^1$ se décompose sous la forme ci-dessus dans le cas de la série principale unitaire, le cas général s'en déduisant par prolongement analytique.

Comme on s'intéresse aux vecteurs localement analytiques, il est commode de changer un peu de point de vue et de privilégier le module $D_{\rm rig}$. On part donc de $\Delta \in \Phi\Gamma^{\rm et}(\mathscr{R})$, de rang 2 ; on note $\mathbf{V}(\Delta)$ (ou souvent, simplement V) la représentation de $\mathscr{G}_{\mathbf{Q}_p}$ qui lui correspond, et on note $\mathbf{\Pi}(\Delta)$ la représentation $\mathbf{\Pi}(\mathbf{V}(\Delta))^{\rm an}$ de telle sorte que la suite exacte ci-dessus devient:

$$0 \to \mathbf{\Pi}(\Delta)^* \otimes \omega_\Delta \to \Delta \boxtimes \mathbf{P}^1 \to \mathbf{\Pi}(\Delta) \to 0.$$

On note aussi $\Delta^{[0]}$ le sous-$\mathscr{E}^\dagger$-module des éléments de Δ d'ordre 0 et $\widehat{\Delta}^{[0]}$ le $\mathscr{E}$-module $\mathscr{E} \otimes_{\mathscr{E}^\dagger} \Delta^{[0]}$ (si $\Delta = D_{\rm rig}$, alors $\Delta^{[0]} = D^\dagger$ et $\widehat{\Delta}^{[0]} = D$). On a alors le résultat suivant.

Théorème 8.1. *$J^*(\mathbf{\Pi}(\Delta))$ est non nul si et seulement si Δ est triangulable.*

Autrement dit, $J^*(\mathbf{\Pi}(\Delta))$ est non nul si et seulement si $\mathbf{\Pi}(\widehat{\Delta}^{[0]})$ est de la série principale unitaire. Si Δ est triangulable, on peut décrire explicitement le module $J^*(\mathbf{\Pi}(\Delta))$ et en déduire la structure de la représentation $\mathbf{\Pi}(\Delta)$, mais cela va demander d'introduire un certain nombre de no(ta)tions.

0.3. (φ, Γ)-modules triangulables de rang 2

La classification des (φ, Γ)-modules triangulables de rang 2 se ramène à celle des (φ, Γ)-modules de rang 1 et de leurs extensions, ce qui fait l'objet d'une bonne partie de [10] et [8].

Si $\delta \in \widehat{\mathscr{T}}(L)$, on note $\mathscr{R}(\delta)$ le (φ, Γ) module obtenu en multipliant l'action de φ sur $\mathscr{R}$ par $\delta(p)$ et celle de $\sigma_a \in \Gamma$ par $\delta(a)$. On note souvent un élément de $\mathscr{R}(\delta)$ sous la forme $x \otimes \delta$, avec $x \in \mathscr{R}$; les actions de φ et σ_a sont alors données par $\varphi(x \otimes \delta) = (\delta(p)\varphi(x)) \otimes \delta$ et $\sigma_a(x \otimes \delta) = (\delta(a)\sigma_a(x)) \otimes \delta$. Le module $\mathscr{R}(\delta)$ admet une base canonique sur $\mathscr{R}$, à savoir $1 \otimes \delta$, qui est propre sous les actions de φ et Γ pour le caractère δ.

On a alors le résultat suivant [10, th. 0.2].

Proposition 8.2. (i) *Si D est un (φ, Γ)-module de rang 1 sur $\mathscr{R}$, il existe un unique $\delta \in \widehat{\mathscr{T}}(L)$ tel que $D \cong \mathscr{R}(\delta)$.*

(ii) *Si $\delta_1, \delta_2 \in \widehat{\mathscr{T}}(L)$, alors* $\mathrm{Ext}^1(\mathscr{R}(\delta_2), \mathscr{R}(\delta_1))$ *est un L-espace vectoriel de dimension* 1 *sauf si $\delta_1\delta_2^{-1}$ est de la forme x^{-i}, avec i entier ≥ 0, ou de la forme $|x|x^i$, avec i entier ≥ 1 ; dans ces deux cas,* $\mathrm{Ext}^1(\mathscr{R}(\delta_2), \mathscr{R}(\delta_1))$ *est de dimension* 2 *et l'espace projectif associé est naturellement isomorphe à* $\mathbf{P}^1(L)$.

On note $\mathscr{S}$ l'ensemble des $(\delta_1, \delta_2, \mathscr{L})$, où (δ_1, δ_2) est un élément de $\widehat{\mathscr{T}}(L) \times \widehat{\mathscr{T}}(L)$ et $\mathscr{L} \in \mathrm{Proj}(\mathrm{Ext}^1(\mathscr{R}(\delta_2), \mathscr{R}(\delta_1)))$, cet espace étant identifié à $\mathbf{P}^0(L) = \{\infty\}$ (resp. $\mathbf{P}^1(L)$) si $\mathrm{Ext}^1(\mathscr{R}(\delta_2), \mathscr{R}(\delta_1))$ est de dimension 1 (resp. 2). Si $s = (\delta_1, \delta_2, \mathscr{L})$, on note $\Delta(s)$ le (φ, Γ)-module extension de $\mathscr{R}(\delta_2)$ par $\mathscr{R}(\delta_1)$ déterminé, à isomorphisme près, par $\mathscr{L}$. On a donc une suite exacte de (φ, Γ)-modules

$$0 \to \mathscr{R}(\delta_1) \to \Delta(s) \to \mathscr{R}(\delta_2) \to 0.$$

On note $\mathscr{S}_*$ le sous-ensemble de $\mathscr{S}$ constitué des s vérifiant les conditions

$$v_p(\delta_1(p)) + v_p(\delta_2(p)) = 0 \quad \text{et} \quad v_p(\delta_1(p)) > 0.$$

Si $s \in \mathscr{S}_*(L)$, on associe à s les invariants $u(s) \in \mathbf{Q}_+$ et $w(s) \in L$ définis par

$$u(s) = v_p(\delta_1(p)) = -v_p(\delta_2(p)) \quad \text{et} \quad w(s) = w(\delta_1) - w(\delta_2).$$

On partitionne $\mathscr{S}_*$ sous la forme $\mathscr{S}_* = \mathscr{S}_*^{\mathrm{ng}} \coprod \mathscr{S}_*^{\mathrm{cris}} \coprod \mathscr{S}_*^{\mathrm{st}} \coprod \mathscr{S}_*^{\mathrm{ord}} \coprod \mathscr{S}_*^{\mathrm{ncl}}$, où

- $\mathscr{S}_*^{\mathrm{ng}}$ est l'ensemble des s tels que $w(s)$ ne soit pas un entier ≥ 1 ;
- $\mathscr{S}_*^{\mathrm{cris}}$ est l'ensemble des s tels que $w(s)$ soit un entier ≥ 1, $u(s) < w(s)$ et $\mathscr{L} = \infty$;
- $\mathscr{S}_*^{\mathrm{st}}$ est l'ensemble des s tels que $w(s)$ soit un entier ≥ 1, $u(s) < w(s)$ et $\mathscr{L} \neq \infty$;
- $\mathscr{S}_*^{\mathrm{ord}}$ est l'ensemble des s tels que $w(s)$ soit un entier ≥ 1, $u(s) = w(s)$;
- $\mathscr{S}_*^{\mathrm{ncl}}$ est l'ensemble des s tels que $w(s)$ soit un entier ≥ 1, $u(s) > w(s)$.

Enfin, on pose $\mathscr{S}_{\mathrm{irr}} = \mathscr{S}_*^{\mathrm{ng}} \coprod \mathscr{S}_*^{\mathrm{cris}} \coprod \mathscr{S}_*^{\mathrm{st}}$, et on note $\mathscr{S}_*^{\mathrm{HT}}$ (resp. $\mathscr{S}_*^{\mathrm{dR}}$) l'ensemble des $s \in \mathscr{S}_{\mathrm{irr}}$ tels que $w(s) \in \mathbf{Z} - \{0\}$ (resp. $w(s) \in \mathbf{N} - \{0\}$). Alors $\mathscr{S}_*^{\mathrm{dR}} = \mathscr{S}_*^{\mathrm{st}} \coprod \mathscr{S}_*^{\mathrm{cris}}$ et $\mathscr{S}_*^{\mathrm{HT}}$ est la réunion disjointe de $\mathscr{S}_*^{\mathrm{dR}}$ et de $\mathscr{S}_*^{\mathrm{HT}} \cap \mathscr{S}_*^{\mathrm{ng}}$ qui est l'image de $\mathscr{S}_*^{\mathrm{ncl}}$ par l'application $(\delta_1, \delta_2, \mathscr{L}) \mapsto (x^{-w(s)}\delta_1, x^{w(s)}\delta_2, \mathscr{L})$ (on a d'ailleurs $\mathscr{L} = \infty$ dans cette situation).

Le résultat suivant [10, th. 0.5] rassemble les principales propriétés de $\Delta(s)$.

Proposition 8.3. *Soit $s \in \mathscr{S}$.*

(i) $\Delta(s) \in \Phi\Gamma^{\mathrm{et}}(\mathscr{R})$ *et est irréductible si et seulement si* $s \in \mathscr{S}_{\mathrm{irr}}$ (si $s \in \mathscr{S}_*^{\mathrm{ncl}}$, alors $\Delta(s)$ n'est pas étale et si $s \in \mathscr{S}_*^{\mathrm{ord}}$, alors $\Delta(s)$ est étale mais pas irréductible).

(ii) *Si $s = (\delta_1, \delta_2, \mathscr{L})$ et $s' = (\delta_1', \delta_2', \mathscr{L}')$ sont deux éléments distincts de $\mathscr{S}_{\mathrm{irr}}$, alors $\Delta(s) \cong \Delta(s')$ si et seulement si $s, s' \in \mathscr{S}_*^{\mathrm{cris}}$ et $\delta_1' = x^{w(s)}\delta_2$, $\delta_2' = x^{-w(s)}\delta_1$.*

Remarque 8.4. L'application $s = (\delta_1, \delta_2, \infty) \mapsto s' = (x^{w(s)}\delta_2, x^{-w(s)}\delta_1, \infty)$ est une involution de $\mathscr{S}_*^{\mathrm{cris}}$. Les points fixes de cette involution sont dits *exceptionnels* ; ce sont les s tels que $w(s) \in \mathbf{N} - \{0\}$ et $\delta_1 = x^{w(s)}\delta_2$.

Si $s \in \mathscr{S}_{\mathrm{irr}}$, on note $V(s)$ la L-représentation de $\mathscr{G}_{\mathbf{Q}_p}$ qui lui est associée. Cette représentation est trianguline par construction ; réciproquement, si V est trianguline, de dimension 2, et irréductible, alors il existe $s \in \mathscr{S}_{\mathrm{irr}}$ tel que $V \cong V(s)$. Le résultat suivant [10, th. 0.8] donne une bonne idée du genre de propriétés de $V(s)$ que l'on peut lire sur s.

Proposition 8.5. *Soit $s \in \mathscr{S}_{\mathrm{irr}}$.*

(i) *Les poids de Hodge-Tate de $V(s)$ sont $w(\delta_1)$ et $w(\delta_2)$.*

(ii) *Si $w(\delta_1) \in \mathbf{Z}$* (i.e. si δ_1 est localement algébrique), *alors*

- $V(s)$ *est cristabéline* (i.e. devient cristalline sur une extension abélienne de $\mathbf{Q}_p$) *si et seulement si* $s \in \mathscr{S}_*^{\text{cris}}$,
- $V(s)$ *est la tordue d'une représentation semi-stable* (non cristalline) *par un caractère d'ordre fini si et seulement si* $s \in \mathscr{S}_*^{\text{st}}$,
- $V(s)$ *est de Hodge-Tate si et seulement si* $s \in \mathscr{S}_*^{\text{HT}}$ *et elle est de Rham si et seulement si* $s \in \mathscr{S}_*^{\text{dR}}$.

0.4. La série principale localement analytique

Remarquons que $J^*(\Pi)$ est stable par B ; c'est donc un B-module sur lequel N agit trivialement. Le th. 8.6 ci-dessous décrit ce module dans le cas de $\mathbf{\Pi}(\Delta(s))$. Si $\eta_1, \eta_2 \in \widehat{\mathscr{T}}(L)$, on note $\eta_1 \otimes \eta_2$ le caractère de B défini par

$$(\eta_1 \otimes \eta_2)\left(\begin{smallmatrix} a & b \\ 0 & d \end{smallmatrix}\right) = \eta_1(a)\eta_2(d).$$

Si $s \in \mathscr{S}_{\text{irr}}$, on note $J^*(s)$ le B-module $J^*(\mathbf{\Pi}(\Delta(s)))$.

Théorème 8.6. *Soit* $s \in \mathscr{S}_{\text{irr}}$.

(i) *Si* $s \in \mathscr{S}_*^{\text{ng}} - (\mathscr{S}_*^{\text{ng}} \cap \mathscr{S}_*^{\text{HT}})$ *ou si* $s \in \mathscr{S}_*^{\text{st}}$, *alors* $J^*(s) \cong \delta_1^{-1} \otimes \delta_2^{-1}\chi$.

(ii) *Si* $s \in \mathscr{S}_*^{\text{ng}} \cap \mathscr{S}_*^{\text{HT}}$, *alors* $J^*(s) = (\delta_1^{-1} \otimes \delta_2^{-1}\chi) \oplus (x^{w(s)}\delta_1^{-1} \otimes x^{-w(s)}\delta_2^{-1}\chi)$.

(iii) *Si* $s \in \mathscr{S}_*^{\text{cris}}$, *alors* $J^*(s) = (\delta_1^{-1} \otimes \delta_2^{-1}\chi) \oplus (x^{-w(s)}\delta_2^{-1} \otimes x^{w(s)}\delta_1^{-1}\chi)$ *sauf si s est exceptionnel où ces deux caractères sont égaux et* $J^*(s) = (\delta_1^{-1} \otimes \delta_2^{-1}\chi) \otimes \left(\begin{smallmatrix} 1 & v_p(a/d) \\ 0 & 1 \end{smallmatrix}\right)$.

Si η est un caractère localement analytique de B, on note $\text{Ind}^{\text{an}}\eta$ l'induite localement analytique de B à G de η. Si Π est une représentation localement analytique de G, et si $\mu \in J^*(\Pi)$ est un vecteur propre pour l'action de B pour le caractère η (i.e. $g \cdot \mu = \eta(g)\,\mu$, si $g \in B$), alors $v \mapsto \tilde{\phi}_v$, où $\tilde{\phi}_v : G \to L$ est définie par $\tilde{\phi}_v(g) = \langle \mu, g \cdot v \rangle$, est un morphisme G-équivariant de Π dans $\text{Ind}^{\text{an}}\eta$. On peut utiliser cette propriété et le th. 8.6 pour décrire complètement la représentation $\mathbf{\Pi}(\Delta(s))$.

Si $\eta_1, \eta_2 \in \widehat{\mathscr{T}}(L)$, on note $B^{\text{an}}(\eta_1, \eta_2)$ la représentation localement analytique $\text{Ind}^{\text{an}}(\eta_2 \otimes \eta_1\chi^{-1})$ de G. On remarque que, dans tous les cas, $\delta_1^{-1} \otimes \delta_2^{-1}\chi$ est un caractère de B apparaissant comme sous-module de $J(s)$; il existe donc un morphisme G-équivariant p_s, non identiquement nul, de $\mathbf{\Pi}(\Delta(s))$ dans $B^{\text{an}}(\delta_2, \delta_1)$. On a alors le résultat suivant, où l'on a noté δ_s le caractère $\chi^{-1}\delta_1\delta_2^{-1}$ (on dit que s est *générique* si $\delta_s \notin \{x^k,\ k \in \mathbf{N}\}$ et *spécial* dans le cas contraire, ce qui inclut le cas $s \in \mathscr{S}_*^{\text{st}}$).

Théorème 8.7. *Si $s = (\delta_1, \delta_2, \mathscr{L}) \in \mathscr{S}_{\rm irr}$, alors p_s est une surjection. De plus:*

(i) *Si s est générique, le noyau de p_s est isomorphe à $B^{\rm an}(\delta_1, \delta_2)$; on a donc une suite exacte $0 \to B^{\rm an}(\delta_1, \delta_2) \to \mathbf{\Pi}(\Delta(s)) \to B^{\rm an}(\delta_2, \delta_1) \to 0$.*

(ii) *Si s est spécial, alors $B^{\rm an}(\delta_1, \delta_2)$ contient une sous-représentation $W(\delta_1, \delta_2)$ de dimension $k+1$. Si $\mathrm{St}^{\rm an}(\delta_1, \delta_2)$ est le quotient de $B^{\rm an}(\delta_1, \delta_2)$ par $W(\delta_1, \delta_2)$, on a un isomorphisme naturel*

$$\mathrm{Ext}^1(\mathscr{R}(\delta_2), \mathscr{R}(\delta_1)) \cong \mathrm{Ext}^1(W(\delta_1, \delta_2), \mathrm{St}^{\rm an}(\delta_1, \delta_2)),$$

et une suite exacte $0 \to E_{\mathscr{L}} \to \mathbf{\Pi}(\Delta(s)) \to B^{\rm an}(\delta_2, \delta_1) \to 0$, où $E_{\mathscr{L}}$ est l'extension de $W(\delta_1, \delta_2)$ par $\mathrm{St}^{\rm an}(\delta_1, \delta_2)$ correspondant à l'extension de $\mathscr{R}(\delta_2)$ par $\mathscr{R}(\delta_1)$ déterminée par $\mathscr{L}$.

Remarque 8.8. (i) D'après [24], on a $\dim_L \mathrm{Ext}^1(B^{\rm an}(\delta_2, \delta_1), B^{\rm an}(\delta_1, \delta_2)) = 1$, si $\delta_s \notin \{x^k,\ k \in \mathbf{N}\}$. Il en résulte que $\mathbf{\Pi}(\Delta(s))$ est l'unique extension non triviale de $B^{\rm an}(\delta_2, \delta_1)$ par $B^{\rm an}(\delta_1, \delta_2)$.

(ii) Si $\delta_s = x^k$, avec $k \in \mathbf{N}$, alors $\mathrm{Ext}^1(\mathscr{R}(\delta_2), \mathscr{R}(\delta_1))$ est naturellement isomorphe à $\mathrm{Hom}(\mathbf{Q}_p^*, L)$. L'extension de $W(\delta_1, \delta_2)$ par $\mathrm{St}^{\rm an}(\delta_1, \delta_2)$ correspondant à un élément non nul ℓ de $\mathrm{Hom}(\mathbf{Q}_p^*, L)$ est celle considérée par Breuil [5, 6].

(iii) Dans tous les cas, la représentation unitaire $\mathbf{\Pi}(V(s))$ (dont $\mathbf{\Pi}(\Delta(s))$ est l'ensemble des vecteurs localement analytiques) est la complétion universelle unitaire de $\mathrm{Ker}\, p_s$ (c'est la conjonction de résultats éparpillés dans la littérature, à savoir [2] pour le cas $s \in \mathscr{S}_*^{\rm cris}$ non exceptionnel, [28] pour le cas exceptionnel, [5, 6] pour le cas $s \in \mathscr{S}_*^{\rm st}$ et [19, th. 5.3.7] (qui s'appuie sur [13]) pour le reste).

(iv) Le théorème a aussi été démontré par Liu, Xie et Zhang [27]. Le point clé des deux démonstrations est le même, à savoir un dévissage du module $\Delta(s) \boxtimes \mathbf{P}^1$ sous la forme d'une suite exacte de G-modules

$$0 \to \mathscr{R}(\delta_1) \boxtimes \mathbf{P}^1 \to \Delta(s) \boxtimes \mathbf{P}^1 \to \mathscr{R}(\delta_2) \boxtimes \mathbf{P}^1 \to 0,$$

la projection de $\Delta(s) \boxtimes \mathbf{P}^1$ sur $\mathscr{R}(\delta_2) \boxtimes \mathbf{P}^1$ donnant naissance à l'application p_s ci-dessus. Les deux[(1)] stratégies pour aboutir à ce résultat sont par contre complètement orthogonales. Celle de Liu, Xie et Zhang utilise la description de $\mathbf{\Pi}(V(s))$ comme complétion universelle pour identifier un sous-module de $\mathbf{\Pi}(\Delta(s))$ et, par dualité, un quotient de $\mathbf{\Pi}(\Delta(s))^*$.

[1] Dospinescu [16] a trouvé une démonstration directe de la stabilité du sous-module $\mathscr{R}(\delta_1) \boxtimes \mathbf{P}^1$ par G, ce qui fournit une troisième manière d'aboutir au résultat.

La notre utilise la détermination du module de Jacquet dual pour montrer que $B^{\mathrm{an}}(\delta_2, \delta_1)$ est un quotient de $\mathbf{\Pi}(\Delta(s))$; elle est indépendante[(2)] de [2, 5, 6, 13] et permet en fait d'en retrouver certains résultats.

(v) Le théorème répond à des questions de Breuil et Emerton (cf. [5, conj. 4.4.1] et [18, conj. 6.7.3 et 6.7.7]). Dans les cas $s \in \mathscr{S}_*^{\mathrm{cris}}$ et $\mathscr{S}_*^{\mathrm{ng}} \cap \mathscr{S}_*^{\mathrm{HT}}$, on peut utiliser l'autre vecteur propre de $J^*(s)$ pour obtenir des renseignements supplémentaires sur $\mathbf{\Pi}(\Delta(s))$. Dans le cas $s \in \mathscr{S}_*^{\mathrm{cris}}$, cela permet de démontrer (cf. prop. 8.97) une conjecture de Breuil [2, conj. 5.3.7 et 5.4.4] (cf. [25] pour une autre démonstration).

Une analyse des composantes de Jordan-Hölder de $\mathbf{\Pi}(\Delta(s))$ couplée avec des résultats d'Emerton [17] permet de déterminer les vecteurs localement algébriques de $\mathbf{\Pi}(\Delta(s))$, et donc de retrouver le résultat [14, th. VI.6.50] selon lequel la correspondance locale p-adique encode la classique dans le cas de la série principale unitaire.

On note $\mathbf{\Pi}(\Delta(s))^{\mathrm{alg}}$ l'ensemble des vecteurs localement algébriques de $\mathbf{\Pi}(\Delta(s))$ pour l'action de $\mathbf{SL}_2(\mathbf{Q}_p)$ [si le caractère central $\delta_1\delta_2\chi^{-1}$ est localement algébrique, alors $\mathbf{\Pi}(\Delta(s))^{\mathrm{alg}}$ l'ensemble des vecteurs localement algébriques de $\mathbf{\Pi}(\Delta(s))$]. Si $k \in \mathbf{N}$, on note Sym^k la représentation algébrique de G, puissance symétrique k-ième de la représentation standard de $G \subset \mathbf{GL}_2(L)$ sur L^2 ; si η est un caractère localement constant de B, on note $\mathrm{Ind}^{\mathrm{lisse}}\eta$ l'induite lisse de B à G de η.

Théorème 8.9. *Soit* $s = (\delta_1, \delta_2, \mathscr{L}) \in \mathscr{S}_{\mathrm{irr}}$.

(i) $\mathbf{\Pi}(\Delta(s))^{\mathrm{alg}} \neq 0$ *si et seulement si* $s \in \mathscr{S}_*^{\mathrm{dR}}$.

(ii) *Si* $s \in \mathscr{S}_*^{\mathrm{st}}$, *alors* $\mathbf{\Pi}(\Delta(s))^{\mathrm{alg}} = \mathrm{St} \otimes \mathrm{Sym}^{w(s)-1} \otimes \delta_2$ *et est irréductible.*

(iii) *Si* $s \in \mathscr{S}_*^{\mathrm{cris}}$, *alors*

$$\mathbf{\Pi}(\Delta(s))^{\mathrm{alg}} = \big(\mathrm{Ind}^{\mathrm{lisse}}(|x|\delta_s x^{1-w(s)} \otimes |x|^{-1})\big) \otimes \mathrm{Sym}^{w(s)-1} \otimes \delta_2$$

et est irréductible, sauf si $\delta_s = x^{w(s)-1}$ *ou si* $\delta_s = |x|^{-2}x^{w(s)-1}$ *auquel cas* $\mathbf{\Pi}(\Delta(s))^{\mathrm{alg}}$ *est une extension de* $\mathrm{Sym}^{w(s)-1} \otimes \delta_2$ *par* $\mathrm{St} \otimes \mathrm{Sym}^{w(s)-1} \otimes \delta_2$.

0.5. Céoukonfaikoi

Cet article contient quatre chapitres de natures assez distinctes et un appendice complétant la première partie.

2 Pas totalement car l'existence de la correspondance de Langlands locale p-adique se démontre par prolongement analytique à partir du cas de la série principale unitaire…

- Le premier chapitre est consacré à une nouvelle démonstration de la prop. 8.2. Cette démonstration reprend les techniques d'analyse fonctionnelle p-adique déjà utilisées par Chenevier [8] pour une version en famille. Nous dévissons la situation complètement de manière à obtenir une description des extensions de $\mathscr{R}(\delta_2)$ par $\mathscr{R}(\delta_1)$ la mieux adaptée possible à la description des vecteurs localement analytiques des représentations de la série principale. Nous avons aussi un peu modifié le point de vue en remplaçant le complexe $0 \to D \to D \oplus D \to D \to 0$ habituel par la cohomologie du semi-groupe A^+ et sa décomposition issue de la suite spectrale de Hochshild-Serre déduite de la suite exacte $1 \to \Phi^+ \to A^+ \to A^0 \to 1$. (L'appendice contient un certain nombre de sorites sur ce sujet ; en particulier des formules explicites (lemme 8.104) concernant la dualité.) C'est relativement cosmétique comme modification, mais cela a l'avantage de mettre sur le même plan les cas $p = 2$ et $p \neq 2$; cela supprime aussi des problèmes de normalisation induits par le choix d'un générateur de $\mathbf{Z}_p^*$.
- Le second chapitre regroupe des résultats de [30] concernant la série principale analytique. Ils sont complétés par la détermination des extensions de $W(\delta_1, \delta_2)$ par $\mathrm{St}^{\mathrm{an}}(\delta_1, \delta_2)$.
- Le troisième chapitre est consacré à l'étude du module $\Delta \boxtimes \{0\} = \cap_{n\in\mathbf{N}}\varphi^n(\Delta)$, si Δ est un (φ, Γ)-module étale sur $\mathscr{R}$. Ceci est crucial pour la détermination[3] du module de Jacquet dual $J^*(\mathbf{\Pi}(\Delta))$ de $\mathbf{\Pi}(\Delta)$, si Δ est de dimension 2. Notons que l'étude de $D^{\mathrm{nr}} = \cap_{n\in\mathbf{N}}\varphi^n(D)$, si $D \in \Phi\Gamma^{\mathrm{et}}(\mathscr{E})$, joue un grand rôle dans [12].
- Dans le dernier chapitre, on utilise les résultats du troisième pour déterminer le module de Jacquet dual de $\mathbf{\Pi}(\Delta)$, puis on utilise ce que l'on obtient pour dévisser le module $\Delta(s) \boxtimes \mathbf{P}^1$. Enfin, on utilise les résultats du second pour décrire le module $\mathbf{\Pi}(\Delta(s))$.

0.6. Remerciements

Cet article a pour origine une question que m'a posée Matthew Emerton pendant l'École d'été de 2008. Je le remercie de me l'avoir mentionnée. Je remercie le programme CEFIPRA d'avoir financé le séjour au Tata Institute en juillet-août 2010 au cours duquel une première version de cet article a été terminée. Je remercie aussi les universités UCLA et UC Berkeley (et l'Institut Miller qui a financé le séjour à Berkeley) pour leur hospitalité en janvier-février et à l'automne 2011 ; des parties conséquentes de cet article y ont été écrites.

[3] Dospinescu [16] a remarqué que l'on pouvait aussi utiliser l'action infinitésimale [15] de N au lieu de φ pour calculer $J^*(\mathbf{\Pi}(\Delta))$.

1. Cohomologie de $\mathscr{R}(\delta)$

Ce chapitre est consacré à la démonstration d'une version renforcée (th. 8.47) du (ii) de la prop. 8.2 qui décrit l'espace $\mathrm{Ext}^1(\mathscr{R}(\delta_2), \mathscr{R}(\delta_1))$ des extensions de $\mathscr{R}(\delta_2)$ par $\mathscr{R}(\delta_1)$, si δ_1, δ_2 sont des caractères de $\mathbf{Q}_p^*$. Cette démonstration repose sur le dévissage $0 \to \mathscr{D}(\mathbf{Z}_p) \to \mathscr{R} \to \mathrm{LA}(\mathbf{Z}_p) \otimes \chi^{-1} \to 0$ de $\mathscr{R}$ fourni par le dictionnaire d'analyse fonctionnelle p-adique (n° 1.2), et sur l'interprétation de $\mathrm{Ext}^1(\mathscr{R}(\delta_2), \mathscr{R}(\delta_1))$ en terme de cohomologie du semi-groupe A^+ que l'on dévisse en utilisant la suite spectrale de Hochschild-Serre (n° 1.3).

1.1. L'anneau de Robba $\mathscr{R}$

Soit $(\zeta_{p^n})_{n\geq 0}$ un système compatible de racines p^n-ièmes de l'unité (i.e. $\zeta_{p^{n+1}}^p = \zeta_{p^n}$, $\zeta_1 = 1$ et $\zeta_p \neq 1$; si $p = 2$, on part de $\zeta_1 = -1$ au lieu de $\zeta_1 = 1$). Si $n \in \mathbf{N}$, on note F_n le corps $\mathbf{Q}_p(\zeta_{p^n})$ (et donc $F_0 = \mathbf{Q}_p$), et on pose $L_n = L \otimes_{\mathbf{Q}_p} F_n$.

Si $n \geq 1$, on pose $r_n = v_p(\zeta_{p^n} - 1) = \frac{1}{(p-1)p^{n-1}}$ (si $p = 2$, on pose $r_n = \frac{1}{2^n}$), et on note $\mathscr{E}^{]0,r_n]}$ l'anneau des fonctions L-analytiques sur la couronne $C_n = \{z \in \mathbf{C}_p,\ 0 < v_p(z) \leq r_n\}$ que l'on voit comme un sous-ensemble des séries de Laurent $\sum_{k\in\mathbf{Z}} a_k T^k$ à coefficients dans L. Alors $\mathscr{E}^{]0,r_n]}$ est un anneau de Fréchet-Stein (tout idéal fermé est principal), la topologie de Fréchet étant définie par la famille de valuations $v^{[r_b,r_n]}$, pour $b \geq n$, avec, si $f = \sum_{k\in\mathbf{Z}} a_k T^k$,

$$v^{[r_b,r_n]} = \inf_{r_b \leq v_p(z) \leq r_n} v_p(f(z)) = \min\big(\inf_{k\in\mathbf{Z}}(v_p(a_k)+r_b k), \inf_{k\in\mathbf{Z}}(v_p(a_k)+r_n k)\big).$$

On définit l'anneau de Robba $\mathscr{R}$ comme la réunion des $\mathscr{E}^{]0,r_n]}$ pour $n \geq 1$; on le munit de la topologie de la limite inductive. On note $\mathscr{R}^+$ l'intersection de $\mathscr{R}$ et $L[[T]]$ (c'est l'anneau des fonctions L-analytiques sur le disque unité $\{z,\ v_p(z) > 0\}$).

On munit l'anneau $\mathscr{R}$ d'un frobenius φ : c'est un endomorphisme de L-algèbres, continu, envoyant T sur $(1+T)^p - 1$. On munit aussi $\mathscr{R}$ d'une action continue de Γ, respectant les structures de L-algèbres, σ_a envoyant T sur $(1+T)^a - 1$. Les actions de φ et Γ commutent entre elles.

On dispose d'un inverse à gauche ψ de φ, défini par $\psi(\sum_{i=0}^{p-1} (1+T)^i \varphi(f_i)) = f_0$. Cet inverse commute à Γ, et on a $\psi(f\varphi(g)) = g\psi(f)$, pour tous f, g. En particulier,

$$\psi^k\big((1+T)^b\varphi^k(z)\big) = \psi^k((1+T)^b)z = \begin{cases} (1+T)^{b/p^k} z & \text{si } b \in p^k\mathbf{Z}_p, \\ 0 & \text{si } b \notin p^k\mathbf{Z}_p. \end{cases}$$

On note ∂ l'opérateur $(1+T)\frac{d}{T}$. On a $\partial\circ\varphi = p\,\varphi\circ\partial$, $\partial\circ\psi = p^{-1}\,\psi\circ\partial$ et $\partial\circ\sigma_a = a\,\sigma_a\circ\partial$, si $a\in\mathbf{Z}_p^*$.

Soit $t=\log(1+T)$. C'est un élément de $\mathscr{R}^+$ vérifiant $\partial t = 1$, $\varphi(t) = pt$, $\psi(t) = p^{-1}t$ et $\sigma_a(t) = a\,t$ si $a\in\mathbf{Z}_p^*$. On note $L\{\{t\}\}$ le sous-anneau de $\mathscr{R}^+$ des $\sum_{n\in\mathbf{N}} a_n t^n$, où $(a_n)_{n\in\mathbf{N}}$ est une suite d'éléments de L telle que $\sum_{n\in\mathbf{N}} a_n z^n$ converge pour tout $z\in\mathbf{C}_p$; ce sous-anneau est stable par ∂, φ, ψ et Γ.

Si $n\geq 1$, on note ι_n l'injection de $\mathscr{E}^{]0,r_n]}$ dans $L_n[[t]]$ envoyant f sur $f(\zeta_{p^n}e^{t/p^n}-1)$. Alors ι_n commute à Γ, et on a $\iota_{n+1}\circ\varphi = \iota_n$ et $\iota_n\circ\psi = \mathrm{Tr}_{F_{n+1}/F_n}\circ\iota_{n+1}$.

Si $f=\sum_{n\in\mathbf{Z}} a_n T^n\in\mathscr{R}$, on pose $\mathrm{rés}_0(f\,dT) = a_{-1}$.

Proposition 8.10. *Si $f\in\mathscr{R}$, on a :*

- $\mathrm{rés}_0\big(\sigma_a(f)\frac{dT}{1+T}\big) = a^{-1}\mathrm{rés}_0\big(f\frac{dT}{1+T}\big)$, *pour tout $a\in\mathbf{Z}_p^*$,*
- $\mathrm{rés}_0\big(\varphi(f)\frac{dT}{1+T}\big) = \mathrm{rés}_0\big(\psi(f)\frac{dT}{1+T}\big) = \mathrm{rés}_0\big(f\frac{dT}{1+T}\big)$,
- $\mathrm{rés}_0\big(\partial f\frac{dT}{1+T}\big) = \mathrm{rés}_0(df) = 0$.

Démonstration. Pour les deux premiers points, cf. [12, prop. I.2.2] ; le dernier est immédiat. □

1.2. Dictionnaire d'analyse fonctionnelle p-adique

1.2.1. Quelques faisceaux P^+-équivariants sur $\mathbf{Z}_p$

On note $\mathrm{LA}(\mathbf{Z}_p)$ l'espace des fonctions localement analytiques sur $\mathbf{Z}_p$ à valeurs dans L, et $\mathscr{D}(\mathbf{Z}_p)$ le dual topologique de $\mathrm{LA}(\mathbf{Z}_p)$; c'est l'espace des distributions sur $\mathbf{Z}_p$ à valeurs dans L.

Si $\mu\in\mathscr{D}(\mathbf{Z}_p)$, on définit sa transformée d'Amice A_μ par la formule

$$A_\mu(T) = \int_{\mathbf{Z}_p}(1+T)^x\,\mu = \sum_{n\in\mathbf{N}}\Big(\int_{\mathbf{Z}_p}\binom{x}{n}\mu\Big)T^n\in L[[T]].$$

Si $f\in\mathscr{R}$, on définit $\phi_f:\mathbf{Z}_p\to L$ par la formule

$$\phi_f(x) = \mathrm{rés}_0\big((1+T)^{-x}f(T)\tfrac{dT}{1+T}\big).$$

Proposition 8.11. (i) *$\mu\mapsto A_\mu(T)$ induit un isomorphisme $\mathscr{D}(\mathbf{Z}_p)\cong\mathscr{R}^+$.*

(ii) *Le sous-anneau $L\{\{t\}\}$ de $\mathscr{R}^+$ est l'image de l'espace des distributions à support $\{0\}$ par $\mu\mapsto A_\mu$.*

(iii) *$f\mapsto\phi_f$ induit un isomorphisme $\mathscr{R}/\mathscr{R}^+\cong\mathrm{LA}(\mathbf{Z}_p)$.*

(iv) *On a $\int_{\mathbf{Z}_p}\phi_f\mu = \mathrm{rés}_0\big(\sigma_{-1}(f)A_\mu\frac{dT}{1+T}\big)$ si $f\in\mathscr{R}$ et $\mu\in\mathscr{D}(\mathbf{Z}_p)$.*

Démonstration. C'est une combinaison du th. I.1.4 de [14] et de la prop. II.4.6 de [11], à ceci près que l'on a changé le signe de x dans la définition de ϕ_f.

Soit P^+ le semi-groupe $\left(\begin{smallmatrix} \mathbf{Z}_p-\{0\} & \mathbf{Z}_p \\ 0 & 1 \end{smallmatrix}\right)$. On munit $\mathscr{R}$ d'une action de P^+ en posant

$$\left(\begin{smallmatrix} p^k a & b \\ 0 & 1 \end{smallmatrix}\right) \cdot f = (1+T)^b \varphi^k \circ \sigma_a(f), \quad \text{si } k \in \mathbf{N}, a \in \mathbf{Z}_p^* \text{ et } b \in \mathbf{Z}_p.$$

On munit $\mathrm{LA}(\mathbf{Z}_p)$ et $\mathscr{D}(\mathbf{Z}_p)$ d'actions de P^+ en posant :

$$\left(\left(\begin{smallmatrix} p^k a & b \\ 0 & 1 \end{smallmatrix}\right) \cdot \phi\right)(x) = \begin{cases} \phi\left(\frac{x-b}{p^k a}\right), & \text{si } x \in b + p^k \mathbf{Z}_p, \\ 0 & \text{si } x \notin b + p^k \mathbf{Z}_p, \end{cases}$$

$$\int_{\mathbf{Z}_p} \phi \left(\begin{smallmatrix} p^k a & b \\ 0 & 1 \end{smallmatrix}\right) \cdot \mu = \int_{\mathbf{Z}_p} \phi(p^k a x + b)\, \mu.$$

Si $\delta \in \widehat{\mathscr{T}}(L)$, on peut voir δ comme un caractère de P^+ en posant $\delta\left(\begin{smallmatrix} a & b \\ 0 & 1 \end{smallmatrix}\right) = \delta(a)$. Si M est un P^+-module, on note $M \otimes \delta$ le P^+ module M avec action de P^+ tordue par δ. (Le P^+-module $\mathscr{R} \otimes \delta$ est en général noté $\mathscr{R}(\delta)$.)

Proposition 8.12. *Les applications $\mu \mapsto A_\mu$ et $f \mapsto \phi_f$ induisent la suite exacte suivante de P^+-modules topologiques :*

$$0 \to \mathscr{D}(\mathbf{Z}_p) \to \mathscr{R} \to \mathrm{LA}(\mathbf{Z}_p) \otimes \chi^{-1} \to 0.$$

Démonstration. La P^+-équivariance est une conséquence des calculs suivants qui utilisent à plein la prop. 8.10 (remarquons que $\chi^{-1}(a) = a^{-1}$ si $a \in \mathbf{Z}_p^*$ et $\chi^{-1}(p) = 1$):

$$\begin{aligned} \int_{\mathbf{Z}_p} (1+T)^x \left(\begin{smallmatrix} p^k a & b \\ 0 & 1 \end{smallmatrix}\right) \cdot \mu &= \int_{\mathbf{Z}_p} (1+T)^{p^k a x + b}\, \mu \\ &= (1+T)^b \varphi^k \circ \sigma_a\Big(\int_{\mathbf{Z}_p} (1+T)^x\, \mu\Big), \end{aligned}$$

$$\phi_{\left(\begin{smallmatrix} p^k a & b \\ 0 & 1 \end{smallmatrix}\right) z}(x) = \mathrm{rés}_0\Big(\left(\begin{smallmatrix} 1 & -x \\ 0 & 1 \end{smallmatrix}\right)\left(\begin{smallmatrix} p^k a & b \\ 0 & 1 \end{smallmatrix}\right) z \, \tfrac{dT}{1+T}\Big) = \mathrm{rés}_0\Big(\sigma_a((1+T)^{(b-x)/a} \varphi^k(z)) \, \tfrac{dT}{1+T}\Big)$$
$$= a^{-1} \mathrm{rés}_0\Big(\psi^k((1+T)^{(b-x)/a} \varphi^k(z)) \, \tfrac{dT}{1+T}\Big)$$
$$= \begin{cases} a^{-1} \mathrm{rés}_0\Big((1+T)^{(b-x)/p^k a} z) \, \tfrac{dT}{1+T}\Big) = \chi^{-1}(p^k a) \phi_z\left(\frac{b-x}{p^k a}\right) & \text{si } x \in b + p^k \mathbf{Z}_p, \\ 0 & \text{si } x \in b + p^k \mathbf{Z}_p. \end{cases}$$

Si U est un ouvert compact de $\mathbf{Z}_p$, on note $\mathbf{1}_U$ sa fonction caractéristique. On peut décomposer U comme une réunion disjointe finie $\coprod_{i \in I} i + p^n \mathbf{Z}_p$, et on définit Res_U sur $\mathscr{R}$ par la formule $\mathrm{Res}_U = \sum_{i \in I} \mathrm{Res}_{i + p^n \mathbf{Z}_p}$, où l'on a posé $\mathrm{Res}_{i+p^n\mathbf{Z}_p} = \left(\begin{smallmatrix} 1 & i \\ 0 & 1 \end{smallmatrix}\right) \circ \varphi^n \circ \psi^n \circ \left(\begin{smallmatrix} 1 & -1 \\ 0 & 1 \end{smallmatrix}\right)$. (Ceci ne dépend pas du choix de la décomposition [12, § III.1].) On note $\mathscr{R} \boxtimes U$ l'image de Res_U.

La restriction Res_U, les opérateurs ψ, ∂, et la multiplication par t s'interprètent aussi simplement en termes du dictionnaire.

Proposition 8.13. (i) *Si $f \in \mathscr{R}$, on a :*

- $\phi_{\psi(f)}(x) = (\psi(\phi_f))(x)$, *avec* $(\psi(\phi))(x) = \phi(px)$,
- $\phi_{\mathrm{Res}_U f} = \mathrm{Res}_U \phi_f$, *avec* $\mathrm{Res}_U \phi = \mathbf{1}_U \phi$,
- $\phi_{\partial f}(x) = x\phi_f(x)$,
- $\phi_{tf}(x) = -\phi'_f(x)$.

(ii) *Si $\mathscr{D}(\mathbf{Z}_p)$, on a:*

- $\psi(A_\mu) = A_{\psi(\mu)}$, *où* $\int_{\mathbf{Z}_p} \phi\, \psi(\mu) = \int_{p\mathbf{Z}_p} \phi\big(\frac{x}{p}\big)\, \mu$ *et donc* $\big(\begin{smallmatrix} p & 0 \\ 0 & 1 \end{smallmatrix}\big) \cdot \psi(\mu) = \mathrm{Res}_{p\mathbf{Z}_p} \mu$,
- $\mathrm{Res}_U A_\mu = A_{\mathrm{Res}_U \mu}$, *avec* $\int_{\mathbf{Z}_p} \phi\, \mathrm{Res}_U \mu = \int_U \phi\, \mu = \int_{\mathbf{Z}_p} \mathbf{1}_U \phi\, \mu$,
- $\partial A_\mu = A_{x\mu}$.
- *Si* $\mu \in \mathscr{D}(\mathbf{Z}_p)$, *alors* $t A_\mu = A_{\mathrm{d}\mu}$, *où* $\int_{\mathbf{Z}_p} \phi\, \mathrm{d}\mu = \int_{\mathbf{Z}_p} \phi'(x)\, \mu$.

Démonstration. Pour le (ii), voir [11, § II.4]. En ce qui concerne le (i), on a:

- $$\begin{aligned}\phi_{\psi(f)}(x) &= \mathrm{rés}_0\big((1+T)^{-x}\psi(f)\,\tfrac{dT}{1+T}\big) = \mathrm{rés}_0\big(\psi((1+T)^{-px} f)\,\tfrac{dT}{1+T}\big)\\ &= \mathrm{rés}_0\big((1+T)^{-px} f\,\tfrac{dT}{1+T}\big) = \phi_f(px).\end{aligned}$$
- $$\begin{aligned}\phi_{\mathrm{Res}_{p^n\mathbf{Z}_p} f}(x) = \phi_{\varphi^n\psi^n(f)}(x) &= \begin{cases}\phi_{\psi^n(f)}(x/p^n) & \text{si } x \in p^n\mathbf{Z}_p,\\ 0 & \text{si } x \notin p^n\mathbf{Z}_p,\end{cases}\\ &= (\mathrm{Res}_{p^n\mathbf{Z}_p}\phi)(x).\end{aligned}$$

Ceci démontre le résultat si $U = p^n\mathbf{Z}_p$. Le cas $U = i + p^n\mathbf{Z}_p$ s'en déduit, grâce à la prop. 8.12, en conjuguant par $\big(\begin{smallmatrix} 1 & i \\ 0 & 1 \end{smallmatrix}\big)$, et le cas général par linéarité.

- $\phi_{\partial f}(x) = \mathrm{rés}_0\big((\partial((1+T)^{-x} f) + x(1+T)^{-x} f)\,\frac{dT}{1+T}\big) = x\phi_f(x)$.
- $\phi_{tf}(x) = \mathrm{rés}_0\big((1+T)^{-x} t f\,\frac{dT}{1+T}\big) = -\frac{d}{dx}\mathrm{rés}_0\big((1+T)^{-x} f\,\frac{dT}{1+T}\big) = -\phi'_f(x)$.

Remarque 8.14. Il ressort de la prop. 8.13 que $U \mapsto \mathscr{R} \boxtimes U$ est un faisceau sur $\mathbf{Z}_p$, équivariant pour l'action de P^+ (agissant sur $\mathbf{Z}_p$ par $\big(\begin{smallmatrix} a & b \\ 0 & 1 \end{smallmatrix}\big) \cdot x = ax + b$), et que la suite exacte de la prop. 8.12 est une suite exacte de faisceaux P^+-équivariants sur $\mathbf{Z}_p$.

Exercise 8.15. (i) Montrer que $\phi_f(x) = \mathbf{1}_{\mathbf{Z}_p^*} x^{-1}$ si $f = -\frac{1}{p}\log\frac{\varphi(T)}{T^p}$ (calculer ∂f).

(ii) On rajoute à $\mathscr{R}$ un élément $\log T$ vérifiant $\varphi(\log T) = p \log T + \log \frac{\varphi(T)}{T^p}$, $\psi(\log T) = \frac{1}{p} \log T$ et $\partial \log T = 1 + \frac{1}{T}$. Vérifier que la formule $\phi_{\log T} = \mathbf{1}_{\mathbf{Z}_p} x^{-1}$ est compatible avec le (i) et les formules de la prop. 8.13.

1.2.2. Dualité

On définit un accouplement $\{\ ,\ \}$ sur $\mathscr{R}(\delta) \times \mathscr{R}(\chi\delta^{-1})$ par la formule

$$\{f \otimes \delta, g \otimes \chi\delta^{-1}\} = \text{rés}_0\big(\sigma_{-1}(f) g \, \tfrac{dT}{1+T}\big).$$

Proposition 8.16. *L'accouplement $\{\ ,\ \}$ identifie $\mathscr{R}(\delta)$ au dual topologique de $\mathscr{R}(\chi\delta^{-1})$. De plus, si $z \in \mathscr{R}(\delta)$ et $z' \in \mathscr{R}(\chi\delta^{-1})$, alors :*

- $\{g \cdot z, g \cdot z'\} = \{z, z'\}$ *pour tout* $g \in P^+$,
- $\{\psi(z), z'\} = \{z, \varphi(z')\}$ *et* $\{z, \psi(z')\} = \{\varphi(z), z'\}$,
- $\{\text{Res}_U z, z'\} = \{\text{Res}_U z, \text{Res}_U z'\} = \{z, \text{Res}_U z'\}$ *pour tout ouvert compact* U *de* $\mathbf{Z}_p$,
- $\{\partial z, z'\} = \{z, \partial z'\}$, *où l'on a posé* $\partial(f \otimes \eta) = (\partial f) \otimes \eta$ *si* $\eta \in \widehat{\mathscr{T}}(L)$.
- $\{tz, z'\} = -\{z, tz'\}$.

Démonstration. Pour l'identification avec le dual et les deux premiers points, cf. [12, prop. III.2.3] et [14, § I.1]. La propriété d'adjonction de Res_U pour U quelconque suit du cas particulier $U = i + p^n\mathbf{Z}_p$, où l'on a $\text{Res}_U = \left(\begin{smallmatrix} 1 & i \\ 0 & 1 \end{smallmatrix}\right) \circ \varphi^n \circ \psi^n \circ \left(\begin{smallmatrix} 1 & -i \\ 0 & 1 \end{smallmatrix}\right)$, pour lequel on peut appliquer le premier point puis deux fois le second puis de nouveau le premier ; la formule $\{\text{Res}_U z, z'\} = \{\text{Res}_U z, \text{Res}_U z'\}$ suit de la propriété d'adjonction et de ce que Res_U est un projecteur $[\,\text{Res}_U(\text{Res}_U(z)) = \text{Res}_U(z)\,]$.

Le quatrième point suit de ce que $\partial \circ \sigma_{-1} = -\sigma_{-1} \circ \partial$ et $(\partial f)\, g + f\,(\partial g) = \partial(fg)$, ce qui fait que l'on a, si $z = f \otimes \delta$ et $z' = g \otimes \chi\delta^{-1}$,

$$-\{\partial z, z'\} + \{z, \partial z'\} = \text{rés}_0\big(\partial(\sigma_{-1}(f) g)\, \tfrac{dT}{1+T}\big) = 0.$$

Enfin, le cinquième est immédiat sur la formule définissant $\{\ ,\ \}$ si on remarque que $\sigma_{-1}(t) = -t$.

Remarque 8.17. L'accouplement $\{\ ,\ \}$ est nul sur $\mathscr{R}^+(\delta) \times \mathscr{R}^+(\chi\delta^{-1})$. Il induit donc un accouplement sur $\mathscr{R}^+(\delta) \times (\mathscr{R}/\mathscr{R}^+)(\chi\delta^{-1}) \cong (\mathscr{D}(\mathbf{Z}_p) \otimes \delta) \times (\text{LA}(\mathbf{Z}_p) \otimes \delta^{-1})$. Le (iv) de la prop. 8.11 montre que l'accouplement ainsi obtenu est l'accouplement naturel donné par l'intégration. Nous laissons au lecteur le plaisir d'expliciter ce que deviennent les formules de la prop. 8.16 quand on les traduit en termes d'intégration sur $\mathbf{Z}_p$.

1.2.3. Le théorème d'Amice

Terminons ce § par une démonstration du théorème d'Amice (cor. 8.20) par dualité. Si $h \in \mathbf{N}$, soit $\widehat{\mathrm{PD}}_h$ le sous-$\mathscr{O}_L$-module de $L[[T]]$ des $\sum_{n\in\mathbf{N}} a_n T^n/[\frac{n}{p^h}]!$, où les a_n appartiennent à $\mathscr{O}_L$.

Lemme 8.18. *$\widehat{\mathrm{PD}}_h$ est le sous-$\mathscr{O}_L[[T]]$-module de $\widehat{\mathrm{PD}}_0$ engendré par $\varphi^h(\widehat{\mathrm{PD}}_0)$.*

Démonstration. Comme $\varphi^h(T) - T^{p^h} \in p\mathscr{O}_L[[T]]$, la matrice exprimant les $\frac{\varphi^h(T)^n}{n!}$ en termes des $\frac{T^{np^h}}{n!}$ est triangulaire supérieure, à coefficients dans $\mathscr{O}_L[[T]]$, avec des 1 sur la diagonale. Le résultat s'en déduit.

Si $h \in \mathbf{N}$, on note $\mathrm{LA}_h(\mathbf{Z}_p)$ l'ensemble des $\phi : \mathbf{Z}_p \to L$, analytiques sur $a + p^h\mathbf{Z}_p$ pour tout $a \in \mathbf{Z}_p$, et on note LA_h^0 la boule unité de $\mathrm{LA}_h(\mathbf{Z}_p)$: on a $\phi \in \mathrm{LA}_h^0$ si et seulement si, pour tout $a \in \mathbf{Z}_p$ (il suffit de le vérifier pour a décrivant un système de représentants modulo p^h), la fonction $x \mapsto \phi(a + p^h x)$ est de la forme $\sum_{i=0}^{+\infty} \alpha_i x^i$, avec $\alpha_i \in \mathscr{O}_L$ pour tout i et $\alpha_i \to 0$ quand $i \to +\infty$.

Proposition 8.19. *Si $f = \sum_{n\geq 1} a_n T^{-1}$, les conditions suivantes sont équivalentes :*

(i) $v_p(a_n) \geq v_p([\frac{n}{p^h}]!)$ *et* $v_p(a_n) - v_p([\frac{n}{p^h}]!) \to +\infty$ *quand* $n \to +\infty$.

(ii) $z \mapsto \{z, f\}$ *s'étend en une forme linéaire continue de $\widehat{\mathrm{PD}}_h$ dans $\mathscr{O}_L$.*

(iii) $\phi_f \in \mathrm{LA}_h^0$.

Démonstration. L'équivalence entre (i) et (ii) est immédiate. Dans le cas $h = 0$, l'équivalence entre (ii) et (iii) vient de ce que la matrice des $n!\binom{x}{n}$ en fonction des x^n est triangulaire supérieure, à coefficients dans $\mathbf{Z}_p$, avec des 1 sur la diagonale. Dans le cas général, le lemme 8.18 implique que (ii) équivaut à $z \mapsto \{(1+T)^i\varphi^h(z), f\} \in \mathscr{O}_L$ et donc à ce que $z \mapsto \{z, \psi^h((1+T)^{-i} f)\} \in \mathscr{O}_L$, pour tout $z \in \widehat{\mathrm{PD}}_0$ et tout $i \in \mathbf{Z}_p$. Or le cas $h = 0$ montre que ceci équivaut à ce que $\phi_{\psi^h((1+T)^{-i} f)} \in \mathrm{LA}_0$, pour tout $i \in \mathbf{Z}_p$, et comme $\phi_{\psi^h((1+T)^{-i} f)}(x) = \phi_f(p^h x + i)$, cela équivaut à $\phi_f \in \mathrm{LA}_h^0$. Ceci permet de conclure.

Corollaire 8.20. *Les $[\frac{n}{p^h}]!\binom{x}{n}$, pour $n \in \mathbf{N}$, forment une base orthonormale de $\mathrm{LA}_h(\mathbf{Z}_p)$.*

Démonstration. C'est une traduction de l'équivalence entre les (i) et (iii) de la prop. 8.19.

1.3. Extensions de (φ, Γ)-modules et cohomologie de A^+

On s'intéresse aux (φ, Γ)-modules extensions de $\mathscr{R}(\delta_2)$ par $\mathscr{R}(\delta_1)$. Le calcul du groupe $\mathrm{Ext}^1(\mathscr{R}(\delta_2), \mathscr{R}(\delta_1)) \cong \mathrm{Ext}^1(\mathscr{R}, \mathscr{R}(\delta))$, avec $\delta = \delta_2^{-1}\delta_1$, a été effectué dans [10] ; ces résultats ont été complétés par Liu [26] et repris par Chenevier [8] en utilisant le dictionnaire d'analyse p-adique pour dévisser le module $\mathscr{R}(\delta)$. Dans ce qui suit, nous poussons la méthode de Chenevier un peu plus loin, ce qui nous permet de donner une description plus précise du groupe $\mathrm{Ext}^1(\mathscr{R}(\delta_2), \mathscr{R}(\delta_1))$, description qui nous sera utile pour déterminer les vecteurs localement analytiques dans le cas spécial.

Pour ce faire, remarquons qu'un (φ, Γ)-module est un A^+-module, où A^+ est le sous-semi-groupe $\left(\begin{smallmatrix} \mathbf{Z}_p - \{0\} & 0 \\ 0 & 1 \end{smallmatrix}\right)$ de P^+, et qu'on a un isomorphisme $\mathrm{Ext}^1(\mathscr{R}, \mathscr{R}(\delta)) \cong H^1(A^+, \mathscr{R}(\delta))$: si D est une extension de $\mathscr{R}$ par $\mathscr{R}(\delta)$, et si $e \in D$ est un relèvement de $1 \in \mathscr{R}$, alors $g \mapsto (g-1)\cdot e$ est un 1-cocycle sur A^+ qui est un cobord si et seulement si l'extension est scindée. Le calcul de $H^i(A^+, \mathscr{R}(\delta))$ se fait en utilisant la suite exacte $0 \to \mathscr{D}(\mathbf{Z}_p) \otimes \delta \to \mathscr{R}(\delta) \to \mathrm{LA}(\mathbf{Z}_p) \otimes \chi^{-1}\delta \to 0$ de A^+-modules et les résultats suivants pour lesquels nous renvoyons à l'appendice A.

On a $A^+ = \Phi^+ \times A^0$, où Φ^+ est le semi-groupe $\left(\begin{smallmatrix} p^{\mathbf{N}} & 0 \\ 0 & 1 \end{smallmatrix}\right)$ et A^0 le groupe $\left(\begin{smallmatrix} \mathbf{Z}_p^* & 0 \\ 0 & 1 \end{smallmatrix}\right)$. On note $\varphi = \left(\begin{smallmatrix} p & 0 \\ 0 & 1 \end{smallmatrix}\right)$ le générateur de Φ et σ_a, si $a \in \mathbf{Z}_p^*$, l'élément $\left(\begin{smallmatrix} a & 0 \\ 0 & 1 \end{smallmatrix}\right)$. Tout élément g de A^+ s'écrit alors, de manière unique, sous la forme $\varphi^{k(g)}\sigma_{a(g)} = \sigma_{a(g)}\varphi^{k(g)}$, avec $k(g) \in \mathbf{N}$ et $a(g) \in \mathbf{Z}_p^*$. Enfin, on peut décomposer A^0 sous la forme $\Delta \times A^1$, où Δ est un groupe fini et A^1 est un groupe topologiquement isomorphe à $\mathbf{Z}_p$ et dont on note σ_u un générateur (alors u est un générateur de $1 + 2p\mathbf{Z}_p$).

Si M est un L-espace vectoriel topologique muni d'une action continue de A^+, la suite spectrale de Hochschild-Serre nous fournit des isomorphismes

$$H^0(A^+, M) \cong H^0(A^0, H^0(\Phi^+, M)) \;\text{ et }\; H^2(A^+, M) \cong H^1(A^0, H^1(\Phi^+, M)),$$

et une suite exacte

$$0 \to H^1(A^0, H^0(\Phi^+, M)) \to H^1(A^+, M) \to H^0(A^0, H^1(\Phi^+, M)) \to 0.$$

De plus, $H^i(A^0, M_0) = H^0(\Delta, H^i(A^1, M_0))$, si $M_0 = H^j(\Phi^+, M)$ et $j = 0, 1$. Comme $H^0(\Phi^+, M)$ et $H^1(\Phi^+, M)$ s'identifient respectivement au noyau et conoyau de $\varphi - 1$ agissant sur M, et $H^0(A^1, M_0)$ et $H^1(A^1, M_0)$ à ceux de $\sigma_u - 1$ agissant sur M_0, cela donne une description parfaitement concrète des groupes $H^i(A^+, M)$.

Par ailleurs, si M et M^* sont en dualité, M^* est muni d'actions des semi-groupes opposés $\Phi^- = \psi^{\mathbf{N}}$ et A^- de Φ^+ et A^+, et on a le même dévissage de $H^j(A^-, M^*)$ que ci-dessus en remplaçant A^+, Φ^+, M par A^-, Φ^-, M^*.

Proposition 8.21. *On suppose que :*

- $(\varphi - 1) \cdot M$ *et* $(\psi - 1) \cdot M^*$ *sont fermés dans* M *et* M^*,
- $(\sigma_u - 1) \cdot M_0$ *est fermé dans* M_0, *si* M_0 *est un des modules* $H^j(\Phi^+, M)$, *pour* $j = 0, 1$, *et* $H^j(\Phi^-, M^*)$, *pour* $j = 0, 1$.

Alors on a des identifications naturelles :

$$H^j(A^-, M^*) \cong H^{2-j}(A^+, M)^* \quad \text{et} \quad H^j(A^+, M) \cong H^{2-j}(A^-, M^*)^*, \quad \textit{pour } j = 0, 1, 2.$$

1.4. Cohomologie de Φ^+ et Φ^-

Soit $\delta \in \widehat{\mathscr{T}}(L)$. On a déjà expliqué comment voir δ comme un caractère de $A^+ \subset P^+$; on peut aussi le voir comme un caractère de A^- en posant $\delta(\psi^n \sigma_a) = \delta(p)^{-n} \delta(a)$. Le dual du A^+-module $M \otimes \delta$ est alors le A^--module $M^* \otimes \delta^{-1}$.

1.4.1. Les groupes $H^i(\Phi^\pm, \mathrm{LA}(\mathbf{Z}_p) \otimes \delta)$ et $H^i(\Phi^\pm, \mathscr{D}(\mathbf{Z}_p) \otimes \delta^{-1})$

La proposition suivante détermine la structure de $H^i(X, M)$, comme A^0 module, si $X = \Phi^+, \Phi^-$ et $M = \mathrm{LA}(\mathbf{Z}_p) \otimes \delta$ ou $M = \mathscr{D}(\mathbf{Z}_p) \otimes \delta$ (si η est un caractère de $\mathbf{Z}_p^*$, on note $L(\eta)$ le L-espace vectoriel de dimension 1 sur lequel $\sigma_a \in A^0$ agit par multiplication par $\eta(a)$). On a bien évidemment $H^i(X, M) = 0$, si $i \neq 0, 1$.

Proposition 8.22. *Les images de* $\varphi - 1$ *et* $\psi - 1$ *sont fermées dans* $\mathrm{LA}(\mathbf{Z}_p) \otimes \delta$ *et* $\mathscr{D}(\mathbf{Z}_p) \otimes \delta^{-1}$. *De plus :*

(i) $H^0(\Phi^+, \mathrm{LA}(\mathbf{Z}_p) \otimes \delta) = 0$ *et* $H^1(\Phi^-, \mathscr{D}(\mathbf{Z}_p) \otimes \delta^{-1}) = 0$.

(ii) *Si* $\delta(p) \notin \{p^i,\ i \in \mathbf{N}\}$, *alors*

- $H^0(\Phi^+, \mathscr{D}(\mathbf{Z}_p) \otimes \delta^{-1}) = H^1(\Phi^+, \mathscr{D}(\mathbf{Z}_p) \otimes \delta^{-1}) = 0$,
- $H^1(\Phi^+, \mathrm{LA}(\mathbf{Z}_p) \otimes \delta) \cong \mathrm{LA}(\mathbf{Z}_p^*) \otimes \delta$ *et* $H^0(\Phi^-, \mathscr{D}(\mathbf{Z}_p) \otimes \delta^{-1}) \cong \mathscr{D}(\mathbf{Z}_p^*) \otimes \delta^{-1}$,
- $H^0(\Phi^-, \mathrm{LA}(\mathbf{Z}_p) \otimes \delta) = H^1(\Phi^-, \mathrm{LA}(\mathbf{Z}_p) \otimes \delta) = 0$.

(iii) *Si* $\delta(p) = p^i$, *avec* $i \in \mathbf{N}$, *alors*

- $H^0(\Phi^+, \mathscr{D}(\mathbf{Z}_p) \otimes \delta^{-1}) \cong H^1(\Phi^+, \mathscr{D}(\mathbf{Z}_p) \otimes \delta^{-1}) \cong L(x^i \delta^{-1})$,

- $H^1(\Phi^+, \mathrm{LA}(\mathbf{Z}_p)\otimes\delta)$ *et* $H^0(\Phi^-, \mathscr{D}(\mathbf{Z}_p)\otimes\delta^{-1})$ *vivent dans des suites exactes*

$$0 \to L(x^{-i}\delta) \longrightarrow \mathrm{LA}(\mathbf{Z}_p^*)\otimes\delta \longrightarrow H^1(\Phi^+, \mathrm{LA}(\mathbf{Z}_p)\otimes\delta) \longrightarrow L(x^{-i}\delta) \longrightarrow 0$$
$$0 \to L(x^i\delta^{-1}) \longrightarrow H^0(\Phi^-, \mathscr{D}(\mathbf{Z}_p)\otimes\delta^{-1}) \longrightarrow \mathscr{D}(\mathbf{Z}_p^*)\otimes\delta^{-1} \longrightarrow L(x^i\delta^{-1}) \to 0,$$

- $H^0(\Phi^-, \mathrm{LA}(\mathbf{Z}_p)\otimes\delta) \cong H^1(\Phi^-, \mathrm{LA}(\mathbf{Z}_p)\otimes\delta) \cong L(x^{-i}\delta)$.

Démonstration. $H^0(\Phi^+, M\otimes\delta)$ et $H^0(\Phi^-, M\otimes\delta)$ sont les noyaux de $\delta(p)\varphi-1$ et $\delta(p)^{-1}\psi-1$ agissant sur M, et $H^1(\Phi^+, M\otimes\delta)$ et $H^1(\Phi^-, M\otimes\delta)$ sont les conoyaux de $\delta(p)\varphi-1$ et $\delta(p)^{-1}\psi-1$ agissant sur M. La proposition est donc la combinaison des lemmes 8.25, 8.27, 8.28 et 8.29 ci-dessous.

Remarque 8.23. (i) Dans le cas $\delta(p) = p^i$, avec $i \in \mathbf{N}$, le sous-espace $L(x^{-i}\delta)$ de $\mathrm{LA}(\mathbf{Z}_p^*)\otimes\delta$ est la droite engendrée par $\mathbf{1}_{\mathbf{Z}_p^*}x^i$, l'application de $H^1(\Phi^+, \mathrm{LA}(\mathbf{Z}_p)\otimes\delta)$ dans $L(x^{-i}\delta)$ est celle qui envoie le cocyle $g\mapsto\phi_g$ sur $\phi_g^{(i)}(0)$. Le sous-espace $L(x^i\delta^{-1})$ de $H^0(\Phi^-, \mathscr{D}(\mathbf{Z}_p)\otimes\delta^{-1})$ est la droite engendrée par $\mathrm{d}^i\,\mathrm{Dir}_0$, l'application de $\mathscr{D}(\mathbf{Z}_p^*)\otimes\delta^{-1}$ dans $L(x^i\delta^{-1})$ est $\mu\mapsto\int_{\mathbf{Z}_p^*}x^i\mu$.

(ii) Les modules $\mathrm{LA}(\mathbf{Z}_p)\otimes\delta$ et $\mathscr{D}(\mathbf{Z}_p)\otimes\delta^{-1}$ sont duaux l'un de l'autre. Comme les images de $\varphi-1$ et $\psi-1$ sont fermées, on sait a priori que $H^j(\Phi^\pm, \mathrm{LA}(\mathbf{Z}_p)\otimes\delta)$ et $H^{1-j}(\Phi^\mp, \mathscr{D}(\mathbf{Z}_p)\otimes\delta^{-1})$ sont duaux l'un de l'autre. Cela peut se vérifier sur la description qu'en donne la proposition.

1.4.2. Un peu de topologie

Dans tous les énoncés qui suivent, « isomorphisme » signifie « isomorphisme d'espaces vectoriels topologiques ». Grâce au théorème de l'image ouverte pour les banachs et les fréchets, il suffit en général de vérifier la bijectivité (la continuité est le plus souvent claire) pour prouver qu'un morphisme L-linéaire est un isomorphisme.

Lemme 8.24. (i) *Soit E un L-espace vectoriel topologique. Si X est un sous-espace fermé, de codimension finie, de E, alors tout sous-espace de E qui contient X est fermé dans E.*

(ii) *Si E est un L-banach (ou, plus généralement, un L-fréchet), et si u : $E\to E$ est continue, alors* $\mathrm{Im}(u)$ *est fermée dans E si elle est de codimension finie.*

Démonstration. Le (i) suit juste de ce que E/X est séparé (puisque X est fermé) et de dimension finie, et donc tout sous-espace de E/X est fermé.

Pour démontrer le (ii), on choisit un supplémentaire F de $\mathrm{Im}(u)$ dans E. L'hypothèse implique que F est de dimension finie et donc est un banach (ou un fréchet si l'on préfère). L'application $\mathrm{id}\oplus u : F\oplus E \to E$ est surjective, et si on note G le noyau de u, elle induit un isomorphisme de $F\oplus E/G \cong F\oplus \mathrm{Im}(u)$ sur E, ce qui permet de conclure puisque $\mathrm{Im}(u)$ est fermée dans $F \oplus \mathrm{Im}(u)$.

1.4.3. Actions de φ et ψ sur $\mathrm{LA}(\mathbf{Z}_p)$

Lemme 8.25. *Soit $\alpha \in L^*$.*

(i) *Si $\alpha \notin \{p^{-i},\ i \in \mathbf{N}\}$, alors $1 - \alpha\psi : \mathrm{LA}(\mathbf{Z}_p) \to \mathrm{LA}(\mathbf{Z}_p)$ est un isomorphisme.*

(ii) *Si $\alpha = p^{-i}$, avec $i \in \mathbf{N}$, le noyau de $1 - \alpha\psi : \mathrm{LA}(\mathbf{Z}_p) \to \mathrm{LA}(\mathbf{Z}_p)$ est $L\cdot \mathbf{1}_{\mathbf{Z}_p}x^i$ et $1-\alpha\psi$ induit un isomorphisme de $\{\phi \in \mathrm{LA}(\mathbf{Z}_p),\ \phi^{(i)}(0) = 0\}$ sur lui-même.*

Démonstration. Si ϕ est dans le noyau de $1 - \alpha\psi$, c'est que $\phi(x) = \alpha\phi(px)$, pour tout x. Alors $\phi(x) = \alpha^n\phi(p^n x)$, pour tout x, ce qui montre que ϕ est analytique sur $\mathbf{Z}_p$ tout entier, et que si $\phi(x) = \sum_{i\geq 0} a_i x^i$ sur $\mathbf{Z}_p$, alors $\sum_{i=0}^{+\infty} a_i(1 - \alpha p^i)x^i$ est identiquement nul sur $\mathbf{Z}_p$. Ceci implique que $a_i(1 - \alpha p^i) = 0$, pour tout i ; on en déduit que le noyau est nul si $\alpha \notin \{p^{-i}$, $i \in \mathbf{N}\}$, et que c'est $L \cdot \mathbf{1}_{\mathbf{Z}_p}x^i$, si $\alpha = p^{-i}$ avec $i \in \mathbf{N}$.

Maintenant, soit $\phi \in \mathrm{LA}(\mathbf{Z}_p)$, et soit $k \in \mathbf{N}$ tel que $k + v_p(\alpha) > 0$. On peut écrire ϕ sous la forme $\sum_{i=0}^{k-1} a_i x^i + \phi_k$, où ϕ_k a un zéro d'ordre $\geq k$ en zéro. La série $\sum_{n=0}^{+\infty} \alpha^n \phi_k(p^n x)$ converge dans $\mathrm{LA}(\mathbf{Z}_p)$ et la somme g vérifie $(1 - \alpha\psi)\cdot g = \phi_k$. Par ailleurs, si $p^i\alpha \neq 1$, on a $(1 - \alpha\psi)(\frac{a_i}{1-p^i\alpha}\mathbf{1}_{\mathbf{Z}_p}x^i) = \mathbf{1}_{\mathbf{Z}_p}x^i$. Il s'ensuit que $1 - \alpha\psi$ est surjectif si $\alpha \notin \{p^{-i},\ i \in \mathbf{N}\}$ et que ϕ est dans l'image de $1 - \alpha\psi$ si et seulement si $a_i = 0$ si $\alpha = p^{-i}$. On en déduit le résultat.

Lemme 8.26. *Soit $\alpha \in L^*$.*

(i) *Si $\alpha \notin \{p^i,\ i \in \mathbf{N}\}$, il existe $u : \mathrm{LA}(\mathbf{Z}_p) \to \mathrm{LA}(\mathbf{Z}_p)$, linéaire continu, tel que $u \circ (1 - \alpha\varphi)$ soit l'identité sur $\mathrm{LA}(\mathbf{Z}_p)$.*

(ii) *Si $\alpha = p^i$, avec $i \in \mathbf{N}$, il existe $u : \mathrm{LA}(\mathbf{Z}_p) \to \mathrm{LA}(\mathbf{Z}_p)$, linéaire continu, tel que l'image de $u \circ (1 - \alpha\varphi) - 1$ soit contenue dans $L \cdot \mathbf{1}_{\mathbf{Z}_p}x^i$.*

Démonstration. Soit k tel que $k - v_p(\alpha) > 0$. Si $\phi \in \mathrm{LA}$ (resp. $\phi \in \mathrm{LA}$ vérifie $\phi^{(i)}(0)$, si $\alpha = p^i$, avec $i \in \mathbf{N}$), on peut écrire ϕ, de manière unique, sous la forme $\sum_{j=0}^{k-1} a_j(1-\alpha\varphi)\mathbf{1}_{\mathbf{Z}_p}x^i + \phi_k$, où ϕ_k a un zéro d'ordre k en 0 (avec $a_i = 0$ si $\alpha = p^i$). La série $-\sum_{n\geq 1}\alpha^{-n}\psi^n(\phi_k)$ converge dans $\mathrm{LA}(\mathbf{Z}_p)$, ce qui permet de définir u par la formule $u(\phi) = \sum_{j=0}^{k-1} a_j\mathbf{1}_{\mathbf{Z}_p}x^i - \sum_{n\geq 1}\alpha^{-n}\psi^n(\phi_k)$. Pour vérifier que $u\circ(1-\alpha\varphi)$ vérifie les propriétés désirées, il suffit alors de prouver que $u\circ(1-\alpha\varphi)\cdot f = f$ si f a un zéro d'ordre au moins k car le résultat est évident pour un polynôme de degré $\leq k-1$. Or, dans ce cas, $\phi = (1-\alpha\varphi)f$ a un zéro d'ordre au moins k en 0, et donc $\phi = \phi_k$. Comme $(1-\alpha\varphi)f(x) = f(x) - \alpha\mathbf{1}_{p\mathbf{Z}_p}f(\frac{x}{p})$, et donc $\psi^n((1-\alpha\varphi)f)(x) = f(p^n x) - \alpha f(p^{n-1}x)$ si $n \geq 1$, on obtient:

$$\begin{aligned} u(\phi)(x) &= -\Big(\sum_{n\geq 1}\alpha^{-n}(f(p^n x) - \alpha f(p^{n-1}x))\Big) \\ &= \sum_{n\geq 0}\alpha^{-n}f(p^n x) - \sum_{n\geq 1}\alpha^{-n}f(p^n x) = f(x). \end{aligned}$$

On en déduit le résultat.

Lemme 8.27. *Soit $\alpha \in L^*$. Alors $1-\alpha\varphi : \mathrm{LA}(\mathbf{Z}_p) \to \mathrm{LA}(\mathbf{Z}_p)$ est injectif et son image est fermée.*

De plus:

- *si $\alpha \notin \{p^i,\ i \in \mathbf{N}\}$, alors $\mathrm{LA}(\mathbf{Z}_p^*)$ est un supplémentaire de $(1-\alpha\varphi)\mathrm{LA}(\mathbf{Z}_p)$ dans $\mathrm{LA}(\mathbf{Z}_p)$.*
- *Si $\alpha = p^i$, avec $i \in \mathbf{N}$, alors $\mathrm{LA}(\mathbf{Z}_p^*) \cap (1-\alpha\varphi)\mathrm{LA}(\mathbf{Z}_p)$ est la droite $L\cdot\mathbf{1}_{\mathbf{Z}_p^*}x^i$ et $\mathrm{LA}(\mathbf{Z}_p^*) + (1-\alpha\varphi)\mathrm{LA}(\mathbf{Z}_p)$ est l'orthogonal de $\mathrm{d}^i\mathrm{Dir}_0$.*

Démonstration. Si ϕ est dans le noyau de $1-\alpha\varphi$, on a $\phi = \alpha^n\varphi^n(\phi)$, pour tout n. Par suite ϕ est à support dans $p^n\mathbf{Z}_p$ pour tout n, et donc est identiquement nulle par continuité en 0. D'où l'injectivité de $1-\alpha\varphi$.

Que l'image de $1-\alpha\varphi$ soit fermée résulte du lemme précédent (si $(1-\alpha\varphi)\phi_n$ tend vers une limite, il en est de même de $u((1-\alpha\varphi)\phi_n) = \phi_n$; si $\alpha = p^i$, cet argument montre que l'image de $\{\phi,\ \phi^{(i)}(0) = 0\}$ est fermée, et comme elle est de codimension 1 dans $(1-\alpha\varphi)\mathrm{LA}(\mathbf{Z}_p)$, cela prouve que $(1-\alpha\varphi)\mathrm{LA}(\mathbf{Z}_p)$ est aussi fermé).

Si $(1-\alpha\varphi)\phi$ est à support dans $\mathbf{Z}_p^*$, on a $\psi((1-\alpha\varphi)\phi) = 0$, et donc $(\psi-\alpha)\cdot\phi = 0$. D'après le lemme 8.25, cela implique que $\phi = 0$ sauf si $\alpha = p^i$ ou cela implique que $\phi \in L\cdot\mathbf{1}_{\mathbf{Z}_p}x^i$.

Par ailleurs, si $\phi \in \mathrm{An}(p^n\mathbf{Z}_p)$ (et vérifie $\phi^{(i)}(0) = 0$ dans le cas $\alpha = p^i$), on peut écrire ϕ sous la forme $\sum_{j=0}^{+\infty} a_j\mathbf{1}_{p^n\mathbf{Z}_p}x^j$ (avec $a_i = 0$ si $\alpha = p^i$) et alors la fonction $\phi - (1-\alpha\varphi)\big(\sum_{j=0}^{+\infty}\frac{a_j}{1-\alpha p^{-j}}\mathbf{1}_{p^n\mathbf{Z}_p}x^j\big)$ est identiquement nulle sur $p^{n+1}\mathbf{Z}_p$. Il s'ensuit que l'on peut écrire tout élément ϕ de $\mathrm{LA}(\mathbf{Z}_p)$

(vérifiant $\phi^{(i)}(0) = 0$ dans le cas $\alpha = p^i$), sous la forme $\phi_1 + (1 - \alpha\varphi)\phi_2$, avec ϕ_1 nulle dans un voisinage de 0 et donc de la forme $\sum_{n=0}^{N} \phi_{1,n}$, avec $\phi_{1,n}$ à support dans $p^n\mathbf{Z}_p^*$. On peut alors écrire $\phi_{1,n}$ sous la forme $\varphi^n\psi^n(\phi_{1,n}) = ((1-\alpha\varphi)-1)^n\psi^n(\alpha^{-1}\phi_{1,n})$, et un développement de $((1-\alpha\varphi)-1)^n$ exprime $\phi_{1,n}$ comme la somme d'un élément de $(1 - \alpha\varphi)\mathrm{LA}(\mathbf{Z}_p)$ et de $\psi^n(\alpha^{-1}\phi_{1,n})$ qui est à support dans $\mathbf{Z}_p^*$.

On en déduit le résultat.

1.4.4. Actions de φ et ψ sur $\mathscr{D}(\mathbf{Z}_p)$

Lemme 8.28. *Soit* $\alpha \in L^*$.

(i) *Si* $\alpha \notin \{p^{-i},\ i \in \mathbf{N}\}$, *alors* $1 - \alpha\varphi : \mathscr{D}(\mathbf{Z}_p) \to \mathscr{D}(\mathbf{Z}_p)$ *est un isomorphisme.*

(ii) *Si* $\alpha = p^{-i}$, *avec* $i \in \mathbf{N}$, *le noyau de* $1 - \alpha\varphi : \mathscr{D}(\mathbf{Z}_p) \to \mathscr{D}(\mathbf{Z}_p)$ *est* $L\cdot \mathrm{d}^i\,\mathrm{Dir}_0$ *et* $1-\alpha\varphi$ *induit un isomorphisme de* $\{\mu \in \mathscr{D}(\mathbf{Z}_p),\ \int_{\mathbf{Z}_p} x^i\,\mu = 0\}$ *sur lui-même.*

Démonstration. Comme $\langle(1 - \alpha\varphi)\mu, \phi\rangle = \langle\mu, (1 - \alpha\psi)\phi\rangle$, l'énoncé est une traduction, par dualité, du lemme 8.25.

Lemme 8.29. *Soit* $\alpha \in L^*$.

(i) $1 - \alpha\psi : \mathscr{D}(\mathbf{Z}_p) \to \mathscr{D}(\mathbf{Z}_p)$ *est surjectif.*

(ii) *Si* $\alpha \notin \{p^i,\ i \in \mathbf{N}\}$, *alors* $\mathrm{Res}_{\mathbf{Z}_p^*}$ *induit un isomorphisme du noyau de* $1 - \alpha\psi$ *sur* $\mathscr{D}(\mathbf{Z}_p^*)$.

(iii) *Si* $\alpha = p^i$, *avec* $i \in \mathbf{N}$, *le noyau de* $\mathrm{Res}_{\mathbf{Z}_p^*}$ *sur* $\mathrm{Ker}(1 - \alpha\psi)$ *est* $L \cdot \mathrm{d}^i\,\mathrm{Dir}_0$ *et l'image est l'espace* $\{\mu \in \mathscr{D}(\mathbf{Z}_p^*),\ \int_{\mathbf{Z}_p^*} x^i\,\mu = 0\}$, *qui est de codimension* 1 *dans* $\mathscr{D}(\mathbf{Z}_p^*)$.

Démonstration. Le (i) se déduit du lemme 8.26 par dualité, si $\alpha \notin \{p^i,\ i \in \mathbf{N}\}$ (on a $\mu = (1 - \alpha\varphi)\cdot {}^t u(\mu)$). Dans le cas $\alpha = p^i$, avec $i \in \mathbf{N}$, le lemme 8.26 permet de montrer que l'image de $1 - \alpha\psi$ contient le sous-espace de codimension 1 des μ vérifiant $\int_{\mathbf{Z}_p} x^i\,\mu = 0$. Comme $(1 - p^i\psi)(\mathrm{d}^i\,\mathrm{Dir}_1) = \mathrm{d}^i\,\mathrm{Dir}_1$ n'appartient pas à ce sous-espace, il s'ensuit que l'image de $1 - \alpha\psi$ est $\mathscr{D}(\mathbf{Z}_p)$ tout entier, ce qui prouve le (i) dans ce cas aussi.

Les (ii) et (iii) sont des traductions du lemme 8.27.

1.5. Cohomologie de A^0

1.5.1. Sur $\mathscr{R}(\delta)$

On rappelle que A^n est le sous-groupe $\left(\begin{smallmatrix} 1+2p^n\mathbf{Z}_p & 0 \\ 0 & 1 \end{smallmatrix}\right)$ de A^0, si $n \geq 1$. On note X_n le groupe des caractères de A^0/A^n (que l'on identifie aussi au groupe des caractères de $\mathbf{Z}_p^*$, constants modulo $1+2p^n\mathbf{Z}_p$).

Lemme 8.30. *Soient $n \in \mathbf{N}$ et $\delta \in \widehat{\mathscr{T}}(L)$. Alors $L_n(\delta) \cong \oplus_{\eta \in X_n} L(\eta\delta)$.*

Démonstration. Un élément de A^0/A^n est d'ordre $(p-1)p^{n-1}$ (2^n si $p=2$), et comme F_n contient les racines de l'unité d'ordre $(p-1)p^{n-1}$ (resp. 2^n), l'action d'un élément de A^0/A^n sur L_n se diagonalise. On a donc $L_n(\delta) \cong \oplus_{\eta\in X_n} L(\eta\delta)^{m_\eta}$, avec $m_\eta \in \mathbf{N}$. Or la théorie de Galois implique que $m_\eta \leq 1$ (le quotient de deux éléments de copies de $L(\eta\delta)$ est fixe par $A^0/A^n \cong \mathrm{Gal}(L_n/L)$, et donc appartient à L) et une comparaison des dimensions implique donc que $m_\eta = 1$ pour tout η, ce qui permet de conclure.

Lemme 8.31. (i) *Si $\eta \neq 1$, alors $H^i(A^0, L(\eta)) = 0$, si $i = 0, 1$.*

(ii) *$H^0(A^0, L) = L$ et $H^1(A^0, L)$ est le L-espace vectoriel de dimension* 1 *engendré par $\sigma_a \mapsto \log a$.*

Démonstration. Le résultat concernant H^0 est immédiat. En ce qui concerne H^1, on a $H^1(A^0, L(\eta)) = H^1(A^1, H^0(\Delta, L(\eta))) \cong H^0(\Delta, L(\eta))/(\sigma_u - 1)$. Il en résulte que $H^1(A^0, L(\eta)) = 0$ sauf si Δ et σ_u agissent trivialement sur $L(\eta)$ (et donc si $\eta = 1$) où $H^1(A^0, L(\eta))$ est de dimension 1. On en déduit le résultat.

Lemme 8.32. *Soient $n \in \mathbf{N}$ et $\delta \in \widehat{\mathscr{T}}(L)$.*

(i) *Si $-w(\delta) \notin \mathbf{N}$, ou si $x^{-w(\delta)}\delta$ n'est pas constant modulo $1+2p^n\mathbf{Z}_p$, alors $H^i(A^0, L_n[[t]] \otimes \delta) = 0$, pour $i = 0, 1$.*

(ii) *Si $-w(\delta) \in \mathbf{N}$, et si $x^{-w(\delta)}\delta$ est constant modulo $1+2p^n\mathbf{Z}_p$, alors $\dim_L H^i(A^0, L_n[[t]] \otimes \delta) = 1$, pour $i = 0, 1$.*

Démonstration. Par dévissage, on se ramène à calculer $H^i(A^0, t^j L_n \otimes \delta)$, et comme $t^j L_n \otimes \delta \cong L_n(x^j\delta)$, on conclut en utilisant les lemmes 8.30 et 8.31.

Proposition 8.33. $H^0(A^0, \mathscr{R}(\delta)) = \begin{cases} 0 & si\ \delta_{|\mathbf{Z}_p^*} \notin \{x^{-i},\ i \in \mathbf{N}\}, \\ L \cdot t^i, & si\ \delta_{|\mathbf{Z}_p^*} = x^{-i},\ avec\ i \in \mathbf{N}. \end{cases}$

Démonstration. Soit $x \in \mathscr{R}(\delta)$, fixe par A^0, et soit $n \in \mathbf{N}$ tel que $x \in \mathscr{E}^{]0,r_{n-1}]}(\delta)$. Alors $\iota_n(x)$ et $\iota_n(\varphi(x))$ appartiennent au L-espace vectoriel $H^0(A^0, L_n[[t]] \otimes \delta)$ qui est de dimension ≤ 1, d'après le lemme 8.32. Comme ι_n est une injection, on en déduit qu'il existe $\alpha \in \mathbf{N}$ tel que $(\alpha\varphi - 1)x = 0$.

D'après le lemme 8.27, cela implique que x est dans l'image de $\mathscr{D}(\mathbf{Z}_p) \otimes \delta$, et donc appartient à $L \cdot t^i$, avec $i \in \mathbf{N}$, d'après le lemme 8.28 (la transformée d'Amice de $\mathrm{d}^i\mathrm{Dir}_0$ est t^i). On conclut[(4)] en utilisant le fait que $\sigma_a(t^i) = a^i t^i$, si $a \in \mathbf{Z}_p^*$.

1.5.2. Sur $\mathscr{R}(\delta) \boxtimes \mathbf{Z}_p^*$.

Soit $\eta : \mathbf{Z}_p^* \to L^*$ un caractère continu. On voit η comme un caractère de A^0.

Proposition 8.34. *L'image de $\sigma_u - 1$ est fermée dans $\mathrm{LA}(\mathbf{Z}_p^*) \otimes \eta$ et $\mathscr{D}(\mathbf{Z}_p^*) \otimes \eta$. De plus :*

(i) $H^0(A^0, \mathrm{LA}(\mathbf{Z}_p^*) \otimes \eta)$ *est le L-espace vectoriel de dimension* 1 *engendré par $\mathbf{1}_{\mathbf{Z}_p^*}\eta$, et $H^1(A^0, \mathrm{LA}(\mathbf{Z}_p^*) \otimes \eta) = 0$.*

(ii) $H^0(A^0, \mathscr{D}(\mathbf{Z}_p^*) \otimes \eta) = 0$*; si $a \mapsto \mu_a$ est un* 1*-cocycle, il existe $\lambda \in L$ tel que $\int_{\mathbf{Z}_p^*} \eta^{-1}\mu_a = \lambda \log a$, pour tout $a \in \mathbf{Z}_p^*$, et l'application $(a \mapsto \mu_a) \mapsto \lambda$ induit un isomorphisme de $H^1(A^0, \mathscr{D}(\mathbf{Z}_p^*) \otimes \eta)$ sur L.*

Démonstration. On a $H^i(A^0, M) = H^0(A^0/A^1, H^i(A^1, M))$. Le résultat est donc une traduction du cor. 8.39 à part la description explicite de $H^1(A^0, \mathscr{D}(\mathbf{Z}_p^*) \otimes \eta)$. Celle-ci résulte de ce que $\mu \mapsto \int_{\mathbf{Z}_p^*} \eta^{-1}\mu$ est A^0-équivariante de $\mathscr{D}(\mathbf{Z}_p^*) \otimes \eta$ dans L, et induit une application surjective, et donc bijective pour des raisons de dimension, de $H^1(A^0, \mathscr{D}(\mathbf{Z}_p^*) \otimes \eta)$ sur $H^1(A^0, L)$ (engendré par $\sigma_a \mapsto \log a$ d'après le lemme 8.31).

Corollaire 8.35. *On a $H^0(A^0, \mathscr{R}(\delta) \boxtimes \mathbf{Z}_p^*) = H^1(A^0, \mathscr{R}(\delta) \boxtimes \mathbf{Z}_p^*) = 0$.*

Démonstration. Pour H^0, c'est une conséquence directe de la prop. 8.33 (en effet, $t^i \notin \mathscr{R}(\delta) \boxtimes \mathbf{Z}_p^*$). Pour H^1, on utilise la suite exacte longue de cohomologie déduite de la suite exacte

$$0 \to \mathscr{D}(\mathbf{Z}_p^*) \otimes \delta \to \mathscr{R}(\delta) \boxtimes \mathbf{Z}_p^* \to \mathrm{LA}(\mathbf{Z}_p^*) \otimes \chi^{-1}\delta \to 0.$$

La nullité de $H^0(A^0, \mathscr{R}(\delta) \boxtimes \mathbf{Z}_p^*)$ et le fait que $H^0(A^0, \mathrm{LA}(\mathbf{Z}_p^*) \otimes \chi^{-1}\delta)$ et $H^1(A^0, \mathscr{D}(\mathbf{Z}_p^*) \otimes \delta)$ sont tous deux de dimension 1 impliquent que $H^1(A^0, \mathscr{R}(\delta) \boxtimes \mathbf{Z}_p^*)$ s'injecte dans $H^1(A^0, \mathrm{LA}(\mathbf{Z}_p^*) \otimes \chi^{-1}\delta)$, qui est nul d'après le (i) de la prop. 8.34.

Remarque 8.36. On a, plus généralement, $H^i(A^0, D \boxtimes \mathbf{Z}_p^*) = 0$ pour tout (φ, Γ)-module D de rang fini sur $\mathscr{R}$ (cela résulte de [14, prop. V.1.19]).

[4] Cette démonstration est assez détournée ; des techniques différentielles [16] mènent directement au résultat.

1.5.3. Cohomologie de A^1

Soit $\delta : \mathbf{Z}_p^* \to \mathscr{O}_L^*$ un caractère continu. On choisit un générateur topologique u de $1+2p\mathbf{Z}_p$.

Lemme 8.37. *$\delta(u)\sigma_u - 1$ est surjectif sur $\mathrm{LA}(\mathbf{Z}_p^*)$ et admet une section continue ; son noyau est le L-espace vectoriel de base les $\mathbf{1}_{i+2p\mathbf{Z}_p}\delta$, pour $i \in \Delta$.*

Démonstration. Remarquons que σ_u laisse stable $\mathrm{LA}(i + 2p\mathbf{Z}_p)$ pour tout $i \in \Delta$, et que le changement de variable $x = iu^{-y}$ permet de ramener l'étude de $\delta(u)\sigma_u - 1$ sur $\mathrm{LA}(i + 2p\mathbf{Z}_p)$ à celle de $\alpha\tau - 1$ sur $\mathrm{LA}(\mathbf{Z}_p)$, où $\alpha = \delta(u)$ vérifie $v_p(\alpha - 1) > 0$ car $1 + 2p\mathbf{Z}_p$ est un pro-p-groupe, et $\tau : \mathrm{LA}(\mathbf{Z}_p) \to \mathrm{LA}(\mathbf{Z}_p)$ est défini par $(\tau(\phi))(y) = \phi(y+1)$.

L'étude de $\alpha\tau-1$ peut se faire en utilisant le développement de Mahler de ϕ : si $\phi = \sum_{n\geq 0} a_n\binom{y}{n}$ et $(\alpha\tau-1)\cdot\phi = \sum_{n\geq 0} b_n\binom{y}{n}$, on a $b_n = (\alpha-1)a_n+\alpha a_{n+1}$ puisque $\tau\big(\binom{y}{n}\big) = \binom{y}{n} + \binom{y}{n-1}$. Si les b_n sont donnés ce système d'équation a une unique solution une fois a_0 fixé, à savoir celle définie par récurrence par $a_{n+1} = \alpha^{-1}(b_n - (\alpha-1)a_n)$; par ailleurs, il existe $s > 0$ tel que $v_p(b_n) \geq ns + C$, avec $C \in \mathbf{R}$, pour tout $n \in \mathbf{N}$, et une récurrence immédiate montre que $v_p(a_n) \geq (n-1)s' + C'$, avec $s' = \inf(s, v_p(\alpha-1))$ et $C' = \inf(C, v_p(a_0))$, ce qui implique que $\sum_{n\geq 0} a_n\binom{y}{n} \in \mathrm{LA}(\mathbf{Z}_p)$. En résumé, $\alpha\tau - 1$ est surjectif, admet une section continue (celle pour laquelle $a_0 = 0$), et son noyau est de dimension 1 (et donc est engendré par $y \mapsto \alpha^{-y}$). On en déduit le résultat car $\alpha^{-y} = \delta(i)^{-1}\delta(x)$.

Lemme 8.38. *$\delta(u)\sigma_u - 1$ est injectif sur $\mathscr{D}(\mathbf{Z}_p^*)$ et l'image de $\delta(u)\sigma_u - 1$ est constituée des μ vérifiant $\int_{i+2p\mathbf{Z}_p} \delta^{-1}\mu = 0$, pour tout $i \in \Delta$; en particulier, elle est fermée dans $\mathscr{D}(\mathbf{Z}_p^*)$.*

Démonstration. C'est une traduction, par dualité, du lemme 8.37.

Corollaire 8.39. (i) *$H^0(A^1, \mathrm{LA}(\mathbf{Z}_p^*)\otimes\delta)$ et $H^1(A^1, \mathscr{D}(\mathbf{Z}_p^*)\otimes\delta^{-1})$ sont de dimension $[A^0 : A^1]$ et duaux l'un de l'autre (isomorphes à $L[\Delta]$ comme A^0-modules).*
(ii) *$H^1(A^1, \mathrm{LA}(\mathbf{Z}_p^*)\otimes\delta)$ et $H^0(A^1, \mathscr{D}(\mathbf{Z}_p^*)\otimes\delta^{-1})$ sont nuls.*

Démonstration. C'est une simple traduction des lemmes 8.37 et 8.38.

Remarque 8.40. Il ne faudrait pas croire que l'image de $\sigma_u - 1$ est toujours fermée. Par exemple, on vérifie facilement, en utilisant le lemme 8.37, qu'un élément ϕ de $\mathrm{LA}(\mathbf{Z}_p)$ est dans l'image de $\delta(a)\sigma_a - 1$ si et seulement si on a $\phi(x) = \sum_{j\geq 0} a_j x^j$ au voisinage de 0 et la série $\sum_{j\geq 0} \frac{a_j}{\delta(a)a^{-j}-1}x^j$ a un rayon de convergence non nul (ce qui inclut la condition $a_i = 0$ si $\delta(x) = x^{-i}$).

On voit donc que l'image de $\sigma_u - 1$ sur $\mathrm{LA}(\mathbf{Z}_p)\otimes\delta$ est fermée si et seulement si $w(\delta)$ n'est pas un nombre de Liouville p-adique (i.e. si $\limsup \frac{1}{n} v_p(w(\delta) -n) < +\infty$). Si $w(\delta)$ est de Liouville, cette image est dense mais $\sigma_u - 1$ n'est pas surjectif.

1.6. Cohomologie de A^+

1.6.1. A valeurs dans $\mathrm{LA}(\mathbf{Z}_p) \otimes \delta$ et $\mathscr{D}(\mathbf{Z}_p) \otimes \delta^{-1}$

Proposition 8.41. *L'image de $\sigma_u - 1$ est fermée dans $H^j(\Phi^\pm, \mathrm{LA}(\mathbf{Z}_p) \otimes \delta)$ et $H^j(\Phi^\pm, \mathscr{D}(\mathbf{Z}_p) \otimes \delta^{-1})$, si $j = 0, 1$. De plus:*

(i) *Si $\delta \notin \{x^i,\ i \in \mathbf{N}\}$, alors*

- $H^j(A^+, \mathscr{D}(\mathbf{Z}_p) \otimes \delta^{-1}) = 0$, *pour* $j = 0, 1, 2$.
- $\dim_L H^j(A^+, \mathrm{LA}(\mathbf{Z}_p) \otimes \delta) = 0, 1, 0$, *si* $j = 0, 1, 2$.
- $H^j(A^-, \mathrm{LA}(\mathbf{Z}_p) \otimes \delta) = 0$, *pour* $j = 0, 1, 2$.
- $\dim_L H^1(A^-, \mathscr{D}(\mathbf{Z}_p) \otimes \delta^{-1}) = 0, 1, 0$, *si* $j = 0, 1, 2$.

(ii) *Si $\delta = x^i$, avec $i \in \mathbf{N}$, alors*

- $\dim_L H^j(A^+, \mathscr{D}(\mathbf{Z}_p) \otimes \delta^{-1}) = 1, 2, 1$ *si* $j = 0, 1, 2$.
- $\dim_L H^j(A^+, \mathrm{LA}(\mathbf{Z}_p) \otimes \delta) = 0, 2, 1$, *si* $j = 0, 1, 2$.
- $\dim_L H^j(A^-, \mathrm{LA}(\mathbf{Z}_p) \otimes \delta) = 1, 2, 1$, *si* $j = 0, 1, 2$.
- $\dim_L H^j(A^-, \mathscr{D}(\mathbf{Z}_p) \otimes \delta^{-1}) = 1, 2, 0$, *si* $j = 0, 1, 2$.

Démonstration. On a :

- $\dim_L H^0(A^\pm, M) = \dim_L H^0(A^0, H^0(\Phi^\pm, M))$,
- $\dim_L H^1(A^\pm, M) = \dim_L H^0(A^0, H^1(\Phi^\pm, M)) + \dim_L H^1(A^0, H^0(\Phi^\pm, M))$,
- $\dim_L H^2(A^\pm, M) = \dim_L H^1(A^0, H^1(\Phi^\pm, M))$.

La proposition est donc une conséquence immédiate des prop. 8.22 et 8.34 sauf si $\delta(p) = p^i$, avec $i \in \mathbf{N}$, où il faut travailler un peu plus pour calculer $H^j(A^0, H^1(\Phi^+, \mathrm{LA}(\mathbf{Z}_p) \otimes \delta))$ et $H^j(A^0, H^0(\Phi^-, \mathscr{D}(\mathbf{Z}_p) \otimes \delta^{-1}))$ et pour montrer que l'image de $\sigma_u - 1$ est fermée. Supposons donc que $\delta(p) = p^i$, avec $i \in \mathbf{N}$ et notons simplement LA et $\mathscr{D}$ les modules $\mathrm{LA}(\mathbf{Z}_p)\otimes\delta$ et $\mathscr{D}(\mathbf{Z}_p)\otimes\delta^{-1}$.

- Soit $X = \big(\mathrm{LA}(\mathbf{Z}_p^*)/L \cdot (\mathbf{1}_{\mathbf{Z}_p^*} x^i)\big) \otimes \delta$. La prop. 8.22 fournit une suite exacte $0 \to X \to H^1(\Phi^+, \mathrm{LA}) \to L(x^{-i}\delta) \to 0$. Comme $\sigma_u - 1$ est surjectif sur $\mathrm{LA}(\mathbf{Z}_p^*) \otimes \delta$, il l'est sur X, et donc l'image de $\sigma_u - 1$ est fermée dans $H^1(\Phi^+, \mathrm{LA})$ puisqu'elle contient X qui est fermé et de codimension finie. Comme $H^1(A^0, \mathrm{LA}(\mathbf{Z}_p^*) \otimes \delta) = 0$ (prop. 8.34), la suite exacte longue de cohomologie déduite de la suite exacte $0 \to L(x^{-i}\delta) \to \mathrm{LA}(\mathbf{Z}_p^*) \otimes \delta \to X \to 0$ et le lemme 8.31 montrent que $H^0(A^0, X)$ a même dimension que $H^0(A^0, \mathrm{LA}(\mathbf{Z}_p^*) \otimes \delta)$, c'est-à-dire 1, et que $H^1(A^0, X) = 0$. On déduit alors de la suite exacte longue de cohomologie déduite de la suite exacte $0 \to X \to H^1(\Phi^+, \mathrm{LA}) \to L(x^{-i}\delta) \to 0$, et de la nullité de $H^1(A^0, X)$, que :

$$\dim H^0(A^0, H^1(\Phi^+, \mathrm{LA})) = 1 + \dim H^0(A^0, L(x^{-i}\delta)) = \begin{cases} 1 & \text{si } \delta \neq x^i, \\ 2 & \text{si } \delta = x^i, \end{cases}$$

$$\dim H^1(A^0, H^1(\Phi^+, \mathrm{LA})) = \dim H^1(A^0, L(x^{-i}\delta)) = \begin{cases} 0 & \text{si } \delta \neq x^i, \\ 1 & \text{si } \delta = x^i. \end{cases}$$

- Soit $Y = \{\mu \in \mathscr{D}(\mathbf{Z}_p^*),\ \int_{\mathbf{Z}_p^*} x^i \mu = 0\} \otimes \delta^{-1}$, de telle sorte que la suite $0 \to L(x^i\delta^{-1}) \to H^0(\Phi^-, \mathscr{D}) \to Y \to 0$ soit exacte (c'est la suite exacte duale de la suite $0 \to X \to H^1(\Phi^+, \mathrm{LA}) \to L(x^{-i}\delta) \to 0$ du point précédent). On a $H^0(A^0, Y) = 0$ puisque $H^0(A^0, \mathscr{D}(\mathbf{Z}_p^*) \otimes \delta^{-1}) = 0$ et $\dim_L H^1(A^0, Y) = 1$ (cela suit de la prop. 8.34, de la suite exacte $0 \to Y \to \mathscr{D}(\mathbf{Z}_p^*) \otimes \delta^{-1} \to L(x^i\delta^{-1})$, et du lemme 8.31). On en déduit les formules

$$\dim H^0(A^0, H^0(\Phi^-, \mathscr{D})) = \dim H^0(A^0, L(x^{-i}\delta)) = \begin{cases} 0 & \text{si } \delta \neq x^i, \\ 1 & \text{si } \delta = x^i, \end{cases}$$

$$\dim H^1(A^0, H^0(\Phi^-, \mathscr{D})) = 1 + \dim H^1(A^0, L(x^{-i}\delta)) = \begin{cases} 1 & \text{si } \delta \neq x^i, \\ 2 & \text{si } \delta = x^i. \end{cases}$$

Pour montrer que l'image de $\sigma_u - 1$ agissant sur $H^0(\Phi^-, \mathscr{D})$ est fermée, il suffit alors d'utiliser le (ii) du lemme 8.24.

Remarque 8.42. Les modules $\mathscr{D}(\mathbf{Z}_p) \otimes \delta^{-1}$ et $\mathrm{LA}(\mathbf{Z}_p) \otimes \delta$ sont duaux l'un de l'autre. Il en résulte, d'après la prop. 8.21 dont les hypothèses sont vérifiées (prop. 8.22 et 8.41), que $H^j(A^-, \mathscr{D}(\mathbf{Z}_p) \otimes \delta^{-1})$ est le dual de $H^{2-j}(A^+, \mathrm{LA}(\mathbf{Z}_p) \otimes \delta)$; en particulier ces deux modules ont la même dimension sur L, ce qui est compatible avec les résultats de la prop. 8.41.

Proposition 8.43. *(i) Si $\delta \notin \{x^i,\ i \in \mathbf{N}\}$, alors $H^1(A^+, \mathrm{LA}(\mathbf{Z}_p) \otimes \delta)$ est de dimension 1, engendré par la classe du cocycle $g \mapsto (g-1)\cdot((\mathbf{1}_{\mathbf{Z}_p}\delta)\otimes\delta)$.*

(ii) Si $\delta = x^i$, avec $i \in \mathbf{N}$, alors l'application envoyant ℓ sur la classe du cocycle $g \mapsto (g-1)\cdot((\mathbf{1}_{\mathbf{Z}_p}x^i\ell)\otimes\delta)$ induit un isomorphisme de l'espace $\mathrm{Hom}(\mathbf{Q}_p^*, L)$, de dimension 2, sur $H^1(A^+, \mathrm{LA}(\mathbf{Z}_p) \otimes \delta)$.

•

Démonstration. Il est clair sur la construction que l'on a bien défini des cocycles. Comme les espaces au départ et à l'arrivée ont la même dimension, il suffit donc, pour conclure, de vérifier que les cocycles ainsi construits ne sont pas des cobords, autrement dit que l'on ne peut pas avoir $(g-1)\cdot((\phi-f)\otimes\delta) = 0$, pour tout $g \in A^+$, avec $\phi \in \mathrm{LA}$ et $f = \mathbf{1}_{\mathbf{Z}_p}\delta$ ou $f = \mathbf{1}_{\mathbf{Z}_p}x^i\ell$, ce qui est clair car la condition $(\varphi - 1)\cdot((\phi - f)\otimes\delta) = 0$ implique que ϕ et f coïncident sur $\mathbf{Z}_p - \{0\}$.

1.6.2. A valeurs dans $\mathscr{R}(\delta)$

Lemme 8.44. (i) *$H^0(A^+, \mathscr{R}(\delta)) = 0$ si $\delta \notin \{x^{-i},\ i \in \mathbf{N}\}$ et $H^0(A^0, \mathscr{R}(\delta)) = L \cdot t^i$, si $\delta = x^{-i}$, avec $i \in \mathbf{N}$.*

(ii) *$H^0(A^-, \mathscr{R}(\delta)) = 0$ si $\delta \notin \{x^{-i},\ i \in \mathbf{N}\}$ et $H^0(A^-, \mathscr{R}(\delta)) = L \cdot t^i$, si $\delta = x^{-i}$, avec $i \in \mathbf{N}$.*

Démonstration. Cela suit de la prop. 8.33 et des identités $\varphi(t^i \otimes \delta) = p^i\delta(p)\, t^i \otimes \delta$ et $\psi(t^i \otimes \delta) = p^{-i}\delta(p)^{-1}\, t^i \otimes \delta$.

Lemme 8.45. *Les images de $\varphi - 1$ et $\psi - 1$ sont fermées dans $\mathscr{R}(\delta)$.*

Démonstration. On note $\mathscr{D}$ et LA les modules $\mathscr{D}(\mathbf{Z}_p) \otimes \delta$ et $\mathrm{LA}(\mathbf{Z}_p) \otimes \chi^{-1}\delta$. On a donc une suite exacte $0 \to \mathscr{D} \to \mathscr{R}(\delta) \to \mathrm{LA} \to 0$ de Φ^+ et Φ^--modules. Si $u = \varphi, \psi$, notons X_u l'image inverse de $(u - 1)\cdot \mathrm{LA}$ dans $\mathscr{R}(\delta)$. Comme $(u - 1)\cdot \mathrm{LA}$ est fermé dans LA, on voit que X_u est fermé dans $\mathscr{R}(\delta)$.

- Si $u = \psi$ ou si $u = \varphi$ et $\delta(p) \notin \{p^{-i},\ i \in \mathbf{N}\}$, alors $u - 1$ est surjectif sur $\mathscr{D}$ et donc $(u - 1)\cdot \mathscr{R}(\delta) = X_u$, ce qui permet de conclure dans ce cas.
- Si $u = \varphi$ et $\delta(p) = p^{-i}$, avec $i \in \mathbf{N}$, alors $(u - 1)\cdot \mathscr{R}(\delta)$ est de codimension ≤ 1 dans X_u car $(u - 1)\cdot \mathscr{D}$ est de codimension 1 dans $\mathscr{D}$. Par ailleurs, $(u - 1)\cdot \mathscr{R}(\delta)$ est orthogonal au noyau de $\psi - 1$ sur $\mathscr{R}(\delta^{-1})$, et donc, en particulier, à $z = (\partial^i \frac{1}{T}) \otimes \delta^{-1}$ (on a $\psi(z) = \delta(p)p^i\partial^i(\psi(\frac{1}{T})) \otimes \delta^{-1} = z$, puisque $\delta(p)p^i = 1$ et $\psi(\frac{1}{T}) = \frac{1}{T}$). Comme l'orthogonal de z ne contient pas $\mathscr{R}^+(\delta) = \mathscr{D}$, on a $(u - 1)\cdot \mathscr{R}(\delta) = X_u \cap z^\perp$, ce qui prouve que $(u - 1)\cdot \mathscr{R}(\delta)$ est fermé dans $\mathscr{R}(\delta)$ comme intersection de deux fermés.

Ceci termine la démonstration du lemme.

Lemme 8.46. *$\sigma_u - 1$ est d'image fermée dans $H^j(\Phi^\pm, \mathscr{R}(\delta))$, si $j = 0, 1$.*

Démonstration. On déduit de la suite exacte $0 \to \mathscr{D} \to \mathscr{R}(\delta) \to \mathrm{LA} \to 0$ et de la nullité de $H^0(\Phi^+, \mathrm{LA})$ et $H^1(\Phi^-, \mathscr{D})$ (prop. 8.22), des isomorphismes

$$H^0(\Phi^+, \mathscr{R}(\delta)) \cong H^0(\Phi^+, \mathscr{D}) \quad \text{et} \quad H^1(\Phi^-, \mathscr{R}(\delta)) \cong H^1(\Phi^-, \mathrm{LA}),$$

et des suites exactes

$$0 \to H^1(\Phi^+, \mathscr{D}) \to H^1(\Phi^+, \mathscr{R}(\delta)) \to H^1(\Phi^+, \mathrm{LA}) \to 0,$$
$$0 \to H^0(\Phi^-, \mathscr{D}) \to H^0(\Phi^-, \mathscr{R}(\delta)) \to H^0(\Phi^-, \mathrm{LA}) \to 0.$$

Le résultat pour $H^0(\Phi^+, \mathscr{R}(\delta))$ et $H^1(\Phi^-, \mathscr{R}(\delta))$ se déduit donc directement de la prop. 8.41.

Le résultat pour $H^0(\Phi^-, \mathscr{R}(\delta))$ se déduit de la prop. 8.41 et du (i) du lemme 8.24 car $H^0(\Phi^-, \mathrm{LA})$ est de dimension finie.

Le résultat pour $H^1(\Phi^+, \mathscr{R}(\delta))$ se déduit de la prop. 8.41 et du (ii) du lemme 8.24 car $H^1(\Phi^+, \mathscr{D})$ est de dimension finie (on ne peut pas appliquer directement le (ii) du lemme 8.24 car $H^1(\Phi^+, \mathrm{LA})$ n'est pas un fréchet, mais on peut reprendre la preuve du lemme 8.37 en remarquant qu'elle permet d'écrire $\mathrm{LA}(\mathbf{Z}_p^*)$ comme une limite inductive compacte de banachs sur lesquels $\delta(u)\sigma_u - 1$ est surjective ; cela permet d'écrire $H^1(\Phi^+, \mathrm{LA})$ comme une limite inductive compacte de banach auxquels on peut appliquer le résultat du lemme 8.24).

Théorème 8.47. *$H^j(A^-, \mathscr{R}(\chi\delta^{-1}))$ et $H^{2-j}(A^+, \mathscr{R}(\delta))$ sont duaux l'un de l'autre, si $j = 0, 1, 2$. De plus,*

- *Si $\delta \notin \{x^{-i},\ i \in \mathbf{N}\} \cup \{x^i\chi,\ i \in \mathbf{N}\}$, alors* $\dim H^j(A^+, \mathscr{R}(\delta)) = 0, 1, 0$, *si $j = 0, 1, 2$.*
- *Si $\delta = x^{-i}$, avec $i \in \mathbf{N}$, alors* $\dim H^j(A^+, \mathscr{R}(\delta)) = 1, 2, 0$, *si $j = 0, 1, 2$.*
- *Si $\delta = x^i\chi$, avec $i \in \mathbf{N}$, alors* $\dim H^j(A^+, \mathscr{R}(\delta)) = 0, 2, 1$, *si $j = 0, 1, 2$.*
- $\dim H^j(A^+, \mathscr{R}(\delta)) = \dim H^j(A^-, \mathscr{R}(\delta))$, *pour tous δ et $j \in \{0, 1, 2\}$.*

Démonstration. La dualité entre $H^j(A^-, \mathscr{R}(\chi\delta^{-1}))$ et $H^{2-j}(A^+, \mathscr{R}(\delta))$ est une conséquence de la prop. 8.21 dont les hypothèses sont vérifiées grâce aux lemmes 8.45 et 8.46.

La dimension des H^0 se calcule en utilisant la prop. 8.33 ; celle des H^2 s'en déduit.

Enfin, la dimension de $H^1(A^+, \mathscr{R}(\delta))$ se calcule en utilisant la suite exacte longue de cohomologie déduite de la suite exacte $0 \to \mathscr{D}(\mathbf{Z}_p)\otimes\delta \to \mathscr{R}(\delta) \to \mathrm{LA}(\mathbf{Z}_p) \otimes \chi^{-1}\delta \to 0$.

- Si $\delta \notin \{x^{-i},\ i \in \mathbf{N}\}$, on a $H^j(A^+, \mathscr{D}(\mathbf{Z}_p) \otimes \delta) = 0$, pour tout j, ce qui nous fournit un isomorphisme

$$H^1(A^+, \mathscr{R}(\delta)) \cong H^1(A^+, \mathrm{LA}(\mathbf{Z}_p) \otimes \chi^{-1}\delta),$$

ce qui prouve que $H^1(A^+, \mathscr{R}(\delta))$ est de dimension 1 sauf si $\delta \in \{x^i\chi,\ i \in \mathbf{N}\}$ où il est de dimension 2.

- Si $\delta = x^{-i}$, avec $i \in \mathbf{N}$, on utilise la nullité de $H^2(A^+, \mathscr{R}(\delta))$ établie ci-dessus. On en déduit que l'application $H^1(A^+, \mathrm{LA}(\mathbf{Z}_p) \otimes \chi^{-1}\delta) \to H^2(A^+, \mathscr{D}(\mathbf{Z}_p) \otimes \delta)$ est un isomorphisme pour des raisons de dimension ; il en est donc de même de l'application $H^1(A^+, \mathscr{D}(\mathbf{Z}_p) \otimes \delta) \to H^1(A^+, \mathscr{R}(\delta))$ puisque $H^0(A^+, \mathscr{D}(\mathbf{Z}_p) \otimes \delta) = 0$. Il s'ensuit que $H^1(A^+, \mathscr{R}(\delta))$ est de dimension 2.

La dimension de $H^1(A^-, \mathscr{R}(\delta))$ s'en déduisant par dualité, cela permet de conclure.

Remarque 8.48. (i) Il résulte de la démonstration que l'application naturelle de $\mathrm{Ext}^1(\mathscr{R}, \mathscr{R}(\delta)) = H^1(A^+, \mathscr{R}(\delta))$ dans $H^1(A^+, \mathrm{LA}(\mathbf{Z}_p) \otimes \chi^{-1}\delta)$ est un isomorphisme sauf si $\delta = x^{-i}$, avec $i \in \mathbf{N}$, où cette application est identiquement nulle. Il semble donc que ce cas soit assez pathologique ce qui est corroboré par la remarque de Chenevier [8, 3.10].

(ii) Si G^+ est un semi-groupe, et si M est muni d'actions de G^+ et de son semi-groupe opposé G^- telles que $g^{-1} \cdot g = \mathrm{id}$ si $g \in G^+$ (auquel cas $g \cdot g^{-1}$ est un projecteur), on dispose d'un foncteur cohomologique ι envoyant $c \in C^n(G^+, M)$ sur $\iota(c) \in C^n(G^-, M)$ défini par

$$\iota(c)(g_1, \ldots, g_n) = (-1)^{n(n+1)/2}(g_1 \cdots g_n) \cdot c(g_n^{-1}, \ldots, g_1^{-1})$$

(un petit calcul montre que $d_{n+1} \circ \iota = \iota \circ d_{n+1}$). Cela nous fournit des applications naturelles $H^n(G^+, M) \to H^n(G^-, M)$, pour $n \in \mathbf{N}$. Si G est un groupe, on a $G^+ = G^-$ et ces applications sont l'identité des groupes considérés. Le cas de A^+ et $M = \mathrm{LA}(\mathbf{Z}_p) \otimes \delta$ montre que $H^n(G^+, M) \to H^n(G^-, M)$ n'est, en général, pas un isomorphisme. Par contre, si D est un (φ, Γ)-module sur $\mathscr{R}$, on a $H^n(A^+, D) \cong H^n(A^-, D)$ pour tout n (c'est dû au fait que $H^n(A^0, \mathrm{Ker}\,\psi) = 0$, pour tout n), ce qui explique que $\dim H^j(A^+, \mathscr{R}(\delta)) = \dim H^j(A^-, \mathscr{R}(\delta))$, pour tous δ et j.

2. La série principale localement analytique

Ce chapitre est consacré à des rappels au sujet de la représentation localement analytique $B^{\mathrm{an}}(\delta_2, \delta_1)$, en particulier en ce qui concerne sa suite de Jordan-Hölder (prop. 8.54). Il contient aussi une construction d'un isomorphisme naturel

$$\mathrm{Ext}^1(W(\delta_1, \delta_2), \mathrm{St}^{\mathrm{an}}(\delta_1, \delta_2)) \cong \mathrm{Ext}^1(\mathscr{R}(\delta_2), \mathscr{R}(\delta_1)), \qquad \text{si } \delta_1 = x^k\delta_2\chi \text{ avec } k \in \mathbf{N},$$

qui joue un rôle important dans la description de $\mathbf{\Pi}(\Delta(s))$ si s est spécial.

2.1. Construction de représentations localement analytiques

Soit $\mathscr{S}$ l'espace défini au § 0.3. Si $s = (\delta_1, \delta_2, \mathscr{L}) \in \mathscr{S}$, on note :

- δ_s et ω_s les caractères $\delta_1\delta_2^{-1}\chi^{-1}$ et $\delta_1\delta_2\chi^{-1}$,
- $B^{\rm an}(\delta_1, \delta_2)$ l'espace des $\phi : \mathbf{Q}_p \to L$ localement analytiques, telles que $x \mapsto \delta_s(x)\phi(1/x)$ se prolonge en une fonction analytique dans un voisinage de 0.

On munit $B^{\rm an}(\delta_1, \delta_2)$ d'une action à gauche de G définie par $g \cdot \phi = \phi \star g$, où

$$(\phi \star \left(\begin{smallmatrix} a & b \\ c & d \end{smallmatrix}\right))(x) = \delta_1^{-1}\chi(ad - bc)\delta_s(cx + d)\phi(\frac{ax + b}{cx + d}).$$

Remarque 8.49. (i) On a aussi

$$(\phi \star \left(\begin{smallmatrix} a & b \\ c & d \end{smallmatrix}\right))(x) = \delta_2(\frac{1}{cx + d})\chi^{-1}\delta_1(\frac{cx + d}{ad - bc})\phi(\frac{ax + b}{cx + d}).$$

(ii) Le caractère central de $B^{\rm an}(\delta_1, \delta_2)$ est ω_s.
(iii) Si $\eta \in \widehat{\mathscr{T}}(L)$, alors $B^{\rm an}(\eta\delta_1, \eta\delta_2) = B^{\rm an}(\delta_1, \delta_2) \otimes \eta$.

La représentation $B^{\rm an}(\delta_1, \delta_2)$ s'interprète, plus conceptuellement, comme une induite analytique. Si W est une représentation localement analytique de B, on note ${\rm Ind}^{\rm an} W$ *l'induite analytique de B à G* de W, c'est-à-dire l'espace des fonctions $\tilde{\phi} : G \to W$, localement analytiques, telles que $\tilde{\phi}(bg) = b \cdot \tilde{\phi}(g)$, pour tous $g \in G$ et $b \in B$, muni de l'action à gauche de G définie par $(h \cdot \tilde{\phi})(g) = \tilde{\phi}(gh)$.

Remarque 8.50. (i) Si $\tilde{\phi} \in {\rm Ind}^{\rm an} W$, on définit $\phi : \mathbf{Q}_p \to W$ par $\phi(x) = \tilde{\phi}(\left(\begin{smallmatrix} 0 & 1 \\ -1 & x \end{smallmatrix}\right))$. On obtient de la sorte un isomorphisme de ${\rm Ind}^{\rm an} W$ sur l'ensemble des $\phi : \mathbf{Q}_p \to W$, localement analytiques, telles que $x \mapsto \left(\begin{smallmatrix} x^{-1} & -1 \\ 0 & -x \end{smallmatrix}\right) \cdot \phi(x^{-1})$ se prolonge en une fonction analytique sur un voisinage de 0. Les formules $\left(\begin{smallmatrix} a & b \\ c & d \end{smallmatrix}\right)^{-1} = \frac{1}{ad-bc}\left(\begin{smallmatrix} d & -b \\ -c & a \end{smallmatrix}\right)$ et

$$\begin{aligned}\tfrac{1}{ad-bc}\left(\begin{smallmatrix} 0 & 1 \\ -1 & x \end{smallmatrix}\right)\left(\begin{smallmatrix} d & -b \\ -c & a \end{smallmatrix}\right) &= \tfrac{1}{ad-bc}\left(\begin{smallmatrix} -c & a \\ -cx-d & ax+b \end{smallmatrix}\right) \\ &= \left(\begin{smallmatrix} 1/(cx+d) & c/(ad-bc) \\ 0 & (cx+d)/(ad-bc) \end{smallmatrix}\right)\left(\begin{smallmatrix} 0 & 1 \\ -1 & (ax+b)/(cx+d) \end{smallmatrix}\right)\end{aligned}$$

montrent que l'action de G sur ${\rm Ind}^{\rm an} W$ est alors donnée par $g \cdot \phi = \phi \star g^{-1}$, avec

$$(\phi \star \left(\begin{smallmatrix} a & b \\ c & d \end{smallmatrix}\right))(x) = \left(\begin{smallmatrix} 1/(cx+d) & c/(ad-bc) \\ 0 & (cx+d)/(ad-bc) \end{smallmatrix}\right) \cdot \phi(\frac{ax + b}{cx + d}).$$

(ii) On en déduit, grâce au (i) de la rem. 8.49 que $\tilde{\phi} \mapsto \phi$, où $\phi(x) = \tilde{\phi}\big(\big(\begin{smallmatrix} 0 & 1 \\ -1 & x \end{smallmatrix}\big)\big)$, est un isomorphisme G-équivariant

$$\mathrm{Ind}^{\mathrm{an}}(\delta_2 \otimes \chi^{-1}\delta_1) \cong B^{\mathrm{an}}(\delta_1, \delta_2).$$

On note N le sous-groupe $\big(\begin{smallmatrix} 1 & \mathbf{Q}_p \\ 0 & 1 \end{smallmatrix}\big)$ de G. Si Π est une représentation localement analytique de G, on note $J^*(\Pi) = (\Pi^*)^N$ le *module de Jacquet dual* de Π. Ce module est stable par B, puisque B normalise N, et donc est muni d'une action de B (déterminée par celle de $T = \big\{\big(\begin{smallmatrix} a & 0 \\ 0 & d \end{smallmatrix}\big),\ a, d \in \mathbf{Q}_p^*\big\}$ puisque N agit trivialement). Si $\chi_1, \chi_2 \in \widehat{\mathscr{T}}(L)$, on note $\chi_1 \otimes \chi_2$ le caractère de B défini par $(\chi_1 \otimes \chi_2)\big(\big(\begin{smallmatrix} a & b \\ 0 & d \end{smallmatrix}\big)\big) = \chi_1(a)\chi_2(d)$.

Remarque 8.51. (i) Si $\mu \in J^*(\Pi) - \{0\}$ est propre sous l'action de B pour le caractère δ, alors $v \mapsto \tilde{\phi}_v$, où $\tilde{\phi}_v : G \to L$ est définie par $\tilde{\phi}_v(g) = \langle \mu, g \cdot v \rangle$, est un morphisme G-équivariant de Π dans $\mathrm{Ind}^{\mathrm{an}}\delta^{-1}$. En effet, si $h \in B$, on a $\tilde{\phi}_v(hg) = \langle \mu, hg \cdot v \rangle = \langle h^{-1} \cdot \mu, g \cdot v \rangle = \delta^{-1}(h)\tilde{\phi}_v(g)$.

(ii) L'orthogonal $\Pi \boxtimes \mathbf{Q}_p$ de $J^*(\Pi)$ est l'adhérence de l'espace engendré par les $(u-1) \cdot v$, pour $v \in \Pi$ et $u \in N$, et $J^*(\Pi)$ est le dual du *module de Jacquet* $J(\Pi) = \Pi/(\Pi \boxtimes \mathbf{Q}_p)$ de Π. Comme d'habitude, l'application $v \mapsto \phi_v(g) = \overline{g \cdot v}$ (image de $g \cdot v$ dans $J(\Pi)$), induit un morphisme G-équivariant de Π dans $\mathrm{Ind}^{\mathrm{an}} J(\Pi)$.

2.2. Composantes de Jordan-Hölder

La représentation $B^{\mathrm{an}}(\delta_1, \delta_2)$ n'est, de manière visible, pas toujours irréductible. Ses composantes de Jordan-Hölder ont été déterminées par Schneider et Teitelbaum [30]. L'énoncé du résultat (prop. 8.54 ci-dessous) va demander un peu de préparation.

- Si $\chi_1, \chi_2 \in \widehat{\mathscr{T}}(L)$ sont localement constants, on note $\mathrm{Ind}^{\mathrm{lisse}}(\chi_1 \otimes \chi_2)$ l'espace des $\phi : G \to L$, localement constantes, telles que $\phi\big(\big(\begin{smallmatrix} a & b \\ 0 & d \end{smallmatrix}\big)g\big) = \delta_1(a)\delta_2(d)\phi(g)$, pour tous $g \in G$ et $\big(\begin{smallmatrix} a & b \\ 0 & d \end{smallmatrix}\big) \in B$. On munit $\mathrm{Ind}^{\mathrm{lisse}}(\chi_1 \otimes \chi_2)$ d'une action de G grâce à la formule $(h \cdot \phi)(g) = \phi(gh)$. Les résultats suivants sont parfaitement classiques [21, th. 3.3].

Proposition 8.52. (i) $\mathrm{Ind}^{\mathrm{lisse}}(\chi_1 \otimes \chi_2)$ *est une représentation irréductible de* G *sauf si* $\chi_1 = \chi_2$ *ou si* $\chi_1 = |x|^2\chi_2$ *: dans le premier cas,* $\mathrm{Ind}^{\mathrm{lisse}}(\chi_1 \otimes \chi_2)$ *est une extension de* $\mathrm{St} \otimes \chi_1$ *par* χ_1*, où* St *est la steinberg, dans le second, c'est une extension de* χ_1 *par* $\mathrm{St} \otimes \chi_1$.

(ii) *Si* $\delta_1 \neq |x|^{\pm 1}\delta_2$*, alors* $\mathrm{Ind}^{\mathrm{lisse}}(\delta_1 \otimes |x|^{-1}\delta_2) \cong \mathrm{Ind}^{\mathrm{lisse}}(\delta_2 \otimes |x|^{-1}\delta_1)$ *(et les deux représentations sont irréductibles).*

- Si $w(s)-1 \in \mathbf{N}$, on note $B^{\text{alg}}(\delta_1, \delta_2)$ l'ensemble des fonctions $\phi : \mathbf{Q}_p \to L$, localement polynomiales de degré $\leq w(s)-1$, telles que $\delta_s(x)\phi(1/x)$ soit polynomiale, de degré $\leq w(s)-1$, dans un voisinage de 0. Alors, muni de l'action de G ci-dessus, $B^{\text{alg}}(\delta_1, \delta_2)$ est une sous-représentation de $B^{\text{an}}(\delta_1, \delta_2)$.
- Si $\delta_1 = \delta_2 \chi x^{w(s)-1}$ (cela équivaut à $\delta_s = x^{w(s)-1}$), on pose

$$W(\delta_1, \delta_2) = \text{Sym}^{w(s)-1} \otimes \chi^{-1} \delta_1 x^{1-w(s)} \quad \text{et} \quad \text{St}^{\text{alg}}(\delta_1, \delta_2) = \text{St} \otimes W(\delta_1, \delta_2).$$

Alors $W(\delta_1, \delta_2)$ est une représentation de dimension finie dont la duale est $W(\delta_1, \delta_2) \otimes \chi \delta_1^{-1} \delta_2^{-1}$ (cela suit de ce que la duale de Sym^k est $\text{Sym}^k \otimes x^{-k}$).

Remarque 8.53. (i) On a $B^{\text{alg}}(\delta_1\eta, \delta_2\eta) = B^{\text{alg}}(\delta_1, \delta_2) \otimes \eta$.

(ii) Si δ_2 est localement constant, il en est de même de $\delta_1 \chi^{-1} x^{1-w(s)}$, et on a $B^{\text{alg}}(\delta_1, \delta_2) \cong \big(\text{Ind}^{\text{lisse}} \delta_2 \otimes \delta_1 \chi^{-1} x^{1-w(s)}\big) \otimes \text{Sym}^{w(s)-1}$.

(iii) Si $s = (\delta_1, \delta_2, \infty) \in \mathscr{S}_*^{\text{cris}}$ et si on note s' l'élément de $\mathscr{S}_*^{\text{cris}}$ défini par $s' = (\delta_1', \delta_2', \infty)$ et $\delta_1' = x^{w(s)}\delta_2$, $\delta_2' = x^{-w(s)}\delta_1$, alors $B^{\text{alg}}(\delta_1, \delta_2)$ et $B^{\text{alg}}(\delta_1', \delta_2')$ ont les mêmes composantes de Jordan-Hölder ; en particulier, si $\delta_s \neq x^{w(s)-1}, |x|^{-2} x^{w(s)-1}$, ces deux représentations sont isomorphes. En effet, on peut, quitte à tordre par un caractère, supposer que δ_2 est localement constant (il en est alors de même de δ_2'), et le résultat suit du (ii) et de la prop. 8.52.

Proposition 8.54. (i) *Si $w(s) - 1 \notin \mathbf{N}$, alors $B^{\text{an}}(\delta_1, \delta_2)$ est irréductible.*

(ii) *Si $w(s)-1 \in \mathbf{N}$, alors $(\frac{d}{dx})^{w(s)}$ induit un morphisme équivariant surjectif de $B^{\text{an}}(\delta_1, \delta_2)$ sur $B^{\text{an}}(x^{-w(s)}\delta_1, x^{w(s)}\delta_2)$ dont le noyau est $B^{\text{alg}}(\delta_1, \delta_2)$.*

(iii) *La représentation $B^{\text{alg}}(\delta_1, \delta_2)$ est irréductible sauf si $\delta_s = x^{w(s)-1}$ où $B^{\text{alg}}(\delta_1, \delta_2)$ est une extension de $\text{St}^{\text{alg}}(\delta_1, \delta_2)$ par $W(\delta_1, \delta_2)$ qui sont toutes deux irréductibles, ou si $\delta_s = |x|^{-2} x^{w(s)-1}$, où $B^{\text{alg}}(\delta_1, \delta_2)$ est une extension de $W(|x|\delta_1, |x|^{-1}\delta_2)$ par $\text{St}^{\text{alg}}(|x|\delta_1, |x|^{-1}\delta_2)$.*

2.3. Extensions de $W(\delta_1, \delta_2)$ par $\text{St}^{\text{an}}(\delta_1, \delta_2)$

Si $\delta_s = x^k$, avec $k \in \mathbf{N}$, on note $\text{St}^{\text{an}}(\delta_1, \delta_2)$ le quotient de $B^{\text{an}}(\delta_1, \delta_2)$ par $W(\delta_1, \delta_2)$. On a donc une suite exacte

$$0 \to \text{St}^{\text{alg}}(\delta_1, \delta_2) \to \text{St}^{\text{an}}(\delta_1, \delta_2) \to B^{\text{an}}(x^{-1-k}\delta_1, x^{k+1}\delta_2) \to 0.$$

On cherche à décrire les extensions de $W(\delta_1, \delta_2)$ par $\text{St}^{\text{an}}(\delta_1, \delta_2)$ (admettant un caractère central). Si $k \in \mathbf{N}$, notons W_k la sous-G-représentation de $B^{\text{an}}(\chi, x^{-k})$ définie par les polynômes de degré $\leq k$ (c'est la duale de Sym^k)

et X_k le quotient de $B^{\rm an}(\chi, x^{-k})$ par W_k. En tant que B représentation, X_k s'identifie au sous-espace des $\phi \in B^{\rm an}(\chi, x^{-k})$ qui tendent vers 0 en ∞.

Si $\delta_s = x^k$, on a $\mathrm{St}^{\rm an}(\delta_1, \delta_2) = X_k \otimes \delta_1\chi^{-1}$ et $W(\delta_1, \delta_2) = W_k \otimes \delta_1\chi^{-1}$. On en déduit un isomorphisme naturel

$$\mathrm{Ext}^1(W(\delta_1, \delta_2), \mathrm{St}^{\rm an}(\delta_1, \delta_2)) \cong \mathrm{Ext}^1(W_k, X_k).$$

Si $\ell \in \mathrm{Hom}(\mathbf{Q}_p^*, L)$, on note ℓ^+ la fonction valant 0 sur $\mathbf{Z}_p$ et coïncidant avec ℓ en dehors de $\mathbf{Z}_p$ (notons que les ℓ^+ forment un L-espace vectoriel de dimension 2 de base v_p^+ et $\log^+$, où log est le logarithme normalisé par $\log p = 0$). On note Y_k l'espace $X_k \oplus W_k v_p^+ \oplus W_k \log^+$, que l'on munit d'une action de G, étendant celle sur X_k, en posant

$$(P\ell^+) \star \left(\begin{smallmatrix} a & b \\ c & d \end{smallmatrix}\right) = (cx+d)^k P\left(\frac{ax+b}{cx+d}\right)\left(\ell^+\left(\frac{ax+b}{cx+d}\right) + \ell(cx+d)\right).$$

(Le membre de droite est dans Y_k car $\ell^+\left(\frac{ax+b}{cx+d}\right) + \ell(cx+d) - \ell^+(x)$ est localement analytique sur $\mathbf{P}^1$.) Ceci fait de Y_k une extension de $\mathrm{Hom}(\mathbf{Q}_p^*, L) \otimes W_k = W_k \oplus W_k$ par X_k.

Théorème 8.55. *Si $\ell \in \mathrm{Hom}(\mathbf{Q}_p^*, L)$ est non nul, le sous espace $E_\ell = X_k \oplus W_k\ell^+$ de Y_k est une extension de W_k par X_k, et l'application $\ell \mapsto E_\ell$ induit un isomorphisme*

$$\mathrm{Ext}^1(W_k, X_k) \cong \mathrm{Hom}(\mathbf{Q}_p^*, L).$$

Démonstration. Il est immédiat que E_ℓ est une extension de W_k par X_k (ces extensions ont d'ailleurs été considérées par Breuil [5, 6]) ; le problème est donc de prouver que toute extension non triviale est de cette forme.

On utilise l'isomorphisme standard

$$\mathrm{Ext}^1(W_k, X_k) \cong \mathrm{Ext}^1(\mathbf{1}, X_k \otimes W_k^*) \cong H^1(G, X_k \otimes W_k^*).$$

On note $e_1^i e_2^{k-i}$, pour $0 \le i \le k$, la base standard de $\mathrm{Sym}^k = W_k^*$ (on a $\left(\begin{smallmatrix} a & b \\ c & d \end{smallmatrix}\right) \cdot e_1^i e_2^{k-i} = (ae_1 + ce_2)^i (be_1 + de_2)^{k-i}$). Comme W_k est irréductible, le lemme de Schur montre que $H^0(G, W_k \otimes W_k^*)$ est de dimension 1, et un petit calcul montre que $(xe_1 + e_2)^k$ en est une base $\big[$il suffit de vérifier que $(xe_1 + e_2)^k$ est stable par $\left(\begin{smallmatrix} a & 0 \\ 0 & 1 \end{smallmatrix}\right)$, pour $a \in \mathbf{Q}_p^*$, par $\left(\begin{smallmatrix} 1 & b \\ 0 & 1 \end{smallmatrix}\right)$ pour $b \in \mathbf{Q}_p$, et par $\left(\begin{smallmatrix} 0 & 1 \\ 1 & 0 \end{smallmatrix}\right)\big]$. L'isomorphisme $\mathrm{Ext}^1(W_k, X_k) \cong \mathrm{Ext}^1(\mathbf{1}, X_k \otimes W_k^*)$ s'obtient de la manière suivante.

- Si $0 \to X_k \to E \to W_k \to 0$ est une extension de W_k par X_k, l'image inverse de $L \cdot (xe_1 + e_2)^k \subset W_k \otimes W_k^*$ dans $E \otimes W_k^*$ est une extension de $L \cdot (xe_1 + e_2)^k \cong \mathbf{1}$ par $X_k \otimes W_k^*$.
- Si $0 \to X_k \otimes W_k^* \to E \to \mathbf{1} \to 0$ est une extension de $\mathbf{1}$ par $X_k \otimes W_k^*$, le quotient de $E \otimes W_k$ par $X_k \otimes (W_k^* \otimes W_k)_0$ est une extension de W_k par X_k, où

l'on a noté $(W_k^* \otimes W_k)_0$ le noyau de l'application naturelle $W_k^* \otimes W_k \to L$, envoyant $\mu \otimes v$ sur $\langle \mu, v \rangle$.

Ces deux opérations sont inverses l'une de l'autre et fournissent l'isomorphisme souhaité.

Pour décrire $H^1(G, X_k \otimes W_k^*)$, nous aurons besoin d'un peu de préparation.

Lemme 8.56. (i) *Si $g \mapsto c_g$ est un* 1*-cocycle continu sur B, à valeurs dans $Y_k \otimes W_k^*$, qui est nul sur A, alors $c_g = 0$ pour tout $g \in B$.*

(ii) *Si $g \mapsto c_g$ est un* 1*-cocycle continu sur G, à valeurs dans $Y_k \otimes W_k^*$, qui est nul sur A, alors $c_g = 0$ pour tout $g \in G$.*

Démonstration. Si $c_\alpha = 0$ pour tout $\alpha \in A$, on a $c_{\alpha\beta\alpha^{-1}} = \alpha \cdot c_\beta$, pour tous $\alpha \in A$ et $\beta \in B$. Si $a \in \mathbf{Q}_p^*$ et si $b \in \mathbf{Q}_p$, posons $\alpha(a) = \left(\begin{smallmatrix} a & 0 \\ 0 & 1 \end{smallmatrix}\right)$ et $\beta(b) = \left(\begin{smallmatrix} 1 & b \\ 0 & 1 \end{smallmatrix}\right)$. On peut appliquer ce qui précède à $\beta = \beta(b)$ et $\alpha = \alpha(p^n)$, avec $n \in \mathbf{N}$, de telle sorte que $\alpha\beta\alpha^{-1} = \beta(p^n b) = \beta(b)^{p^n}$. On obtient alors les relations

$$c_{\beta(p^n b)} = \alpha(p^n) \cdot c_{\beta(b)} \quad \text{et} \quad c_{\beta(p^n b)} = \sum_{j=0}^{p^n-1} \beta(jb) \cdot c_{\beta(b)}.$$

Écrivons $c_{\beta(b)}$ sous la forme $\sum_{i=0}^k \phi_i e_1^i e_2^{k-i}$, où $\phi_i = \phi_i' + P_i v_p^+ + Q_i \log^+$, avec $\phi_i' \in B^{\mathrm{an}}(\chi, x^{-k})$ tendant vers 0 en ∞ et P_i, Q_i des polynômes de degré $\leq k$; écrivons de même $c_{\beta(p^n b)}$ sous la forme $\sum_{i=0}^k \phi_{n,i} e_1^i e_2^{k-i}$. La seconde relation ci-dessus nous donne

$$\sum_{i=0}^{k} \phi_{n,i}(x) e_1^i e_2^{k-i} = \sum_{j=0}^{p^n-1} \sum_{i=0}^{k} \phi_i(x - jb) e_1^i (jbe_1 + e_2)^{k-i}.$$

Comme il existe $r \in \mathbf{N}$ tel que les ϕ_i soient analytiques sur $a + p^r\mathbf{Z}_p$, pour tout $a \in \mathbf{Q}_p$, l'identité ci-dessus montre que $\phi_{n,i}$ est analytique sur $p^r\mathbf{Z}_p$, pour tout $n \in \mathbf{N}$. Maintenant, la première relation ci-dessus nous donne $\phi_{n,i}(x) = p^{ni}\phi_i\left(\frac{x}{p^n}\right)$. Il en résulte que ϕ_i est analytique sur $p^{r-n}\mathbf{Z}_p$ pour tout $n \in \mathbf{N}$; c'est donc la restriction à $\mathbf{Q}_p$ d'une fonction analytique sur $\mathbf{C}_p$. Par ailleurs, $\phi_i - P_i v_p - Q_i \log$ est la restriction d'une fonction analytique dans un voisinage de ∞, nulle en ∞ ; On en déduit que $R_i = P_i v_p + Q_i \log$ est analytique sur la couronne $v_p(x) \leq -N$, si $N \in \mathbf{N}$ est assez grand, ce qui implique que $P_i = Q_i = 0$ (dériver permet de se ramenener au cas où P_i et Q_i sont des constantes, et on montre que $P_i = Q_i = 0$ en considérant le développement de Laurent de $R_i(ax) - R_i(x)$, pour $a \in \mathbf{Q}_p^*$). Il en résulte que ϕ_i est la restriction d'une fonction analytique sur $\mathbf{P}^1(\mathbf{C}_p)$ (et donc une constante), nulle à l'infini et donc partout puisque constante. En résumé, on a $c_{\beta(b)} = 0$ pour tout $b \in \mathbf{Q}_p$. On en déduit le (i) puisque B est engendré par A et les $\beta(b)$, pour $b \in \mathbf{Q}_p$.

Si maintenant $g \mapsto c_g$ est 1-cocycle continu sur G, à valeurs dans $X_k \otimes W_k^*$, qui est nul sur A, le (i) prouve que $c_g = 0$ pour tout $g \in B$, et que $w \cdot c_{wgw} = 0$ pour tout $g \in B$. On en déduit que c_g est nul pour tout g dans les borels inférieur et supérieur, ce qui permet de conclure puisque G est engendré par ces borels.

Lemme 8.57. *$H^1(A, X_k \otimes W_k^*)$ est un L-espace vectoriel de dimension 2 $(k+1)$, isomorphe à $\mathrm{Hom}(\mathbf{Q}_p^*, L) \otimes W_k^*$ par l'application qui envoie $\ell \otimes x^i$ sur le cocycle $g \mapsto (g-1) \cdot \ell^+ f_i$, avec $f_i = \mathbf{1}_{\mathbf{P}^1 - \mathbf{Z}_p} x^i e_1^i e_2^{k-i}$.*

Démonstration. On a $X_k \otimes W_k^* = \oplus_{i=0}^k X_k e_1^i e_2^{k-i}$, et $X_k e_1^i e_2^{k-i} \cong X_k \otimes x^i$ comme A-module ; on est donc ramené à prouver que l'application envoyant ℓ sur le cocycle $g \mapsto (g-1) \cdot (\ell \mathbf{1}_{\mathbf{P}^1 - \mathbf{Z}_p} x^i) \otimes x^i$ induit un isomorphisme $\mathrm{Hom}(\mathbf{Q}_p^*, L) \cong H^1(A, X_k \otimes x^i)$.

Notons Φ le sous-groupe $\left(\begin{smallmatrix} p^{\mathbf{Z}} & 0 \\ 0 & 1 \end{smallmatrix}\right)$ de A. Pour les raisons habituelles, on a une suite exacte :

$$\begin{aligned} 0 \to H^1(A^0, H^0(\Phi, X_k \otimes x^i)) &\to H^1(A, X_k \otimes x^i) \\ &\to H^0(A^0, H^1(\Phi, X_k \otimes x^i)) \to 0. \end{aligned}$$

On dispose d'un isomorphisme $\phi \mapsto (\phi_1, \phi_2)$ de X_k sur $\mathrm{LA}(\mathbf{Z}_p) \oplus \mathrm{LA}(\mathbf{Z}_p)$, où ϕ_1 est la restriction de ϕ à $\mathbf{Z}_p$ et $\phi_2(x) = \frac{1}{px}\phi(\frac{1}{px})$. Un petit calcul montre que l'action de $1 - \left(\begin{smallmatrix} p & 0 \\ 0 & 1 \end{smallmatrix}\right)$ sur X_k devient, via cet isomorphisme, l'application

$$(\phi_1, \phi_2) \mapsto \big((1 - p^i\varphi) \cdot \phi_1 - p^{i+1}\mathrm{Res}_{\mathbf{Z}_p^*} \frac{1}{x}\phi_2(\frac{1}{x}), (1 - p^{i+1}\psi) \cdot \phi_2\big).$$

Or $1 - p^{i+1}\psi$ induit un isomorphisme de $\mathrm{LA}(\mathbf{Z}_p)$ d'après le lemme 8.25. On en déduit que l'injection de $\mathrm{LA}(\mathbf{Z}_p)$ dans X_k induit des isomorphismes

$$H^j(\Phi^+, \mathrm{LA}(\mathbf{Z}_p) \otimes x^i) \cong H^j(\Phi, X_k \otimes x^i), \quad \text{pour } j = 0, 1,$$

et $H^1(A^+, \mathrm{LA}(\mathbf{Z}_p) \otimes x^i) \cong H^1(A, X_k \otimes x^i)$. On conclut en utilisant la prop. 8.43 et le fait que les cocycles $g \mapsto (g-1) \cdot ((\mathbf{1}_{\mathbf{P}^1 - \mathbf{Z}_p} x^i \ell) \otimes x^i)$ et $g \mapsto (g-1) \cdot ((\mathbf{1}_{\mathbf{Z}_p} x^i \ell) \otimes x^i)$ sont opposés dans $X_k \otimes x^i$ (en effet, si $g = \left(\begin{smallmatrix} a & 0 \\ 0 & 1 \end{smallmatrix}\right)$, on a $(g-1) \cdot ((\mathbf{1}_{\mathbf{P}^1} x^i \ell) \otimes x^i) = \ell(a)((\mathbf{1}_{\mathbf{P}^1} x^i) \otimes x^i)$ qui est nul dans $X_k \otimes x^i$ puisque $\mathbf{1}_{\mathbf{P}^1} x^i = 0$ dans X_k).

Lemme 8.58. *L'application qui envoie ℓ sur le cocycle*

$$g \mapsto (g-1) \cdot ((xe_1 + e_2)^k \ell^+) = ((g-1) \cdot \ell^+)(xe_1 + e_2)^k$$

induit un isomorphisme de $\mathrm{Hom}(\mathbf{Q}_p^*, L)$ sur $H^1(G, X_k \otimes W_k^*)$ qui est donc de dimension 2.

Démonstration. Soit $g \mapsto c_g$ un 1-cocycle continu, trivial sur Z (on s'intéresse aux extensions admettant un caractère central). Quitte à modifier ce cocycle par un cobord on peut, d'après le lemme 8.57, supposer que la restriction de $g \mapsto c_g$ à A est de la forme $g \mapsto (g-1)\cdot\sum_{i=0}^k \ell_i^+ f_i$, où les ℓ_i sont des éléments de $\mathrm{Hom}(\mathbf{Q}_p^*, L)$. Mais alors le cocycle $c'_g = c_g - (g-1)\cdot\sum_{i=0}^k \ell_i^+ f_i$, à valeurs dans $Y_k \otimes W_k^*$, est identiquement nul sur A. Il est donc identiquement nul sur G, d'après le (i) du lemme 8.56. En résumé, on a $c_g = (g-1)\cdot\sum_{i=0}^k \ell_i^+ f_i$, pour tout $g \in G$.

Maintenant, c_g est à valeurs dans X_k ; son image dans $\mathrm{Hom}(\mathbf{Q}_p^*, L) \otimes W_k \otimes W_k^*$ est donc nulle. Il s'ensuit que $\sum_{i=0}^k \ell_i \otimes f_i$ est fixe par G et donc appartient à $\mathrm{Hom}(\mathbf{Q}_p^*, L) \otimes (xe_1 + e_2)^k$, ce qui permet de conclure.

Remarque 8.59. (i) Soit $\ell \in \mathrm{Hom}(\mathbf{Q}_p^*, L)$, et soit E_ℓ l'extension de $\mathbf{1}$ par $X_k \otimes W_k^*$ qui lui est attachée par le lemme 8.58. On peut identifier E_ℓ au sous-espace $(X_k \otimes W_k^*) \oplus L\cdot((xe_1+e_2)^k\ell^+)$ de $Y_k \otimes W_k^*$. L'extension de W_k par X_k qui lui correspond est alors l'image de $E_\ell \otimes W_k \subset Y_k \otimes W_k^* \otimes W_k$ par $\phi \otimes \mu \otimes v \mapsto \langle \mu, v\rangle \phi$; c'est donc le sous-espace $X_k \oplus W_k \ell^+$ de Y_k. Ceci termine la démonstration du th. 8.55.

(ii) Si on regarde de plus près la démonstration, on obtient un isomorphisme naturel

$$\mathrm{Ext}^1_G(W_k, X_k) \cong H^1(A^+, \mathrm{LA}(\mathbf{Z}_p) \otimes x^k).$$

(Ce dernier espace est naturellement isomorphe à $\mathrm{Hom}(\mathbf{Q}_p^*, L)$ d'après la prop. 8.43, d'où l'isomorphisme $\mathrm{Ext}^1_G(W_k, X_k) \cong \mathrm{Hom}(\mathbf{Q}_p^*, L)$ du théorème). De manière précise, la restriction de G à A fournit une application naturelle de $\mathrm{Ext}^1_G(W_k, X_k)$ dans $\mathrm{Ext}^1_A(W_k, X_k)$. Or $W_k = \oplus_{i=0}^k L \otimes x^{-i}$ en tant que A-module, et donc $\mathrm{Ext}^1_A(W_k, X_k) = \oplus_{i=0}^k \mathrm{Ext}^1_A(L \otimes x^{-i}, X_k) = \oplus_{i=0}^k H^1(A, X_k \otimes x^i)$; la projection sur le terme correspondant à $i = k$ fournit donc une application de $\mathrm{Ext}^1_G(W_k, X_k)$ dans $H^1(A, X_k \otimes x^k)$, et le lemme 8.57 montre que cette application est un isomorphisme. Par ailleurs, on a démontré au cours de la preuve du lemme 8.57 que l'injection de $\mathrm{LA}(\mathbf{Z}_p)$ dans X_k induit un isomorphisme $H^1(A^+, \mathrm{LA}(\mathbf{Z}_p) \otimes x^k) \cong H^1(A, X_k \otimes x^k)$, d'où l'isomorphisme $\mathrm{Ext}^1_G(W_k, X_k) \cong H^1(A^+, \mathrm{LA}(\mathbf{Z}_p) \otimes x^k)$.

(iii) On déduit du (ii) et de la prop. 8.43 des isomorphismes naturels

$$\begin{aligned}\mathrm{Ext}^1_G(W(\delta_1,\delta_2), \mathrm{St}^{\mathrm{an}}(\delta_1,\delta_2)) &\cong H^1(A^+, \mathrm{LA}(\mathbf{Z}_p) \otimes x^k)\\ &\cong \mathrm{Ext}^1(\mathscr{R}(\delta_2), \mathscr{R}(\delta_1)),\end{aligned}$$

si $\delta_1 = x^k \delta_2 \chi$.

3. Le module $\Delta \boxtimes \{0\}$

Ce chapitre est consacré à l'étude du module $\Delta \boxtimes \{0\} = \cap_{n\in\mathbf{N}}\varphi^n(\Delta)$, si Δ est un (φ, Γ)-module sur $\mathscr{R}$. Il est clair que $\Delta \boxtimes \{0\}$ est muni d'une structure de (φ, Γ)-module sur $\cap_{n\in\mathbf{N}}\varphi^n(\mathscr{R})$, et on prouve que ce dernier anneau est égal à $L\{\{t\}\}$ et que $\Delta \boxtimes \{0\}$ est libre sur $L\{\{t\}\}$, de rang inférieur ou égal à celui de Δ sur $\mathscr{R}$ (et même que l'application naturelle $\mathscr{R} \otimes_{L\{\{t\}\}} (\Delta \boxtimes \{0\}) \to \Delta$ est injective). La démonstration repose sur un théorème de structure pour les φ-modules de type fini sur $L\{\{t\}\}$ (th. 8.65) et le fait que $\cap_{n\in\mathbf{N}}\varphi^n(\mathrm{Fr}(\mathscr{R})) = \mathrm{Fr}(L\{\{t\}\})$ (th. 8.77).

3.1. (φ, Γ)-module sur $L\{\{t\}\}$

3.1.1. L'anneau $L\{\{t\}\}$

On rappelle que $L\{\{t\}\}$ est le sous-anneau de $\mathscr{R}^+$ des séries $\sum_{n\in\mathbf{N}} a_n t^n$ à coefficients dans L, qui convergent pour toute valeur de t. C'est la limite projective des anneaux de fonctions L-analytiques sur les disques $v_p(t) \geq -n$, pour $n \in \mathbf{N}$, et comme chacun de ces anneaux est un anneau principal de Banach, $L\{\{t\}\}$ est un anneau de Fréchet–Stein (tout sous-module fermé d'un module libre de rang d est libre et de rang $\leq d$; un sous-module de type fini d'un module libre de type fini est fermé (et donc libre)). La théorie des polygones de Newton montre que $f \in L\{\{t\}\}$ ne s'annule pas si et seulement si $f \in L^*$; en particulier, on a $(L\{\{t\}\})^* = L^*$.

On peut aussi considérer $L\{\{t\}\}$ comme un sous-anneau de $\mathscr{E}^{]0,r_a]}$, et la topologie sur $L\{\{t\}\}$ induite par celle de $\mathscr{E}^{]0,r_a]}$ coïncide avec celle décrite ci-dessus : l'injection de $L\{\{t\}\}$ dans $\mathscr{E}^{]0,r_a]}$ est continue et son image est fermée (elle est fermée dans $\mathscr{R}^+$ car c'est l'intersection des noyaux des $1-\varphi^n\psi^n$ (cf. prop. 8.68), et $\mathscr{R}^+$ est fermé dans $\mathscr{E}^{]0,r_a]}$) ; le théorème de l'image ouverte pour les fréchets montre donc que cette injection est un isomorphisme de fréchets.

Un *diviseur* sur $\mathbf{C}_p$ est une expression de la forme $D = \sum_{\alpha\in\mathbf{C}_p} n_\alpha[\alpha]$, où les n_α sont des éléments de $\mathbf{Z}$. Un tel diviseur est *effectif* si $n_\alpha \geq 0$ pour tout α, *localement fini* si $\sum_{v_p(\alpha)\geq -M} |n_\alpha| < +\infty$ pour tout $M \in \mathbf{N}$; il est *défini sur* L si $n_{g(\alpha)} = n_\alpha$, pour tout $\alpha \in \mathbf{C}_p$ et $g \in \mathrm{Gal}(\overline{\mathbf{Q}}_p/L)$. En particulier, si D est localement fini et défini sur L, on a $n_\alpha = 0$ pour tout $\alpha \notin \overline{\mathbf{Q}}_p$.

Si $f \in L\{\{t\}\}$ est non nul, alors $\mathrm{Div}(f) = \sum_{\alpha\in\mathbf{C}_p} v_\alpha(f)[\alpha]$ est un diviseur effectif sur $\mathbf{C}_p$, localement fini et défini sur L. Si I est un idéal non nul de type fini (et donc principal) de $L\{\{t\}\}$, on note $\mathrm{Div}(I)$ le diviseur $\mathrm{Div}(f)$, pour n'importe quel générateur f de I. L'application $I \mapsto \mathrm{Div}(I)$, ainsi définie, induit une bijection de l'ensemble des idéaux non nuls et de type fini de $L\{\{t\}\}$

sur celui des diviseurs effectifs, localement finis et définis sur L ; de plus, $g \in I - \{0\}$ si et seulement si $\mathrm{Div}(g) - \mathrm{Div}(I)$ est un diviseur effectif.

Les actions de φ et Γ sur $\mathscr{R}^+$ induisent des automorphismes de $L\{\{t\}\}$: on a $\varphi\big(\sum_{n\in\mathbf{N}} a_n t^n\big) = \sum_{n\in\mathbf{N}} p^n a_n t^n$ et $\sigma_b(\sum_{n\in\mathbf{N}} a_n t^n) = \sum_{n\in\mathbf{N}} a_n b^n t^n$.

Lemme 8.60. *Un idéal de $L\{\{t\}\}$, de type fini et stable par Γ ou par φ, est de la forme (t^k), avec $k \in \mathbf{N}$.*

Démonstration. Soit I un idéal de $L\{\{t\}\}$, de type fini et stable par Γ, et soit $D = \sum_{\alpha\in\mathbf{C}_p} n_\alpha[\alpha]$ le diviseur associé.

- Si I est stable par $\sigma_a \in \Gamma$, alors D est fixe par σ_a et donc $n_{a\alpha} = n_\alpha$, quels que soient $a \in \mathbf{Z}_p^*$ et $\alpha \in \mathbf{C}_p$. Comme D est localement fini, cela implique $n_\alpha = 0$ si $\alpha \neq 0$.
- Si I est stable par φ, alors D est fixe par φ et donc $n_{p\alpha} = n_\alpha$, quel que soit $\alpha \in \mathbf{C}_p$. Comme D est localement fini, cela implique $n_\alpha = 0$ si $\alpha \neq 0$.

  Dans les deux cas, cela implique que $I = (t^{n_0})$, ce que l'on cherchait à prouver.

Lemme 8.61. *Si $\alpha \in L$, alors $\varphi - \alpha$ induit un isomorphisme sur $L\{\{t\}\}$ sauf si $\alpha = p^i$, avec $i \in \mathbf{N}$, auquel cas le noyau de $\varphi - \alpha$ est $L\,t^i$ et $\varphi - \alpha$ induit un isomorphisme du supplémentaire $\{\sum_{k\in\mathbf{N}} a_k t^k,\ a_i = 0\}$ du noyau.*

Démonstration. Cela suit de ce que $(\varphi - \alpha)\big(\sum_{k\in\mathbf{N}} a_k t^k\big) = \sum_{k\in\mathbf{N}}(p^k - \alpha)a_k t^k$.

Lemme 8.62. *Si $A, B \in \mathbf{GL}_d(L\{\{t\}\})$, et si $U \in \mathbf{M}_d(L\{\{t\}\})$ a un coefficient constant inversible et vérifie $A\varphi(U) = UB$, alors $U \in \mathbf{GL}_d(L\{\{t\}\})$.*

Démonstration. Soit $\Delta = \det U$. Comme A et B sont inversibles leurs déterminants appartiennent à L^* et la relation satisfaite par U se traduit par $\Delta(pt) = \alpha\Delta(t)$, avec $\alpha \in L^*$. Il s'ensuit que si a est un zéro de Δ, alors $p^k a$ aussi, et comme Δ n'a qu'un nombre fini de zéro dans toute boule, cela implique que Δ ne s'annule pas en dehors de 0. Comme Δ ne s'annule pas non plus en 0 par hypothèse, il s'ensuit que $\Delta \in L^*$, ce qui permet de conclure.

3.1.2. φ-modules

Un φ-module M sur $L\{\{t\}\}$ est un $L\{\{t\}\}$-module libre de type fini, muni d'une action semi-linéaire bijective de φ (on remarquera qu'il suffit que l'action soit surjective pour qu'elle soit bijective).

Lemme 8.63. *Soit M un φ-module de rang d sur $L\{\{t\}\}$, et soit $v \in M$ tel qu'il existe $\alpha \in L^*$ avec $\varphi(v) = \alpha v$. Alors il existe $k \in \mathbf{N}$ tel que $t^{-k}v \in M$ et $M/L\{\{t\}\}t^{-k}v$ soit libre (de rang $d-1$) sur $L\{\{t\}\}$.*

Démonstration. Soit $e_1, \ldots, e_d$ une base de M sur $L\{\{t\}\}$ et soient $v = v_1 e_1 + \cdots + v_d e_d$ la décomposition de v dans cette base et I l'idéal $(v_1, \ldots, v_d)$. Alors I est stable par φ et donc de la forme (t^k), avec $k \in \mathbf{N}$, d'après le lemme 8.60. On en déduit le résultat. □

Soit M un φ-module sur $L\{\{t\}\}$, et soit $\overline{M} = M/tM$; c'est un L-espace vectoriel de dimension finie muni d'un isomorphisme φ. Si $P \in L[X]$ est irréductible et unitaire, on note M_P (resp. $\overline{M}_P$) l'ensemble des $v \in M$ (resp. $v \in \overline{M}$) tels que $P(\varphi)^n \cdot v = 0$, pour $n \gg 0$, et si $k \in \mathbf{N}$, on note $P[k]$ le polynôme $p^{kd} P(X/p^k)$ (ses racines sont celles de P multipliées par p^k).

Lemme 8.64. *L'application naturelle $M_P \to \overline{M}_P$ est surjective.*

Démonstration. Quitte à étendre les scalaires, on peut supposer que toutes les valeurs propres de φ sur $\overline{M}$ appartiennent à L. Soit α une valeur propre de φ sur $\overline{M}$ de valuation minimale, et soient v un vecteur propre pour cette valeur propre et $\tilde{v}$ un relèvement de v dans M. Si $n \in \mathbf{N}$, soit $u_n = \alpha^{-n}\varphi^n(\tilde{v})$; on a $u_{n+1} - u_n = \alpha^{-n}\varphi^n(x)$, où $x = \alpha^{-1}\varphi(\tilde{v}) - \tilde{v} \in tM$. L'hypothèse selon laquelle α est de valuation minimale implique que l'on peut trouver une base de $\overline{M}$ dans laquelle la matrice de $\alpha^{-1}\varphi$ soit à coefficients dans $\mathscr{O}_L$. On peut relever cette base en une base de M sur $L\{\{t\}\}$ (en appliquant un élément de $\mathbf{GL}_d(L)$ bien choisi à une base quelconque de M), et dans la base $e_1, \ldots, e_d$ obtenue, la matrice B de $\alpha^{-1}\varphi$ est à coefficients dans $\mathscr{O}_L + tL\{\{t\}\}$. Comme la matrice de $\alpha^{-1}\varphi$ dans la base $te_1, \ldots, te_d$ de tM est pB, et comme celle de $\alpha^{-n}\varphi^n = (\alpha^{-1}\varphi)^n$ est $p^n B(t)B(pt)\cdots B(p^{n-1}t)$, on en déduit que $\alpha^{-n}\varphi^n(x) \to 0$ quand $n \to +\infty$, et donc que u_n a une limite u quand $n \to +\infty$. Comme u_n a pour image v dans $\overline{M}$, pour tout n, il en est de même de u. Par ailleurs, on a $\varphi(u) = \alpha u$ par construction. On en déduit, grâce au lemme 8.63, que $M' = M/(L\{\{t\}\}u)$ est un φ-module libre de rang $d-1$ sur $L\{\{t\}\}$. On peut donc lui appliquer l'hypothèse de récurrence et relever tout élément x de $\overline{M}'_\beta = (\overline{M}/Lv)_\beta$, si β est une valeur propre de φ sur M'/tM', en un élément $\tilde{x}$ de M tel que $(\varphi - \beta)^n \tilde{x} \in tL\{\{t\}\}u$, pour n assez grand. Il existe alors $a \in tL\{\{t\}\}$ tel que $(\varphi - \beta)^n \tilde{x} = au$, et il résulte du lemme 8.61 que l'on peut trouver $b \in tL\{\{t\}\}$ tel que $(\alpha\varphi - \beta)b = a$ (resp. $(\alpha\varphi - \beta)b - a \in Lt^i$), si $\beta \notin p^{\mathbf{N}-\{0\}}\alpha$ (resp. si $\beta = p^i\alpha$, avec $i \geq 1$). Alors $\tilde{x} - bu$ est un relèvement de x dans M_β. On en déduit le résultat.

Théorème 8.65. *Si M est un φ-module de rang d sur $L\{\{t\}\}$, il existe une base $e_1, \ldots, e_d$ de M dans laquelle la matrice de φ est $A + N$, où $A \in \mathrm{GL}_d(L)$ est semi-simple, inversible, et N est nilpotente, commute à A, et se décompose sous la forme $N = N_0 + tN_1 + t^2N_2 + \cdots$, où $N_i \in \mathrm{M}_d(L)$ envoie le noyau*

M_P de $P(A)$ dans celui $M_{P[-i]}$ de $P(p^i A)$, pour tout P (en particulier, la somme est finie).

Démonstration. On choisit une section $\iota_P : \overline{M}_P \to M_P$ de l'application naturelle (c'est possible d'après le lemme 8.64), et on note $\iota : \overline{M} \to M$ la somme directe des ι_P (on a $\overline{M}_P = 0$ sauf pour un nombre fini de P et $\overline{M} = \oplus_P \overline{M}_P$). Il est alors plus ou moins immediat que $M_P = \iota(\overline{M}_P) \oplus t\iota(\overline{M}_{P[-1]}) \oplus \cdots$ (la somme est finie). Il en résulte que si l'on choisit une base de $\overline{M}_P$, pour tout P, et l'on considère la famille des $\iota(e_i)$, où les e_i parcourent la réunion des bases des $\overline{M}_P$, les polynômes étant rangés dans l'ordre croissant pour la valuation de leurs racines, alors la matrice de φ est triangulaire supérieure par blocs, chacun des blocs diagonaux étant la matrice de φ sur un des $\overline{M}_P$, les blocs au-dessus de la diagonale étant tous nuls sauf celui correspondant à P en horizontal et $P[k]$ en vertical, qui est à coefficients dans $L\,t^k$. En particulier, cette matrice est inversible (ce qui prouve, d'après le lemme 8.62, que les $\iota(e_i)$ forment une base de M sur $L\{\{t\}\}$), et elle a la forme voulue.

3.1.3. (φ, Γ)-modules

Un *(φ, Γ)-module sur* $L\{\{t\}\}$ est un φ-module muni en plus d'une action semi-linéaire de Γ commutant à celle de φ.

On dit que v est *propre sous l'action de φ et* Γ s'il existe $\delta \in \widehat{\mathscr{T}}(L)$ tel que l'on ait $\varphi(v) = \delta(p)v$ et $\sigma_a(v) = \delta(a)v$, pour tout $a \in \mathbf{Z}_p^*$ (on dit alors que v est *propre pour le caractère* δ).

Exemple 8.66. (i) Un $L\{\{t\}\}$-module de rang 1 possède, à multiplication près par un élément de $(L\{\{t\}\})^* = L^*$, une unique base e. Il en résulte que si M est un (φ, Γ)-module de rang 1 sur $L\{\{t\}\}$, et si e en est une base, alors il existe $\delta \in \widehat{\mathscr{T}}(L)$ tel que e soit propre pour δ.

(ii) Si M est un (φ, Γ)-module de rang 2 sur $L\{\{t\}\}$, il résulte du th. 8.65 que, quitte à faire une extension quadratique de L, il existe e_1 propre pour l'action de φ et Γ (pour un caractère δ_1) tel que $L\{\{t\}\}e_1$ soit saturé dans M. On a alors $M/L\{\{t\}\}e_1 = L\{\{t\}\}e_2$, où e_2 est propre pour un caractère δ_2, et on est dans un des deux cas exclusifs suivants :

- e_2 se relève dans M en e_2' propre pour δ_2,
- il existe $k \in \mathbf{N}$ et $(\alpha, \beta) \in L^2 - \{(0,0)\}$, tels que $\delta_2 = x^k\delta_1$ et e_2 se relève dans M en e_2' vérifiant

$$\varphi(e_2') = p^k\delta_1(p)e_2' + \alpha t^k e_1 \text{ et } \sigma_a(e_2') = a^k\delta_1(a)e_2' + \beta t^k e_1, \text{ si } a \in \mathbf{Z}_p^*.$$

Dans les deux cas, e_1, e_2' forment une base de M sur $L\{\{t\}\}$. De plus, dans le premier cas ou si $k = 0$ dans le second, le L-espace vectoriel $M_0 = Le_1 + Le_2'$

est stable par φ et Γ et c'est le seul sous-L-espace vectoriel de M, stable par φ et Γ, tel que l'application naturelle $L\{\{t\}\} \otimes_L M_0 \to M$ soit un isomorphisme. (Dans le second cas, si $k \geq 1$, les sous-L-espaces vectoriels de dimension 2, stables par φ et Γ, et non inclus dans $L\{\{t\}\}e_1$, sont de la forme $Lt^{k+i}e_1 + Lt^i e_2'$, avec $i \in \mathbf{N}$; ils n'engendrent donc pas M.)

3.2. L'action de φ sur $\mathrm{Fr}(\mathscr{R})$

3.2.1. L'action de φ sur $\mathscr{R}$

Lemme 8.67. (i) $\varphi^n(\mathscr{R}) \cap \mathscr{R}^+ = \varphi^n(\mathscr{R}^+)$.
(ii) $\cap_{n\in\mathbf{N}}\varphi^n(\mathscr{R}) = \cap_{n\in\mathbf{N}}\varphi^n(\mathscr{R}^+)$.

Démonstration. $\phi_{\varphi^n(f)}(x) = \begin{cases} 0, & \text{si } x \notin p^n\mathbf{Z}_p, \\ \phi_f(\frac{x}{p^n}), & \text{si } x \in p^n\mathbf{Z}_p. \end{cases}$

On en déduit que :

- $\phi_f(x) = 0$ si $\phi_{\varphi^n(f)} = 0$, ce qui démontre le (i) car $\phi_f = 0$ équivaut à $f \in \mathscr{R}^+$,
- l'appartenance de f à $\varphi^n(\mathscr{R})$ entraîne la nullité de ϕ_f en dehors de $p^n\mathbf{Z}_p$. Comme ϕ_f est continue (en particulier en 0), on en déduit que $f \in \cap_{n\in\mathbf{N}}\varphi^n(\mathscr{R})$ implique $\phi_f = 0$, et donc $f \in \mathscr{R}^+$. Le (ii) est donc une conséquence du (i).

Proposition 8.68. $\cap_{n\in\mathbf{N}}\varphi^n(\mathscr{R}) = L\{\{t\}\}$.

Démonstration. D'après le lemme 8.67, on a $\cap_{n\in\mathbf{N}}\varphi^n(\mathscr{R}) = \cap_{n\in\mathbf{N}}\varphi^n(\mathscr{R}^+)$. Maintenant, l'application qui à une distribution μ associe sa transformée d'Amice A_μ induit un isomorphisme de $\mathscr{D}(\mathbf{Z}_p)$ sur $\mathscr{R}^+$. L'image inverse de $\varphi^n(\mathscr{R}^+)$ par cet isomorphisme est l'ensemble des distributions à support dans $p^n\mathbf{Z}_p$ et donc celle de $\cap_{n\in\mathbf{N}}\varphi^n(\mathscr{R}^+)$ est l'ensemble des distributions de support $\{0\}$. L'isomorphisme $\cap_{n\in\mathbf{N}}\varphi^n(\mathscr{R}) = L\{\{t\}\}$ est donc une traduction du (ii) de la prop. 8.11.

3.2.2. Le corps $\mathrm{Fr}(\mathscr{R})$

Si a est un entier ≥ 1, on pose $r_a = \frac{1}{(p-1)p^{a-1}}$, et on note D_a le disque $v_p(T) > r_a$ et C_a la couronne $0 < v_p(T) \leq r_a$ complémentaire de D_a dans le disque unité. La théorie des polygones de Newton permet de montrer que $f \in \mathscr{E}^{]0,r_a]}$ est inversible si et seulement si f ne s'annule pas sur C_a.

Un *diviseur* sur C_a est une expression de la forme $D = \sum_{\alpha\in C_a} n_\alpha[\alpha]$, où les n_α sont des éléments de $\mathbf{Z}$. Un tel diviseur est *localement fini*, si $\sum_{v_p(\alpha)\geq r_n}|n_\alpha| < +\infty$, pour tout $n \geq a$; il est *défini sur* L si $n_{g(\alpha)} = n_\alpha$, pour tout $\alpha \in \mathbf{C}_p$ et $g \in \mathrm{Gal}(\overline{\mathbf{Q}}_p/L)$.

Si $f \in \mathscr{E}^{]0,r_a]}$ est non nul, alors $\mathrm{Div}(f) = \sum_{\alpha\in C_a} v_\alpha(f)[\alpha]$ est un diviseur sur C_a localement fini et défini sur L. Si I est un idéal non nul de type fini (et donc principal) de $\mathscr{E}^{]0,r_a]}$, on note $\mathrm{Div}(I)$ le diviseur $\mathrm{Div}(f)$, pour n'importe quel générateur f de I. L'application $I \mapsto \mathrm{Div}(I)$, ainsi définie, induit une bijection de l'ensemble des idéaux non nuls et de type fini de $\mathscr{E}^{]0,r_a]}$ sur celui des diviseurs effectifs, localement finis et définis sur L ; de plus, $g \in I - \{0\}$ si et seulement si $\mathrm{Div}(g) - \mathrm{Div}(I)$ est un diviseur effectif.

Soit $D = \sum_{\alpha\in C_a} n_\alpha[\alpha]$ un diviseur localement fini. On dit que D est C_a-stable par $\boldsymbol{\mu}_{p^n}$ ($n = \infty$ inclus) si $n_{\zeta(1+\alpha)-1} = n_\alpha$ pour tous $\alpha \in C_a$ et $\zeta \in \boldsymbol{\mu}_{p^n}$ tels que $\zeta(1+\alpha) - 1 \in C_a$.

Lemme 8.69. *Si* $D = \sum_{\alpha\in C_a} n_\alpha[\alpha]$ *est effectif, localement fini, défini sur* L *et* C_a*-stable par* $\boldsymbol{\mu}_{p^\infty}$*, il existe* $g \in L\{\{t\}\} \subset \mathscr{E}^{]0,r_a]}$ *tel que* $D = \mathrm{Div}(g)$ (en tant que diviseur sur C_a).

Démonstration. Soit $D' = \sum_{\alpha\in\mathbf{C}_p} n_\beta[\beta]$ le diviseur sur $\mathbf{C}_p$ défini par $n_\beta = n_\alpha$, si $\log(1+\alpha) = \beta$. Comme deux solutions de l'équation $\log(1+\alpha) = \beta$ s'obtiennent à partir de l'une d'entre elles en faisant agir $\boldsymbol{\mu}_{p^\infty}$, la condition selon laquelle D est C_a-stable par $\boldsymbol{\mu}_{p^\infty}$ implique que D' est bien défini. Il est alors facile de voir que D' est localement fini et défini sur L, et donc est le diviseur d'un élément g de $L\{\{t\}\}$ (vu comme une fonction analytique de t sur $\mathbf{C}_p$). Par construction, le diviseur de g vu comme une fonction analytique sur C_a via le changement de variable $t = \log(1+T)$ est alors D. On en déduit le résultat.

Soient $U = \{f \in \mathscr{R}^+,\ f(0) = 1\}$ et $U_a \subset U$ l'ensemble des f ne s'annulant pas sur D_a.

Lemme 8.70. *Tout élément non nul* f *de* $\mathscr{E}^{]0,r_a]}$ *peut s'écrire sous la forme* $f = f^+ f^{\{0\}}$, *avec* $f^+ \in U_a$, *et* $f^{\{0\}} \in (\mathscr{E}^{]0,r_a]})^*$.

Démonstration. Soit $D = \sum_{0<v_p(z)\leq r_a} v_z(f)\, z$ le diviseur de f. Si $b \geq a$, l'ensemble $\{z,\ r_b < v_p(z) \leq r_a \text{ et } v_z(f) \neq 0\}$ est fini et on a $v_{\sigma(z)}(f) = v_z(f)$, si $\sigma \in \mathrm{Gal}(\overline{\mathbf{Q}}_p/L)$. Ces conditions impliquent l'existence de $f^+ \subset U_a$ dont le diviseur est D, et alors $g = (f^+)^{-1} f$ ne s'annule pas sur C_a et donc appartient à $(\mathscr{E}^{]0,r_a]})^* = (\mathscr{E}^{(0,r_a]})^*$. Ceci prouve l'existence d'une décomposition sous la forme souhaitée.

Corollaire 8.71. *Tout élément non nul* f *de* $\mathrm{Fr}(\mathscr{E}^{]0,r_a]})$ *peut s'écrire sous la forme* $f = f^{\{0\}}\frac{f^+}{f^-}$, *avec* $f^+, f^- \in U_a$ *sans zéros communs, et* $f^{\{0\}} \in (\mathscr{E}^{]0,r_a]})^*$.

Si $f = f^{\{0\}}\frac{f^+}{f^-} \in \mathrm{Fr}(\mathscr{E}^{]0,r_a]})$, son diviseur $\mathrm{Div}(f) = \sum_{\alpha\in C_a} v_\alpha(f)[\alpha]$ est aussi égal à $\mathrm{Div}(f^+) - \mathrm{Div}(f^-)$.

Lemme 8.72. *Si $f \in \mathscr{E}^{]0,r_a]}$ est non nul, il existe $g \in L\{\{t\}\}$ tel que $f^{-1}\mathscr{E}^{]0,r_a]} \cap \mathrm{Fr}(L\{\{t\}\}) = g^{-1}L\{\{t\}\}$.*

Démonstration. Soit D le diviseur $\sum_{v_p(T)>0} n_\alpha[\alpha]$, où

$$n_\alpha = \inf\big(\{v_{(1+\alpha)\zeta-1}(f),\ \zeta \in \boldsymbol{\mu}_{p^\infty},\ v_p((1+\alpha)\zeta - 1) \le r_a\}\big).$$

Par construction, D est invariant par $\boldsymbol{\mu}_{p^\infty}$ (où $\zeta \in \boldsymbol{\mu}_{p^\infty}$ agit par $\alpha \mapsto (1+\alpha)\zeta - 1$), et est localement fini car on a $n_\alpha \le v_\alpha(f)$, si $v_p(\alpha) \le r_a$. Il existe donc $g \in L\{\{t\}\}$ dont D est le diviseur, et il est facile de voir que tout élément de $L\{\{t\}\}$ divise g s'il divise f. On en déduit le résultat.

3.2.3. L'action de φ sur $\mathrm{Fr}(\mathscr{R})$

Lemme 8.73. (i) *$\mathscr{R}^+$ est un $\varphi(\mathscr{R}^+)$-module libre de rang p de base les $(1+T)^i$, pour $0 \le i \le p-1$.*

(ii) *Si $f \in \mathscr{R}^+$, alors $\mathrm{N}_{\mathscr{R}^+/\varphi(\mathscr{R}^+)} f = \prod_{\eta^p=1} f((1+T)\eta - 1)$.*

(iii) *Si $f \in \mathscr{R}^+$, alors f divise $\mathrm{N}_{\mathscr{R}^+/\varphi(\mathscr{R}^+)} f$ dans $\mathscr{R}^+$.*

Démonstration. Les (i) et (ii) suivent de ce que $\varphi(T) = (1+T)^p - 1$; le (iii) est une conséquence du (ii).

Lemme 8.74. (i) *Tout élément f de* $\mathrm{Fr}(\mathscr{R})$ *peut s'écrire sous la forme $f = \frac{a}{\varphi(b)}$, avec $a, b \in \mathscr{R}$.*

(ii) *$\psi(f) = \frac{\psi(a)}{b}$ ne dépend pas de l'écriture choisie.*

Démonstration. Comme $\mathrm{Fr}(\mathscr{R}) = \cup_{a\ge 1}\mathrm{Fr}(\mathscr{E}^{]0,r_a]})$, on peut écrire f sous la forme $f = f^{\{0\}}\frac{f^+}{f^-}$, et le (iii) du lemme 8.73 fournit une écriture sous la forme voulue avec $a = f^{\{0\}} f^+ \mathrm{N}_{\mathscr{R}^+/\varphi(\mathscr{R}^+)} f^- / f^-$ et $b = \varphi^{-1}(\mathrm{N}_{\mathscr{R}^+/\varphi(\mathscr{R}^+)} f^-)$.

Si $f = \frac{a}{\varphi(b)} = \frac{a'}{\varphi(b')}$, on a $a\varphi(b') = a'\varphi(b)$ et donc $\psi(a)b' = \psi(a')b$, ce qui démontre le (ii).

Lemme 8.75. *Si $n \in \mathbf{N}$, alors $\mathscr{R} \cap \varphi^n(\mathrm{Fr}(\mathscr{R})) = \varphi^n(\mathscr{R})$.*

Démonstration. Soit $y \in \mathrm{Fr}(\mathscr{R})$ tel que $\varphi^n(y) \in \mathscr{R}$. Il existe $a \ge 1$ tel que $y \in \mathrm{Fr}(\mathscr{E}^{(0,r_a]})$ et, d'après le cor. 8.71, on peut écrire y sous la forme $y = y^{\{0\}}\frac{y^+}{y^-}$, avec $y^+, y^- \in U_a$ sans zéros communs, et $y^{\{0\}} \in (\mathscr{E}^{]0,r_a]})^*$. Maintenant, si $\varphi^n(y) \in \mathscr{R}$, il existe b (que l'on peut supposer $\ge a+n$) tel que $\varphi^n(y)$ n'ait pas de pôle dans C_b. Ceci implique que $\varphi^n(y^-)$ n'a pas de zéro dans C_b, et donc que y^- n'a pas de zéro dans C_{b-n}. On en déduit l'appartenance de y^- à $(\mathscr{E}^{]0,r_{b-n}]})^*$ et celle de y à $\mathscr{R}$, ce qui permet de conclure.

Si $a \in \mathbf{Z}$ est ≤ 1, on pose $r_a = r_1 + 1 - a$. Alors $\psi(\mathscr{E}^{]0,r_a]}) \subset \mathscr{E}^{]0,r_{a+1}]}$ pour tout $a \in \mathbf{Z}$ (cela résulte de [13, prop. I.13]), et $\varphi(\mathscr{E}^{]0,r_{a+1}]}) \subset \mathscr{E}^{]0,r_a]}$ pour tout $a \in \mathbf{Z}$ car $z \in C_{a+1}$ implique $(1+z)^p - 1 \in C_a$.

Lemme 8.76. *Si* $f \in \mathrm{Fr}(\mathscr{E}^{]0,r_a]}) \cap \varphi^n(\mathrm{Fr}(\mathscr{R}))$*, alors* $\mathrm{Div}(f)$ *est* C_a*-stable par* $\boldsymbol{\mu}_{p^n}$.

Démonstration. D'après le lemme 8.73, on peut écrire f sous la forme $f = \frac{g}{\varphi^n(h)}$, avec $\varphi^n(h) = \mathrm{N}_{\mathscr{R}^+/\varphi^n(\mathscr{R}^+)}(f^-) \in \mathscr{R}^+$ et $g \in \mathscr{E}^{]0,r_a]}$ (cf. cor. 8.71 pour la définition de f^-). L'hypothèse $f \in \varphi^n(\mathrm{Fr}(\mathscr{R}))$ implique que $f = \varphi^n(\psi^n(f))$. Or $\psi^n(f) = \frac{\psi^n(g)}{h}$, et $\psi^n(g) \in \mathscr{E}^{]0,r_{a-n}]}$. Il s'ensuit que si $\sum_{\alpha\in C_{a-n}} n_\alpha[\alpha]$ est le diviseur de $\psi^n(f)$ sur C_{a-n} (avec $n_\alpha = v_\alpha(\psi^n(g)) - v_\alpha(h)$), celui de f est $\sum_{\alpha\in C_{a-n}} \sum_{\beta\in C_a,\ (1+\beta)^{p^n}=1+\alpha} n_\alpha[\beta]$. Il est clair sur cette expression que ce diviseur est C_a-stable par $\boldsymbol{\mu}_{p^n}$.

Théorème 8.77. $\cap_{n\in\mathbf{N}}\varphi^n(\mathrm{Fr}(\mathscr{R})) = \mathrm{Fr}(L\{\{t\}\})$.

Démonstration. Soit $f \in \cap_{n\in\mathbf{N}}\varphi^n(\mathrm{Fr}(\mathscr{R}))$. Il existe $a \geq 1$ tel que $f \in \mathrm{Fr}(\mathscr{E}^{]0,r_a]})$, et il résulte du lemme 8.76 que $\mathrm{Div}(f)$ est C_a-stable par $\boldsymbol{\mu}_{p^\infty}$. Il existe donc $g \in \mathrm{Fr}(L\{\{t\}\})$ tel que $\mathrm{Div}(g) = \mathrm{Div}(f)$ (cf. lemme 8.69). Cela signifie que $\frac{f}{g} \in (\mathscr{E}^{]0,r_a]})^* \subset \mathscr{R}$. Par ailleurs, φ est bijectif sur $\mathrm{Fr}(L\{\{t\}\})$ puisqu'il l'est sur $L\{\{t\}\}$. Il s'ensuit, grâce au lemme 8.75, que $\frac{f}{g} \in \cap_{n\in\mathbf{N}}\varphi^n(\mathscr{R})$, et donc, grâce à la prop. 8.68, que $\frac{f}{g} \in L\{\{t\}\}$. On en déduit le résultat.

3.3. Le $L\{\{t\}\}$-module $\Delta \boxtimes \{0\}$

3.3.1. Structure de $\Delta \boxtimes \{0\}$

Soit $\Delta \in \Phi\Gamma^{\mathrm{et}}(\mathscr{R})$, de rang d. D'après [1, th. I.3.3], il existe $m(\Delta) \in \mathbf{N}$ et, pour $a \geq m(\Delta)$, un sous-$\mathscr{E}^{]0,r_a]}$-module $\Delta^{]0,r_a]}$ de Δ vérifiant les propriétés suivantes (qui les caractérisent) :

- $\Delta^{]0,r_{m(D)}]}$ est de rang d sur $\mathscr{E}^{]0,r_{m(D)}]}$ et les applications naturelles $\mathscr{E}^{]0,r_a]} \otimes \Delta^{]0,r_{m(D)}]} \to \Delta^{]0,r_a]}$ et $\mathscr{R} \otimes \Delta^{]0,r_{m(D)}]} \to \Delta$ (les produits tensoriels sont au-dessus de $\mathscr{E}^{]0,r_{m(D)}]}$) sont des isomorphismes (et donc $\Delta^{]0,r_a]}$ est de rang d sur $\mathscr{E}^{]0,r_a]}$ pour tout $a \geq m(D)$).
- $\varphi(\Delta^{]0,r_a]}) \subset \Delta^{]0,r_{a+1}]}$, pour tout $a \geq m(D)$.

On peut déduire de ces propriétés que :

- $\psi(\Delta^{]0,r_{a+1}]}) \subset \Delta^{]0,r_a]}$, pour tout $a \geq m(D)$ et que $\varphi(x) \in \Delta^{]0,r_{a+1}]}$ si et seulement si $x \in \Delta^{]0,r_a]}$.

On note $\Delta \boxtimes \{0\}$, le L-espace vectoriel $\cap_{n\in\mathbf{N}}\varphi^n(\Delta)$. C'est un $L\{\{t\}\} = \cap_{n\in\mathbf{N}}\varphi^n(\mathscr{R})$-module.

Lemme 8.78. *Soient* $x_1, \dots, x_d \in \Delta \boxtimes \{0\}$. *Si* $x_1, \dots, x_d$ *sont linéairement dépendants sur* $\mathscr{R}$, *ils le sont déjà sur* $L\{\{t\}\}$.

Démonstration. Soit $\sum_{i=1}^{r} f_i x_i = 0$ une relation de longueur minimale sur $\mathscr{R}$. En divisant tout par f_1, on obtient une relation $\sum_{i=1}^{r} g_i x_i = 0$ de longueur minimale sur $\mathrm{Fr}(\mathscr{R})$, avec $g_1 = 1$. Comme $x_i \in \varphi^n(\Delta)$, on a $\varphi^n(\psi^n(x_i)) = x_i$ et $\psi^n(g_i x_i) = \psi^n(g_i \varphi^n(\psi^n(x_i))) = \psi^n(g_i)\psi^n(x_i)$; on en déduit la relation $\sum_{i=1}^{r} \varphi^n(\psi^n(g_i))x_i = 0$. Comme $\varphi^n(\psi^n(g_1)) = 1$, la minimalité de la relation entraîne la relation $g_i = \varphi^n\psi^n(g_i))$, pour tout i et tout $n \in \mathbf{N}$. Il s'ensuit que $g_i \in \cap_{n\in\mathbf{N}}\varphi^n(\mathrm{Fr}(\mathscr{R})) = \mathrm{Fr}(L\{\{t\}\})$. On en déduit le résultat.

Théorème 8.79. *Si* $\Delta \in \Phi\Gamma^{\mathrm{et}}(\mathscr{R})$ *est de rang* d, *alors* $\Delta \boxtimes \{0\}$ *est un* (φ, Γ)-*module de rang* $\leq d$ *sur* $L\{\{t\}\}$.

Démonstration. Soient j la dimension du sous-$\mathrm{Fr}(L\{\{t\}\})$-espace vectoriel de $\mathrm{Fr}(\mathscr{R}) \otimes_{\mathscr{R}} \Delta$ engendré par $M = \Delta \boxtimes \{0\}$ (d'après le lemme 8.78, on a $j \leq d$), et $e_1, \dots, e_j$ une base de cet espace vectoriel constituée d'éléments de $\Delta \boxtimes \{0\}$.

Soit $a \in \mathbf{N}$ tel que $\Delta^{]0,r_a]}$ contienne $e_1, \dots, e_j$ et, si $n \geq a$, soit $M_n = \Delta^{]0,r_n]} \cap M$. Alors φ induit une bijection de M_n sur M_{n+1} (car φ est bijectif sur M et $\varphi(x) \in \Delta^{]0,r_{n+1}]}$ si et seulement si $x \in \Delta^{]0,r_n]}$) et on a $M = \cup_{n\geq a} M_n$.

Soit X l'adhérence du $\mathscr{E}^{]0,r_a]}$-module engendré par M_a dans $\Delta^{]0,r_a]}$. Alors X est un $\mathscr{E}^{]0,r_a]}$-module de rang j [il est libre en tant que sous-module fermé du $\mathscr{E}^{]0,r_a]}$-module libre $\Delta^{]0,r_a]}$; il contient $e_1, \dots, e_j$ qui sont libres sur $\mathrm{Fr}(\mathscr{R})$, et donc son rang est $\geq j$; il est contenu dans le $\mathrm{Fr}(\mathscr{R})$-espace vectoriel F engendré par $e_1, \dots, e_j$ ce qui implique qu'une base de X sur $\mathscr{E}^{]0,r_a]}$ (une telle base peut se compléter en une base de $\Delta^{]0,r_a]}$) est aussi une base de F sur $\mathrm{Fr}(\mathscr{R})$] et $e_1, \dots, e_j$ est une base de $\mathrm{Fr}(\mathscr{E}^{]0,r_a]}) \otimes X$ sur $\mathrm{Fr}(\mathscr{E}^{]0,r_a]})$. Si δ est le déterminant de $e_1, \dots, e_j$ dans une base de M_a sur $\mathscr{E}^{]0,r_a]}$, les coordonnées de δx par rapport à $e_1, \dots, e_j$ sont dans $\mathscr{E}^{]0,r_a]}$, si $x \in M_a$. Par ailleurs, $M_a = \Delta^{]0,r_a]} \cap \big(\oplus_{i=1}^{j} \mathrm{Fr}(L\{\{t\}\})\, e_i\big)$. On en déduit, en utilisant le lemme 8.72, l'existence de $g \in L\{\{t\}\}$ tel que M_a soit inclus dans $\oplus_{i=1}^{j} g^{-1} L\{\{t\}\}\, e_i$. Comme M_a est fermé dans $\Delta^{]0,r_a]}$ (c'est l'intersection des noyaux des $1 - \varphi^n\psi^n$, pour $n \in \mathbf{N}$), son image par l'application qui à x associe les coordonnées de gx dans la base $e_1, \dots, e_j$ est fermée dans $(\mathscr{E}^{]0,r_a]})^j$ et est incluse dans $(L\{\{t\}\})^j$ par construction de g. Comme la topologie de $L\{\{t\}\}$ induite par celle de $\mathscr{E}^{]0,r_a]}$ est sa topologie naturelle, on en déduit que cette image est fermée dans $(L\{\{t\}\})^j$, et comme c'est un $L\{\{t\}\}$-module, il en résulte qu'elle est libre de rang j sur $L\{\{t\}\}$. On en déduit que M_a est libre de rang j sur $L\{\{t\}\}$ et on peut donc supposer que $e_1, \dots, e_j$ est une base de M_a sur $L\{\{t\}\}$.

Maintenant, M_{a+n} est stable par Γ et engendré par $\varphi^n(e_1), \dots, \varphi^n(e_j)$. Il en résulte que l'idéal de $L\{\{t\}\}$ engendré par le déterminant de $e_1, \dots, e_j$ dans la base $\varphi^n(e_1), \dots, \varphi^n(e_j)$ est stable par Γ et donc que ce déterminant est de la forme αt^k, avec $a \in L^*$ et $k \in \mathbf{N}$. Par ailleurs, le déterminant $b(\gamma)$ de $\gamma(e_1), \dots, \gamma(e_j)$ dans la base $e_1, \dots, e_j$ appartient à $(L\{\{t\}\})^* = L^*$ puisque M_a est stable par Γ. La relation $(\det \sigma_a)\sigma_a(\det \varphi) = (\det \varphi)\varphi(\det \sigma_a)$ se traduit alors par la relation $b(\sigma_a)\alpha^{-1}a^{-k}t^{-k} = \alpha^{-1}t^{-k}b(\sigma_a)$, dont on déduit que $k = 0$. Il en résulte que $M_{a+n} = M_a$, pour tout n, et donc $\Delta \boxtimes \{0\} = M_a$. Ceci permet de conclure.

3.3.2. Le module $\Delta \boxtimes \{0\}$ dans le cas de rang 2

Si $s \in \mathscr{S}$, le (φ, Γ)-module $\Delta(s)$ vit, par construction, dans une suite exacte $0 \to \mathscr{R}(\delta_1) \to \Delta(s) \to \mathscr{R}(\delta_2) \to 0$. On note e_1 la base canonique $1 \otimes \delta_1$ de $\mathscr{R}(\delta_1)$ et e_2 celle de $\mathscr{R}(\delta_2)$. On note aussi p_s la projection $\Delta(s) \to \mathscr{R}(\delta_2)$ et on choisit $\hat{e}_2 \in \Delta(s)$ tel que $p_s(\hat{e}_2) = e_2$.

Si $s \in \mathscr{S}_*^{\rm cris}$ n'est pas exceptionnel, alors $t^{w(s)}e_2$ a un (unique) relèvement e_2' dans $\Delta(s)$, vérifiant $\varphi(e_2') = p^{w(s)}\delta_2(p)e_2'$ et $\sigma_a(e_2') = a^{w(s)}\delta_2(a)e_2'$, pour tout $a \in \mathbf{Z}_p^*$.

Si $x^{w(s)}\delta_2 = \delta_1$, alors $t^{w(s)}e_2$ a un relèvement e_2' dans $\Delta(s)$ (unique à addition près d'un multiple de e_1), vérifiant $\varphi(e_2') = p^{w(s)}\delta_2(p)e_2' + e_1$ et $\sigma_a(e_2') = a^{w(s)}\delta_2(a)e_2'$, pour tout $a \in \mathbf{Z}_p^*$.

Remarque 8.80. Si $s \in \mathscr{S}_*^{\rm cris}$ n'est pas exceptionnel, l'isomorphisme de la prop. 8.3 entre $\Delta(s)$ et $\Delta(s')$ se voit de la manière suivante. Le sous-$\mathscr{R}$-module $\mathscr{R}e_2'$ de $\Delta(s)$ est stable par φ et Γ (isomorphe à $\mathscr{R}(x^{w(s)}\delta_2)$) et est saturé. On en déduit la suite exacte $0 \to \mathscr{R}(x^{w(s)}\delta_2) \to \Delta(s) \to \mathscr{R}(x^{-w(s)}\delta_1) \to 0$, et l'image de e_1 dans $\mathscr{R}(x^{-w(s)}\delta_1)$ est $t^{w(s)}e_1'$, où l'on a noté e_1' la base canonique de $\mathscr{R}(x^{-w(s)}\delta_1)$.

Lemme 8.81. *Soit $\Delta \in \Phi\Gamma^{\rm et}(\mathscr{R})$, irréductible de rang 2.*

(i) *Si Δ possède un vecteur propre pour les actions de φ et Γ, alors Δ est triangulable.*

(ii) *Si $s \in \mathscr{S}_*^{\rm ng} \cup \mathscr{S}_*^{\rm st}$ les seuls vecteurs propres* (à multiplication près par un élément de L^*) *de $\Delta(s)$ pour les actions de φ et Γ sont les $t^k e_1$* (propre pour $x^k\delta_1$)*, pour $k \in \mathbf{N}$; c'est aussi le cas si $s \in \mathscr{S}_*^{\rm cris}$ est exceptionnel.*

(iii) *Si $s \in \mathscr{S}_*^{\rm cris}$ n'est pas exceptionnel, les seuls vecteurs propres de $\Delta(s)$ pour les actions de φ et Γ sont les $t^k e_1$* (propre pour $x^k\delta_1$)*, pour $k \in \mathbf{N}$, et les $t^k e_2'$* (propre pour $x^{k+w(s)}\delta_2$)*, pour $k \in \mathbf{N}$.*

Démonstration. Si Δ_1 est un sous-(φ,Γ)-module de rang 1 d'un (φ,Γ)-module Δ sur $\mathscr{R}$, il existe $k \in \mathbf{N}$ tel que $t^{-k}\Delta_1$ soit saturé dans Δ (cf. dém. de [10, lemme 3.2]). Ceci permet de démontrer le (i).

Maintenant, les seuls sous-(φ,Γ)-modules de $\mathscr{R}(\delta)$ sont les $t^k\mathscr{R}(\delta)$, pour $k \in \mathbf{N}$ (cf. prop. 8.33). On en déduit que les seuls vecteurs propres (à multiplication près par un élément de L^*) de $\mathscr{R}(\delta_1)$ pour les actions de φ et Γ sont les $t^k e_1$, pour $k \in \mathbf{N}$. De plus, si $v \notin \mathscr{R}e_1$ est propre pour φ et Γ, il en est de même de son image dans $\mathscr{R}(\delta_2)$ qui est donc de la forme $\alpha t^i e_2$, avec $\alpha \in L^*$; autrement dit, il existe $i \in \mathbf{N}$, tel que $t^i e_2$ admette un relèvement propre pour φ et Γ. D'après [10, lemme 4.9], cela implique que $s \in \mathscr{S}_*^{\rm cris}$ n'est pas exceptionnel et que v est un multiple de $t^k e_2'$, avec $k \in \mathbf{N}$.

On en déduit le résultat.

Théorème 8.82. *Soit $\Delta \in \Phi\Gamma^{\rm et}(\mathscr{R})$, irréductible de dimension 2. Alors $\Delta \boxtimes \{0\}$ possède un unique sous-L-espace vectoriel $(\Delta \boxtimes \{0\})_0$, stable par φ et Γ, tel que l'application naturelle $L\{\{t\}\} \otimes_L (\Delta(s) \boxtimes \{0\})_0 \to \Delta \boxtimes \{0\}$ soit un isomorphisme. Plus précisément :*

(i) *Si Δ n'est pas triangulable, alors $\Delta \boxtimes \{0\} = 0$ et donc $(\Delta \boxtimes \{0\})_0 = 0$.*
(ii) *Si $s = (\delta_1,\delta_2,\mathscr{L}) \in \mathscr{S}_*^{\rm ng} \coprod \mathscr{S}_*^{\rm st}$, alors $(\Delta(s) \boxtimes \{0\})_0 = L\, e_1$.*
(iii) *Si $s = (\delta_1,\delta_2,\mathscr{L}) \in \mathscr{S}_*^{\rm cris}$, alors $(\Delta(s) \boxtimes \{0\})_0 = L\, e_1 \oplus L\, e_2'$.*

Démonstration. Si $M = \Delta \boxtimes \{0\}$ n'est pas nul, alors il est de rang 1 ou 2 et quitte à faire une extension quadratique de L, il existe $v \in M$, propre pour les actions de φ et Γ (cf. ex. 8.66), ce qui, d'après le (i) du lemme 8.81, implique que Δ est triangulable.

Maintenant, soit $s = (\delta_1,\delta_2,\mathscr{L}) \in \mathscr{S}_{\rm irr}$. On a une suite exacte $0 \to \mathscr{R}e_1 \boxtimes \{0\} \to \Delta(s) \boxtimes \{0\} \to \mathscr{R}e_2 \boxtimes \{0\}$, et $\{\mathscr{R}e_1 \boxtimes \{0\} = \cap_{n\in\mathbf{N}}\varphi^n(\mathscr{R})e_1 = L\{\{t\}\}e_1$. De même $\mathscr{R}e_2 \boxtimes \{0\} = L\{\{t\}\}e_2$, et si l'image de $\Delta(s) \boxtimes \{0\}$ dans $\mathscr{R}e_2 \boxtimes \{0\}$ est non nulle, elle est de la forme $t^i L\{\{t\}\}e_2$, avec $i \in \mathbf{N}$. D'après l'ex. 8.66, on a alors deux possibilités :

- $t^i e_2$ admet un relèvement propre pour φ et Γ, et d'après [10, lemme 4.9], cela implique que $s \in \mathscr{S}_*^{\rm cris}$ et est non exceptionnel, auquel cas un tel relèvement de $t^i e_2$ existe si et seulement si $i \geq w(s)$ (où l'on peut prendre $t^{i-w(s)}e_2'$ comme relèvement).
- il existe $k \in \mathbf{N}$ et $(\alpha,\beta) \in L^2$ non nul, et un relèvement f de $t^i e_2$ tels que

$$\varphi(f) = p^k\delta_1(p)f + \alpha t^k e_1 \ \text{ et } \ \sigma_a(f) = a^k\delta_1(a)f + \beta t^k e_1, \ \text{ si } a \in \mathbf{Z}_p^*.$$

On en déduit que $x^k\delta_1 = x^i\delta_2$. Comme $v_p(\delta_1(p)) + v_p(\delta_2(p)) = 0$ et $v_p(\delta_1(p)) > 0$, cela implique $i > k$, et donc que $s \in \mathscr{S}_*^{\rm cris}$ est exceptionnel, et que $w(s) = i - k$. En particulier, on a $i \geq w(s)$, et comme on peut relever $t^{w(s)}e_2$ en e_2', on en déduit que $i = w(s)$ et que $k = 0$.

Cela prouve que M est de rang 2 si et seulement si $s \in \mathscr{S}_*^{\rm cris}$, auquel cas il est engendré par e_1 et e_2'. Comme M est de rang ≥ 1, si $s \in \mathscr{S}_{\rm irr}$, cela prouve qu'il est de rang 1, engendré par e_1, si $s \in \mathscr{S}_*^{\rm ng} \cup \mathscr{S}_*^{\rm st}$.

Enfin, l'existence et l'unicité de $(\Delta(s) \boxtimes \{0\})_0$ se déduisent de la description de M_0 dans l'ex. 8.66 (on remarquera que le second cas de cet exemple avec $k \geq 1$ a été exclu par la discussion précédente).

Ceci permet de conclure.

Remarque 8.83. Si $D \in \Phi\Gamma^{\rm et}(\mathscr{E})$, à côté de $D \boxtimes \{0\} = \cap_{n\in\mathbf{N}}\varphi^n(D)$ (qui est noté $D^{\rm nr}$ dans [12]), on dispose [12] de sous-modules $D^\natural \subset D^\sharp$ qui jouent un grand rôle dans l'établissement et l'étude [13, 2, 14] de la correspondance de Langlands locale p-adique pour $\mathbf{GL}_2(\mathbf{Q}_p)$. Par analogie avec les définitions et les propriétés de $D^\natural$ et $D^\sharp$, on est amené à définir, pour $\Delta \in \Phi\Gamma^{\rm et}(\mathscr{E})$, les sous-modules suivants :

- $\Delta^\sharp = \cap_{n\in\mathbf{N}}\psi^n(\Delta^{]0,r_a]})$, où $a \geq m(D)$ est quelconque.
- $\Delta^\natural$ est l'orthogonal de $\check{\Delta} \boxtimes \{0\}$ dans $\Delta^\sharp$.

Nous laissons au lecteur le soin de montrer, en utilisant le dictionnaire d'analyse fonctionnelle, que dans le cas $\Delta = \mathscr{R}$, alors $\Delta^\natural = \mathscr{R}^+$ et $\Delta^\sharp/\Delta^\natural$ est le sous-espace de $\mathrm{LA}(\mathbf{Z}_p)$ des ϕ qui sont la restriction à $\mathbf{Z}_p$ d'une fonction analytique sur $\mathbf{C}_p$.

4. Les vecteurs localement analytiques de la série principale unitaire

Ce chapitre est consacré à la description (prop. 8.94) de $\mathbf{\Pi}(\Delta(s))$. L'ingrédient principal en est la détermination (th. 8.87) du module de Jacquet dual $J^*(\mathbf{\Pi}(\Delta))$ qui permet (th. 8.89) de dévisser le module $\Delta(s) \boxtimes \mathbf{P}^1$.

4.1. La représentation localement analytique $\mathbf{\Pi}(\Delta)$

4.1.1. Le faisceau $U \mapsto \Delta \boxtimes U$

Soit $\Delta \in \Phi\Gamma^{\rm et}(\mathscr{R})$. Alors Δ est muni d'une action de P^+ donnée par $\left(\begin{smallmatrix} p^k a & b \\ 0 & 1\end{smallmatrix}\right) \cdot z = (1+T)^b\varphi^k \circ \sigma_a(z)$, si $k \in \mathbf{N}$, $a \in \mathbf{Z}_p^*$ et $b \in \mathbf{Z}_p$, et d'un inverse à gauche ψ de φ qui commute à l'action de Γ et qui est donné par la formule

$\psi(\sum_{i=0}^{p-1}(1+T)^i\varphi(x_i)) = x_0$. On utilise ces données pour associer à Δ un faisceau $U \mapsto \Delta \boxtimes U$ sur $\mathbf{Z}_p$ (où U décrit les ouverts compacts de $\mathbf{Z}_p$), équivariant sous l'action de P^+, où P^+ agit sur $\mathbf{Z}_p$ par la formule $\left(\begin{smallmatrix} a & b \\ 0 & 1 \end{smallmatrix}\right) \cdot x = ax + b$ habituelle. De manière précise :

- $\Delta \boxtimes \mathbf{Z}_p = \Delta$ et $\Delta \boxtimes \emptyset = 0$,
- $\Delta \boxtimes (i + p^k\mathbf{Z}_p) = \left(\begin{smallmatrix} p^k & i \\ 0 & 1 \end{smallmatrix}\right)\Delta \subset \Delta$,
- $\mathrm{Res}_{i+p^k\mathbf{Z}_p} : \Delta \boxtimes \mathbf{Z}_p \to \Delta \boxtimes (i + p^k\mathbf{Z}_p)$ est donné par $\mathrm{Res}_{i+p^k\mathbf{Z}_p} = \left(\begin{smallmatrix} 1 & i \\ 0 & 1 \end{smallmatrix}\right) \circ \varphi^k \circ \psi^k \circ \left(\begin{smallmatrix} 1 & -i \\ 0 & 1 \end{smallmatrix}\right)$.

On remarquera que l'on a une suite exacte $0 \to \mathscr{D} \to \Delta \to \mathrm{LA} \boxtimes \chi^{-1} \to 0$ de faisceaux P^+-équivariants sur $\mathbf{Z}_p$ si $\Delta = \mathscr{R}$.

Ce qui précède est valable pour un élément de $\Phi\Gamma^{\mathrm{et}}(\mathscr{R})$ de rang arbitraire. Si Δ est de rang 2, on peut lui associer un faisceau $U \mapsto \Delta \boxtimes U$ sur $\mathbf{P}^1 = \mathbf{P}^1(\mathbf{Q}_p)$ (où U décrit les ouverts compacts de $\mathbf{P}^1$), équivariant sous l'action de G, où G agit sur $\mathbf{P}^1 = \mathbf{P}^1(\mathbf{Q}_p)$ par la formule $\left(\begin{smallmatrix} a & b \\ c & d \end{smallmatrix}\right) \cdot x = \frac{ax+b}{cx+d}$ habituelle. (Si $a \in \mathbf{Q}_p^*$, l'élément $\left(\begin{smallmatrix} a & 0 \\ 0 & a \end{smallmatrix}\right)$ du centre de G agit par multiplication par $\omega_\Delta(a)$, où ω_Δ est le caractère $\chi^{-1}\det\Delta$, mais la définition de l'action de G repose [14, § II.1] sur des formules nettement plus compliquées que celle de P^+). On a comme ci-dessus $\Delta \boxtimes \mathbf{Z}_p = \Delta$, et le faisceau ainsi obtenu sur $\mathbf{Z}_p$, muni de la restriction de l'action de G à P^+, n'est autre que le faisceau P^+-équivariant sur $\mathbf{Z}_p$ défini ci-dessus. Par ailleurs, si U est un ouvert compact de $\mathbf{P}^1$, on dispose d'une application de prolongement par 0 qui permet de voir $\Delta \boxtimes U$ comme un sous-module de l'espace $\Delta \boxtimes \mathbf{P}^1$ des sections globales.

Comme $\mathbf{P}^1$ est obtenu en recollant $\mathbf{Z}_p$ et $w \cdot \mathbf{Z}_p$ le long de $\mathbf{Z}_p^*$, et comme G est engendré par P^+ et w, le faisceau $U \mapsto \Delta \boxtimes U$ sur $\mathbf{P}^1 = \mathbf{P}^1(\mathbf{Q}_p)$ est complètement décrit par sa restriction à $\mathbf{Z}_p$ (avec l'action de P^+) et par l'action de w sur $\Delta \boxtimes \mathbf{Z}_p^*$. Si on note w_Δ cette action, l'application $z \mapsto (\mathrm{Res}_{\mathbf{Z}_p} z, \mathrm{Res}_{\mathbf{Z}_p} w \cdot z)$ permet de décrire $\Delta \boxtimes \mathbf{P}^1$ comme l'ensemble des $(z_1, z_2) \in \Delta \times \Delta$ vérifiant $\mathrm{Res}_{\mathbf{Z}_p^*} z_2 = w_\Delta(\mathrm{Res}_{\mathbf{Z}_p^*} z_1)$. L'action de G sur $\Delta \boxtimes \mathbf{P}^1$ est alors décrite par les formules du *squelette d'action* (avec $z = (z_1, z_2)$) :

- $\left(\begin{smallmatrix} 0 & 1 \\ 1 & 0 \end{smallmatrix}\right) \cdot z = (z_2, z_1)$.
- Si $a \in \mathbf{Q}_p^*$, alors $\left(\begin{smallmatrix} a & 0 \\ 0 & a \end{smallmatrix}\right) \cdot z = (\omega_\Delta(a) z_1, \omega_\Delta(a) z_2)$.
- Si $a \in \mathbf{Z}_p^*$, alors $\left(\begin{smallmatrix} a & 0 \\ 0 & 1 \end{smallmatrix}\right) \cdot z = (\left(\begin{smallmatrix} a & 0 \\ 0 & 1 \end{smallmatrix}\right) z_1, \omega_\Delta(a)\left(\begin{smallmatrix} a^{-1} & 0 \\ 0 & 1 \end{smallmatrix}\right) z_2)$.
- Si $z' = \left(\begin{smallmatrix} p & 0 \\ 0 & 1 \end{smallmatrix}\right) \cdot z$, alors $\mathrm{Res}_{p\mathbf{Z}_p} z' = \left(\begin{smallmatrix} p & 0 \\ 0 & 1 \end{smallmatrix}\right) \cdot z_1$ et $\mathrm{Res}_{\mathbf{Z}_p} w z' = \omega_\Delta(p)\psi(z_2)$.

- Si $b \in p\mathbf{Z}_p$, et si $z' = \left(\begin{smallmatrix} 1 & b \\ 0 & 1 \end{smallmatrix}\right) \cdot z$, alors[5]

$$\mathrm{Res}_{\mathbf{Z}_p} z' = \left(\begin{smallmatrix} 1 & b \\ 0 & 1 \end{smallmatrix}\right) \cdot z_1 \quad \text{et} \quad \mathrm{Res}_{p\mathbf{Z}_p} w z' = u_b(\mathrm{Res}_{p\mathbf{Z}_p}(z_2)),$$

$$\text{où } u_b = \omega_\Delta(1+b) \left(\begin{smallmatrix} 1 & -1 \\ 0 & 1 \end{smallmatrix}\right) \circ w_\Delta \circ \left(\begin{smallmatrix} (1+b)^{-2} & b(1+b)^{-1} \\ 0 & 1 \end{smallmatrix}\right)$$
$$\circ\, w_\Delta \circ \left(\begin{smallmatrix} 1 & 1/(1+b) \\ 0 & 1 \end{smallmatrix}\right) \text{ sur } \Delta \boxtimes p\mathbf{Z}_p.$$

Le G-module $\Delta \boxtimes \mathbf{P}^1$ vit alors dans une suite exacte de G-modules :

$$0 \to \mathbf{\Pi}(\Delta)^* \otimes \omega_\Delta \to \Delta \boxtimes \mathbf{P}^1 \to \mathbf{\Pi}(\Delta) \to 0,$$

où $\mathbf{\Pi}(\Delta)$ est une représentation localement analytique de G (topologiquement, c'est une limite inductive compacte de L-banach et son dual $\mathbf{\Pi}(\Delta)^*$ est donc un L-fréchet). Si V est la représentation de $\mathscr{G}_{\mathbf{Q}_p}$ qui est associée à Δ par l'équivalence de catégories de Fontaine, et si $\mathbf{\Pi}(V)$ est la représentation unitaire associée à V par la correspondance de Langlands locale p-adique, alors $\mathbf{\Pi}(\Delta)$ est la sous-représentation $\mathbf{\Pi}(V)^{\mathrm{an}}$ des vecteurs localement analytiques de $\mathbf{\Pi}(V)$.

4.1.2. Dualité

Soit $\check{\Delta} = \mathrm{Hom}_{\mathscr{R}}(\Delta, \mathscr{R}\frac{dT}{1+T})$ le dual de Tate de Δ. On définit un accouplement G-équivariant $\{\ ,\ \}_{\mathbf{P}^1}$ sur $(\check{\Delta} \boxtimes \mathbf{P}^1) \times (\Delta \boxtimes \mathbf{P}^1)$, en posant

$$\{x, y\}_{\mathbf{P}^1} = \{\mathrm{Res}_{\mathbf{Z}_p} x, \mathrm{Res}_{\mathbf{Z}_p} y\} + \{\mathrm{Res}_{\mathbf{Z}_p}\left(\begin{smallmatrix} 0 & 1 \\ p & 0 \end{smallmatrix}\right) x, \mathrm{Res}_{\mathbf{Z}_p}\left(\begin{smallmatrix} 0 & 1 \\ p & 0 \end{smallmatrix}\right) y\},$$

où $\{\ ,\ \} : \check{\Delta} \times \Delta \to L$ est l'accouplement $(x, y) \mapsto \mathrm{rés}_0(\langle \sigma_{-1} x, y\rangle)$, et $\langle\ ,\ \rangle$ est l'accouplement tautologique sur $\check{\Delta} \times \Delta$. Cet accouplement identifie $\check{\Delta} \boxtimes \mathbf{P}^1$ au dual topologique de $\Delta \boxtimes \mathbf{P}^1$. Plus généralement, si U est un ouvert compact de $\mathbf{P}^1$, alors $\{\ ,\ \}_{\mathbf{P}^1}$ identifie $\check{\Delta} \boxtimes U$ au dual topologique de $\Delta \boxtimes U$, et $\check{\Delta} \boxtimes U$ et $\Delta \boxtimes V$ sont orthogonaux si $U \cap V = \emptyset$.

On choisit un isomorphisme de $\wedge^2 \Delta$ sur $\mathscr{R}(\det \Delta)$, ce qui nous fournit un isomorphisme de (φ, Γ)-modules de $\Delta \otimes \omega_\Delta^{-1}$ sur $\check{\Delta}$. Cela permet, si $z \in \Delta$, de voir $z \otimes \omega_\Delta^{-1}$ comme un élément de $\check{\Delta}$. Si v_1, v_2 est une base de Δ sur $\mathscr{R}$, telle que $v_1 \wedge v_2 = 1 \otimes \det \Delta$, alors $v_1 \otimes \omega_\Delta^{-1}, v_2 \otimes \omega_\Delta^{-1}$ est une base de $\check{\Delta}$ sur $\mathscr{R}$, et on a

$$\{\sigma_{-1}\big(x_1(v_1 \otimes \omega_\Delta^{-1}) + x_2(v_2 \otimes \omega_\Delta^{-1})\big), y_1 v_1 + y_2 v_2\} = \mathrm{rés}_0\big((x_1 y_2 - x_2 y_1)\,\tfrac{dT}{1+T}\big),$$

pour tous $x_1, x_2, y_1, y_2 \in \mathscr{R}$. L'isomorphisme P^+-équivariant $\Delta \otimes \omega_\Delta^{-1} \cong \check{\Delta}$ s'étend en un isomorphisme G-équivariant $(\Delta \boxtimes \mathbf{P}^1) \otimes \omega_\Delta^{-1} \cong \check{\Delta} \boxtimes \mathbf{P}^1$. On a donc une suite exacte

[5] La formule pour u_b de [14] comporte plusieurs fautes de frappe comme me l'a fait remarquer G. Dospinescu.

$$0 \to \mathbf{\Pi}(\Delta)^* \to \check{\Delta} \boxtimes \mathbf{P}^1 \to \mathbf{\Pi}(\Delta) \otimes \omega_\Delta^{-1} \to 0,$$

et alors $\mathbf{\Pi}(\Delta)^* \subset \check{\Delta} \boxtimes \mathbf{P}^1$ et $\mathbf{\Pi}(\Delta)^* \otimes \omega_\Delta \subset \Delta \boxtimes \mathbf{P}^1$ sont les orthogonaux l'un de l'autre ([14, rem. V.2.21]).

Lemme 8.84. *Si $v \in \Delta = \Delta \boxtimes \mathbf{Z}_p \subset \Delta \boxtimes \mathbf{P}^1$ est propre pour l'action de φ et Γ pour le caractère δ, alors $w \cdot (v \otimes \omega_\Delta^{-1}) \in \check{\Delta} \boxtimes \mathbf{P}^1$ est propre pour l'action de $T = \left(\begin{smallmatrix} \mathbf{Q}_p^* & 0 \\ 0 & \mathbf{Q}_p^* \end{smallmatrix}\right)$ pour le caractère $\delta^{-1} \otimes \delta\omega_\Delta^{-1}$.*

Démonstration. On a

$$\begin{aligned} \left(\begin{smallmatrix} a & 0 \\ 0 & d \end{smallmatrix}\right) \cdot (w \cdot (v \otimes \omega_\Delta^{-1})) &= \omega_\Delta^{-1}(a) \left(\begin{smallmatrix} 1 & 0 \\ 0 & d/a \end{smallmatrix}\right) w \cdot (v \otimes \omega_\Delta^{-1}) \\ &= \omega_\Delta^{-1}(a) w \cdot \left(\begin{smallmatrix} d/a & 0 \\ 0 & 1 \end{smallmatrix}\right) \cdot (v \otimes \omega_\Delta^{-1}) \\ &= \omega_\Delta^{-1}(a) w \cdot (\delta\omega_\Delta^{-1}(d/a)\, v \otimes \omega_\Delta^{-1}) \\ &= \delta^{-1}(a)\, \delta\omega_\Delta^{-1}(d) (w \cdot (v \otimes \omega_\Delta^{-1})), \end{aligned}$$

ce qui permet de conclure.

4.2. Le foncteur de Jacquet dual

Notre but est de déterminer le module de Jacquet dual de $\mathbf{\Pi}(\Delta)$. Pour cela nous allons avoir besoin d'introduire certaines notations.

Si $s \in \mathscr{S}_{\mathrm{irr}}$, on définit les éléments suivants de $(\Delta(s) \otimes \omega_s^{-1}) \boxtimes \mathbf{P}^1 = (\Delta(s) \boxtimes \mathbf{P}^1)^*$ (cf. n° 3.3.2 pour la définition de e_1 et e_2'):

- $f_1 = w \cdot (e_1 \otimes \omega_s^{-1})$ (pour tout $s \in \mathscr{S}_{\mathrm{irr}}$),
- $f_2 = w \cdot (e_2' \otimes \omega_s^{-1})$ (si $s \in \mathscr{S}_*^{\mathrm{cris}}$),
- $f_1' = w \cdot (t^{-w(s)} e_1 \otimes \omega_s^{-1})$ (si $s \in \mathscr{S}_*^{\mathrm{HT}} \cap \mathscr{S}_*^{\mathrm{ng}}$).

Remarque 8.85. Il résulte du lemme 8.84 et du lemme 8.88 ci-dessous, que:

- f_1 est propre sous l'action de B pour le caractère $\delta_1^{-1} \otimes \delta_2^{-1}\chi$,
- f_1' est propre sous l'action de B pour le caractère $x^{w(s)}\delta_1^{-1} \otimes x^{-w(s)}\delta_2^{-1}\chi$ (si $s \in \mathscr{S}_*^{\mathrm{HT}} \cap \mathscr{S}_*^{\mathrm{ng}}$),
- f_2 est propre sous l'action de B pour le caractère $x^{-w(s)}\delta_2^{-1} \otimes x^{w(s)}\delta_1^{-1}\chi$ (si $s \in \mathscr{S}_*^{\mathrm{cris}}$ n'est pas exceptionnel).

Lemme 8.86. *Les éléments f_1, f_1' et f_2 appartiennent au sous-module $\Pi(s)^*$ de $\check{\Delta} \boxtimes \mathbf{P}^1$.*

Démonstration. Il est équivalent de prouver que e appartient à $\Pi(s)^* \otimes \omega_\Delta$, si $e \in \{e_1, e_1', e_2'\}$ (avec $e_1' = t^{-w(s)}e_1$), ou encore que e est orthogonal au sous-module $\Pi(s)^*$ de $\check{\Delta} \boxtimes \mathbf{P}^1$. Or $\Pi(s)^*$ contient un sous-$\mathscr{O}_L$-module compact W, stable par G, engendrant un sous-L-espace dense, à savoir le $\mathscr{O}_L$-dual de la boule unité de $\mathbf{\Pi}(D)$, si $D = \widehat{\Delta}^{[0]}$ de telle sorte que $\Pi(s) = \mathbf{\Pi}(D)^{\mathrm{an}}$; il suffit donc de prouver que e est orthogonal à W. Comme W est stable par $\left(\begin{smallmatrix} p^{-n} & 0 \\ 0 & 1 \end{smallmatrix}\right)$, et comme $\{\ ,\ \}_{\mathbf{P}^1}$ est G-équivariant, cela implique que $\left\{\left(\begin{smallmatrix} p^n & 0 \\ 0 & 1 \end{smallmatrix}\right) \cdot e, W\right\}_{\mathbf{P}^1} = \{e, W\}_{\mathbf{P}^1}$ pour tout n. Or $\left(\begin{smallmatrix} p^n & 0 \\ 0 & 1 \end{smallmatrix}\right) \cdot e = \alpha_e e$, avec $v_p(e) > 0$ (on a $\alpha_e = \delta_1(p)$ si $e = e_1$, $\alpha_e = p^{-w(s)}\delta_1(p)$ et $w(s) < 0$ si $s \in \mathscr{S}_*^{\mathrm{HT}} \cap \mathscr{S}_*^{\mathrm{ng}}$, et $\alpha_e = p^{w(s)}\delta_2(p)$ si $s \in \mathscr{S}_*^{\mathrm{cris}}$). On a donc $\{e, W\}_{\mathbf{P}^1} = \alpha_e^n\{e, W\}_{\mathbf{P}^1}$, pour tout n, et comme W est compact, $\{e, W\}_{\mathbf{P}^1}$ est borné et $\cap_{n\in\mathbf{N}}\alpha_e^n\{e, W\}_{\mathbf{P}^1} = 0$, ce qui permet de conclure.

Théorème 8.87. (i) *Si Δ n'est pas triangulable, alors $J^*(\mathbf{\Pi}(\Delta)) = 0$.*
(ii) *Si $s \in \mathscr{S}_*^{\mathrm{ng}} - (\mathscr{S}_*^{\mathrm{ng}} \cap \mathscr{S}_*^{\mathrm{HT}})$ ou si $s \in \mathscr{S}_*^{\mathrm{st}}$, alors $J^*(\Pi(s)^{\mathrm{an}}) = L\, f_1$.*
(iii) *Si $s \in \mathscr{S}_*^{\mathrm{ng}} \cap \mathscr{S}_*^{\mathrm{HT}}$, alors $J^*(\Pi(s)^{\mathrm{an}}) = L\, f_1 \oplus L\, f_1'$.*
(iv) *Si $s \in \mathscr{S}_*^{\mathrm{cris}}$, alors $J^*(\Pi(s)^{\mathrm{an}}) = L\, f_1 \oplus L\, f_2$.*

Démonstration. D'après le lemme 8.86, f_1, f_1' et f_2 appartiennent à $\Pi(s)^*$. Le théorème est donc une conséquence du résultat suivant (en tordant par ω_Δ^{-1}).

Lemme 8.88. (i) *Si Δ n'est pas triangulable, alors $(\Delta \boxtimes \mathbf{P}^1)^N = 0$.*
(ii) *Si $s \in \mathscr{S}_*^{\mathrm{ng}} - (\mathscr{S}_*^{\mathrm{ng}} \cap \mathscr{S}_*^{\mathrm{HT}})$ ou si $s \in \mathscr{S}_*^{\mathrm{st}}$, alors $(\Delta(s) \boxtimes \mathbf{P}^1)^N = L\, w \cdot e_1$.*
(iii) *Si $s \in \mathscr{S}_*^{\mathrm{ng}} \cap \mathscr{S}_*^{\mathrm{HT}}$, alors $(\Delta(s) \boxtimes \mathbf{P}^1)^N = L\, w \cdot e_1 \oplus L\, w \cdot e_1'$.*
(iv) *Si $s \in \mathscr{S}_*^{\mathrm{cris}}$, alors $(\Delta(s) \boxtimes \mathbf{P}^1)^N = L\, w \cdot e_1 \oplus L\, w \cdot e_2$.*

Démonstration. On a $\left(\begin{smallmatrix} p^n & 0 \\ 0 & 1 \end{smallmatrix}\right)\mathrm{Res}_{p^{-n}\mathbf{Z}_p}\left(\left(\begin{smallmatrix} 1 & 1 \\ 0 & 1 \end{smallmatrix}\right)z\right) = (1+T)^{p^n}\left(\begin{smallmatrix} p^n & 0 \\ 0 & 1 \end{smallmatrix}\right)\mathrm{Res}_{p^{-n}\mathbf{Z}_p}z$. On ne déduit que si $\left(\begin{smallmatrix} 1 & 1 \\ 0 & 1 \end{smallmatrix}\right)z = z$, alors $\mathrm{Res}_{p^{-n}\mathbf{Z}_p}z = 0$ pour tout n, et donc que $w \cdot z \in \Delta \boxtimes p^{n+1}\mathbf{Z}_p = \varphi^{n+1}(\Delta)$, pour tout n. Il en résulte que z est dans l'image de $\Delta \boxtimes \{0\}$ par $x \mapsto w \cdot x$. On en déduit le (i) en utilisant le (i) du th. 8.82. □

Supposons maintenant que $\Delta = \Delta(s)$. On a a priori une inclusion $(\Delta \boxtimes \mathbf{P}^1)^N \subset (\Delta \boxtimes \mathbf{P}^1)^{\mathfrak{n}}$, où $\mathfrak{n}$ est l'algèbre de Lie de N, avec égalité si $(\Delta \boxtimes \mathbf{P}^1)^{\mathfrak{n}}$ est de dimension finie (en effet, il existe alors un sous-groupe ouvert de N qui agit trivialement sur $(\Delta \boxtimes \mathbf{P}^1)^{\mathfrak{n}}$ et donc N agit trivialement car $\left(\begin{smallmatrix} p & 0 \\ 0 & 1 \end{smallmatrix}\right)$ induit un isomorphisme de $(\Delta \boxtimes \mathbf{P}^1)^{\mathfrak{n}}$). Il s'agit donc d'étudier l'action infinitésimale de G sur le module $w \cdot (\Delta \boxtimes \{0\})$ ou, ce qui revient au même en conjuguant par w, sur $\Delta \boxtimes \{0\}$. L'algèbre de Lie $\mathfrak{g}$ de G est engendrée par $u^+ = \left(\begin{smallmatrix} 0 & 1 \\ 0 & 0 \end{smallmatrix}\right)$, $u^- = \left(\begin{smallmatrix} 0 & 0 \\ 1 & 0 \end{smallmatrix}\right)$, $a^+ = \left(\begin{smallmatrix} 1 & 0 \\ 0 & 0 \end{smallmatrix}\right)$ et $a^- = \left(\begin{smallmatrix} 0 & 0 \\ 0 & 1 \end{smallmatrix}\right)$, et on a $u^+u^- - u^-u^+ = a^+ - a^-$.

On cherche à déterminer le noyau de u^- agissant[6] sur $\Delta \boxtimes \{0\}$. Commençons par remarquer que, si $v \in \Delta \boxtimes \{0\}$ est un vecteur propre pour Γ pour le caractère δ_v (i.e. $\sigma_a(v) = \delta_v(a)v$, si $a \in \mathbf{Z}_p^*$), alors v est vecteur propre de a^+ pour la valeur propre $w(\delta_v)$. Par ailleurs,

$$\begin{aligned}\left(\begin{smallmatrix} b & 0 \\ 0 & 1 \end{smallmatrix}\right)u^- &= \lim_{x\to 0} \frac{1}{x}\left(\begin{smallmatrix} b & 0 \\ 0 & 1 \end{smallmatrix}\right)\left(\left(\begin{smallmatrix} 1 & 0 \\ x & 1 \end{smallmatrix}\right) - 1\right) = b^{-1} \lim_{x\to 0} \frac{1}{b^{-1}x}\left(\left(\begin{smallmatrix} 1 & 0 \\ b^{-1}x & 1 \end{smallmatrix}\right) - 1\right)\left(\begin{smallmatrix} b & 0 \\ 0 & 1 \end{smallmatrix}\right) \\ &= b^{-1}u^-\left(\begin{smallmatrix} b & 0 \\ 0 & 1 \end{smallmatrix}\right).\end{aligned}$$

Il en résulte que $u^-(v)$ est propre sous l'action de Γ pour le caractère $x^{-1}\delta_v$, si v l'est pour le caractère δ_v. En revenant à la description explicite de $M = (\Delta \boxtimes \{0\})_0$ et de $(\Delta \boxtimes \{0\})$, donnée dans le th. 8.82, on en déduit que u^- tue M.

Par ailleurs, on a $u^+(v) = tv$, pour tout $v \in \Delta \boxtimes \{0\}$ et $a^+ + a^-$ est la multiplication par $w(\delta_\Delta) = 2w(\delta_1) - w(s) - 1$. Maintenant, $\sigma_a(v) = \delta_1(a)v$, pour tout a assez proche de 1 et tout $v \in M$, et donc $a^+(v) = w(\delta_1)v$. On déduit de la relation $u^+u^- - u^-u^+ = a^+ - a^-$ que

$$\begin{aligned} tu^-(t^j v) - u^-(t^{j+1}v) &= (2w(\delta_1) + 2j - (2w(\delta_1) - w(s) - 1))t^j v \\ &= (2j + w(s) + 1)t^j v, \end{aligned}$$

si $v \in M$ et $j \in \mathbf{N}$. Comme $u^-(v) = 0$, une récurrence immédiate permet d'en déduire que $u^-(t^j v) = -(j(w(s)+1) + j(j-1))t^{j-1}v$.

Maintenant, si $j \geq 1$, la nullité de $u^-(t^j v)$ équivaut à celle de $w(s) + 1 + j - 1 = w(s) + j$. Il s'ensuit que, dans le cas $s \in \mathscr{S}_*^{\mathrm{cris}}$, $u^-(t^j v) = 0$ n'est possible que si $j = 0$ (car $w(s) \in \mathbf{N} - \{0\}$) et donc le noyau de u^- sur $\Delta \boxtimes \{0\}$ est réduit à $(\Delta \boxtimes \{0\})_0$.

Dans les cas restant, cela n'est possible que si $w(s)$ est un entier < 0, et si $j = -w(s)$. On en déduit le résultat.

4.3. Dévissage de $\Delta(s) \boxtimes \mathbf{P}^1$

Nous allons prouver que $\Delta(s) \boxtimes \mathbf{P}^1$ se dévisse comme $\Delta(s)$. Plus précisément, on a le résultat suivant qui permet, en utilisant la prop. 8.54, de déterminer les composantes de Jordan-Hölder de $\Delta(s) \boxtimes \mathbf{P}^1$.

Théorème 8.89. (i) $\mathscr{R}e_1 \boxtimes \mathbf{P}^1 = \{(z_1, z_2) \in \Delta(s) \boxtimes \mathbf{P}^1,\ z_1, z_2 \in \mathscr{R}e_1\}$ *est stable par G, et le G-module* $\Delta(s) \boxtimes \mathbf{P}^1$ *vit dans une suite exacte de G-modules*[7]

6 On pourrait utiliser la formule explicite [15] pour l'action de u^- pour simplifier ce qui suit.

7 Le (i) du lemme 8.91 permet de munir $\mathscr{R}e_2 \boxtimes \mathbf{Z}_p^*$ de l'action quotient de $w_{\Delta(s)}$ et donc de définir $\mathscr{R}e_2 \boxtimes \mathbf{P}^1$; les formules du squelette d'action munissent alors ce module d'une action de G.

$$0 \to \mathscr{R}e_1 \boxtimes \mathbf{P}^1 \to \Delta(s) \boxtimes \mathbf{P}^1 \to \mathscr{R}e_2 \boxtimes \mathbf{P}^1 \to 0.$$

(ii) *Si $i = 1, 2$, on a une suite exacte de G-modules :*

$$0 \to B^{\rm an}(\delta_{3-i}, \delta_i)^* \otimes \omega_s \to \mathscr{R}e_i \boxtimes \mathbf{P}^1 \to B^{\rm an}(\delta_i, \delta_{3-i}) \to 0.$$

Démonstration. La démonstration va demander un peu de préparation ; le résultat est une traduction de la prop. 8.92 ci-dessous.

Soit $f_1 = w\left(\begin{smallmatrix} -1 & 0 \\ 0 & 1 \end{smallmatrix}\right) \cdot (e_1 \otimes \omega_s^{-1}) = \omega_s^{-1}\delta_1(-1)w \cdot (e_1 \otimes \omega_s^{-1}) \in (\Delta(s) \otimes \omega_s^{-1}) \boxtimes \mathbf{P}^1$. D'après la rem. 8.85, f_1 est propre sous l'action de B, pour le caractère $\delta_1^{-1} \otimes \delta_2^{-1}\chi$. On en déduit, grâce aux rem. 8.51 et 8.50, que Φ_1, envoyant z sur $\phi_{1,z} : \mathbf{Q}_p \to L$ définie par $\phi_{1,z}(x) = \left\{f_1, \left(\begin{smallmatrix} 0 & 1 \\ -1 & x \end{smallmatrix}\right) \cdot z\right\}_{\mathbf{P}^1}$, est G-équivariante de $\Delta(s) \boxtimes \mathbf{P}^1$ dans $B^{\rm an}(\delta_2, \delta_1)$.

Lemme 8.90. *L'application $\Phi_1 : \Delta(s) \boxtimes \mathbf{P}^1 \to B^{\rm an}(\delta_2, \delta_1)$ est surjective et son noyau est l'ensemble des z vérifiant $p_s(\mathrm{Res}_{\mathbf{Z}_p} z) \in \mathscr{R}^+e_2$ et $p_s(\mathrm{Res}_{\mathbf{Z}_p}\left(\begin{smallmatrix} 0 & 1 \\ p & 0 \end{smallmatrix}\right)z) \in \mathscr{R}^+e_2$.*

Démonstration. On a $\left(\begin{smallmatrix} 0 & 1 \\ -1 & x \end{smallmatrix}\right) = w\left(\begin{smallmatrix} -1 & 0 \\ 0 & 1 \end{smallmatrix}\right)\left(\begin{smallmatrix} 1 & -x \\ 0 & 1 \end{smallmatrix}\right)$, et donc

$$\begin{aligned}\phi_{1,z}(x) &= \left\{w\left(\begin{smallmatrix} -1 & 0 \\ 0 & 1 \end{smallmatrix}\right) \cdot (e_1 \otimes \omega_s^{-1}), w\left(\begin{smallmatrix} -1 & 0 \\ 0 & 1 \end{smallmatrix}\right)\left(\begin{smallmatrix} 1 & -x \\ 0 & 1 \end{smallmatrix}\right) \cdot z\right\}_{\mathbf{P}^1} \\ &= \left\{e_1 \otimes \omega_s^{-1}, \left(\begin{smallmatrix} 1 & -x \\ 0 & 1 \end{smallmatrix}\right) \cdot z\right\}_{\mathbf{P}^1}.\end{aligned}$$

Maintenant, $e_1 \otimes \omega_s^{-1}$ étant à support $\{0\}$, on a $\phi_{1,\mathrm{Res}_U z} = \mathrm{Res}_U \phi_{1,z}$, et comme $\mathbf{P}^1$ est la réunion disjointe de $\mathbf{Z}_p$ et $\left(\begin{smallmatrix} 0 & 1 \\ p & 0 \end{smallmatrix}\right)\mathbf{Z}_p$, on est ramené à prouver que Φ_1 induit une surjection de $\Delta(s) \boxtimes \mathbf{Z}_p = \Delta(s)$ sur $B^{\rm an}(\delta_2, \delta_1) \boxtimes \mathbf{Z}_p = \mathrm{LA}(\mathbf{Z}_p)$, dont le noyau est $p_s^{-1}(\mathscr{R}^+e_2)$ (la partie concernant $\left(\begin{smallmatrix} 0 & 1 \\ p & 0 \end{smallmatrix}\right)\mathbf{Z}_p$ s'en déduit alors par G-équivariance). L'application $(z_1, z_2) \mapsto z_1e_1 + z_2\hat{e}_2$ est un isomorphisme de $\mathscr{R}$-modules de $\mathscr{R} \times \mathscr{R}$ sur $\Delta(s)$, et si $z = z_1e_1 + z_2\hat{e}_2$, on a $\phi_{1,z}(x) = \phi_{z_2}(x)$. On peut donc déduire la surjectivité de $\Delta(s) \mapsto \mathrm{LA}(\mathbf{Z}_p)$ du (iii) de la prop. 8.11 ainsi que le fait que z est dans le noyau de Φ_1 si et seulement si $z_2 \in \mathscr{R}^+e_2$.

Lemme 8.91. (i) *$\mathscr{R}e_1 \boxtimes \mathbf{Z}_p^*$ est stable par $w_{\Delta(s)}$.*

(ii) *Le sous-module $\mathscr{R}e_1 \boxtimes \mathbf{P}^1 = \{(z_1, z_2) \in \Delta(s) \boxtimes \mathbf{P}^1,\ z_1, z_2 \in \mathscr{R}e_1\}$ de $\Delta(s) \boxtimes \mathbf{P}^1$ est stable par G.*

Démonstration. $\Delta(s) \boxtimes \mathbf{Z}_p^*$ est stable par w et contenu dans $\mathrm{Ker}\, \Phi_1$. On déduit donc du lemme 8.90 que l'image par p_s de $w(\mathscr{R}e_1 \boxtimes \mathbf{Z}_p^*)$ est incluse dans $\mathscr{R}^+e_2 \boxtimes \mathbf{Z}_p^*$. Or $\mathscr{R}e_1 \boxtimes \mathbf{Z}_p^*$ est un sous-$\mathscr{R}(\Gamma)$-module de $\Delta(s) \boxtimes \mathbf{Z}_p^*$, et $w_{\Delta(s)}$ étant $\mathscr{R}(\Gamma)$-semi-linéaire (car $w_{\Delta(s)} \circ \sigma_a = \omega_s(a)\sigma_{a^{-1}} \circ w_{\Delta(s)}$, traduction de $w\left(\begin{smallmatrix} a & 0 \\ 0 & 1 \end{smallmatrix}\right) = \left(\begin{smallmatrix} a & 0 \\ 0 & a \end{smallmatrix}\right)\left(\begin{smallmatrix} a^{-1} & 0 \\ 0 & 1 \end{smallmatrix}\right)w$), $w_{\Delta(s)}(\mathscr{R}e_1 \boxtimes \mathbf{Z}_p^*)$ est aussi un $\mathscr{R}(\Gamma)$-module. Il en est donc de même de son image par p_s et comme celle-ci est incluse dans le

$\mathscr{R}^+(\Gamma)$-module $\mathscr{R}^+e_2 \boxtimes \mathbf{Z}_p^*$, qui est de rang fini (et même de rang 1), elle est nulle, ce qui démontre le (i).

Enfin, comme $\mathscr{R}e_1$ est stable par P^+ et ψ et $\mathscr{R}e_1 \boxtimes \mathbf{Z}_p^*$ est stable par $w_{\Delta(s)}$, il résulte des formules du squelette d'action que $\mathscr{R}e_1 \boxtimes \mathbf{P}^1$ est stable par G (cf. [14, prop. V.2.8] pour des résultats du même genre).

Ceci permet de conclure.

le (i) du lemme permet de munir $\mathscr{R}e_2 \boxtimes \mathbf{Z}_p^*$ d'une action de $w_{\Delta(s)}$, et donc de définir un module $\mathscr{R}e_2 \boxtimes \mathbf{P}^1$. Les formules du squelette d'action passent alors au quotient modulo $\mathscr{R}e_1 \boxtimes \mathbf{P}^1$, et donc munissent $\mathscr{R}e_2 \boxtimes \mathbf{P}^1$ d'une action de G, rendant G-équivariante l'application naturelle $p_s : \Delta(s) \boxtimes \mathbf{P}^1 \to \mathscr{R}e_2 \boxtimes \mathbf{P}^1$. Ceci prouve le (i) du th. 8.90.

Si $i = 1, 2$, on note $\mathscr{R}^+e_i \boxtimes \mathbf{P}^1$ l'ensemble des $z \in \mathscr{R}e_i \boxtimes \mathbf{P}^1$ vérifiant $\mathrm{Res}_{\mathbf{Z}_p} z \in \mathscr{R}^+e_i$ et $\mathrm{Res}_{\mathbf{Z}_p}\left(\begin{smallmatrix} 0 & 1 \\ p & 0 \end{smallmatrix}\right)z \in \mathscr{R}^+e_i$.

Proposition 8.92. *Si $i = 1, 2$, le sous-module $\mathscr{R}^+e_i \boxtimes \mathbf{P}^1$ de $\mathscr{R}e_i \boxtimes \mathbf{P}^1$ est stable par G et on a des isomorphismes de G-modules :*

$$\mathscr{R}^+e_i \boxtimes \mathbf{P}^1 \cong B^{\mathrm{an}}(\delta_{3-i}, \delta_i)^* \otimes \omega_s \text{ et } (\mathscr{R}e_i \boxtimes \mathbf{P}^1)/(\mathscr{R}^+e_i \boxtimes \mathbf{P}^1) \cong B^{\mathrm{an}}(\delta_i, \delta_{3-i}).$$

Démonstration. $\mathscr{R}^+e_2 \boxtimes \mathbf{P}^1$ est l'image de $\mathrm{Ker}\,\Phi_1$ par p_s ; on en déduit sa stabilité par G. De plus, $(\mathscr{R}e_2 \boxtimes \mathbf{P}^1)/(\mathscr{R}^+e_2 \boxtimes \mathbf{P}^1)$ est isomorphe à l'image de Φ_1, c'est-à-dire $B^{\mathrm{an}}(\delta_2, \delta_1)$.

Maintenant, $w \cdot e_2$ est stable par N. En effet, on a $\mathrm{Res}_U w \cdot e_2 = 0$ pour tout ouvert compact U de $\mathbf{Q}_p$, et donc $\mathrm{Res}_U \left(\begin{smallmatrix} 1 & b \\ 0 & 1 \end{smallmatrix}\right) w \cdot e_2 = 0$, pour tout ouvert compact U de $\mathbf{Q}_p$. On en déduit que $\left(\begin{smallmatrix} 1 & b \\ 0 & 1 \end{smallmatrix}\right) w \cdot e_2$ est de la forme $\sum_{k \in \mathbf{N}} \alpha_k(b) w \cdot t^k e_2$, avec $\alpha_k(b) \in L$ et $\sum_{k \in \mathbf{N}} \alpha_k(b) X^k$ convergeant sur $\mathbf{C}_p$ tout entier. Par ailleurs, $\left(\begin{smallmatrix} a & 0 \\ 0 & 1 \end{smallmatrix}\right) w \cdot t^k e_2 = \omega_s(a) w \left(\begin{smallmatrix} a^{-1} & 0 \\ 0 & 1 \end{smallmatrix}\right) \cdot t^k e_2 = \omega_s(a)\delta_2(a)^{-1} a^{-k} w \cdot t^k e_2$. En caculant de deux manières $\left(\begin{smallmatrix} a & 0 \\ 0 & 1 \end{smallmatrix}\right)\left(\begin{smallmatrix} 1 & b \\ 0 & 1 \end{smallmatrix}\right) w \cdot e_2 = \left(\begin{smallmatrix} 1 & ab \\ 0 & 1 \end{smallmatrix}\right)\left(\begin{smallmatrix} a & 0 \\ 0 & 1 \end{smallmatrix}\right) w \cdot e_2$, on obtient $\alpha_k(ab) = a^{-k}\alpha_k(b)$, et donc $\alpha_k(b) = c_k b^{-k}$. Or $b \mapsto \left(\begin{smallmatrix} 1 & b \\ 0 & 1 \end{smallmatrix}\right) w \cdot e_2$ est analytique dans un voisinage de 0, et donc $c_k = 0$ pour tout $k \neq 0$ et $c_0 = 1$.

On déduit de ce qui précède que $w \cdot \hat{e}_2$, qui est un relèvement de $w \cdot e_2$, est fixe par N modulo $\mathscr{R}e_1 \boxtimes \mathbf{P}^1$. Ceci permet, comme pour le lemme 8.90, de définir une application G-équivariante $\Phi_2 : \mathscr{R}e_1 \boxtimes \mathbf{P}^1 \to B^{\mathrm{an}}(\delta_1, \delta_2)$, en envoyant z sur la fonction $\phi_{2,z}$ définie par $\tilde{\phi}_{2,z}(x) = \{w \cdot (\hat{e}_2 \otimes \omega_s^{-1}), \left(\begin{smallmatrix} 0 & 1 \\ -1 & x \end{smallmatrix}\right) \cdot z\}_{\mathbf{P}^1}$. Le noyau de Φ_2 est $\mathscr{R}^+e_1 \boxtimes \mathbf{P}^1$ pour les mêmes raisons que précédemment ; on en déduit sa stabilité par G ainsi que l'isomorphisme $(\mathscr{R}e_1 \boxtimes \mathbf{P}^1)/(\mathscr{R}^+e_1 \boxtimes \mathbf{P}^1) \cong B^{\mathrm{an}}(\delta_1, \delta_2)$.

Enfin, dans la dualité entre $\Delta(s) \boxtimes \mathbf{P}^1$ et $\check{\Delta}(s) \boxtimes \mathbf{P}^1$ les espaces $\mathscr{R}e_1 \boxtimes \mathbf{P}^1$ et $\mathscr{R}e_1' \boxtimes \mathbf{P}^1$ sont les orthogonaux l'un de l'autre (on a $\{z', z\}_{\mathbf{P}^1} = \{\mathrm{Res}_{\mathbf{Z}_p} z', \mathrm{Res}_{\mathbf{Z}_p} z\} + \{\mathrm{Res}_{\mathbf{Z}_p}\left(\begin{smallmatrix} 0 & 1 \\ p & 0 \end{smallmatrix}\right)z', \mathrm{Res}_{\mathbf{Z}_p}\left(\begin{smallmatrix} 0 & 1 \\ p & 0 \end{smallmatrix}\right)z\}$, et si $\hat{e}_2' = \hat{e}_2 \otimes \omega_s^{-1}$, on a

$\{x_1'e_1' + x_2'\hat{e}_2', x_1e_1 + x_2e_2\} = \text{rés}_0\big(x_1'x_2 - x_2'x_1 \frac{dT}{1+T}\big)$ si $x_1, x_2, x_1', x_2' \in \mathscr{R}$). Cette dualité induit donc des dualités G-équivariantes entre $\mathscr{R}e_i \boxtimes \mathbf{P}^1$ et $\mathscr{R}e_{3-i}' \boxtimes \mathbf{P}^1$ dans laquelle $\mathscr{R}^+e_i \boxtimes \mathbf{P}^1$ et $\mathscr{R}^+e_{3-i}' \boxtimes \mathbf{P}^1$ sont les orthogonaux l'un de l'autre (car $\mathscr{R}^+$ est l'orthogonal de $\mathscr{R}^+$ dans $\mathscr{R}$). On en déduit que $\mathscr{R}^+e_i \boxtimes \mathbf{P}^1$ est le dual de $(\mathscr{R}e_{3-i}' \boxtimes \mathbf{P}^1)/(\mathscr{R}^+e_{3-i}' \boxtimes \mathbf{P}^1)$ est donc est isomorphe à $B^{\text{an}}(\chi\delta_i^{-1}, \chi\delta_{3-i}^{-1})^* \cong B^{\text{an}}(\delta_{3-i}, \delta_i)^* \otimes \omega_s$.

Cela permet de conclure.

Remarque 8.93. (i) Si δ, ω sont deux caractères continus de $\mathbf{Q}_p^*$, on peut construire des représentations $\mathscr{E}(\delta) \boxtimes_\omega \mathbf{P}^1$ et $\mathscr{R}(\delta) \boxtimes_\omega \mathbf{P}^1$ en reprenant la stratégie menant à la construction de $D \boxtimes \mathbf{P}^1$ et $D_{\text{rig}} \boxtimes \mathbf{P}^1$ dans le cas où D est de rang 2 sur $\mathscr{E}$. La représentation $\mathscr{R}e_i \boxtimes \mathbf{P}^1$ de la prop. 8.92 est alors isomorphe à $\mathscr{R}(\delta_i) \boxtimes_{\omega_s} \mathbf{P}^1$.

(ii) Si D est irréductible de rang ≥ 3, on peut encore définir le G-module $D \boxtimes_\omega \mathbf{P}^1$ (cf. [14, § II.1]), mais la stratégie menant à la construction de $D_{\text{rig}} \boxtimes_\omega \mathbf{P}^1$ est en défaut. Les résultats de Paskunas [29] laissent à penser qu'il n'est pas possible de munir $D_{\text{rig}} \boxtimes \mathbf{P}^1$ d'une action de G.

4.4. Dévissage de $\mathbf{\Pi}(\Delta(s))$

On note simplement $\Pi(s)$ la représentation $\mathbf{\Pi}(\Delta(s))$. On a donc une suite exacte $0 \to \Pi(s)^* \otimes \omega_s \to \Delta(s) \boxtimes \mathbf{P}^1 \to \Pi(s) \to 0$ de G modules.

Proposition 8.94. *Soit* $s = (\delta_1, \delta_2, \mathscr{L}) \in \mathscr{S}_{\text{irr}}$.

(i) *Si s est générique, on a des suites exactes*

$$0 \to \mathscr{R}^+e_1 \boxtimes \mathbf{P}^1 \to \Pi(s)^* \otimes \omega_s \to \mathscr{R}^+e_2 \boxtimes \mathbf{P}^1 \to 0,$$
$$0 \to B^{\text{an}}(\delta_1, \delta_2) \to \Pi(s) \to B^{\text{an}}(\delta_2, \delta_1) \to 0.$$

(ii) *Si s est spécial,* $\mathscr{R}^+e_2 \boxtimes \mathbf{P}^1$ *contient un sous-espace fermé* $(\mathscr{R}^+e_2 \boxtimes \mathbf{P}^1)_0$, *stable par G, de codimension* $w(s)$ *et* $\mathscr{R}e_1 \boxtimes \mathbf{P}^1$ *contient un sous-espace fermé* $(\mathscr{R}e_1 \boxtimes \mathbf{P}^1)^0$, *stable par G, dans lequel* $\mathscr{R}^+e_1 \boxtimes \mathbf{P}^1$ *est de codimension* $w(s)$, *et on a des suites exactes*

$$0 \to (\mathscr{R}e_1 \boxtimes \mathbf{P}^1)^0 \to \Pi(s)^* \otimes \omega_s \to (\mathscr{R}^+e_2 \boxtimes \mathbf{P}^1)_0 \to 0,$$
$$0 \to E_{\mathscr{L}} \to \Pi(s) \to B^{\text{an}}(\delta_2, \delta_1) \to 0,$$

où $E_{\mathscr{L}}$ *est l'extension de* $W(\delta_1, \delta_2)$ *par* $\text{St}^{\text{an}}(\delta_1, \delta_2)$ *correspondant à celle de* $\mathscr{R}(\delta_2)$ *par* $\mathscr{R}(\delta_1)$ *via l'isomorphisme du (iii) de la rem. 8.59.*

Démonstration. La démonstration consiste à analyser comment se répartissent les composantes de Jordan-Hölder entre $\Pi(s)^* \otimes \omega_s$ et $\Pi(s)$.

Dans le cas (i):

- Si $\delta_s \neq |x|^{-2}x^{w(s)-1}$, ces composantes sont de dimension infinie et sont de deux types bien distincts : des limites inductives compactes de banach qui font partie des composantes de $\Pi(s)$ et des fréchets (non banach) qui interviennent dans $\Pi(s)^* \otimes \omega_s$. Cela permet de démontrer la première suite exacte en regardant les fréchets ; la seconde s'en déduit.
- Si $\delta_s = |x|^{-2}x^{w(s)-1}$, alors $B^{\rm an}(\delta_1, \delta_2)$ a une composante de Jordan-Hölder qui est de dimension finie, mais celle-ci n'est pas un sous-objet, ce qui fait que le plus grand sous-objet de $\mathscr{R}e_1 \boxtimes \mathbf{P}^1$ qui est de type fréchet est quand-même $\mathscr{R}^+e_1 \boxtimes \mathbf{P}^1$. De même, $\mathscr{R}^+e_2 \boxtimes \mathbf{P}^1$ a une composante de dimension finie, mais celle-ci n'apparaît pas en quotient, et donc $\mathscr{R}^+e_2 \boxtimes \mathbf{P}^1$ est le plus petit sous-objet de $\mathscr{R}e_2 \boxtimes \mathbf{P}^1$ contenant les composantes de type fréchet de dimension infinie. On en déduit le résultat dans ce cas aussi.

Dans le cas (ii), la situation est un peu plus délicate à cause des deux composantes de dimension finie qui peuvent faire partie des composantes de l'un ou de l'autre. Or $\Pi(s)$ ne contient pas de sous-représentation de dimension finie (cela résulte, par exemple, des cor. I.3.2 et II.2.9 de [14] dont on déduit que $\Pi(s)$ ne contient pas de sous-L-espace de dimension finie stable par B). Il en résulte que le sous-quotient $W(\delta_1, \delta_2)$ de $\mathscr{R}e_1 \boxtimes \mathbf{P}^1$ apparaît dans $\Pi(s)^* \otimes \omega_s$, et comme il n'intervient pas dans $\mathscr{R}^+e_1 \boxtimes \mathbf{P}^1$, cela prouve que $(\mathscr{R}e_1 \boxtimes \mathbf{P}^1) \cap \Pi(s)^* \otimes \omega_s$ est l'extension $(\mathscr{R}e_1 \boxtimes \mathbf{P}^1)^0$ de $W(\delta_1, \delta_2)$ par $\mathscr{R}^+e_1 \boxtimes \mathbf{P}^1$ apparaissant dans $\mathscr{R}e_1 \boxtimes \mathbf{P}^1$. La suite exacte $0 \to \Pi(s)^* \otimes \omega_s \to \Delta(s) \boxtimes \mathbf{P}^1 \to \Pi(s) \to 0$ prouve alors que la composante $W(\delta_1, \delta_2)$ de $\mathscr{R}^+e_2 \boxtimes \mathbf{P}^1$ n'apparaît pas dans $\Pi(s)^* \otimes \omega_s$ (sinon, $\Pi(s)$ n'aurait pas de composante de Jordan-Hólder de dimension finie contrairement à $\Pi(s)^*$). Il s'ensuit que l'image de $\Pi(s)^* \otimes \omega_s$ dans $\mathscr{R}e_2 \boxtimes \mathbf{P}^1$ est strictement incluse dans $\mathscr{R}^+e_2 \boxtimes \mathbf{P}^1$ et que le quotient est isomorphe à $W(\delta_1, \delta_2)$; cette image est donc le sous-module $(\mathscr{R}^+e_2 \boxtimes \mathbf{P}^1)_0$ [qui est l'orthogonal du sous-module $W(\chi\delta_2^{-1}, \chi\delta_1^{-1})$ de $B^{\rm an}(\chi\delta_2^{-1}, \chi\delta_1^{-1})$]. On en déduit la première suite exacte. Le lemme du serpent nous fournit alors une suite exacte $0 \to B^{\rm an}(\delta_1, \delta_2)/W(\delta_1, \delta_2) \to \Pi(s) \to E \to 0$, où $E = (\mathscr{R}e_2 \boxtimes \mathbf{P}^1)/(\mathscr{R}^+e_2 \boxtimes \mathbf{P}^1)_0$ est une extension de $B^{\rm an}(\delta_2, \delta_1)$ par $W(\delta_1, \delta_2)$. Il en résulte que $\Pi(s)$ contient une extension E' de $W(\delta_1, \delta_2)$ par ${\rm St}^{\rm an}(\delta_1, \delta_2)$ et que l'on a une suite exacte $0 \to E' \to \Pi(s) \to B^{\rm an}(\delta_2, \delta_1) \to 0$. Il reste à déterminer la classe de l'extension E', et pour cela, nous allons utiliser les isomorphismes

$$\operatorname{Ext}^1_G(W(\delta_1, \delta_2), {\rm St}^{\rm an}(\delta_1, \delta_2)) \cong H^1(A^+, {\rm LA}(\mathbf{Z}_p) \otimes x^k) \cong \operatorname{Ext}^1(\mathscr{R}(\delta_2), \mathscr{R}(\delta_1))$$

de la rem. 8.59 dont nous reprenons les notations. Commençons par remarquer que notre $W(\delta_1, \delta_2)$ est $(\mathscr{R}^+e_2 \boxtimes \mathbf{P}^1)/(\mathscr{R}^+e_2 \boxtimes \mathbf{P}^1)_0$; les t^ie_2, pour $0 \leq i \leq k$

en forment donc une base sur L. Par ailleurs, $W(\delta_1, \delta_2) \cong W_k \otimes \delta_1\chi^{-1} = W_k \otimes x^k\delta_2$, et comme $\left(\begin{smallmatrix} a & 0 \\ 0 & 1 \end{smallmatrix}\right)$ agit par multiplication par $\delta_2(a)$ sur e_2, on voit que e_2 correspond à l'élément $x^{-k} \otimes x^k$ de $W_k \otimes x^k$. L'image de l'extension de $W(\delta_1, \delta_2)$ par $\mathrm{St}^{\mathrm{an}}(\delta_1, \delta_2)$ dans $H^1(A^+, \mathrm{LA}(\mathbf{Z}_p) \otimes x^k)$ est donc, d'après la rem. 8.59, obtenue de la manière suivante : on relève e_2 en $\tilde{e}_2$ dans $\Pi(s)$, et on se débrouille pour que le cocycle $\left(\begin{smallmatrix} a & 0 \\ 0 & 1 \end{smallmatrix}\right) \mapsto c_a = \left(\delta_2(a)^{-1}\left(\begin{smallmatrix} a & 0 \\ 0 & 1 \end{smallmatrix}\right) - 1\right) \cdot \tilde{e}_2$ sur A^+ soit à valeurs dans les fonctions à support dans $\mathbf{Z}_p$; l'image de ce cocycle dans $H^1(A^+, \mathrm{LA}(\mathbf{Z}_p) \otimes x^k)$ est alors la classe qui nous intéresse. Une manière d'assurer que c_a est à valeurs dans $\mathbf{Z}_p$ est de choisir un relèvement $\hat{e}_2$ de e_2 dans $\Delta(s) = \Delta(s) \boxtimes \mathbf{Z}_p \subset \Delta(s) \boxtimes \mathbf{P}^1$, et de prendre pour $\tilde{e}_2$ l'image de $\hat{e}_2$ dans $\mathrm{LA}(\mathbf{Z}_p) \otimes x^k$. Or cette description du cocycle $a \mapsto c_a$ montre que son image dans $H^1(A^+, \mathrm{LA}(\mathbf{Z}_p) \otimes x^k)$ est celle de l'extension $\Delta(s)$ de $\mathscr{R}(\delta_2)$ par $\mathscr{R}(\delta_1)$.

Ceci permet de conclure.

4.5. Vecteurs localement algébriques

On note $\Pi(s)^{\mathrm{alg}}$ l'ensemble des vecteurs localement algébriques de $\Pi(s)$ pour l'action de $\mathbf{SL}_2(\mathbf{Q}_p)$ (si δ_1 est localement algébrique, c'est l'ensemble des vecteurs localement algébriques de $\Pi(s)$).

Théorème 8.95. *Soit $s = (\delta_1, \delta_2, \mathscr{L}) \in \mathscr{S}_{\mathrm{irr}}$.*

(i) $\Pi(s)^{\mathrm{alg}} \neq 0$ *si et seulement si $s \in \mathscr{S}_*^{\mathrm{dR}}$.*

(ii) *Si $s \in \mathscr{S}_*^{\mathrm{st}}$, alors* $\Pi(s)^{\mathrm{alg}} = \mathrm{St} \otimes \mathrm{Sym}^{w(s)-1} \otimes \delta_2$.

(iii) *Si $s \in \mathscr{S}_*^{\mathrm{cris}}$, alors*

$$\Pi(s)^{\mathrm{alg}} = \left(\mathrm{Ind}^{\mathrm{lisse}}(|x|\delta_s x^{1-w(s)} \otimes |x|^{-1})\right) \otimes \mathrm{Sym}^{w(s)-1} \otimes \delta_2$$

et est irréductible, sauf si $\delta_s = x^{w(s)-1}$ ou si $\delta_s = |x|^{-2}x^{w(s)-1}$ auquel cas $\Pi(s)^{\mathrm{alg}}$ est une extension de $\mathrm{Sym}^{w(s)-1} \otimes \delta_2$ *par* $\mathrm{St} \otimes \mathrm{Sym}^{w(s)-1} \otimes \delta_2$.

Démonstration. Il y a plusieurs cas :

- Si s n'est pas spécial (i.e. si $\delta_s \neq x^{w(s)-1}$, avec $w(s)$ entier ≥ 1), l'énoncé est équivalent à $\Pi(s)^{\mathrm{alg}} = B^{\mathrm{alg}}(\delta_1, \delta_2)$, ce qui est immédiat si $w(s)$ n'est pas un entier < 0 au vu des composantes de Jordan-Hölder de $B(\delta_1, \delta_2)$ et $B(\delta_2, \delta_1)$ (prop. 8.54) et de la suite exacte $0 \to B(\delta_1, \delta_2) \to \Pi(s) \to B(\delta_2, \delta_1) \to 0$.
- Si $w(s)$ est un entier < 0, on cherche à prouver que $\Pi(s)^{\mathrm{alg}} = 0$, et donc que $B^{\mathrm{alg}}(\delta_2, \delta_1)$ n'est pas un sous-objet de $\Pi(s)$. Dans le cas contraire, $B^{\mathrm{alg}}(\delta_2, \delta_1)$ serait une sous-représentation de la représentation unitaire $\mathbf{\Pi}(V(s))$, et son complété universel serait non nul, ce qui contredit un résultat d'Emerton [17]. (On a $\sum_{i=0}^{p^n-1} \left(\begin{smallmatrix} p^n & i \\ 0 & 1 \end{smallmatrix}\right) \cdot \mathbf{1}_{\mathbf{Z}_p} =$

$\delta_2(p)^n \sum_{i=0}^{p^n-1} \mathbf{1}_{i+p^n\mathbf{Z}_p} = \delta_2(p)^n \mathbf{1}_{\mathbf{Z}_p}$; il en résulte, puisque $v_p(\delta_2) < 0$, que l'image de $\mathbf{1}_{\mathbf{Z}_p}$ dans le complété universel de $B^{\text{alg}}(\delta_2, \delta_1)$ est nulle, et donc que ce complété universel est nul puisque $B^{\text{alg}}(\delta_2, \delta_1)$ est irréductible.)

- Si $\delta_s = x^k$, avec $k \in \mathbf{N}$ (ce qui couvre le cas $s \in \mathscr{S}^{\text{st}}$ et le cas $s \in \mathscr{S}^{\text{cris}}$ spécial), alors $\Pi(s)$ est une extension de $B(\delta_2, \delta_1)$ par $E_\ell \otimes \delta_1 \chi^{-1}$, où E_ℓ est l'extension de W_k par X_k correspondant à $\ell \in \text{Hom}(\mathbf{Q}_p^*, L)$ (cf. th. 8.55). On est donc ramené à démontrer le résultat suivant (on rappelle que $\text{Sym}^k \otimes \delta_2 = W_k \otimes \delta_2 x^k = W_k \otimes \delta_1 \chi^{-1}$).

Lemme 8.96. *$E_\ell^{\text{alg}} = \text{St} \otimes W_k$ sauf si $\ell = v_p$ où E_ℓ^{alg} est une extension non triviale de W_k par* $\text{St} \otimes W_k$ *et donc est isomorphe à* $(\text{Ind}^{\text{lisse}}(|x| \otimes |x|^{-1})) \otimes W_k$.

Démonstration. On reprend les notations du th. 8.55.

Si $\ell = v_p$, le sous-espace de Y_k engendré par $W_k \ell^+$ et les fonctions localement polynomiales sur $\mathbf{P}^1$ de degré $\leq k$ (isomorphe à $\text{St} \otimes W_k$) est stable par G, et l'extension de W_k par $\text{St} \otimes W_k$ ainsi obtenue est non scindée car $\Pi(s)$ ne contient pas de sous-représentation de dimension finie. Ceci permet de conclure dans le cas $\ell = v_p$.

Si $\ell \neq v_p$, on peut, quitte à multiplier ℓ par une constante ce qui ne change pas E_ℓ, supposer que $\ell = \log + \alpha v_p$, avec $\alpha \in L$. Montrons qu'une sous représentation de G qui se surjecte sur W_k contient une fonction qui n'est pas localement polynomiale sur $\mathbf{P}^1$ de degré $\leq k$. Pour cela, il suffit de prouver que si $\phi \in X_k$, alors $\left(\left(\begin{smallmatrix} 1 & 1 \\ 0 & 1 \end{smallmatrix}\right) - 1\right) \cdot (\ell^+ + \phi)$ ne peut pas être nulle dans un voisinage de l'infini. Dans le cas contraire, il existerait $b \in \mathbf{Z}$ tel que $\phi(x+1) - \phi(x) = \log(x+1) - \log x$, si $v_p(x) \leq b$. Par densité de $\mathbf{N}$ dans $\mathbf{Z}_p$, on en déduit que $\phi - \log$ est constante sur $i + \mathbf{Z}_p$, si $v_p(i) \leq b$, ce qui est en contradiction avec le fait que ϕ est analytique au voisinage de ∞ (i.e. $\phi(x) = \sum_{n \in \mathbf{N}} a_n x^{-n}$, si $v_p(x) \ll 0$).

Ceci permet de conclure.

4.6. Cas particuliers

Ce n° est consacré à l'étude des deux cas particuliers où le module de Jacquet est de dimension 2. Le résultat obtenu, dans le cas cristallin, répond aux conjectures 5.3.7 et 5.4.4 de [2].

4.6.1. Le cas cristallin

Soit $s = (\delta_1, \delta_2, \infty) \in \mathscr{S}_*^{\text{cris}}$. On note s' l'élément de $\mathscr{S}_*^{\text{cris}}$ défini par $s' = (\delta_1', \delta_2', \infty)$ et $\delta_1' = x^{w(s)} \delta_2$, $\delta_2' = x^{-w(s)} \delta_1$. Alors s est exceptionnel si et seulement si $s = s'$.

Il résulte de la démonstration du th. 8.95 que :

- si $\delta_s \neq x^{w(s)-1}, |x|^2 x^{w(s)-1}$, alors $\Pi(s)^{\mathrm{alg}} = B^{\mathrm{alg}}(\delta_1, \delta_2)$ (qui est isomorphe à $B^{\mathrm{alg}}(\delta'_1, \delta'_2)$, d'après le (iii) de la rem. 8.53),
- si $\delta_s = |x|^2 x^{w(s)-1}$, alors $\Pi(s)^{\mathrm{alg}} = B^{\mathrm{alg}}(\delta_1, \delta_2)$ (extension de $W(\delta'_1, \delta'_2)$ par $\mathrm{St}^{\mathrm{alg}}(\delta'_1, \delta'_2)$),
- si $\delta_s = x^{w(s)-1}$, alors $\Pi(s)^{\mathrm{alg}}$ est une extension de $W(\delta_1, \delta_2)$ par $\mathrm{St}^{\mathrm{alg}}(\delta_1, \delta_2)$ isomorphe à $B^{\mathrm{alg}}(\delta'_1, \delta'_2) = \Pi(s')^{\mathrm{alg}}$.

On remarque que, dans tous les cas, $\Pi(s)^{\mathrm{alg}} \cong \Pi(s')^{\mathrm{alg}}$. Cela traduit l'existence d'un isomorphisme $\Pi(s) \cong \Pi(s')$.

Proposition 8.97. *Si $s \in \mathscr{S}_*^{\mathrm{cris}}$ n'est pas exceptionnel, on a des suites exactes*

$$0 \to \Pi(s)^{\mathrm{alg}} \to B^{\mathrm{an}}(\delta_1, \delta_2) \oplus B^{\mathrm{an}}(\delta'_1, \delta'_2) \to \Pi(s) \to 0$$
$$0 \to \Pi(s)^{\mathrm{alg}} \to \Pi(s) \to B^{\mathrm{an}}(\delta_2, \delta_1) \oplus B^{\mathrm{an}}(\delta'_2, \delta'_1) \to 0.$$

Démonstration. Comme $\Pi(s) \cong \Pi(s')$, il résulte de la prop. 8.94 que $B^{\mathrm{an}}(\delta_1, \delta_2)$ et $B^{\mathrm{an}}(\delta'_1, \delta'_2)$ s'identifient à des sous-objets de $\Pi(s)$. La première suite exacte s'en déduit en examinant les composantes de Jordan-Hölder des représentations considérées. La seconde en résulte en utilisant les isomorphismes $B^{\mathrm{an}}(s)/B^{\mathrm{alg}}(s) \cong B^{\mathrm{an}}(\delta'_2, \delta'_1)$ et $B^{\mathrm{an}}(s')/B^{\mathrm{alg}}(s') \cong B^{\mathrm{an}}(\delta_2, \delta_1)$.

Remarque 8.98. (i) La flèche $\Pi(s) \to B^{\mathrm{an}}(\delta_2, \delta_1) \oplus B^{\mathrm{an}}(\delta'_2, \delta'_1)$ peut s'induire à partir de l'application $z \mapsto \left(\{f_1, \left(\begin{smallmatrix} 0 & 1 \\ -1 & x \end{smallmatrix}\right) \cdot z\}_{\mathbf{P}^1}, \{f_2, \left(\begin{smallmatrix} 0 & 1 \\ -1 & x \end{smallmatrix}\right) \cdot z\}_{\mathbf{P}^1}\right)$, définie sur $\Delta(s) \boxtimes \mathbf{P}^1$, et qui se factorise à travers $\Pi(s)$.

(ii) Dans le cas exceptionnel, on a $\delta'_1 = \delta_1$ et $\delta'_2 = \delta_2$. Les suites exactes ci-dessus deviennent

$$0 \to B^{\mathrm{alg}}(\delta_1, \delta_2) \to \mathrm{Ind}^{\mathrm{an}}\left((\delta_2 \otimes \chi^{-1}\delta_1) \otimes \left(\begin{smallmatrix} 1 & v_p(a/d) \\ 0 & 1 \end{smallmatrix}\right)\right) \to \Pi(s) \to 0$$
$$0 \to \Pi(s)^{\mathrm{alg}} \to \Pi(s) \to \mathrm{Ind}^{\mathrm{an}}\left((\delta_1 \otimes \chi^{-1}\delta_2) \otimes \left(\begin{smallmatrix} 1 & v_p(a/d) \\ 0 & 1 \end{smallmatrix}\right)\right) \to 0.$$

4.6.2. Le cas Hodge-Tate non de Rham

Si $s \in \mathscr{S}_*^{\mathrm{HT}} \cap \mathscr{S}_*^{\mathrm{ng}}$, alors $J^*(\Pi(s))$ contient un autre vecteur propre que f_1 pour l'action de B, à savoir f'_1, qui est propre pour le caractère $x^{-w(s)}\delta_1^{-1} \otimes x^{w(s)}\delta_2^{-1}\chi$. L'application $z \mapsto \phi'_{1,z}$ définie par $\phi'_{1,z}(x) = \{f'_1, \left(\begin{smallmatrix} 0 & 1 \\ -1 & x \end{smallmatrix}\right) z\}_{\mathbf{P}^1}$ induit donc un morphisme G-équivariant de $\Delta(s) \boxtimes \mathbf{P}^1$ sur $B^{\mathrm{an}}(x^{w(s)}\delta_2, x^{-w(s)}\delta_1)$ qui est surjectif pour les mêmes raisons que $z \mapsto \phi_{1,z}$. Or on a $f'_1 = w \cdot (t^{-w(s)} e_1 \otimes \omega_s^{-1})$, ce qui nous donne

$$\begin{aligned}\phi'_{1,z}(x) &= (-1)^{w(s)}\omega_s^{-1}\delta_1(-1)\{t^{-w(s)}e_1\otimes\omega_s^{-1}, \left(\begin{smallmatrix}1 & -x\\ 0 & 1\end{smallmatrix}\right)z\}_{\mathbf{P}^1}\\ &= \omega_s^{-1}\delta_1(-1)\{e_1\otimes\omega_s^{-1}, t^{-w(s)}\left(\begin{smallmatrix}1 & -x\\ 0 & 1\end{smallmatrix}\right)z\}_{\mathbf{P}^1}\\ &= \omega_s^{-1}\delta_1(-1)\{e_1\otimes\omega_s^{-1}, \left(\frac{d}{dx}\right)^{-w(s)}(\left(\begin{smallmatrix}1 & -x\\ 0 & 1\end{smallmatrix}\right)z)\}_{\mathbf{P}^1}\\ &= \left(\frac{d}{dx}\right)^{-w(s)}\phi_{1,z}(x).\end{aligned}$$

Il s'ensuit que $z\mapsto\phi'_{1,z}$ s'obtient en composant $z\mapsto\phi_{1,z}$ avec le morphisme $\left(\frac{d}{dx}\right)^{-w(s)}: B^{\mathrm{an}}(\delta_2,\delta_1)\to B^{\mathrm{an}}(x^{w(s)}\delta_2, x^{-w(s)}\delta_1)$ de la proposition 8.54. Il est à noter que l'extension de $B^{\mathrm{alg}}(\delta_2,\delta_1)$ (noyau de ce morphisme) par $B^{\mathrm{an}}(\delta_1,\delta_2)$ qui apparaît n'est pas scindée (cf. démonstration du th. 8.95).

4.7. La transformée de Stieljes

Terminons cet article par une curiosité dont je ne sais pas vraiment que penser.

Soit Δ une extension non triviale de $\mathscr{R}$ par $\mathscr{R}(\delta)$. Si $e\in\Delta$ est un relèvement de $1\in\mathscr{R}$, alors $\mathrm{Res}_U e\in\mathscr{R}(\delta)$ pour tout ouvert compact U de $\mathbf{Z}_p$ ne contenant pas 0. Les $\phi_{\mathrm{Res}_U e}$ définissent donc une fonction localement analytique (notée simplement ϕ_e) sur $\mathbf{Z}_p-\{0\}$ (plus exactement, un élément de $\mathrm{LA}(\mathbf{Z}_p-\{0\})\otimes\delta\chi^{-1}$).

Proposition 8.99. (i) *Si $\delta\notin\{x^{-i},\ i\in\mathbf{N}\}\cup\{\chi x^i,\ i\in\mathbf{N}\}$, il existe $\lambda\in L^*$ et $\phi\in\mathrm{LA}(\mathbf{Z}_p)$, uniquement déterminés, tels que $\phi_e=\lambda\delta\chi^{-1}+\phi$, et on peut choisir e de telle sorte que $\phi=0$.*

(ii) *Si $\delta=\chi x^i$, avec $i\in\mathbf{N}$, il existe $\ell\in\mathrm{Hom}(\mathbf{Q}_p^*,L)$ non nul et $\phi\in\mathrm{LA}(\mathbf{Z}_p)$, uniquement déterminés, tels que $\phi_e=x^i\ell+\phi$, et on peut choisir e de telle sorte que $\phi=0$.*

(iii) *Si $\delta=x^{-i}$, avec $i\in\mathbf{N}$, alors $\phi_e\in\mathrm{LA}(\mathbf{Z}_p)$ et on peut choisir e de telle sorte que $\phi_e=0$.*

Démonstration. C'est une traduction du (i) de la rem. 8.48, utilisant la prop. 8.43.

4.7.1. Le cas générique

Soit $s\in\mathscr{S}_{\mathrm{irr}}$ non spécial. On suppose l'extension de $\mathscr{R}(\delta_2)$ par $\mathscr{R}(\delta_1)$ *normalisée* (ce qui signifie que $\lambda=1$ dans le (i) de la prop. 8.99 ; le cas $\delta=x^{-i}$ avec $i\in\mathbf{N}$ est incompatible avec l'hypothèse $s\in\mathscr{S}_{\mathrm{irr}}$).

Si U est un ouvert compact de $\mathbf{P}^1$, on note ${}^cU$ son complémentaire. On définit un morphisme

$$S_{s,U}:(B^{\mathrm{an}}(\delta_1,\delta_2)^*\otimes\omega_s)\boxtimes U\to B^{\mathrm{an}}(\delta_1,\delta_2)\boxtimes{}^cU$$

de la manière suivante : si $\mu \in B^{\mathrm{an}}(\delta_1, \delta_2)^* \boxtimes U$, on peut relever μ en un élément $\tilde{\mu}$ de $\Pi(s)^* \otimes \omega_s$ d'après le (i) de la prop. 8.94. Maintenant, $\Pi(s)^* \otimes \omega_s$ est un sous-espace de $\Delta(s) \boxtimes \mathbf{P}^1$, et donc $\mathrm{Res}_{^cU}\tilde{\mu}$ est un élément de $\Delta(s) \boxtimes {}^cU$ dont l'image dans $\mathscr{R}e_2 \boxtimes {}^cU$ est nulle puisque μ est à support dans U ; il s'ensuit que $\mathrm{Res}_{^cU}\tilde{\mu} \in \mathscr{R}e_1 \boxtimes {}^cU$. On définit alors $S_{s,U}(\mu)$ comme l'image de $\mathrm{Res}_{^cU}\tilde{\mu}$ dans $B^{\mathrm{an}}(\delta_1, \delta_2) \boxtimes {}^cU$ qui est le quotient de $\mathscr{R}e_1 \boxtimes {}^cU$ par $\mathscr{R}^+e_1 \boxtimes {}^cU$. (Ceci ne dépend pas du choix de $\tilde{\mu}$ car deux choix diffèrent par un élément de $\mathscr{R}^+e_1 \boxtimes \mathbf{P}^1$ et donc leurs images par $\mathrm{Res}_{^cU}$ diffèrent par un élément de $\mathscr{R}^+e_1 \boxtimes {}^cU$.)

Proposition 8.100. (i) *Si $g \in G$, on a $g \cdot S_{s,U}(\mu) = S_{s,g\cdot U}(g \cdot \mu)$.*
(ii) *On a $(S_{s,U}(\mu))(x) = \int_U \delta_s(z-x)\,\mu(z)$, si $x \in {}^cU$.*

Démonstration. Le (i) est immédiat : on peut prendre $g \cdot \tilde{\mu}$ comme relèvement de $g \cdot \mu$ et on a alors $\mathrm{Res}_{g\cdot {}^cU}\, g \cdot \tilde{\mu} = g \cdot \mathrm{Res}_{^cU}\tilde{\mu}$.

Pour démontrer le (ii), commençons par prouver la G-équivariance de $\mu \mapsto S'_{s,\mu}$, avec $S'_{s,\mu}(x) = \int_U \delta_s(z-x)\,\mu(z)$. Si $g = \left(\begin{smallmatrix} a & b \\ c & d \end{smallmatrix}\right)$, on a $g^{-1} = \frac{1}{ad-bc}\left(\begin{smallmatrix} d & -b \\ -c & a \end{smallmatrix}\right)$, et comme $\omega_s\delta_1^{-1}\chi = \delta_2$ et $\delta_s\delta_2 = \delta_1\chi^{-1}$, on obtient

$$\begin{aligned}(S'_{s,g\cdot U}(g\cdot\mu))(x) &= \omega_s(ad-bc)\int_{g\cdot U}\delta_s(z-x)\,g\cdot\mu(z)\\ &= \delta_2(ad-bc)\int_U \delta_s(cz+d)\delta_s\big(\tfrac{az+b}{cz+d}-x\big)\,\mu(z)\\ &= \delta_s\delta_2(ad-bc)\delta_s\big(\tfrac{-cx+a}{ad-bc}\big)\int_U \delta_s\big(z-\tfrac{dx-b}{-cx+a}\big)\,\mu(z)\\ &= (S'_{s,U}(\mu)\star g^{-1})(x),\end{aligned}$$

ce que l'on cherchait à vérifier. Maintenant, comme $\mu \mapsto S_{s,U}(\mu)$ et $\mu \mapsto S'_{s,U}(\mu)$ sont G-équivariantes toutes les deux, il suffit de vérifier qu'elles coïncident sur la masse de Dirac en 0 pour montrer qu'elles sont égales : en effet, la G-équivariance implique qu'elles coïncident en toute masse de Dirac et la densité des masses de Dirac dans $B^{\mathrm{an}}(\delta_1, \delta_2)^*$ permet de conclure.

Soit donc $\tilde{e}_2$ un relèvement de e_2 dans $\Pi(s)^* \otimes \omega_s$. Si V est un ouvert compact de $\mathbf{P}^1$ ne contenant pas 0, on a $\mathrm{Res}_V\tilde{e}_2 \in \mathscr{R}e_1 \boxtimes V$; on note ϕ_{V,e_2} l'image de $\mathrm{Res}_V\tilde{e}_2$ dans $B^{\mathrm{an}}(\delta_1, \delta_2) \boxtimes V$, et les ϕ_{V,e_2} se recollent en une fonction ϕ_{e_2}, localement analytique sur $\mathbf{Q}_p^*$, et qui ne dépend pas du choix de $\tilde{e}_2$. On cherche à vérifier que $\phi_{e_2}(x) = \delta_s(x)$, pour tout $x \in \mathbf{Q}_p^*$. Or on a $\left(\begin{smallmatrix} a & 0 \\ 0 & 1 \end{smallmatrix}\right)e_2 = \delta_2(a)e_2$, si $a \in \mathbf{Q}_p^*$. Il s'ensuit que $\left(\begin{smallmatrix} a & 0 \\ 0 & 1 \end{smallmatrix}\right)\tilde{e}_2 - \delta_2(a)\tilde{e}_2 \in (\mathscr{R}e_1 \boxtimes \mathbf{P}^1) \cap (\Pi(s)^* \otimes \omega_s) = (\mathscr{R}^+e_1 \boxtimes \mathbf{P}^1)$; on en déduit que $\left(\begin{smallmatrix} a & 0 \\ 0 & 1 \end{smallmatrix}\right)\phi_{e_2} = \delta_2(a)\phi_{e_2}$, ce qui se traduit par $\delta_1\chi^{-1}(a)\phi_{e_2}\big(\frac{x}{a}\big) = \delta_2(a)\phi_{e_2}(x)$ et donc que $\phi_{e_2}(x) = \delta_s(x)\phi_{e_2}(1)$.

Par ailleurs, $\mathrm{Res}_{\mathbf{Z}_p}\tilde{e}_2$ est un relèvement de e_2 dans $\Delta(s)$, et comme l'extension est supposée normalisée on en déduit, grâce au (i) de la prop. 8.99, qu'il existe $\phi \in \mathrm{LA}(\mathbf{Z}_p)$ telle que l'on ait $\phi_{e_2} = \delta_s + \phi$ sur $\mathbf{Z}_p$. Il en résulte que $\phi = 0$ et $\phi_{e_2}(1) = 1$, ce qui permet de conclure.

4.7.2. Le cas spécial

Supposons maintenant que s est spécial. Alors $\delta_s = x^k$, avec $k \in \mathbf{N}$, et il existe $\ell \in \mathrm{Hom}(\mathbf{Q}_p^*, L)$ décrivant l'extension. La différence avec le cas générique est que l'application de $\Pi(s)^* \otimes \omega_s$ dans $B^{\mathrm{an}}(\delta_1, \delta_2)^* \otimes \omega_s$ n'est pas surjective : son image est l'orthogonal de $W(\delta_1, \delta_2)$ (i.e. de l'espace des polynômes de degré $\leq k$). Si U est un ouvert compact de $\mathbf{P}^1$ on note $(B^{\mathrm{an}}(\delta_1, \delta_2)^* \boxtimes U)_0 \otimes \omega_s$ l'intersection de $(B^{\mathrm{an}}(\delta_1, \delta_2)^* \boxtimes U) \otimes \omega_s$ avec l'orthogonal de $W(\delta_1, \delta_2)$. On définit alors, par le même procédé que dans le cas générique, un morphisme

$$S_{s,U} : (B^{\mathrm{an}}(\delta_1, \delta_2)^* \boxtimes U)_0 \otimes \omega_s \to \mathrm{St}^{\mathrm{an}}(\delta_1, \delta_2) \boxtimes {}^cU,$$

où $\mathrm{St}^{\mathrm{an}}(\delta_1, \delta_2) \boxtimes {}^cU$ est le quotient de $B^{\mathrm{an}}(\delta_1, \delta_2) \boxtimes {}^cU$ par l'espace des polynômes de degré $\leq k$ sur ${}^cU$.

Proposition 8.101. (i) *Si $g \in G$, on a $g \cdot S_{s,U}(\mu) = S_{s,g\cdot U}(g \cdot \mu)$.*
(ii) *On a $(S_{s,U}(\mu))(x) = \int_U (x-z)^k \ell(x-z)\,\mu(z)$, si $x \in {}^cU$.*

Démonstration. Le (i) se démontre exactement comme dans le cas générique. Pour démontrer le (ii), on commence, comme dans le cas générique, par prouver la G-équivariance de $\mu \mapsto S'_{s,\mu}$, avec $S'_{s,\mu}(x) = \int_U (z-x)^k\,\mu(z)$. Si $g = \left(\begin{smallmatrix} a & b \\ c & d \end{smallmatrix}\right)$, on obtient

$$\begin{aligned}
(S'_{s,g\cdot U}(g\cdot\mu))(x) &= \omega_s(ad-bc)\int_{g\cdot U}(z-x)^k\ell(z-x)\,g\cdot\mu(z)\\
&= \delta_2(ad-bc)\int_U (cz+d)^k\left(\tfrac{az+b}{cz+d}-x\right)^k\ell\left(\tfrac{az+b}{cz+d}-x\right)\mu(z)\\
&= (ad-bc)^k\delta_2(ad-bc)\left(\tfrac{-cx+a}{ad-bc}\right)^k\\
&\qquad \int_U\left(z-\tfrac{dx-b}{-cx+a}\right)^k\ell\left(\tfrac{az+b}{cz+d}-x\right)\mu(z)\\
&= (S'_{s,U}(\mu)\star g^{-1})(x),
\end{aligned}$$

la dernière identité venant de ce que $\ell\left(\frac{az+b}{cz+d}-x\right) = \ell\left(z-\frac{dx-b}{-cx+a}\right) + \ell(-cx+a) - \ell(cz+d)$, et de ce que le terme faisant intervenir $\ell(-cx+a)$ disparaît car on intègre un polynôme en z de degré $\leq k$, et que celui faisant intervenir $\ell(cz+d)$ disparaît car le résultat est un polynôme en x de degré $\leq k$.

Il suffit alors, comme dans le cas générique, de prouver que $S_{s,U}(\mu)$ et $S'_{s,U}(\mu)$ coïncident pour $\mu = \mathrm{Dir}_0 - w\cdot(\frac{1}{k!}d^k\mathrm{Dir}_0)$ qui est un élément de

$B^{\mathrm{an}}(\delta_1,\delta_2)_0^*$ dont les translaté sous l'action de G (et même de N) engendrent topologiquement $B^{\mathrm{an}}(\delta_1,\delta_2)_0^*$. Soit donc $\tilde{e}_2$ un relèvement de $e_2 - w\cdot\frac{t^k}{k!}e_2$. On définit, comme ci-dessus, une fonction ϕ_2, localement analytique sur $\mathbf{Q}_p^*$, et on cherche à prouver que son image modulo les polyômes de degré $\leq k$ est $x^k\ell$. Pour les mêmes raisons que ci-dessus, on a $\left(\left(\begin{smallmatrix} a & 0\\ 0 & 1\end{smallmatrix}\right) - \delta_2(a)\right)\cdot\phi_2 = 0$ dans $\mathrm{St}^{\mathrm{an}}(\delta_1,\delta_2)$, ce qui signifie que $\left(\left(\begin{smallmatrix} a & 0\\ 0 & 1\end{smallmatrix}\right) - \delta_2(a)\right)\cdot\phi_2$ est un polyôme de degré $\leq k$, et donc que ϕ_2 est de la forme $x^k\ell'(x)+P(x)$, où $\ell'\in\mathrm{Hom}(\mathbf{Q}_p^*, L)$ et P est un polyôme de degré $\leq k$. On en déduit le résultat, comme dans le cas générique, en utilisant le fait que $\mathrm{Res}_{\mathbf{Z}_p}\tilde{e}_2$ est un relèvement de e_2 dans $\Delta(s)$, et donc que la restriction de ϕ_2 à $\mathbf{Z}_p$ est de la forme $x^k\ell+\phi$, où $\phi\in\mathrm{LA}(\mathbf{Z}_p)$.

Remarque 8.102. La transformation $\mu\mapsto\int_U\frac{1}{z-x}\mu(z)$ (qui correspondrait au cas $\delta_s = x^{-1}$, interdit par l'hypothèse $s\in\mathscr{S}_{\mathrm{irr}}$), est la transformée de Stieljes [23] : son image est l'espace des $\phi : {}^cU\to L$ qui sont la restriction d'une fonction analytique sur $\mathbf{P}^1(\mathbf{C}_p) - U$, nulle en ∞.

A. Cohomologie des semi-groupes

Soit G^+ un semi-groupe topologique. Si M est un L-espace vectoriel topologique muni d'une action continue de G^+ (en bref M est un G^+-module), on note $C^\bullet(G^+,M)$ le complexe habituel

$$0\longrightarrow C^0(G^+,M)\xrightarrow{d_1}C^1(G^+,M)\xrightarrow{d_2}\cdots,$$

où $C^n(G^+,M)$ est le Λ module des fonctions continues de $(G^+)^n$ dans M (et $C^0(G^+,M)=M$ par convention), et d_{n+1} est la différentielle

$$\begin{aligned}d_{n+1}c(g_0,\dots,g_n) = g_0\cdot c(g_1,\dots,g_n) + \sum_{0\leq i\leq n-1}(-1)^{i+1}c(g_0,\dots,g_ig_{i+1},\dots,g_n)\\ +(-1)^{n+1}c(g_0,\dots,g_{n-1}).\end{aligned}$$

Si $n=0,1,2$, on obtient:

$$\begin{aligned}&d_1c(g) = (g-1)\cdot c,\quad \text{si } c\in M = C^0(G^+,M),\\ &d_2c(g_0,g_1) = g_0\cdot c(g_1) - c(g_0g_1) + c(g_0),\\ &d_3(g_0,g_1,g_2) = g_0\cdot c(g_1,g_2) - c(g_0g_1,g_2) + c(g_0,g_1g_2) - c(g_0,g_1).\end{aligned}$$

On note $B^n(G^+,M)$ l'espace des cobords, image de d_n, et $Z^n(G^+,M)$ l'espace de cocycles, noyau de d_{n+1}. Comme $d_{n+1}\circ d_n = 0$, on a $B^n(G^+,M)\subset Z^n(G^+,M)$, et on note $H^n(G^+,M)$ le n-ième groupe de cohomologie de G^+ à valeurs dans M, quotient de $Z^n(G^+,M)$ par $B^n(G^+,M)$.

Notons G^- le semi-groupe opposé de G^+ que l'on voit comme l'ensemble des inverses des éléments de G^+ (et donc $G^+ = \{g^{-1},\ g \in G^-\}$). Si M est un L-espace vectoriel topologique, muni d'une action de G^+, son dual M^* est muni d'une action de G^- donnée par $\langle g \cdot \mu, v\rangle = \langle \mu, g^{-1} \cdot v\rangle$. On dispose donc de groupes de cohomologie $H^n(G^-, M^*)$, pour $n \in \mathbf{N}$.

A.1. Dévissage de la cohomologie de A^+

On a $A^+ = \Phi^+ \times A^0$, où Φ^+ est le semi-groupe $\begin{pmatrix} p^{\mathbf{N}} & 0 \\ 0 & 1 \end{pmatrix}$ et A^0 le groupe $A^0 = \begin{pmatrix} \mathbf{Z}_p^* & 0 \\ 0 & 1 \end{pmatrix}$. On note $\varphi = \begin{pmatrix} p & 0 \\ 0 & 1 \end{pmatrix}$ le générateur de Φ et σ_a, si $a \in \mathbf{Z}_p^*$, l'élément $\begin{pmatrix} a & 0 \\ 0 & 1 \end{pmatrix}$. Tout élément g de A^+ s'écrit alors, de manière unique, sous la forme $\varphi^{k(g)}\sigma_{a(g)} = \sigma_{a(g)}\varphi^{k(g)}$, avec $k(g) \in \mathbf{N}$ et $a(g) \in \mathbf{Z}_p^*$.

On peut décomposer A^0 sous la forme $\Delta \times A^1$, où Δ est le sous-groupe de torsion de A^0 et $A^1 = \begin{pmatrix} 1+2\mathbf{Z}_p & 0 \\ 0 & 1 \end{pmatrix}$. Comme nos coefficients sont tous des L-espaces vectoriels topologiques, on a

$$H^i(A^0, M) = H^i(A^1, H^0(\Delta, M)) = H^0(\Delta, H^i(A^1, M)),$$

pour tout i, et comme $A^1 \cong \mathbf{Z}_p$, on a $H^i(A^1, M) = 0$ et donc $H^i(A^0, M) = 0$, pour tout $i \geq 2$. De plus, si u est un générateur topologique de $1 + 2\mathbf{Z}_p$ (et donc σ_u est un générateur topologique de A^1), alors

$$H^0(A^1, M) = \mathrm{Ker}(\sigma_u - 1) \ \text{ et } \ H^1(A^1, M) \cong M/(\sigma_u - 1) \cdot M,$$

le dernier isomorphisme étant induit par $c \mapsto c(\sigma_u)$ (l'isomorphisme inverse envoie $v \in M$ sur $g \mapsto \frac{g-1}{\sigma_u - 1} \cdot v$, où $\frac{g-1}{\sigma_u-1}$ est vu comme un élément de $\mathbf{Z}_p[[A^1]]$).

La nullité de $H^2(A^0, H^0(\Phi^+, M))$ implique que la suite d'inflation-restriction

$$0 \to H^1(A^0, H^0(\Phi^+, M)) \xrightarrow{\iota_{HS}} H^1(A^+, M)$$
$$\xrightarrow{\pi_{HS}} H^0(A^0, H^1(\Phi^+, M)) \to 0$$

est exacte, pour tout M. Par ailleurs,

$$H^0(\Phi^+, M) = \mathrm{Ker}(\varphi - 1) \ \text{ et } \ H^1(\Phi^+, M) \cong M/(\varphi - 1)M,$$

le dernier isomorphisme étant induit par $c \mapsto c(\varphi)$ (l'isomorphisme inverse envoie $v \in M$ sur le cocycle $g \mapsto \frac{g-1}{\varphi-1} \cdot v$; si $v \in (\varphi - 1) \cdot M$, ce cocycle est un cobord).

On a $H^n(\Phi^+, M) = 0$ si $n \geq 2$. On en déduit, grâce à la suite spectrale de Hochshild-Serre, que $H^i(A^+, M) = 0$, si $i \neq 0, 1, 2$, et que l'on a des isomorphismes

$$\iota_{HS} : H^0(A^0, H^0(\Phi^+, M)) \cong H^0(A^+, M),$$
$$\iota_{HS} : H^1(A^0, H^1(\Phi^+, M)) \cong H^2(A^+, M).$$

Le premier isomorphisme se passe de commentaires, mais nous allons avoir besoin d'expliciter le second.

Soit c un 2-cocycle continu sur A^+, et soit $\alpha : A^+ \times A^+ \to M$ définie par $\alpha(g,h) = c(g,h) - c(h,g)$. En soustrayant de la relation de cocycle pour (h_1, g, h_2) celles pour (h_1, h_2, g) et (g, h_1, h_2), et en utilisant la commutatitivité de A^+, on obtient l'identité

$$d_{2,h}\alpha(g, h_1, h_2) = (1-g) \cdot c(h_1, h_2).$$

Par antisymétrie, on a aussi $d_{2,g}\alpha(g_1, g_2, h) = (h-1) \cdot c(g_1, g_2)$. Par ailleurs, si $c = d_2 c_1$, alors $\alpha(g,h) = (g-1) \cdot c_1(h) - (h-1)c_1(g)$. Maintenant, comme $H^2(\Phi, M) = 0$, on peut écrire c sous la forme $c' + d_2 c_1$, où c' est un 2-cocycle sur A^+ vérifiant $c(h_1, h_2) = 0$, pour tous $h_1, h_2 \in \Phi^+$. Alors les identités ci-dessus montrent que $h \mapsto c'(g,h) - c'(h,g)$ est un élément de $Z^1(A^+, M)$, pour tout $g \in A^+$, et que l'image de $g \mapsto (h \mapsto c'(g,h) - c'(h,g))$ dans $C^1(A^+, H^1(\Phi^+, M))$ est un 1-cocycle trivial sur Φ^+ et donc est l'inflation d'un 1-cocyle sur A^0 ; l'isomorphisme $\iota_{HS}^{-1} : H^2(A^+, M) \cong H^1(A^0, H^1(\Phi^+, M))$ envoie c sur l'image de $g \mapsto (h \mapsto c'(g,h) - c'(h,g))$. (Ceci ne dépend pas de la décomposition $c = c' + d_2 c_1$ choisie car, si $c = c'' + d_2 c_1'$ en est une autre, cela implique que $c_1 - c_1'$ est un 1-cocyle sur Φ, et l'image de $d_2(c_1 - c_1')$ par ι_{HS} est celle de $g \mapsto ((g-1) \cdot (c_1 - c_1'))$, et donc est nulle.)

Si $(\varphi - 1)M$ est fermé dans M, on peut décrire l'isomorphisme inverse $\iota_{HS} : H^1(A^0, H^1(\Phi^+, M)) \to H^2(A^+, M)$ de la manière suivante. Si $\alpha \in Z^1(A^0, H^1(\Phi^+, M))$, on peut relever α en un cocyle sur A^0, à valeurs dans M, car $H^2(A^0, (\varphi-1)M) = 0$ et $H^1(\Phi^+, M) = M/(\varphi-1)M$. Alors $c(g,h) = -\frac{\varphi^{k(g)}-1}{\varphi-1} \cdot \sigma_{a(g)}\alpha(\sigma_{a(h)})$ est un 2-cobord sur A^+, trivial sur $\Phi^+ \times \Phi^+$, dont l'image dans $H^1(A^0, H^1(\Phi^+, M))$ est celle de α.

A.2. Dualité

Dans tout ce qui suit, on suppose que la topologie sur M^* est telle que l'application naturelle $M \to (M^*)^*$ est un isomorphisme.

• *Entre $H^j(\Phi^\pm, M)$ et $H^{1-j}(\Phi^\mp, M^*)$.*

Notons $\Phi^- = \psi^{\mathbf{N}}$ le semi-groupe opposé de Φ^+. Si M est un L-espace vectoriel topologique, muni d'une action de Φ^+, alors $\psi - 1$ est le transposé de $\varphi - 1$. Il s'ensuit que $\mathrm{Ker}(\psi - 1)$ et $\mathrm{Im}(\varphi - 1)$ sont orthogonaux, ainsi que

$\mathrm{Ker}(\varphi - 1)$ et $\mathrm{Im}(\psi - 1)$. L'accouplement naturel $M^* \times M \to L$ induit donc des accouplements

$$\langle\ ,\ \rangle_\Phi : H^i(\Phi^-, M^*) \times H^{1-i}(\Phi^+, M) \to L, \quad \text{pour } i = 0, 1.$$

Si $\mathrm{Im}(\varphi - 1)$ *et* $\mathrm{Im}(\psi - 1)$ *sont fermés,* alors $H^i(\Phi^+, M)$ et $H^i(\Phi^-, M^*)$ sont séparés si $i = 0, 1$, et ces accouplements fournissent les identifications suivantes :

— $H^0(\Phi^-, M^*) \cong H^1(\Phi^+, M)^*$ et $H^0(\Phi^+, M) \cong H^1(\Phi^-, M^*)^*$,
— $H^1(\Phi^+, M) \cong H^0(\Phi^-, M^*)^*$ et $H^1(\Phi^-, M^*) \cong H^0(\Phi^+, M)^*$.

Remarque 8.103. Soient E un L-espace vectoriel topologique, et $u : E \to E$ linéaire continue. On dispose de résultats généraux [7] assurant que l'image de ${}^t u : E^* \to E^*$ est fermée si celle de u l'est (c'est par exemple le cas si E est un fréchet [4, th. IV.4.2.1]). En pratique, la démonstration de la fermeture de $\mathrm{Im}\,{}^t u$ est en général une simple traduction de celle de $\mathrm{Im}\, u$.

- *Entre* $H^j(A^0, M)$ *et* $H^{1-j}(A^0, M^*)$.— Si M_0 est un L-espace vectoriel topologique muni d'une action de A^0, le cup-produit induit un accouplement

$$\langle\ ,\ \rangle_{A^0} : H^i(A^0, M_0^*) \times H^{1-i}(A^0, M_0) \to H^1(A^0, L) = L.$$

De manière explicite, si $x \in H^0(A^0, M_0^*)$ (resp. $H^0(A^0, M_0)$), alors il existe $\lambda \in L$ tel que l'on ait $\langle x, c(\sigma_a)\rangle = \lambda \log a$ (resp. $\langle c(\sigma_a), x\rangle = \lambda \log a$), pour tout $a \in \mathbf{Z}_p^*$, et λ est l'élément de L fourni par l'accouplement ci-dessus.

On suppose que l'application naturelle $M_0 \to (M_0^*)^*$ est un isomorphisme. Soit u un générateur de $1 + 2p\mathbf{Z}_p$. *Si* $(\sigma_u - 1)M_0$ *et* $(\sigma_u - 1)M_0^*$ *sont fermés,* alors $H^i(A^0, M_0)$ et $H^i(A^0, M_0^*)$ sont séparés si $i = 0, 1$, et l'accouplement $\langle\ ,\ \rangle_{A^0}$ induit les identifications suivantes :

— $H^0(A^0, M_0^*) \cong H^1(A^0, M_0)^*$ et $H^0(A^0, M_0) \cong H^1(A^0, M_0^*)^*$,
— $H^1(A^0, M) \cong H^0(A^0, M_0^*)^*$ et $H^1(A^0, M_0^*) \cong H^0(A^0, M_0)^*$.

- *Entre* $H^i(A^+, M)$ *et* $H^{2-i}(A^-, M^*)$.— On note A^- le semi-groupe opposé de A^+. On a alors une décomposition $A^- = \Phi^- \times A^0$.

Lemme 8.104. *Soit* $\langle g, h\rangle = k(h)\log a(g) - k(g)\log a(h)$. *Si* $j = 0, 1, 2$, *il existe un* (unique) *accouplement* $\langle\ ,\ \rangle_{A^+} : H^j(A^-, M^*) \times H^{2-j}(A^+, M) \to L$, *tel que l'on ait, pour tous* $h, g \in A^+$,

$$\langle c, c'(g,h) - c'(h,g)\rangle = \langle c, c'\rangle_{A^+} \langle g, h\rangle,$$
$$\text{si } c \in Z^0(A^-, M^*) \text{ et } c' \in Z^2(A^+, M),$$
$$\langle c(g^{-1}), c'(h)\rangle - \langle c(h^{-1}), c'(g)\rangle = \langle c, c'\rangle_{A^+} \langle g, h\rangle,$$

$$\textit{si } c \in Z^1(A^-, M^*) \textit{ et } c' \in Z^1(A^+, M),$$
$$\langle c(g^{-1}, h^{-1}) - c(h^{-1}, g^{-1}), c'\rangle = \langle c, c'\rangle_{A^+}\langle g, h\rangle,$$
$$\textit{si } c \in Z^2(A^-, M^*) \textit{ et } c' \in Z^0(A^+, M).$$

Démonstration. Soient $c \in Z^0(A^-, M^*)$ et $c' \in Z^2(A^+, M)$. Si $\alpha(g, h) = c'(g, h) - c'(h, g)$, on a $d_{2,h}\alpha(g, h_1, h_2) = (1 - g) \cdot c'(h_1, h_2)$. (Cela a été utilisé pour la description de l'isomorphisme $\iota_{HS} : H^1(A^0, H^1(\Phi^+, M)) \cong H^2(A^+, M)$.) On en déduit que $(g, h) \mapsto \langle c, c'(g, h) - c'(h, g)\rangle$ est bilinéaire antisymétrique et donc de la forme $\lambda(k(h) \log a(g) - k(g) \log a(h))$, avec $\lambda \in L$. On note $\langle c, c'\rangle_{A^+}$ l'élément λ de L ainsi défini. Alors $(c, c') \mapsto \langle c, c'\rangle_{A^+}$ est bilinéaire sur $Z^0(A^-, M^*) \times Z^2(A^+, M)$ et, pour terminer la démonstration du cas $j = 0$, il suffit de vérifier que $\langle c, c'\rangle_{A^+} = 0$ si $c' = d_2c_1$ est un 2-cobord, ce qui suit de l'identité $\alpha(g, h) = (g-1) \cdot c_1(h) - (h-1) \cdot c_1(g)$.

Soient $c \in Z^1(A^\mp, M^*)$ et $c' \in Z^1(A^\pm, M)$. On cherche à prouver que $f(g, h) = \langle c(g^{-1}), c'(h)\rangle - \langle c(h^{-1}), c'(g)\rangle$ est bilinéaire, antisymétrique (ceci est évident), sur A^+. Or $f(g, h_1h_2) - f(g, h_1) - f(g, h_2)$ est égal à

$$\begin{aligned}
&\langle c(g^{-1}), c'(h_1h_2) - c'(h_1) - c'(h_2)\rangle - \langle c(h_2^{-1}h_1^{-1}) - c(h_1^{-1}) - c(h_2^{-1}), c'(g)\rangle \\
&\quad = \langle c(g^{-1}), (h_1 - 1) \cdot c'(h_2)\rangle - \langle (h_2^{-1} - 1) \cdot c(h_1^{-1}), c'(g)\rangle \\
&\quad = \langle (h_1^{-1} - 1) \cdot c(g^{-1}), c'(h_2)\rangle - \langle c(h_1^{-1}), (h_2 - 1) \cdot c'(g)\rangle.
\end{aligned}$$

Comme $A^\pm$ est commutatif, un 1-cocycle sur $A^\pm$ vérifie la relation $(g - 1) \cdot c(h) = c(gh) - c(g) - c(h) = (h - 1) \cdot c(g)$. On en déduit que :

$$\begin{aligned}
&f(g, h_1h_2) - f(g, h_1) - f(g, h_2) \\
&\quad = \langle (g^{-1} - 1) \cdot c(h_1^{-1}), c'(h_2)\rangle - \langle c(h_1^{-1}), (g - 1) \cdot c'(h_2)\rangle \\
&\quad = \langle c(h_1^{-1}), (g - 1) \cdot c'(h_2)\rangle - \langle c(h_1^{-1}), (g - 1) \cdot c'(h_2)\rangle = 0.
\end{aligned}$$

Ceci prouve la linéarité par rapport à h ; celle par rapport à g en découle en utilisant l'antisymétrie. On en déduit l'existence d'un accouplement $\langle\ ,\ \rangle_{A^+} : Z^1(A^-, M^*) \times Z^1(A^+, M)$ tel que

$$f(g, h) = \langle c, c'\rangle_{A^+}(k(h) \log a(g) - k(g) \log a(h)).$$

Par ailleurs, De plus, si c ou c' est un cobord, alors f est identiquement nulle (car $\langle c(g^{-1}), (h-1)c'\rangle = \langle (h^{-1}-1) \cdot c(g^{-1}), c'\rangle = \langle (g^{-1} - 1) \cdot c(h^{-1}), c'\rangle = \langle c(h^{-1}), (g - 1)c'\rangle$ et argument similaire si c' est un cobord). Il en résulte que $\langle\ ,\ \rangle_{A^+}$ se factorise à travers $H^1(A^-, M^*) \times H^1(A^+, M)$.

Ceci démontre le cas $j = 1$, et le cas $j = 2$ s'obtenant en échangeant les rôles de M et M^* et de A^+ et A^- dans le cas $j = 0$, cela permet de conclure.

Proposition 8.105. *On suppose que :*

- $(\varphi - 1) \cdot M$ *et* $(\psi - 1) \cdot M^*$ *sont fermés dans* M *et* M^*,
- $(\sigma_u - 1) \cdot M_0$ *est fermé dans* M_0, *si* M_0 *est un des modules* $H^j(\Phi^+, M)$, *pour* $j = 0, 1$, *et* $H^j(\Phi^-, M^*)$, *pour* $j = 0, 1$.

Alors les accouplements $\langle\ ,\ \rangle_{A^+}$ *induisent des identifications :*

$$H^j(A^-, M^*) \cong H^{2-j}(A^+, M)^* \quad \text{et} \quad H^j(A^+, M) \cong H^{2-j}(A^-, M^*)^*, \quad \textit{pour } j = 0, 1, 2.$$

Démonstration. La première hypothèse implique que $H^j(\Phi^+, M)$ et $H^{1-j}(\Phi^-, M^*)$ sont séparés et duaux l'un de l'autre, si $j = 0, 1$; la seconde implique alors que $H^i(A^0, H^j(\Phi^+, M))$ et $H^{1-i}(A^0, H^{1-j}(\Phi^-, M^*))$ sont duaux l'un de l'autre, si $i = 0, 1$ et $j = 0, 1$, pour l'accouplement $\langle\ ,\ \rangle_{A_0}$.

On en déduit le résultat pour $j = 0$ en utilisant l'identité

$$\langle \iota_{HS}(c), \iota_{HS}(c') \rangle_{A^+} = \langle c, c' \rangle_{A_0},$$

si $c \in H^0(A^0, H^0(\Phi^-, M^*))$ et $c' \in H^1(A^0, H^1(\Phi^+, M))$. (Elle se vérifie en revenant aux formules.) Le résultat pour $j = 2$ s'en déduit en échangeant les rôles de M et M^* et de A^+ et A^-.

Pour vérifier le résultat dans le cas $j = 1$, on utilise les suites exactes

$$0 \to H^1(A^0, H^0(\Phi^+, M)) \to H^1(A^+, M) \to H^0(A^0, H^1(\Phi^+, M)) \to 0,$$
$$0 \to H^1(A^0, H^0(\Phi^-, M^*)) \to H^1(A^-, M^*) \to H^0(A^0, H^1(\Phi^-, M^*)) \to 0.$$

Si $c \in \iota_{HS}(H^1(A^0, H^0(\Phi^-, M^*)))$ et $c' \in \iota_{HS}(H^1(A^0, H^0(\Phi^+, M)))$, on a $c(\varphi^{-1}) = 0$ et $c'(\varphi) = 0$, et donc $\langle c, c' \rangle_{A^+} = 0$. L'accouplement $\langle\ ,\ \rangle_{A^+}$ induit donc des accouplements sur $H^1(A^0, H^0(\Phi^+, M)) \times H^0(A^0, H^1(\Phi^-, M^*))$ et $H^1(A^0, H^0(\Phi^-, M^*)) \times H^0(A^0, H^1(\Phi^+, M))$ dont on vérifie aisément en revenant aux formules qu'ils coïncident, au signe près, avec les accouplements $\langle\ ,\ \rangle_{A_0}$. On en déduit le résultat pour $j = 1$, ce qui permet de conclure.

Références

[1] L. Berger, Équations différentielles p-adiques et (φ, N)-modules filtrés. Astérisque **319** (2008), 13–38.

[2] L. Berger et C. Breuil, Sur quelques représentations potentiellement cristallines de $\mathbf{GL}_2(\mathbf{Q}_p)$, Astérisque **330** (2010), 155–211.

[3] L. Berger et P. Colmez, Familles de représentations de de Rham et monodromie p-adique, Astérisque **319** (2008), 303–337.

[4] N. BOURBAKI, *Espace Vectoriels Topologiques*, chap. I à V, Masson, Paris, 1981.

[5] C. BREUIL, Invariant $\mathscr{L}$ et série spéciale p-adique, Ann. E.N.S. **37** (2004) 559–610.

[6] C. BREUIL, Série spéciale p-adique et cohomologie étale complétée, Astérisque **331** (2010), 65–115.

[7] H. CARTAN, lettre du 4 mars 1940 et A. WEIL, lettre du 9 mars 1940, *Correspondance entre Henri Cartan et André Weil (1928-1991)*, éditée par M. AUDIN, Documents Mathématiques **6**, Société Mathématique de France, 2011.

[8] G. CHENEVIER, Sur la densité des représentations cristallines du groupe de Galois absolu de $\mathbf{Q}_p$, Math. Ann. **355** (2013), 1469–1525.

[9] F. CHERBONNIER et P. COLMEZ, Représentations p-adiques surconvergentes, Invent. Math. **133** (1998), 581–611.

[10] P. COLMEZ, Représentations triangulines de dimension 2, Astérisque **319** (2008), 213–258.

[11] P. COLMEZ, Fonctions d'une variable p-adique, Astérisque **330** (2010), 13–59.

[12] P. COLMEZ, (φ, Γ)-modules et représentations du mirabolique de $\mathbf{GL}_2(\mathbf{Q}_p)$, Astérisque **330** (2010), 61–153.

[13] P. COLMEZ, La série principale unitaire de $\mathbf{GL}_2(\mathbf{Q}_p)$, Astérisque **330** (2010), 213–262.

[14] P. COLMEZ, Représentations de $\mathbf{GL}_2(\mathbf{Q}_p)$ et (φ, Γ)-modules, Astérisque **330** (2010), 281–509.

[15] G. DOSPINESCU, Actions infinitésimales dans la correspondance de Langlands locale p-adique pour $\mathbf{GL}_2(\mathbf{Q}_p)$, Math. Ann. **354** (2012), 627–657.

[16] G. DOSPINESCU, Équations différentielles p-adiques et foncteurs de Jacquet analytiques, ce volume.

[17] M. EMERTON, p-adic L-functions and unitary completions of representations of p-adic reductive groups Duke Math. J. **130** (2005), 353-392.

[18] M. EMERTON, Jacquet modules of locally analytic representations of p-adic reductive groups. I. Construction and first properties, Ann. E.N.S. **39** (2006), 775–839.

[19] M. EMERTON, A local-global compatibility conjecture in the p-adic Langlands programme for $\mathrm{GL}_{2/\mathbb{Q}}$, Pure Appl. Math. Q. **2** (2006), 279–393.

[20] J.-M. FONTAINE, Représentations p-adiques des corps locaux, dans "*The Grothendieck Festschrift*", vol 2, Prog. in Math. **87**, 249–309, Birkhäuser 1991.

[21] H. JACQUET et R. LANGLANDS, Automorphic forms on $\mathbf{GL}(2)$, Lect. Notes in Math. **114**, Springer 1970.

[22] K. KEDLAYA, A p-adic monodromy theorem, Ann. of Math. **160** (2004), 93–184.

[23] N. KOBLITZ, *p-adic analysis : a short course on recent work*, London Math. Soc. Lecture Note Series 46, Cambridge University Press, 1980.

[24] J. KOHLHAASE, The cohomology of locally analytic representations, J. Reine Angew. Math. **651** (2011), 187–240.

[25] R. LIU, Locally Analytic Vectors of some crystabeline representations of $\mathbf{GL}_2(\mathbf{Q}_p)$, Compos. Math. 148 (2012), 28–64.

[26] R. LIU, Cohomology and duality for (φ, Γ)-modules over the Robba ring, Int. Math. Res. Not. IMRN (3) (2008)

[27] R. LIU, B. XIE, Y. ZHANG, Locally Analytic Vectors of Unitary Principal Series of $\mathbf{GL}_2(\mathbf{Q}_p)$, Ann. E.N.S. **45** (2012), 167–190.

[28] V. PASKUNAS, On some crystalline representations of $\mathbf{GL}_2(\mathbf{Q_p})$, Algebra & Number Theory **3** (2009), 411–421.

[29] V. PASKUNAS, The image of Colmez's Montréal functor, Publ. Math. IHES (à paraître).

[30] P. SCHNEIDER et J. TEITELBAUM, Locally analytic distributions and p-adic representation theory, with applications to $\mathbf{GL}_2$, J. Amer. Math. Soc. **15** (2002), 443–468.

9

Equations différentielles p-adiques et modules de Jacquet analytiques

Gabriel Dospinescu

Abstract

Using differential techniques, we compute the Jacquet module of the locally analytic vectors of irreducible admissible unitary representations of $\mathrm{GL}_2(\mathbf{Q}_p)$.

1. Introduction

Le but de cet article est d'étudier le module de Jacquet des $\mathrm{GL}_2(\mathbf{Q}_p)$ représentations de Banach unitaires admissibles, absolument irréductibles. On retrouve les résultats de [7], mais les méthodes sont sensiblement différentes.

1.1. Notations

On fixe une extension finie L de $\mathbf{Q}_p$ et on note $\mathscr{R}$ l'anneau de Robba à coefficients dans L, i.e. l'anneau des séries de Laurent $\sum_{n\in\mathbf{Z}} a_n T^n$, avec $a_n \in L$, qui convergent sur une couronne de type $0 < v_p(T) \leq r$, où $r > 0$ dépend de la série.

Soit $\chi : \mathrm{Gal}(\overline{\mathbf{Q}_p}/\mathbf{Q}_p) \to \mathbf{Z}_p^*$ le caractère cyclotomique. Il induit un isomorphisme de $\Gamma = \mathrm{Gal}(\mathbf{Q}_p(\mu_{p^\infty})/\mathbf{Q}_p)$ sur $\mathbf{Z}_p^*$ et on note $a \to \sigma_a$ son inverse. On munit $\mathscr{R}$ d'une action de Γ et d'un Frobenius φ, en posant $(\sigma_a f)(T) = f((1+T)^a - 1)$ et $(\varphi f)(T) = f((1+T)^p - 1)$. Soit $\nabla = \lim_{a\to 1} \frac{\sigma_a - 1}{a-1}$ l'action infinitésimale de Γ. Explicitement, on a $\nabla(f) = t(1+T)f'(T)$, où $t = \log(1+T)$.

Si $\delta : \mathbf{Q}_p^* \to L^*$ est un caractère continu, on note $\mathscr{R}(\delta)$ le $\mathscr{R}$-module libre de rang 1 ayant une base e (dite canonique) telle que $\varphi(e) = \delta(p)e$ et

Automorphic Forms and Galois Representations, ed. Fred Diamond, Payman L. Kassaei and Minhyong Kim. Published by Cambridge University Press.

$\sigma_a(e) = \delta(a)e$ pour tout $a \in \mathbf{Z}_p^*$. On pose $w(\delta) = \delta'(1)$, la dérivée de δ en 1 (rappelons que δ est automatiquement localement analytique). Si δ est unitaire (i.e. si $\delta(\mathbf{Q}_p^*) \subset \mathscr{O}_L^*$), alors $w(\delta)$ est le poids de Hodge-Tate généralisé du caractère de $\mathrm{Gal}(\overline{\mathbf{Q}_p}/\mathbf{Q}_p)$ attaché à δ par la théorie locale du corps de classes, normalisée de telle sorte que χ corresponde à $x \to x \cdot |x|_p$.

1.2. La correspondance de Langlands locale p-adique

Une L-représentation de $\mathrm{Gal}(\overline{\mathbf{Q}_p}/\mathbf{Q}_p)$ est un L-espace vectoriel de dimension finie, muni d'une action L-linéaire continue de $\mathrm{Gal}(\overline{\mathbf{Q}_p}/\mathbf{Q}_p)$. Les travaux de Fontaine, Berger, Cherbonnier–Colmez et Kedlaya (voir [1, 3, 11, 12]) associent à une L-représentation V un $\mathscr{R}$-module $D_{\mathrm{rig}} = D_{\mathrm{rig}}(V)$ libre de rang $\dim_L V$, muni d'actions semi-linéaires de φ et Γ, qui commutent. Le (φ, Γ)-module D_{rig} est de pente 0, i.e. il contient un (φ, Γ)-module étale $D^\dagger$ sur $\mathscr{E}^\dagger$, tel que $D_{\mathrm{rig}} = \mathscr{R} \otimes_{\mathscr{E}^\dagger} D^\dagger$. Ici $\mathscr{E}^\dagger$ est le sous-anneau des éléments bornés de $\mathscr{R}$ (c'est un corps). D_{rig} est aussi muni d'un inverse à gauche ψ de φ, qui joue un grand rôle dans la théorie.

Dans la suite de cette introduction on suppose que V est de dimension 2 sur L, absolument irréductible. La correspondance de Langlands locale p-adique [5] associe à V un L-espace de Banach $\Pi = \Pi(V)$, muni d'une action continue de $\mathrm{GL}_2(\mathbf{Q}_p)$, qui en fait une représentation unitaire, admissible et topologiquement absolument irréductible. Soit $\delta_D = \chi^{-1} \cdot \det V$, que l'on voit comme caractère de $\mathbf{Q}_p^*$ et comme caractère de $\mathrm{GL}_2(\mathbf{Q}_p)$, en composant avec le déterminant. On peut utiliser les actions de φ, ψ et Γ pour construire un faisceau $\mathrm{GL}_2(\mathbf{Q}_p)$-équivariant sur $\mathbf{P}^1(\mathbf{Q}_p)$, dont D_{rig} est l'espace des sections sur $\mathbf{Z}_p$. Par construction, $D_{\mathrm{rig}}^{\psi=0}$ est l'espace des sections sur $\mathbf{Z}_p^*$ et l'application de restriction à $\mathbf{Z}_p^*$ est donnée par $\mathrm{Res}_{\mathbf{Z}_p^*} = 1 - \varphi \circ \psi$. Soit w_D l'involution de $D_{\mathrm{rig}}^{\psi=0}$ décrivant l'action de $\left(\begin{smallmatrix} 0 & 1 \\ 1 & 0 \end{smallmatrix}\right)$. L'espace des sections globales du faisceau est donc

$$D_{\mathrm{rig}} \boxtimes \mathbf{P}^1 = \{(z_1, z_2) \in D_{\mathrm{rig}} \times D_{\mathrm{rig}} | \quad w_D(\mathrm{Res}_{\mathbf{Z}_p^*}(z_1)) = \mathrm{Res}_{\mathbf{Z}_p^*}(z_2)\}.$$

On montre [5, th. V.2.20] que les vecteurs localement analytiques Π^{an} de Π vivent dans une suite exacte de $\mathrm{GL}_2(\mathbf{Q}_p)$-modules topologiques

$$0 \to (\Pi^{\mathrm{an}})^* \otimes \delta_D \to D_{\mathrm{rig}} \boxtimes \mathbf{P}^1 \to \Pi^{\mathrm{an}} \to 0.$$

1.3. Représentations trianguline

Dans [4], Colmez définit un espace $\mathscr{S}_{\mathrm{irr}}$ de représentations irréductibles de dimension 2 de $\mathrm{Gal}(\overline{\mathbf{Q}_p}/\mathbf{Q}_p)$, appellées trianguline. Un point de $\mathscr{S}_{\mathrm{irr}}$ est un triplet $s = (\delta_1, \delta_2, \mathscr{L})$, où $\delta_1, \delta_2 : \mathbf{Q}_p^* \to L^*$ sont des caractères continus et $\mathscr{L} \in \mathbf{P}^1(L)$ si $\delta_1 = x^k \chi \delta_2$ ($k \in \mathbf{N}$), ou $\mathscr{L} \in \mathbf{P}^0(L) = \{\infty\}$ si $\delta_1 \notin \{x^k \chi \delta_2,\ k \in \mathbf{N}\}$. Si $s \in \mathscr{S}_{\rm irr}$, on note $w(s) = w(\delta_1) - w(\delta_2)$, $V(s)$ la représentation associée et $D_{\rm rig}(s)$ son (φ, Γ)-module sur l'anneau de Robba. On note aussi $\Pi(s) = \Pi(V(s))$. Par construction, on a une suite exacte $0 \to \mathscr{R}(\delta_1) \to D_{\rm rig} \to \mathscr{R}(\delta_2) \to 0$, dont la classe d'isomorphisme est déterminée par $\mathscr{L}$. L'espace $\mathscr{S}_{\rm irr}$ admet une partition $\mathscr{S}_{\rm irr} = \mathscr{S}_*^{\rm ng} \amalg \mathscr{S}_*^{\rm cris} \amalg \mathscr{S}_*^{\rm st}$, où

- $\mathscr{S}_*^{\rm cris} = \{s \in \mathscr{S}_{\rm irr} | w(s) \in \mathbb{N}^*, w(s) > v_p(\delta_1(p)) \quad \text{et} \quad \mathscr{L} = \infty\}$.
- $\mathscr{S}_*^{\rm st} = \{s \in \mathscr{S}_{\rm irr} | w(s) \in \mathbb{N}^*, w(s) > v_p(\delta_1(p)) \quad \text{et} \quad \mathscr{L} \neq \infty\}$.
- $\mathscr{S}_*^{\rm ng} = \{s \in \mathscr{S}_{\rm irr} | w(s) \notin \mathbb{N}^*\}$.

Supposons que δ_1 est localement algébrique et que V est une représentation irréductible de dimension 2. On démontre alors [4] que:

- V correspond à un point de $\mathscr{S}_*^{\rm cris}$ si et seulement si V devient cristalline sur une extension abélienne de $\mathbf{Q}_p$;
- V correspond à un point de $\mathscr{S}_*^{\rm st}$ si et seulement si V est une tordue par un caractère d'ordre fini d'une représentation semi-stable non cristalline.

1.4. Le module de Jacquet analytique

Soit $U = \begin{pmatrix} 1 & \mathbf{Q}_p \\ 0 & 1 \end{pmatrix}$. Si π est une L-représentation localement analytique de $\mathrm{GL}_2(\mathbf{Q}_p)$ (voir [9, 15, 16] pour les bases de la théorie), on note $J(\pi)$ son module de Jacquet naïf, quotient de π par l'adhérence du sous-espace engendré par les vecteurs $(u-1) \cdot v$, où $u \in U$ et $v \in \pi$. Le dual[1] de $J(\pi)$ est $J^*(\pi) = (\pi^*)^U$ et c'est naturellement une représentation localement analytique du tore diagonal de $\mathrm{GL}_2(\mathbf{Q}_p)$. Le premier résultat est l'analogue p-adique d'un résultat classique de la théorie des représentations lisses, et confirme le principe selon lequel les représentations triangulines correspondent aux $\mathrm{GL}_2(\mathbf{Q}_p)$-représentations de la série principale unitaire [6].

Théorème 9.1. *Soient V et Π comme dans 1.2. Alors $J^*(\Pi^{\rm an})$ est un L-espace vectoriel de dimension au plus 2 et il est non nul si et seulement si V est trianguline.*

Ce théorème est aussi démontré dans [7, th. 0.1], en utilisant l'action de φ sur $D_{\rm rig}$. Notre approche est orthogonale (elle utilise l'action infinitésimale de Γ au lieu de celle de φ) et plus directe: si u^+ désigne l'action infinitésimale de

[1] Tous les duaux que l'on considère dans cet article sont topologiques.

U, le résultat principal de [8] montre que le noyau de u^+ sur l'espace $(\Pi^{\mathrm{an}})^*$ s'identifie à l'espace des solutions de l'équation différentielle $(\nabla - a)(\nabla - b)z = 0$, où $z \in D_{\mathrm{rig}}$, a et b sont les poids de Hodge–Tate généralisés de V et ∇ est l'action infinitésimale de Γ sur D_{rig}. Cela ramène l'étude de $J^*(\Pi^{\mathrm{an}})$ à la résolution de cette équation différentielle, ce qui se fait sans mal.

Soit $\delta_1 \otimes \delta_2$ le caractère $(a,d) \to \delta_1(a)\delta_2(d)$ du tore diagonal T de $\mathrm{GL}_2(\mathbf{Q}_p)$. Le résultat suivant précise le théorème 9.1 et correspond à [7, th. 0.6].

Théorème 9.2. *Soit $s = (\delta_1, \delta_2, \mathscr{L}) \in \mathscr{S}_{\mathrm{irr}}$.*

(1) Si $s \in \mathscr{S}_*^{\mathrm{st}}$ ou si $w(s) \notin \mathbf{Z}^*$, alors $J^*(\Pi^{\mathrm{an}}(s)) = \delta_1^{-1} \otimes \delta_2^{-1}\chi$.
(2) Si $w(s) \in \{\dots, -2, -1\}$, alors $J^*(\Pi^{\mathrm{an}}(s)) = (\delta_1^{-1} \otimes \delta_2^{-1}\chi) \oplus (x^{w(s)}\delta_1^{-1} \otimes x^{-w(s)}\delta_2^{-1}\chi)$.
(3) Si $s \in \mathscr{S}_*^{\mathrm{cris}}$, alors $J^*(\Pi^{\mathrm{an}}(s)) = (\delta_1^{-1} \otimes \delta_2^{-1}\chi) \oplus (x^{-w(s)}\delta_2^{-1} \otimes x^{w(s)}\delta_1^{-1}\chi)$ si s est non exceptionnel (i.e. si $\delta_1 \neq x^{w(s)}\delta_2$) et $J^*(\Pi^{\mathrm{an}}(s)) = (\delta_1^{-1} \otimes \delta_2^{-1}\chi) \otimes \begin{pmatrix} 1 & v_p(a/d) \\ 0 & 1 \end{pmatrix}$ dans le cas contraire.

1.5. L'involution w_D et dévissage de $D_{\mathrm{rig}} \boxtimes \mathbf{P}^1$

On suppose que $s = (\delta_1, \delta_2, \mathscr{L}) \in \mathscr{S}_{\mathrm{irr}}$ et on note e_i la base canonique de $\mathscr{R}(\delta_i)$, p_s la projection canonique $D_{\mathrm{rig}}(s) \to \mathscr{R}(\delta_2)$ et $\hat{e}_2 \in D_{\mathrm{rig}}(s)$ tel que $p_s(\hat{e}_2) = e_2$. Si U est un ouvert compact de $\mathbf{Z}_p$, soit $\mathrm{LA}(U)$ l'espace des fonctions localement analytiques sur U à valeurs dans L et soit $\mathscr{D}(U)$ son dual topologique. Si $\delta : \mathbf{Q}_p^* \to L^*$ est un caracère continu, on définit une involution w_δ sur $\mathscr{D}(\mathbf{Z}_p^*)$ en demandant que

$$\int_{\mathbf{Z}_p^*} \phi(w_\delta \mu) = \int_{\mathbf{Z}_p^*} \delta(x)\phi\left(\frac{1}{x}\right)$$

pour tout $\phi \in \mathrm{LA}(\mathbf{Z}_p^*)$. Via l'isomorphisme $(\mathscr{R}^+)^{\psi=0} \simeq \mathscr{D}(\mathbf{Z}_p^*)$ donné par le théorème d'Amice, cela induit une involution w_δ sur $(\mathscr{R}^+)^{\psi=0}$, qui satisfait $w_\delta(\sigma_a f) = \delta(a)\sigma_{\frac{1}{a}}(w_\delta(f))$. Cette involution s'étend de manière unique en une involution de $\mathscr{R}^{\psi=0}$, satisfaisant la même relation que ci-dessus. Le résultat suivant fournit une description plus ou moins explicite de l'involution w_D dans le cas triangulin. Si $s \in \mathscr{S}_*^{\mathrm{cris}}$, cela permet de retrouver[2] et renforcer le délicat lemme II.3.13 de [5].

[2] Notre preuve utilise la stabilité de $D^\natural \boxtimes \mathbf{P}^1$ par G et ne permet pas de la retrouver; dans le chapitre II de [5], cette stabilité est démontrée en utilisant le lemme II.3.13, mais elle peut aussi se déduire des résultats de Berger–Breuil [2] et du ii) de la remarque IV.4.9 de [5].

Théorème 9.3. *Pour tout* $f \in \mathscr{R}^{\psi=0}$ *on a* $w_{D(s)}(f \cdot e_1) = \delta_1(-1)w_{\delta_D \cdot \delta_1^{-2}}(f) \cdot e_1$ *et*

$$p_s(w_{D(s)}(f \cdot \varphi(\hat{e}_2))) = \delta_2(-1)w_{\delta_D \cdot \delta_2^{-2}}(f) \cdot \varphi(e_2).$$

La très mauvaise convergence de la suite[3] définissant w_D rend délicate une preuve directe du théorème 9.3. On déduit du théorème 9.3 le corollaire 9.4 ci-dessous qui est le point de départ [7, 14] pour l'étude des vecteurs localement analytiques de la série principale unitaire.

Soit

$$\mathscr{R} \boxtimes_\delta \mathbf{P}^1 = \{(f_1, f_2) \in \mathscr{R} \times \mathscr{R} | \quad \mathrm{Res}_{\mathbf{Z}_p^*}(f_2) = w_\delta(\mathrm{Res}_{\mathbf{Z}_p^*}(f_1))\}.$$

En copiant les constructions de Colmez, on munit $\mathscr{R} \boxtimes_\delta \mathbf{P}^1$ d'une structure naturelle de $\mathrm{GL}_2(\mathbf{Q}_p)$-module topologique (pour les détails voir 3.1). Ce module $\mathscr{R} \boxtimes_\delta \mathbf{P}^1$ est étroitement lié aux induites paraboliques[4], car on peut montrer qu'il vit dans une suite exacte de $\mathrm{GL}_2(\mathbf{Q}_p)$-modules topologiques

$$0 \to (\mathrm{Ind}_B^G(\delta^{-1} \otimes 1))^* \to \mathscr{R} \boxtimes_\delta \mathbf{P}^1 \to \mathrm{Ind}_B^G(\chi\delta \otimes \chi^{-1}) \to 0.$$

Corollaire 9.4. *Soit* $s \in \mathscr{S}_{\mathrm{irr}}$. *La suite exacte* $0 \to \mathscr{R}(\delta_1) \to D_{\mathrm{rig}} \to \mathscr{R}(\delta_2) \to 0$ *induit une suite exacte de* $\mathrm{GL}_2(\mathbf{Q}_p)$*-modules topologiques*

$$0 \to (\mathscr{R} \boxtimes_{\delta_D \cdot \delta_1^{-2}} \mathbf{P}^1) \otimes \delta_1 \to D_{\mathrm{rig}} \boxtimes \mathbf{P}^1 \to (\mathscr{R} \boxtimes_{\delta_D \cdot \delta_2^{-2}} \mathbf{P}^1) \otimes \delta_2 \to 0.$$

Ce résultat est démontré dans [7, th. 4.6], ainsi que dans [14] (prop. 6.8) par des méthodes différentes. Il permet de donner [7, th.07] une description complète de la représentation Π^{an} (en particulier, de montrer qu'elle est de longueur finie et d'en trouver les constituants de Jordan-Hölder), confirmant ainsi des conjectures de Berger, Breuil et Emerton [2, 10].

[3] Soit $D^\dagger$ le sous-module surconvergent de D_{rig} et D le (φ, Γ)-module sur le corps de Fontaine $\mathscr{E}$ attaché à V. Il découle de la correspondance de Langlands locale p-adique pour $\mathrm{GL}_2(\mathbf{Q}_p)$ ([5], th. II.3.1, prop. V.2.1 et lemme V.2.4) que si $z \in D^{\dagger,\psi=0}$, la suite

$$\sum_{i \in (\mathbb{Z}/p^n\mathbb{Z})^*} \delta_D(i^{-1})(1+T)^i \sigma_{-i^2}\varphi^n \psi^n((1+T)^{-i^{-1}} z)$$

converge dans D (mais pas dans $D^\dagger$) et sa limite $w_D(z)$ appartient à $D^{\dagger,\psi=0}$. L'extension de w_D à $D_{\mathrm{rig}}^{\psi=0}$ se fait en utilisant la densité de $D^\dagger$ dans D_{rig}.

[4] On note dans la suite $\mathrm{Ind}_B^G(\delta_1 \otimes \delta_2)$ l'espace des fonctions localement analytiques $f : \mathrm{GL}_2(\mathbf{Q}_p) \to L$ telles que $f\left(\left(\begin{smallmatrix} a & b \\ 0 & d \end{smallmatrix}\right) g\right) = \delta_1(a)\delta_2(d)f(g)$ pour tous $a, d \in \mathbf{Q}_p^*$, $b \in \mathbf{Q}_p$ et $g \in \mathrm{GL}_2(\mathbf{Q}_p)$.

1.6. Remerciements

Ce travail est une partie de ma thèse de doctorat, réalisée sous la direction de Pierre Colmez et de Gaëtan Chenevier. Je leur suis profondément reconnaissant pour des nombreuses discussions que nous avons eues et pour leurs tout aussi nombreuses suggestions. Je voudrais aussi remercier R. Taylor, X. Caruso et L. Berger pour m'avoir invité à exposer ces résultats à l'I.A.S, dans le cadre du Workshop on Galois Representations and Automorphic Forms, et à l'E.N.S Lyon, dans le cadre de la conférence "Théorie de Hodge p-adique, équations différentielles p-adiques et leurs applications". Merci aussi à Andrea Pulita pour une discussion qui m'a permis de simplifier une démonstration et à Ramla Abdellatif, qui m'a grandement aidé à améliorer la rédaction. Enfin, merci au rapporteur pour des remarques utiles.

2. Le module de Jacquet de Π^{an}

2.1. Un résultat de finitude

On démontre un résultat de finitude pour les équations différentielles p-adiques attachées aux représentations galoisiennes. Rappelons [1, V.1] que si V est une L-représentation de $\mathrm{Gal}(\overline{\mathbf{Q}_p}/\mathbf{Q}_p)$ et si D_{rig} est son (φ, Γ)-module sur $\mathscr{R}$, alors l'action de Γ peut se dériver, d'où une connexion $\nabla = \lim_{a\to 1} \frac{\sigma_a - 1}{a-1}$ sur D_{rig}, au-dessus de la connexion ∇ sur $\mathscr{R}$ (introduite dans 1.1). Si $P \in L[X]$, notons

$$D_{\mathrm{rig}}^{P(\nabla)=0} = \{z \in D_{\mathrm{rig}} | P(\nabla)(z) = 0\}.$$

Proposition 9.5. *Soit V une L-représentation quelconque de* $\mathrm{Gal}(\overline{\mathbf{Q}_p}/\mathbf{Q}_p)$ *et soit D_{rig} son (φ, Γ)-module sur $\mathscr{R}$. Si $P \in L[X]$ est non nul, alors*

$$\dim_L D_{\mathrm{rig}}^{P(\nabla)=0} \leq \dim_L(V) \cdot \deg(P).$$

Démonstration. Quitte à remplacer L par une extension finie, on peut supposer que toutes les racines de P dans $\overline{\mathbf{Q}_p}$ sont dans L. On démontre la proposition par récurrence sur $\deg P$. Pour traiter le cas $\deg(P) = 1$, nous avons besoin de quelques préliminaires.

Lemme 9.6. *On a* $(\mathrm{Frac}(\mathscr{R}))^{\nabla=0} = L$.

Démonstration. Ce résultat est probablement standard, mais faute d'une référence voici une preuve. Rappelons que $\nabla(f) = t \cdot (1+T) f'(T)$ pour $f \in \mathscr{R}$, où $t = \log(1+T)$. En particulier $\mathscr{R}^{\nabla=0} = L$. La condition $\nabla\left(\frac{f}{g}\right) = 0$ équivaut à $f' \cdot g = f \cdot g'$. Soit $r > 0$ tel que f et g soient analytiques sur la couronne $0 < v_p(T) \leq r$. Si $0 < r' \leq r$ et si $z \in \mathbf{C}_p$ est un zéro de g dans la couronne $r' \leq v_p(T) \leq r$, la relation $f' \cdot g = g' \cdot f$ montre que z a la même multiplicité (finie) dans f et dans g. D'après des résultats standard de Lazard (voir par exemple la prop. 4.12 de [1]), le quotient $\frac{f}{g}$ est donc analytique dans la couronne $r' \leq v_p(T) \leq r$. Comme cela vaut pour tout $r' \leq r$, le quotient $\frac{f}{g}$ est analytique sur la couronne $0 < v_p(T) \leq r$ et donc est dans $\mathscr{R}^{\nabla=0} = L$. □

Le cas $\deg(P) = 1$ suit alors du lemme ci-dessus et du fait que le rang de D_{rig} sur $\mathscr{R}$ est $\dim_L(V)$.

Lemme 9.7. *Soit $\alpha \in L$. L'application naturelle $D_{\mathrm{rig}}^{\nabla=\alpha} \otimes_L \mathscr{R} \to D_{\mathrm{rig}}$ est injective.*

Démonstration. Il s'agit de vérifier que si $z_1, z_2, \ldots, z_d \in D_{\mathrm{rig}}^{\nabla=\alpha}$ sont libres sur L, alors ils sont libres sur $\mathscr{R}$. Soit (quitte à renuméroter les z_i) $\sum_{i=1}^k f_i \cdot z_i = 0$ une relation de longueur minimale sur $\mathscr{R}$ et soit $g_i = \frac{f_i}{f_1} \in \mathrm{Frac}(\mathscr{R})$. En appliquant ∇ et en utilisant le fait que $\nabla(z_i) = \alpha \cdot z_i$, on obtient $\sum_{i=1}^k \nabla(g_i) \cdot z_i = 0$. Comme $\nabla(g_1) = 0$, par minimalité on obtient $\nabla(g_i) = 0$ pour tout i. Le lemme 9.6 permet alors de conclure. □

Supposons le résultat démontré pour $\deg P = n$ et montrons-le pour $\deg P = n+1$. Soit $P = (X - \alpha) \cdot Q(X)$, avec $Q \in L[X]$ et notons $W_1 = D_{\mathrm{rig}}^{P(\nabla)=0}$ et $W_2 = D_{\mathrm{rig}}^{Q(\nabla)=0}$. Alors $\nabla - \alpha$ est un opérateur L-linéaire de W_1 dans W_2, dont le noyau est de dimension au plus $\dim_L(V)$ (lemme 9.7) et dont l'image est de dimension au plus $\dim_L W_2 \leq \deg(Q) \cdot \dim_L V$. Le résultat s'en déduit.

2.2. Finitude et annulation du module de Jacquet

Dans la suite on suppose que V et Π sont comme dans 1.2 (donc V est de dimension 2). Pour les autres notations utilisées dans la suite (en particulier ∇ et $D_{\mathrm{rig}} \boxtimes \mathbf{P}^1$), voir 1.2 et 2.1.

L'ingrédient essentiel pour l'étude de $J^*(\Pi^{\mathrm{an}})$ est le résultat suivant, dans lequel u^+ désigne l'action infinitésimale de l'unipotent supérieur U de $\mathrm{GL}_2(\mathbf{Q}_p)$. Soient a et b les poids de Hodge-Tate généralisés de V.

Proposition 9.8. *Pour tout $z = (z_1, z_2) \in D_{\mathrm{rig}} \boxtimes \mathbf{P}^1$ on a*

$$u^+(z) = \left(tz_1, -\frac{(\nabla - a)(\nabla - b)z_2}{t}\right).$$

Démonstration. On peut écrire $z = z_1 + w \cdot (\varphi \circ \psi(z_2))$, où $w = \left(\begin{smallmatrix} 0 & 1 \\ 1 & 0 \end{smallmatrix}\right)$ et où l'on voit D_{rig} comme sous-espace de $D_{\mathrm{rig}} \boxtimes \mathbf{P}^1$ comme dans 1.2. On a donc

$$u^+(z) = u^+(z_1) + w \cdot u^-(\varphi(\psi(z_2)))$$

et le résultat découle alors de [8, th. 1]. □

D'après la proposition 9.5, le L-espace vectoriel

$$X = \{z \in D_{\mathrm{rig}} | (\nabla - a)(\nabla - b)z = 0\}$$

est de dimension au plus 4 (on fera mieux par la suite). Comme X est stable par φ, on a $X \subset \varphi(D_{\mathrm{rig}})$, de telle sorte que $(0, z) \in D_{\mathrm{rig}} \boxtimes \mathbf{P}^1$ pour tout $z \in X$. Rappelons aussi que les vecteurs localement analytiques Π^{an} de Π vivent dans une suite exacte de $\mathrm{GL}_2(\mathbf{Q}_p)$-modules topologiques

$$0 \to (\Pi^{\mathrm{an}})^* \otimes \delta_D \to D_{\mathrm{rig}} \boxtimes \mathbf{P}^1 \to \Pi^{\mathrm{an}} \to 0.$$

Proposition 9.9. *On a une égalité de sous-L-espaces vectoriels de $D_{\mathrm{rig}} \boxtimes \mathbf{P}^1$*

$$J^*(\Pi^{\mathrm{an}}) \otimes \delta_D = \{(0, z) | z \in X\}.$$

En particulier, $J^*(\Pi^{\mathrm{an}})$ est de dimension au plus 4 sur L.

Démonstration. Soit $\pi = \Pi^{\mathrm{an}}$. L'inclusion $\pi^* \otimes \delta_D \subset D_{\mathrm{rig}} \boxtimes \mathbf{P}^1$ induit une inclusion

$$J^*(\pi) \otimes \delta_D \subset (D_{\mathrm{rig}} \boxtimes \mathbf{P}^1)^U \subset (D_{\mathrm{rig}} \boxtimes \mathbf{P}^1)^{u^+=0}$$

et, d'après la proposition 9.8, on a

$$(D_{\mathrm{rig}} \boxtimes \mathbf{P}^1)^{u^+=0} = \{(0, z) | z \in X\}.$$

La proposition 9.5 montre alors que $\dim_L J^*(\pi) \leq 4$.

Montrons maintenant que les inclusions précédentes sont des égalités. Nous aurons besoin du résultat suivant, qui montre en particulier que $(D_{\mathrm{rig}} \boxtimes \mathbf{P}^1)^U = (D_{\mathrm{rig}} \boxtimes \mathbf{P}^1)^{u^+=0}$.

Lemme 9.10. *Soit M une L-représentation localement analytique de $\begin{pmatrix} p^{\mathbf{Z}} & \mathbf{Q}_p \\ 0 & 1 \end{pmatrix}$. Si $M^{u^+=0}$ est de dimension finie sur L, alors $M^{u^+=0} = M^U$.*

Démonstration. Il existe n tel que $M^{u^+=0}$ soit invariant par $\begin{pmatrix} 1 & p^n\mathbf{Z}_p \\ 0 & 1 \end{pmatrix}$. Si $m \in M^{u^+=0}$ et $a \in \mathbf{Q}_p$, on a alors pour tout $k \geq n - v_p(a)$

$$\begin{pmatrix} p^k & 0 \\ 0 & 1 \end{pmatrix} \left(\begin{smallmatrix} 1 & a \\ 0 & 1 \end{smallmatrix}\right) m = \begin{pmatrix} 1 & p^k a \\ 0 & 1 \end{pmatrix} \begin{pmatrix} p^k & 0 \\ 0 & 1 \end{pmatrix} m = \begin{pmatrix} p^k & 0 \\ 0 & 1 \end{pmatrix} m,$$

donc $\left(\begin{smallmatrix} 1 & a \\ 0 & 1 \end{smallmatrix}\right) m = m$. Cela permet de conclure. □

Pour finir la preuve de la proposition 9.9, il reste à voir que $(0, z) \in \pi^* \otimes \delta_D$ pour tout $z \in X$. Les $\varphi^n(z)$ vivent dans X, qui est un L-espace vectoriel de dimension finie. Cela entraîne (voir [1, prop 3.2]) que $z \in \tilde{D}^+_{\mathrm{rig}}$, où[5] $\tilde{D}^+_{\mathrm{rig}} = (\tilde{\mathbf{B}}^+_{\mathrm{rig}} \otimes_{\mathbf{Q}_p} V)^{\mathrm{Ker}\chi}$. Mais d'après [5, lemme V.2.17], le module $\tilde{D}^+_{\mathrm{rig}}$ est inclus dans $\pi^* \otimes \delta_D$. Le résultat en découle (noter que $(0, z) = w \cdot z$). □

Corollaire 9.11. *Si V n'est pas trianguline, alors $J^*(\Pi^{\mathrm{an}}) = 0$.*

Démonstration. Il suffit de vérifier que $X = 0$ si V n'est pas trianguline. Si $X \neq 0$, il existe (après avoir remplacé L par une extension finie) un vecteur propre pour φ et Γ dans X. On en déduit que V est trianguline (utiliser le lemme 3.2 de [4]). □

2.3. Le module de Jacquet dans le cas triangulin

Dans la suite on fixe un point $s = (\delta_1, \delta_2, \mathscr{L}) \in \mathscr{S}_{\mathrm{irr}}$ et on note $V = V(s)$, $\Pi = \Pi(s)$, etc. On note e_i la base canonique de $\mathscr{R}(\delta_i)$ (rappelons que $\varphi(e_i) = \delta_i(p)e_i$ et $\sigma_a(e_i) = \delta_i(a)e_i$) et p_s la projection de D_{rig} sur $\mathscr{R}(\delta_2)$. Noter que les poids de Hodge-Tate généralisés de V sont $w(\delta_1)$ et $w(\delta_2)$ ([4, prop 4.5]). De plus, comme D_{rig} est de pente 0 et que V est irréductible, la théorie des pentes de Kedlaya montre que $v_p(\delta_1(p)) = -v_p(\delta_2(p)) > 0$.

Le but de cette partie est de démontrer le théorème 9.2 de l'introduction. Ce théorème découle de la proposition 9.9 et de la proposition 9.15 ci-dessous, qui détermine l'espace X (défini dans 2.2). Cela va demander quelques préliminaires.

Rappelons que $s \in \mathscr{S}^{\mathrm{cris}}_*$ est dit exceptionnel si $\delta_1 = x^{w(s)}\delta_2$. Si $s \in \mathscr{S}^{\mathrm{cris}}_*$, il existe $e_2' \in D_{\mathrm{rig}}$ tel que $p_s(e_2') = t^{w(s)}e_2$ et $\sigma_a(e_2') = a^{w(s)}\delta_2(a)e_2'$ pour tout $a \in \mathbf{Z}_p^*$. Si s n'est pas exceptionnel, on peut choisir e_2' tel que $\varphi(e_2') = p^{w(s)}\delta_2(p)e_2'$ et alors e_2' est unique à multiplication par un élément de L^* près. Si s est exceptionnel, on peut choisir e_2' tel que $\varphi(e_2') = p^{w(s)}\delta_2(p)e_2' + e_1$, et alors e_2' est unique à addition près d'un élément de Le_1. Ces résultats sont déduits du calcul de la cohomologie de $\mathscr{R}(\delta)$, voir [4, prop. 3.10]. On aura besoin dans la suite du résultat suivant [7, lemme 3.22]:

Lemme 9.12. *Soit V une L-représentation irréductible de dimension 2 de $\mathrm{Gal}(\overline{\mathbf{Q}}_p/\mathbf{Q}_p)$ et soit D_{rig} son (φ, Γ)-module.*

(a) Si D_{rig} possède un vecteur propre pour les actions de φ et Γ, alors V est trianguline.

[5] Rappelons que $\tilde{\mathbf{B}}^+_{\mathrm{rig}} = \cap_{n \geq 1}\varphi^n(\mathbf{B}^+_{\mathrm{cris}})$.

(b) *Si V correspond à un point $s \in \mathscr{S}_*^{\mathrm{st}} \cup \mathscr{S}_*^{\mathrm{ng}}$ ou si $s \in \mathscr{S}_*^{\mathrm{cris}}$ est exceptionnel, alors les vecteurs propres pour l'action de φ et Γ sont dans $\cup_{k\geq 0} L^* \cdot t^k e_1$.*
(c) *Si V correspond à un point $s \in \mathscr{S}_*^{\mathrm{cris}}$ non exceptionnel, les vecteurs propres pour φ et Γ sont dans $L^* \cdot t^k e_1$ ou dans $L^* \cdot t^k e_2'$ pour un $k \geq 0$.*

Le lemme suivant sera utilisé constamment dans la suite. Il fournit aussi une démonstration très directe de la proposition 1.19 de [7].

Lemme 9.13. *Soit $k \in L$. L'espace des solutions de l'équation $\nabla f + kf = 0$ (avec $f \in \mathscr{R}$) est $\{0\}$ si $k \notin \{0, -1, -2, \dots\}$ et $L \cdot t^{-k}$ si $k \in \{0, -1, -2, \dots\}$.*

Démonstration. Soit $f \in \mathscr{R}$ une solution non nulle de l'équation $\nabla f + kf = 0$ et soit j le plus grand entier positif tel que $f \in t^j \cdot \mathscr{R}$. Posons $f = t^j \cdot g$, avec $g \in \mathscr{R} - t \cdot \mathscr{R}$. On a $\nabla g + (k+j)g = 0$. Comme $\nabla(\mathscr{R}) \subset t \cdot \mathscr{R}$, on obtient $k + j = 0$ et $g \in L$. Le résultat s'en déduit. □

Notons $X_2 = \{z \in D_{\mathrm{rig}} | (\nabla - w(\delta_2))z = 0\}$, de telle sorte que $(\nabla - w(\delta_1))X \subset X_2$.

Lemme 9.14. *On a $X_2 = 0$ si $w(s) \notin \{0, -1, -2, \dots\}$ et $X_2 = L \cdot t^{-w(s)} e_1$ si $w(s) \in \{0, -1, -2, \dots\}$.*

Démonstration. Comme $\nabla e_j = w(\delta_j) \cdot e_j$, l'équation $(\nabla - w(\delta_2))(fe_1) = 0$ équivaut à $\nabla f + w(s)f = 0$ et l'équation $(\nabla - w(\delta_2))(fe_2) = 0$ équivaut à $\nabla f = 0$ et donc à $f \in L$.

Supposons que $w(s) \notin \{0, -1, \dots\}$. La suite exacte $0 \to \mathscr{R}(\delta_1) \to D_{\mathrm{rig}} \to \mathscr{R}(\delta_2) \to 0$ et l'observation du premier paragraphe montrent que X_2 s'injecte dans Le_2. Supposons que $X_2 \neq 0$, donc $\dim_L X_2 = 1$. Soit $x \in X_2 - \{0\}$, donc $x \notin \mathscr{R}e_1$ et x est vecteur propre pour φ et Γ. Le lemme 9.12 montre que $s \in \mathscr{S}_*^{\mathrm{cris}}$ n'est pas exceptionnel et qu'il existe $c \in L^*$ et $k \in \mathbb{N}$ tels que $x = ct^k \cdot e_2'$. Comme $x \in X_2$ et $\nabla e_2' = (w(s) + w(\delta_2))e_2'$, on obtient $w(s) = -k \in \{0, -1, -2, \dots\}$, une contradiction. Donc $X_2 = 0$.

Supposons maintenant que $w(s) \in \{0, -1, -2, \dots\}$. Le premier paragraphe nous fournit une suite exacte $0 \to Lt^{-w(s)}e_1 \to X_2 \to Le_2$ et il reste à voir qu'elle n'est pas exacte à droite. Supposons donc qu'il existe $z \in X_2$ qui s'envoie sur e_2. Alors $\varphi(z) - \delta_2(p)z$ est dans $X_2 \cap \mathscr{R}e_1 = Lt^{-w(s)}e_1$ et donc, quitte à travailler[6] avec $z + ct^{-w(s)}e_1$ pour un $c \in L$ bien choisi, on peut supposer que $\varphi(z) = \delta_2(p)z$. Soit $\gamma \in \Gamma$. Il existe $a \in L$ tel que $\gamma(z) = \delta_2(\chi(\gamma))z + at^{-w(s)}e_1$. Comme φ et Γ commutent, un petit calcul

6 Cela utilise le fait que $p^{-w(s)}\delta_1(p) \neq \delta_2(p)$, car $v_p(\delta_1(p)) > 0$ et $v_p(\delta_2(p)) < 0$.

donne $a(p^{-w(s)}\delta_1(p) - \delta_2(p)) = 0$, donc $a = 0$. Mais alors la suite exacte $0 \to \mathscr{R}(\delta_1) \to D_{\rm rig} \to \mathscr{R}(\delta_2) \to 0$ est scindée, contradiction. □

Proposition 9.15. *(1) Si $s \in \mathscr{S}_*^{\rm st}$ ou si $w(s) \notin \mathbf{Z}^*$, alors $X = L \cdot e_1$.*
(2) Si $w(s) \in \{\dots, -2, -1\}$, alors $X = L \cdot e_1 \oplus L \cdot t^{-w(s)} e_1$.
(3) Si $s \in \mathscr{S}_*^{\rm cris}$, alors $X = L \cdot e_1 \oplus L \cdot e_2'$.

Démonstration. Commençons par remarquer que $e_1 \in X$ et que $t^{-w(s)}e_1 \in X$ (resp. $e_2' \in X$) pour $w(s) \in \{\dots, -2, -1\}$ (resp. $s \in \mathscr{S}_*^{\rm cris}$). Pour l'inclusion inverse, on va distinguer deux cas:

- Si $w(s) \notin \{0, -1, -2, \dots\}$, la proposition précédente montre que $(\nabla - w(\delta_1))z = 0$ si $z \in X$, donc[7] la triangulation de $D_{\rm rig}$ induit une suite exacte
$$0 \to Le_1 \to X \to \{fe_2 | \nabla(f) = w(s)f\}.$$
Si $w(s) \notin \mathbf{N}^*$, le lemme 9.13 montre que le terme de droite de cette suite exacte est nul et $X = Le_1$. Si $w(s) \in \mathbf{N}^*$ (ce qui inclut les cas $s \in \mathscr{S}_*^{\rm cris}$ et $s \in \mathscr{S}_*^{\rm st}$), on obtient donc (par le lemme 9.13) une suite exacte $0 \to Le_1 \to X \to L \cdot t^{w(s)}$. Cela montre déjà que $\dim_L(X) \leq 2$ et permet de conclure dans le cas $s \in \mathscr{S}_*^{\rm cris}$. Supposons que $s \in \mathscr{S}_*^{\rm st}$ et montrons que cette suite n'est pas exacte à droite (et donc que $X = L \cdot e_1$). Si ce n'était pas le cas, on trouve comme dans la preuve du lemme 9.14 un $z \in X$ qui s'envoie sur $t^{w(s)}e_2$ et tel que $\varphi(z) = p^{w(s)}\delta_2(p)z$. Comme $\sigma_a(z) - a^{w(s)}\delta_2(a)z \in L \cdot e_1$ et comme φ commute à Γ, on obtient facilement que $\sigma_a(z) = a^{w(s)}\delta_2(a)$, donc z est propre pour φ et Γ, ce qui contredit le lemme 9.12.
- Si $w(s) \in \{0, -1, -2, \dots\}$, le lemme 9.14 montre que $(\nabla - w(\delta_1))X \subset X_2 \subset \mathscr{R}e_1$, donc l'image de X dans $\mathscr{R}(\delta_2)$ est contenue dans $\{fe_2 | \nabla f - w(s)f = 0\}$. Le dernier espace est nul si $w(s) < 0$ et de dimension 1 si $w(s) = 0$. Enfin,
$$X \cap \mathscr{R}e_1 = \{fe_1 | \nabla(\nabla f) + w(s)\nabla f = 0\}$$
et ceci est de dimension 2 (resp. 1) si $w(s) < 0$ (resp. $w(s) = 0$), toujours d'après le lemme 9.13 (si $w(s) = 0$, noter que la relation $\nabla(\nabla f) = 0$ force $\nabla f \in L \cap t \cdot \mathscr{R} = \{0\}$ et donc $f \in L$). Ceci permet de conclure. □

3. L'involution w_D

On décrit l'involution w_D sur $D_{\rm rig}^{\psi=0}$ dans le cas où V est trianguline et on démontre le théorème 9.3.

[7] Noter que $(\nabla - w(\delta_1))(fe_2) = (\nabla - w(s))f \cdot e_2$.

3.1. Le module $\mathscr{R} \boxtimes_\delta \mathbf{P}^1$

Soit $\delta : \mathbf{Q}_p^* \to L^*$ un caractère continu. Rappelons que $\mathscr{R}^{\psi=0}$ est libre de rang 1 sur[8] $\mathscr{R}(\Gamma)$, de base $1+T$ (cela découle de [5, cor. V.1.13]). L'involution i_δ de $L[\Gamma]$ qui envoie σ_a sur $\delta(a)\sigma_{\frac{1}{a}}$ se prolonge de manière unique en une involution i_δ de $\mathscr{R}(\Gamma)$ ([5, lemme V.2.3]) et on définit $w_\delta(f \cdot (1+T)) = i_\delta(f)\cdot(1+T)$ si $f \in \mathscr{R}(\Gamma)$. Cela fournit une involution continue sur $\mathscr{R}^{\psi=0}$ telle que pour toute distribution μ sur $\mathbf{Z}_p^*$ on ait

$$w_\delta\left(\int_{\mathbf{Z}_p^*}(1+T)^x\mu\right) = w_\delta\left(\left(\int_{\mathbf{Z}_p^*}\sigma_x\mu\right)\cdot(1+T)\right) =$$

$$\int_{\mathbf{Z}_p^*}\delta(x)\sigma_{\frac{1}{x}}\mu\cdot(1+T) = \int_{\mathbf{Z}_p^*}\delta(x)(1+T)^{\frac{1}{x}}\mu.$$

Donc l'involution w_δ que l'on vient de définir coïncide avec celle définie dans 1.5.

Soit $\mathrm{LA}(\mathbf{P}^1(\delta))$ l'espace des fonctions localement analytiques $\phi : \mathbf{Q}_p \to L$ telles que $x \to \delta(x)\phi\left(\frac{1}{x}\right)$ se prolonge en une fonction localement analytique, muni de l'action définie par (où $g = \left(\begin{smallmatrix} a & b \\ c & d\end{smallmatrix}\right)$)

$$(g^{-1}\cdot\phi)(x) = \delta(cx+d)\phi\left(\frac{ax+b}{cx+d}\right).$$

Un exercice standard de la théorie des induites paraboliques identifie $\mathrm{Ind}_B^G(1,\delta)\otimes\delta^{-1}$ à $\mathrm{LA}(\mathbf{P}^1(\delta))$ (en tant que $\mathrm{GL}_2(\mathbf{Q}_p)$-modules topologiques). L'application qui envoie μ sur $\left(\mathrm{Res}_{\mathbf{Z}_p}(\mu), \mathrm{Res}_{\mathbf{Z}_p}(w\cdot\mu)\right)$ composée avec l'isomorphisme d'Amice[9] identifie le dual de $\mathrm{LA}(\mathbf{P}^1(\delta))$ (comme espace vectoriel topologique) à

$$\mathscr{R}^+ \boxtimes_\delta \mathbf{P}^1 = \{(f_1, f_2) \in \mathscr{R}^+ \times \mathscr{R}^+ | \mathrm{Res}_{\mathbf{Z}_p^*}(f_2) = w_\delta(\mathrm{Res}_{\mathbf{Z}_p^*}(f_1))\}.$$

L'espace $\mathscr{R}^+ \boxtimes_\delta \mathbf{P}^1$ est ainsi muni d'une action de $\mathrm{GL}_2(\mathbf{Q}_p)$. Cette action est donnée par des formules explicites comme dans [5, II.1] (voir aussi [7, 4.1]). Ces formules permettent de prolonger l'action de $\mathrm{GL}_2(\mathbf{Q}_p)$ à $\mathscr{R} \boxtimes_\delta \mathbf{P}^1$ (défini de la même manière que $\mathscr{R}^+ \boxtimes_\delta \mathbf{P}^1$, en remplaçant $\mathscr{R}^+$ par $\mathscr{R}$).

8 L'anneau $\mathscr{R}(\Gamma)$ est défini par $\mathscr{R}(\Gamma) = \Lambda(\Gamma)\otimes_{\Lambda(\Gamma_2)}\mathscr{R}(\Gamma_2)$, où $\Gamma_2 = \chi^{-1}(1+p^2\mathbf{Z}_p)$, $\Lambda(G)$ est l'algèbre des mesures à valeurs dans $\mathscr{O}_L$ sur le groupe de Lie p-adique G et $\mathscr{R}(\Gamma_2)$ est défini comme l'anneau $\mathscr{R}$, en remplaçant la variable T par $\gamma - 1$, pour n'importe quel générateur topologique γ de Γ_2.

9 Cet isomorphisme identifie $\mathscr{R}^+$ à l'espace des distributions sur $\mathbf{Z}_p$ via la transformée d'Amice $\mu \to \sum_{n\geq 0}\int_{\mathbf{Z}_p}\binom{x}{n}\mu\cdot T^n$.

3.2. Dévissage du module $D_{\mathrm{rig}} \boxtimes \mathbf{P}^1$

On démontre le théorème 9.3 de l'introduction et on l'utilise ensuite pour dévisser le module $D_{\mathrm{rig}} \boxtimes \mathbf{P}^1$. Rappelons que l'on suppose que V est trianguline, correspondant à un point $s = (\delta_1, \delta_2, \mathscr{L})$ de $\mathscr{S}_{\mathrm{irr}}$. Rappelons aussi que D_{rig} est contenu dans $D_{\mathrm{rig}} \boxtimes \mathbf{P}^1$ (voir le paragraphe 1.2). Le résultat suivant est crucial pour la suite:

Lemme 9.16. *On a* $w \cdot e_1 \in (D_{\mathrm{rig}} \boxtimes \mathbf{P}^1)^U$.

Démonstration. C'est une conséquence de la proposition 9.9 et du fait que $w \cdot e_1 = (0, e_1)$ et $e_1 \in X$. □

Proposition 9.17. *Pour tout* $f \in \mathscr{R}^{\psi=0}$ *on a* $w_D(f \cdot e_1) = \delta_1(-1) w_{\delta_D \delta_1^{-2}}(f) \cdot e_1$.

Démonstration. On laisse au lecteur le soin de vérifier l'identité suivante (dans laquelle $w = \left(\begin{smallmatrix} 0 & 1 \\ 1 & 0 \end{smallmatrix}\right)$)

$$\left(\begin{smallmatrix} 1 & 1 \\ 0 & 1 \end{smallmatrix}\right) \cdot \left(\begin{smallmatrix} -1 & 0 \\ 0 & 1 \end{smallmatrix}\right) \cdot w \cdot \left(\begin{smallmatrix} 1 & 1 \\ 0 & 1 \end{smallmatrix}\right) \cdot w = w \cdot \left(\begin{smallmatrix} 1 & 1 \\ 0 & 1 \end{smallmatrix}\right).$$

Appliquons cette identité à $e_1 \in D_{\mathrm{rig}} \boxtimes \mathbf{P}^1$. Le terme de gauche est égal[10] à $\delta_1(-1)(1+T)e_1$, ce qui permet donc d'écrire $w_D((1+T)e_1) = \delta_1(-1)(1+T)e_1$. Comme $\sigma_a(e_1) = \delta_1(a)e_1$ et comme $w_D(\sigma_a(z)) = \delta_D(a)\sigma_{\frac{1}{a}}(w_D(z))$, il est facile de voir que les applications $F(f) = w_D(f e_1)$ et $G(f) = \delta_1(-1) w_{\delta_D \cdot \delta_1^{-2}}(f) \cdot e_1$ sont semi-linéaires pour l'action de $i_{\delta_D \cdot \delta_1^{-1}}$. Comme elles coïncident sur $1+T$, qui est une base de $\mathscr{R}^{\psi=0}$ sur $\mathscr{R}(\Gamma)$, on obtient bien $F = G$, d'où le résultat. □

Soit $\mathscr{R}e_1 \boxtimes \mathbf{P}^1 = (D_{\mathrm{rig}} \boxtimes \mathbf{P}^1) \cap (\mathscr{R}e_1 \times \mathscr{R}e_1)$.

Corollaire 9.18. $\mathscr{R}e_1 \boxtimes \mathbf{P}^1$ *est un sous-module fermé de* $D_{\mathrm{rig}} \boxtimes \mathbf{P}^1$, *stable sous l'action de* $\mathrm{GL}_2(\mathbf{Q}_p)$.

Démonstration. En tant qu'espace vectoriel topologique, $\mathscr{R}e_1 \boxtimes \mathbf{P}^1$ (resp. $D_{\mathrm{rig}} \boxtimes \mathbf{P}^1$) s'identifie à $\mathscr{R}e_1 \times \mathscr{R}e_1$ (resp. $D_{\mathrm{rig}} \times D_{\mathrm{rig}}$). La fermeture de $\mathscr{R}e_1 \boxtimes \mathbf{P}^1$ dans $D_{\mathrm{rig}} \boxtimes \mathbf{P}^1$ suit donc de celle de $\mathscr{R}e_1$ dans D_{rig}. La stabilité de $D_{\mathrm{rig}} \boxtimes \mathbf{P}^1$ sous l'action de $\mathrm{GL}_2(\mathbf{Q}_p)$ découle de la proposition 9.17, de la stabilité de $\mathscr{R}e_1$ par φ et Γ, et des formules donnant l'action de $\mathrm{GL}_2(\mathbf{Q}_p)$ sur $D_{\mathrm{rig}} \boxtimes \mathbf{P}^1$. □

Proposition 9.19. *Pour tout* $B \in \mathscr{R}^{\psi=0}$ *on a*

$$p_s(w_D(B \cdot \varphi(\hat{e}_2))) = \delta_2(-1) w_{\delta_D \cdot \delta_2^{-2}}(B)\varphi(e_2).$$

10 Utiliser le fait que $w \cdot e_1 = (0, e_1)$, le lemme 9.16 et l'égalité $\sigma_{-1}(e_1) = \delta_1(-1)e_1$.

Démonstration. Un argument de semi-linéarité comme dans la preuve de la proposition 9.17 montre que l'on peut supposer que $B = 1 + T$.

Posons $Y = (D_{\mathrm{rig}} \boxtimes \mathbf{P}^1)/(\mathscr{R}e_1 \boxtimes \mathbf{P}^1)$ et notons $z \to [z]$ la projection canonique $D_{\mathrm{rig}} \boxtimes \mathbf{P}^1 \to Y$. Nous aurons besoin du résultat suivant, analogue du lemme 9.16.

Lemme 9.20. *L'élément* $[w \cdot \varphi(\hat{e}_2)]$ *de* Y *est invariant par* U.

Démonstration. Le lemme 9.10 montre qu'il suffit de vérifier que $\dim_L Y^{u^+=0} < \infty$ et que $[w \cdot \varphi(\hat{e}_2)] \in Y^{u^+=0}$. Le L-espace vectoriel

$$W = \{f \in \mathscr{R}(\delta_2) | (\nabla - w(\delta_1))(\nabla - w(\delta_2)) f = 0\}$$

est de dimension finie d'après la proposition 9.5. Si $z = (z_1, z_2) \in D_{\mathrm{rig}} \boxtimes \mathbf{P}^1$ satisfait $[z] \in Y^{u^+=0}$, alors $u^+ z \in \mathscr{R}e_1 \boxtimes \mathbf{P}^1$ et on déduit de la proposition 9.8 que $z_1 \in \mathscr{R}e_1$ et que $p_s(z_2) \in W$. Donc

$$z = z_1 + w \cdot \mathrm{Res}_{p\mathbf{Z}_p}(z_2) \equiv w \cdot \mathrm{Res}_{p\mathbf{Z}_p}(p_s(z_2)) \pmod{\mathscr{R}e_1 \boxtimes \mathbf{P}^1},$$

ce qui montre que $\dim_L Y^{u^+=0} < \infty$.

Pour conclure, il nous reste à vérifier que $u^+(w \cdot \varphi(\hat{e}_2)) \in \mathscr{R}e_1 \boxtimes \mathbf{P}^1$. Cela découle de la proposition 9.8 et du fait que $(\nabla - w(\delta_2))\hat{e}_2 \in \mathscr{R}e_1$ (car $\sigma_a(\hat{e}_2) - \delta_2(a)\hat{e}_2 \in \mathscr{R}e_1$ pour tout $a \in \mathbf{Z}_p^*$). □

Revenons à la preuve de la proposition 9.19. On applique l'identité matricielle du début de la preuve de la proposition 9.17 à $[w \cdot \varphi(\hat{e}_2)]$. Noter que $[w \cdot \varphi(\hat{e}_2)]$ est vecteur propre pour l'opérateur $\left(\begin{smallmatrix} -1 & 0 \\ 0 & 1 \end{smallmatrix}\right)$, de valeur propre $\delta_2 \cdot \delta_D(-1)$. L'identité matricielle s'écrit donc $p_s(w_D((1+T)\varphi(\hat{e}_2))) = \delta_2(-1)(1+T)\varphi(e_2)$, ce qui permet de conclure. □

Corollaire 9.21. *La suite exacte* $0 \to \mathscr{R}e_1 \to D_{\mathrm{rig}} \to \mathscr{R}e_2 \to 0$ *induit une suite exacte de* $\mathrm{GL}_2(\mathbf{Q}_p)$*-modules topologiques*

$$0 \to (\mathscr{R} \boxtimes_{\delta_D \cdot \delta_1^{-2}} \mathbf{P}^1) \otimes \delta_1 \to D_{\mathrm{rig}} \boxtimes \mathbf{P}^1 \to (\mathscr{R} \boxtimes_{\delta_D \cdot \delta_2^{-2}} \mathbf{P}^1) \otimes \delta_2 \to 0.$$

Démonstration. Commençons par définir les morphismes dans cette suite exacte. L'application i de $\mathscr{R} \boxtimes_{\delta_D \cdot \delta_1^{-2}} \mathbf{P}^1$ dans $D_{\mathrm{rig}} \boxtimes \mathbf{P}^1$ envoie (f_1, f_2) sur $(f_1 \cdot e_1, \delta_1(-1) f_2 \cdot e_1)$. L'application de $D_{\mathrm{rig}} \boxtimes \mathbf{P}^1$ dans $\mathscr{R} \boxtimes_{\delta_D \cdot \delta_2^{-2}} \mathbf{P}^1$ envoie $(A_1 \cdot e_1 + B_1 \cdot \varphi(\hat{e}_2), A_2 \cdot e_1 + B_2 \cdot \varphi(\hat{e}_2))$ sur $(B_1, \delta_2(-1) B_2)$, où $A_i, B_i \in \mathscr{R}$ ($\hat{e}_2 \in D_{\mathrm{rig}}$ est un relèvement fixé de e_2). Le fait que ces applications i et pr sont bien définies et induisent une suite exacte d'espaces vectoriels topologiques est une conséquence immédiate des propositions 9.17 et 9.19. La $\mathrm{GL}_2(\mathbf{Q}_p)$-équivariance (à torsion par δ_1, resp. δ_2 près) suit des propositions 9.17 et 9.19, du fait que $f \to f \cdot e_1$ et p_s sont des morphismes de (φ, Γ)-modules et des formules explicites donnant l'action de $\mathrm{GL}_2(\mathbf{Q}_p)$ sur les modules intervenant dans la suite exacte. □

Remarque 9.22. Soit $\delta\colon \mathbf{Q}_p^* \to L^*$ un caractère continu. On a vu dans 3.1 que l'on a un isomorphisme de $\mathrm{GL}_2(\mathbf{Q}_p)$-modules topologiques

$$\mathscr{R}^+ \boxtimes_\delta \mathbf{P}^1 \simeq \mathrm{Ind}_B^G(1,\delta)^* \otimes \delta \simeq (\mathrm{Ind}_B^G(\delta^{-1}\otimes 1))^*.$$

On peut vérifier (en utilisant des arguments identiques à ceux du chapitre II de [5]; voir aussi le paragraphe 4.1 de [7]) que l'application $\mathscr{R} \boxtimes_\delta \mathbf{P}^1 \to \mathrm{Ind}_B^G(\chi\delta\otimes\chi^{-1})$ définie par

$$z \to \phi_z, \quad \phi_z(g) = \mathrm{res}_0\left(\mathrm{Res}_{\mathbf{Z}_p}(wgz)\frac{dT}{1+T}\right)$$

est une surjection $\mathrm{GL}_2(\mathbf{Q}_p)$-équivariante, de noyau $\mathscr{R}^+ \boxtimes_\delta \mathbf{P}^1$. On dispose donc d'une suite exacte de $\mathrm{GL}_2(\mathbf{Q}_p)$-modules topologiques

$$0 \to (\mathrm{Ind}_B^G(\delta^{-1}\otimes 1))^* \to \mathscr{R}\boxtimes_\delta \mathbf{P}^1 \to \mathrm{Ind}_B^G(\chi\delta\otimes\chi^{-1}) \to 0.$$

Combiné au corollaire 9.21 et à la suite exacte

$$0 \to (\Pi^{\mathrm{an}})^* \otimes \delta_D \to D_{\mathrm{rig}} \boxtimes \mathbf{P}^1 \to \Pi^{\mathrm{an}} \to 0,$$

cela permet de montrer que Π^{an} est de longueur finie et que

$$(\Pi^{\mathrm{an}})^{\mathrm{ss}} = (\mathrm{Ind}_B^G(\delta_1\otimes\chi^{-1}\delta_2))^{\mathrm{ss}} \oplus (\mathrm{Ind}_B^G(\delta_2\otimes\chi^{-1}\delta_1))^{\mathrm{ss}}.$$

Il faut travailler un peu plus [7] pour déterminer les extensions entre les constituants de Jordan-Hölder.

Références

[1] L. Berger-Représentations p-adiques et équations différentielles, *Invent. Math.* 148 (2002), 219–284.

[2] L. Berger, C. Breuil-Sur quelques représentations potentiellement cristallines de $G_{\mathbf{Q}_p}$, *Astérisque* 330 (2010), 155–211.

[3] F. Cherbonnier et P. Colmez-Représentations p-adiques surconvergentes, *Invent. Math.* 133 (1998), 581–611.

[4] P. Colmez-Représentations triangulines de dimension 2, *Astérisque* 319 (2008), 213–258.

[5] P. Colmez-Représentations de $\mathrm{GL}_2(\mathbf{Q}_p)$ et (φ,Γ)-modules, *Astérisque* 330 (2010), 281–509.

[6] P. Colmez-La série principale unitaire de $\mathrm{GL}_2(\mathbf{Q}_p)$, *Astérisque* 330 (2010), 213–262.

[7] P. Colmez-La série principale unitaire de $\mathrm{GL}_2(\mathbf{Q}_p)$: vecteurs localement analytiques, ce volume.

[8] G. Dospinescu-Actions infinitésimales dans la correspondance de Langlands locale p-adique, *Math. Ann.* 354 (2012), 627–657.

[9] M. Emerton-Locally analytic vectors in representations of locally p-adic analytic groups, to appear in *Memoirs of the AMS.*

[10] M. Emerton-A local-global compatibility conjecture in the p-adic Langlands programme for $GL_2/\mathbf{Q}$, *Pure Appl. Math. Q.* 2 (2006), 279–393.

[11] J.-M. Fontaine-Représentations p-adiques des corps locaux. I, in *The Grothendieck Festschrift*, Vol II, Progr. Math., vol 87, Birkhauser, 1990, 249–309.

[12] K.S. Kedlaya-A p-adic local monodromy theorem, *Ann. of Math.* 160 (2004), 93–184.

[13] R. Liu, Locally analytic vectors of some crystabeline representations of $\mathrm{GL}_2(\mathbf{Q}_\mathrm{p})$, *Compositio Mathematica*, Volume 148, Issue 01, (2012) 28–64.

[14] R. Liu, B. Xie, Y. Zhang, Locally analytic vectors of unitary principal series of $\mathrm{GL}_2(\mathbf{Q}_\mathrm{p})$, *Annales Scientifiques de l'E.N.S.* Vol. 45, No. 1, (2012) 167–190.

[15] P. Schneider et J. Teitelbaum- Locally analytic distributions and p-adic representation theory, with applications to GL_2, J. *Amer. Math. Soc* 15 (2002), 443–468.

[16] P. Schneider et J. Teitelbaum- Algebras of p-adic distributions and admissible representations, *Invent. Math.* 153 (2003), 145–196.

Printed in the United States
By Bookmasters